Virgilijus Sakalauskas · Anu Bajaj ·
Ajith Abraham · K. Reddy Madhavi ·
Pooja Manghirmalani Mishra
Editors

Bio-Inspired Computing

Proceedings of the 14th International Conference on Innovations in Bio-Inspired Computing and Applications (IBICA 2023) held during December 14–15, 2023
Volume 4: Machine Learning Applications

Springer

Editors
Virgilijus Sakalauskas
Department of Informatics
Vilnius University
Vilnius, Lithuania

Ajith Abraham
School of Artificial Intelligence
Sai University
Chennai, India

Pooja Manghirmalani Mishra
Machine Intelligence Research Labs
Mumbai, Maharashtra, India

Anu Bajaj
Department of Computer Science
and Engineering
Thapar Institute of Engineering
and Technology
Patiala, Punjab, India

K. Reddy Madhavi
School of Computing
Mohan Babu University
Tirupati, Andhra Pradesh, India

ISSN 2367-3370 ISSN 2367-3389 (electronic)
Lecture Notes in Networks and Systems
ISBN 978-3-031-78945-8 ISBN 978-3-031-78946-5 (eBook)
https://doi.org/10.1007/978-3-031-78946-5

This Springer imprint is published by the registered company Springer Nature Switzerland AG
The registered company address is: Gewerbestrasse 11, 6330 Cham, Switzerland

Lecture Notes in Networks and Systems 1231

The series "Lecture Notes in Networks and Systems" publishes the latest developments in Networks and Systems—quickly, informally and with high quality. Original research reported in proceedings and post-proceedings represents the core of LNNS.

Volumes published in LNNS embrace all aspects and subfields of, as well as new challenges in, Networks and Systems.

The series contains proceedings and edited volumes in systems and networks, spanning the areas of Cyber-Physical Systems, Autonomous Systems, Sensor Networks, Control Systems, Energy Systems, Automotive Systems, Biological Systems, Vehicular Networking and Connected Vehicles, Aerospace Systems, Automation, Manufacturing, Smart Grids, Nonlinear Systems, Power Systems, Robotics, Social Systems, Economic Systems and other. Of particular value to both the contributors and the readership are the short publication timeframe and the world-wide distribution and exposure which enable both a wide and rapid dissemination of research output.

The series covers the theory, applications, and perspectives on the state of the art and future developments relevant to systems and networks, decision making, control, complex processes and related areas, as embedded in the fields of interdisciplinary and applied sciences, engineering, computer science, physics, economics, social, and life sciences, as well as the paradigms and methodologies behind them.

Indexed by SCOPUS, EI Compendex, INSPEC, WTI Frankfurt eG, zbMATH, SCImago.

All books published in the series are submitted for consideration in Web of Science.

For proposals from Asia please contact Aninda Bose (aninda.bose@springer.com).

IBICA 2023 Preface – Volume 4

Welcome to the 14th International Conference on Innovations in Bio-Inspired Computing and Applications (IBICA 2023) and 13th World Congress on Information and Communication Technologies (WICT 2023). IBICA-WICT 2023 was held in 5 different cities namely Olten, Switzerland; Porto, Portugal; Kaunas, Lithuania; Greater Noida, India; Kochi, India and in online mode. IBICA-WICT 2023 is hosted and sponsored by the Machine Intelligence Research Labs (MIR Labs), USA and supported by Bennett University, Greater Noida, India.

14th International Conference on Innovations in Bio-Inspired Computing and Applications was organized to provide a platform for world research leaders and practitioners, to discuss the "full spectrum" of current theoretical developments, emerging technologies, and innovative applications of Bio-inspired Computing. Bio-inspired Computing is currently one of the most exciting research areas, and it is continuously demonstrating exceptional strength in solving complex real-life problems. The main driving force of the conference is to further explore the intriguing potential of Bio-inspired Computing.

13th World Congress on Information and Communication Technologies (WICT 2023) was organized to provide an opportunity for the researchers from academia and industry to meet and discuss the latest solutions, scientific results and methods in the usage and applications of ICT in the real world. Last few decades have experienced technologies that are evolving so rapidly, altering the constraints of space and time, and reshaping the way we communicate, learn and think. Rapid advances in information technologies and other digital systems are reshaping our ecosystem. Innovations in ICT allow us to transmit information quickly and widely, propelling the growth of new urban communities, linking distant places and diverse areas of endeavor in productive new ways, which a decade ago was unimaginable. Thus, the theme of this World Congress is "Innovating ICT For Social Revolutions".

IBICA-WICT 2023 received 656 papers from 36 countries and based on the review process we have accepted 263 papers. Each paper was reviewed by at least five or more reviewers, and based on the outcome of the review papers were accepted for inclusion in the conference proceedings. Proceedings were divided into Five Volumes. This Volume is focused on *Machine Learning and Applications.* Machine learning uses algorithms that learn from data to approximate functions. These function approximations can be generated through supervised learning, where algorithms learn patterns from existing data, or unsupervised learning, where they discover general patterns in data. This Volume comprises of 53 articles focused on machine learning and its real-world applications.

First, we would like to thank all the authors for submitting their papers to the conference, for their presentations and discussions during the conference. Our thanks go to program committee members and reviewers, who carried out the most difficult work by carefully evaluating the submitted papers. Our special thanks to the following plenary speakers, for their exciting plenary talks:

- Christine Zarges, Aberystwyth University, UK
- Theresa Schmiedel, University of Applied Sciences and Arts Northwestern Switzerland, Basel, Switzerland
- Ke Feng, Singapore-ETH Centre, The National University of Singapore, Singapore
- Stefka Fidanova, Bulgarian Academy of Sciences, Bulgaria
- Diego Oliva, Universidad de Guadalajara, Mexico
- Sebastian Ventura, University of Cordoba, Spain
- João Pedrosa, INESCTEC, Portugal
- Nuno Bettencourt, Instituto Superior de Engenharia do Porto, Portugal
- Milan Tuba, Singidunum University, Serbia
- Dalia Kriksciuniene, Vilnius University, Lithuania
- Kusum Deep, Indian Institute of Technology Roorkee, India

We express our sincere thanks to the organizing committee chairs for helping us to formulate a rich technical program. Enjoy reading the articles!

Virgilijus Sakalauskas
Thinakaran Perumal
Ajith Abraham
Preethi Thekkath
Dalia Kriksciuniene
Yunchuan Sun
Ryotaro Kamimura
Sreela Sreedhar
Anu Bajaj

IBICA–WICT 2023 Organizing Committees

General Chairs

Virgilijus Sakalauskas	Vilnius University, Lithuania
Thinakaran Perumal	Universiti Putra Malaysia, Malaysia
Ajith Abraham	Sai University, India
Preethi Thekkath	Toc H Institute of Science & Technology, India

Program Chairs

Dalia Kriksciuniene	Vilnius University, Lithuania
Yunchuan Sun	Beijing Normal University, China
Ryotaro Kamimura	Tokai University, Japan
Sreela Sreedhar	Toc H Institute of Science & Technology, India

Web Services Chair

Kun Ma	University of Jinan, China

Publication Chair

Anu Bajaj	Thapar Institute of Engineering and Technology, India

Publicity Chair

Pooja Manghirmalani Mishra	Machine Intelligence Research Labs, India

Local Organizing Committee

Jesna Anver	Toc H Institute of Science & Technology, India
Leda Kamal	Toc H Institute of Science & Technology, India
Elsaba Jacob	Toc H Institute of Science & Technology, India

Mithu Mary George	Toc H Institute of Science & Technology, India
Mima Manual	Toc H Institute of Science & Technology, India
Ashly Jospeh	Toc H Institute of Science & Technology, India
Sajith P. J.	Toc H Institute of Science & Technology, India
Leema Mathew	Toc H Institute of Science & Technology, India
Baazil P. Thampy	Toc H Institute of Science & Technology, India
Sharath P. Raju	Toc H Institute of Science & Technology, India
Philo Sumi	Toc H Institute of Science & Technology, India
Jacob Jose Chemmankuzha	Toc H Institute of Science & Technology, India

International Program Committee

Abdelkrim Haqiq	Hassan 1st University, Morocco
Alberto Bugarín-Diz	Universidade de Santiago de Compostela, Spain
Alberto Guillén	University of Granada, Spain
Alberto Fernandez Hilario	University of Granada, Spain
Alexey Kornaev	Innopolis University, Russia
Alfonso Guarino	University of Foggia, Italy
Alfredo Cuzzocrea	University of Calabria, France
Alicia Troncoso	Pablo de Olavide University, Spain
Ambili P. S.	REVA University, India
Amelia Zafra	University of Cordoba, Spain
Amit Kumar Tyagi	Vellore Institute of Technology, India
Amutha S.	Vellore Institute of Technology, India
Ana de Almeida	Instituto Universitário de Lisboa, Portugal
Ana Madureira	Polytechnique of Porto, Portugal
Anazida Zainal	Universiti Teknologi Malaysia, Malaysia
Andre C. P. L. F. de Carvalho	University of Sao Paulo, Brazil
André S. Santos	ISEP, Portugal
Andrea Soltoggio	Loughborough University, UK
Andries Engelbrecht	University of Stellenbosch, South Africa
Anitha Kumari K.	PSG College of Technology, India
Antonello Florio	Politecnico di Bari, Italy
Antonio Arauzo-Azofra	University of Cordoba, Spain
Antonio Salmeron	Universidad de Almeria, Spain
Anurag Rana	Shoolini University, India
Arijit Ghosal	St. Thomas' College of Engineering and Technology, India
Arthi B.	SRM Institute of Science and Technology, India
Aryan Chaudhary	Bio Tech Sphere Research, India
Aureli Soria-Frisch	Starlab Barcelona, Spain

Azah Kamilah binti Muda	UTeM, Malaysia
Bac Le	University of Science, Vietnam
Bay Vo	Ho Chi Minh City University of Technology, Vietnam
Bernard De Baets	Ghent University, Belgium
Bernardete Ribeiro	CISUC, University of Coimbra, Portugal
Bhawna Goyal	Chandigarh University, India
Carlos Cotta	Universidad de Málaga, Spain
Carlos Pereira	Universidade de Santiago de Compostela, Portugal
Carlos Serro	ISCTE - Instituto Universitário de Lisboa, Portugal
Catarina I. Reis	Polytechnic of Leiria, Portugal
Cesar Hervas-Martinez	University of Cordoba, Spain
Chang-Yong Lee	Kongju National University, Korea
Chao-Chun Chen	National Cheng Kung University, Taiwan
Chun-Hao Chen	National Taipei University of Technology, Taiwan
Corrado Mencar	University of Bari Aldo Moro, Italy
David A. Pelta	Universidad de Granada, Spain
Debdatta Kandar	North-Eastern Hill University, India
Deepika Koundal	UPES, India
Deepti Chaudhary	Kurukshetra University, India
Devi Priya R.	KPR Institute of Engineering and Technology, India
Rajashree Shettar	RV College of Engineering, India
Amit Kumar Mishra	Jain University, India
Iqbal Sarker	Edith Cowan University, Australia
Saira Varghese	Toc H Institute of Science & Technology, India
Durai Raj Vincent P. M.	VIT Vellore, India
Durgesh Srivastava	Chitkara University, India
Eiji Uchino	Yamaguchi University, Japan
Elakkiya R.	BITS Pilani Dubai Campus, UAE
Elizabeth F. G. Goldbarg	Universidade Federal do Rio Grande do Norte, Brazil
Enrique De la Cal	University of Oviedo, Spain
Enrique Herrera-Viedma	University of Granada, Spain
Fabio Zambetta	RMIT University, Australia
Fernando Batista	INESC-ID and ISCTE-IUL, Portugal
Fernando Gomide	University of Campinas, Brazil
Francisco Herrera	Universidad de Granada, Spain
Gabriella Casalino	University of Bari, Italy
Gahangir Hossain	University of North Texas, USA

Gaurav Gupta	Shoolini University, India
George Papakostas	International Hellenic University, Greece
Gianluca Zaza	University of Bari, Italy
Giovanna Castellano	University of Bari, Italy
Gregorio Sainz-Palmero	Universidad de Valladolid, Spain
Guenther Raidl	Vienna University of Technology, Austria
Hajime Nobuhara	University of Tsukuba, Japan
Heba Afify	Biomedical Engineering, Egypt
Hsueh-Chan Lu	National Cheng Kung University, Taiwan
Hudson G. Medeiros	Federal University of Rio Grande do Norte, Brazil
Irwin King	The Chinese University of Hong Kong, Hongkong
Isaac Chairez	Tecnológico de Monterrey, Mexico
Isabel Jesus	Institute of Engineering of Porto, Portugal
Ivo Pereira	University Fernando Pessoa, Portugal
Ja-Hwung Su	National University of Kaohsiung, Taiwan
Janos Botzheim	Eötvös Loránd University Faculty of Informatics, Hungary
Janusz Kacprzyk	Polish Academy of Sciences, Poland
Jayaraman Valadi	FLAME University, India
Jerry Chun-Wei Lin	Western Norway University of Applied Sciences, Norway
Jesus S. Aguilar-Ruiz	Universidad Pablo de Olavide, Spain
Jesus Alcala-Fdez	University of Granada, Spain
Jimmy Ming-Tai Wu	Shandong University of Science and Technology, China
Jolanta Mizera-Pietraszko	Military University of Land Forces, Poland
Jose A. Gamez	Universidad de Castilla-La Mancha, Spain
Jose Jerez	Universidad de Malaga, Spain
Jose A. Lozano	UPV/EHU, Spain
Jose Maria-Luna	University of Cordoba, Spain
Jose Puerta	Universidad de Castilla-La Mancha, Spain
José Everardo B. Maia	UECE, Brazil
Juan Vidal	University of Santiago de Compostela, Spain
Ju-Chin Chen	National Kaohsiung University of Science and Technology, Taiwan
Julia Flores	University of Castilla - La Mancha (UCLM), Spain
K. Reddy Madhavi	Sree Vidyanikethan Engineering College, India
Kawuu W. Lin	National Kaohsiung University of Science and Technology, Taiwan
Kingsley Okoye	Tecnologico de Monterrey, Mexico
Kolla Bhanu Prakash	Koneru Lakshmaiah Education Foundation, India

Radhika S.	Sathyabama Institute of Science and Technology, India
Radu-Emil Precup	Politehnica University of Timisoara, Romania
RajaniKanth Aluvalu	Vardhaman College of Engineering, India
Richard Weber	University of Chile, Chile
Roberto Iglesias Rodriguez	Universidade de Santiago de Compostela, Spain
Rohit Anand	GB Pant Govt. Engineering College, India
Ronald Yager	Machine Learning Institute, Iona College, USA
Rudolf Kruse	University of Magdeburg, Germany
Rutuparna Panda	VSS University of Technology, India
Saman Halgamuge	University of Melbourne, Australia
Sancho Salcedo-Sanz	Universidad de Alcalá, Spain
Sandeep Sreekumar	IndsutryApps, Singapore
Santosh Kumar Srivastava	Galgotias University, India
Sarina Sulaiman	Universiti Teknologi Malaysia, Malaysia
Sebastian Basterrech	VSB-Technical University of Ostrava, Czech Republic
Sebastián Ventura Soto	University of Cordoba, Spain
Shabeer Puthukkudi	British Telecom, United Kingdom
Shali Rani	Chitkara University, India
Shankru Guggari	KLE Technological University, India
Shilpa Gupta	Maharaja Agrasen University, India
Shing Chiang Tan	Multimedia University, Malaysia
Silviu Ionita	University of Pitesti, Romania
Sirine Marrakchi	University of Sfax, Tunisia
Stéphane Grieu	Université de Perpignan, France
Steven D. Prestwich	University College Cork, Ireland
Sujata Dash	Nagaland University, India
Sung-Bae Cho	Yonsei University, Korea
Teresa Ludermir	Federal University of Pernambuco, Brazil
Thillaiarasu Nadesan	Reva University, India
Thinagaran Perumal	Universiti Putra Malaysia, Malaysia
Tho Quan	Ho Chi Minh City University of Technology, Vietnam
Thomas Hanne	University of Applied Sciences Northwestern, Switzerland
Tom Gedeon	The Australian National University, Australia
Tzung-Pei Hong	National University of Kaohsiung, Taiwan
Van Vo	Ho Chi Minh University of Industry, Vietnam
Victor M Rivas	University of Jaen, Spain
Vijay P.	Vellore Institute of Technology, India
Vishwanath N.	Toc H Institute of Science & Technology

Wen-Yang Lin	National University of Kaohsiung, Taiwan
Wilfried Elmenreich	Alpen-Adria-Universitat, Austria
Witold Pedrycz	University of Alberta, Canada
Xiao-Zhi Gao	University of Eastern Finland, Finland
Yaochu Jin	University of Bielefeld, Germany
Yilun Shang	Northumbria University, UK
Yukio Ohsawa	University of Tokyo, Japan

Contents

OSPC: Ontology Synthesis on Peace and Conflict Studies as a Domain of Choice

S. S. Nitin Hariharan[1] and Gerard Deepak[2(✉)]

[1] Department of Computer Science and Engineering, Alagappa Chettiar Government College of Engineering and Technology, Karaikudi, India

[2] Manipal Institute of Technology Bengaluru, Manipal Academy of Higher Education, Manipal, India

gerard.deepak.christuni@gmail.com

Abstract. This paper addresses the lack of a strategic framework for creating and combining ontologies in the peace and conflict studies field. Although this is an important area of study, not much attention has been given to ontology modeling here. To fill this gap, we propose a strategic and automated framework for synthesizing ontologies. Our framework starts by extracting information from the dataset and uses this information to create a feature feedback mechanism by referring to Wiki data and CYC Knowledge Store repositories. This feedback mechanism helps in classification tasks using SVM (Support Vector Machine). We also generate metadata and make the framework more adaptable using the Deep Belief Networks classifier. To measure semantic similarity, we use the SOC-PMI method, and for optimization purposes, we implement the Clonal Selection Principle, which is a meta-optimization strategy. Additionally, we use DBSCAN clustering to generate an initial seed ontology with strong associations. The proposed model achieves high percentages of precision, recall, accuracy, and F-measure, specifically 95.33%, 96.78%, 96.055%, and 96.04952787% respectively. Additionally, the model achieves a low value of FDR (False Discovery Rate) at 0.05% is achieved.

Keywords: CYC · SVM · DBN · SOC-PMI · Clonal Selection Principle · DBSCAN

1 Introduction

Ontology synthesis refers to the process of automatically generating ontologies, which are formal representations of knowledge, from various sources of information. This synthesis involves integrating knowledge-based reasoning techniques to create ontologies that can reason and make inferences based on the underlying knowledge. Automatic modeling of ontologies aims to streamline the process of ontology development by leveraging computational methods and algorithms. Using machine learning and natural language processing techniques, automatic modeling facilitates the extraction and organization of relevant knowledge from diverse data sources. Integrating knowledge-based reasoning within the ontologies enables them to not only represent knowledge but also perform logical deductions and make informed decisions.

V. Sakalauskas et al. (Eds.): IBICA 2023, LNNS 1231, pp. 1–11, 2025.
https://doi.org/10.1007/978-3-031-78946-5_1

***Motivation*:** This research addresses the absence of ontologies specifically designed for peace and conflict studies. Creating an ontology tailored to this field is crucial for applications like search engine indexing and text summarization. Existing models lack automation and effective reasoning. The proposed framework aims to fill this gap by automating ontology synthesis for peace and conflict studies, using diverse knowledge sources from the World Wide Web to ensure relevance. It seeks to provide an intelligent model for ontological modeling, which is currently lacking, especially in niche domains like peace and conflict studies where semantically inclined techniques are needed.

***Contribution*:** The proposed work has several remarkable contributions. It combines the utilization of CYC and WikiData for extracting informative terms from the dataset using TF-IDF and category extraction. These terms are then employed as features in a stream classifier to classify the dataset. This integration of diverse resources and techniques represents a significant contribution. The inclusion of both a lightweight stream classifier and a powerful deep learning classifier for classifying the dataset and metadata, respectively, is an important aspect of the model. This approach not only brings novelty to the research but also effectively manages the computational load within the framework. Another noteworthy contribution is the generation of metadata, which accelerates the incorporation of auxiliary knowledge into the framework at an exponential rate. This process of generating metadata enhances the overall knowledge base and strengthens the synthesis of the ontology. The utilization of DBSCAN clustering to cluster the classified instances from the dataset and metadata is another valuable contribution. This clustering aids in creating a robust upper ontology, while the SOC-PMI measure supports semantic reasoning within the framework. The optimized clonal selection principle plays a significant role in generating an optimal set of solutions. The entities within the framework are ranked based on formalization, axiomatization, and reasoning processes, further enhancing the effectiveness and quality of the ontology synthesis.

Organization: The paper is structured as follows. Section 2 provides an overview of Related Literature, Sect. 3 discusses the Proposed System Architecture, Sect. 4 presents the Results and Performance Evaluation and outcomes, and Sect. 5 concludes with insights.

2 Related Works

Clyde et al. [1] introduce a collaborative approach to ontology design, emphasizing the significance of engaging multiple stakeholders and leveraging their expertise throughout the ontology development process. Soininen et al. [2] strive to create a comprehensive ontology specifically focused on configuration. The study delves into the fundamental concepts and interrelationships that underlie the configuration process, with the aim of establishing a solid foundation for effective configuration management systems. Waralak et al. [3] specifically examine the correlation between ontology modelling and object modelling within the realm of software engineering. The research investigates the advantages of integrating ontologies into the software development process and emphasizes the potential synergies that can arise from combining ontology modelling and object modelling. The paper discusses the implications and benefits of employing ontologies in software engineering applications.

Alobaidi et al. [4] proposed a framework that focuses on generating biomedical ontologies by leveraging Linked Open Data. This framework combines existing biomedical ontologies with external datasets to automatically create novel ontologies. The main goal of this framework is to enhance the ontology development process in the field of biomedicine Huang et al. [5] proposed a data-centric methodology for the development and evolution of ontologies in manufacturing systems. Their framework utilizes machine learning techniques and data analysis to autonomously construct ontologies that facilitate intelligent services within the manufacturing sector. The research highlights the potential of ontology-driven approaches in enhancing the productivity and intelligence of manufacturing systems. Wang et al. [6] presented an approach that demonstrates the application of deep learning algorithms to automatically acquire the structure and semantics of data, enabling the creation of ontologies.

Kumar et al. [7] proposed a method for automatically generating ontologies from plain text by utilizing statistical and natural language processing (NLP) techniques. The framework incorporates statistical methods and NLP tools to extract relevant information and relationships from the text, facilitating the construction of ontologies. The research focuses on automating the process of ontology generation using text-based data. Fraga et al. [8] present a semi-automated methodology for generating ontologies from industrial product data standards. The approach outlines a process that combines manual curation and automated techniques to extract relevant information and build ontologies. The research focuses on generating ontologies specifically from standards in the industrial product domain. Kim et al. [9] introduced an intelligent health diagnosis technique that leverages automatic ontology generation and web-based personal health record services. Leo et al. [10] presented a methodology for creating medical multiple-choice questions (MCQs) using ontologies. The approach harnesses ontological knowledge and reasoning techniques to automatically generate MCQs that cover multiple relevant terms in the medical domain. The research focuses on leveraging ontologies to enhance the process of generating MCQs for medical education and assessment. Rupasingha et al. [11] introduced a specificity-aware methodology for ontology generation, aiming to improve web service clustering. The approach focuses on developing ontologies that capture the unique characteristics and complexities of web services, facilitating more effective clustering of these services. The research endeavors to enhance the performance and accuracy of web service clustering by incorporating specificity-aware techniques in ontology generation. In [12–21] several frameworks in support of the literature of the proposed work have been depicted.

3 Proposed System Architecture

Figure 1. Depicts the system architecture of the proposed Framework for Ontology Synthesis on Peace and Conflict Studies (OSPC), focusing on the selected domain. The architecture commences with a dataset of documents characterized by their categorical nature. The initial step involves extracting categories directly from the dataset. Concurrently, information terms are extracted from the dataset utilizing the term frequency inverse document frequency (TF-IDF) technique. The TFIDF model not only identifies frequently occurring terms across the document corpus but also identifies terms that are infrequent within the corpus. These terms complement the extracted categories and contribute to the generation of an informative term set known as the initial term set. However, the initial term set is relatively limited in depth and lacks enough information. Hence, additional steps are necessary to augment its content.

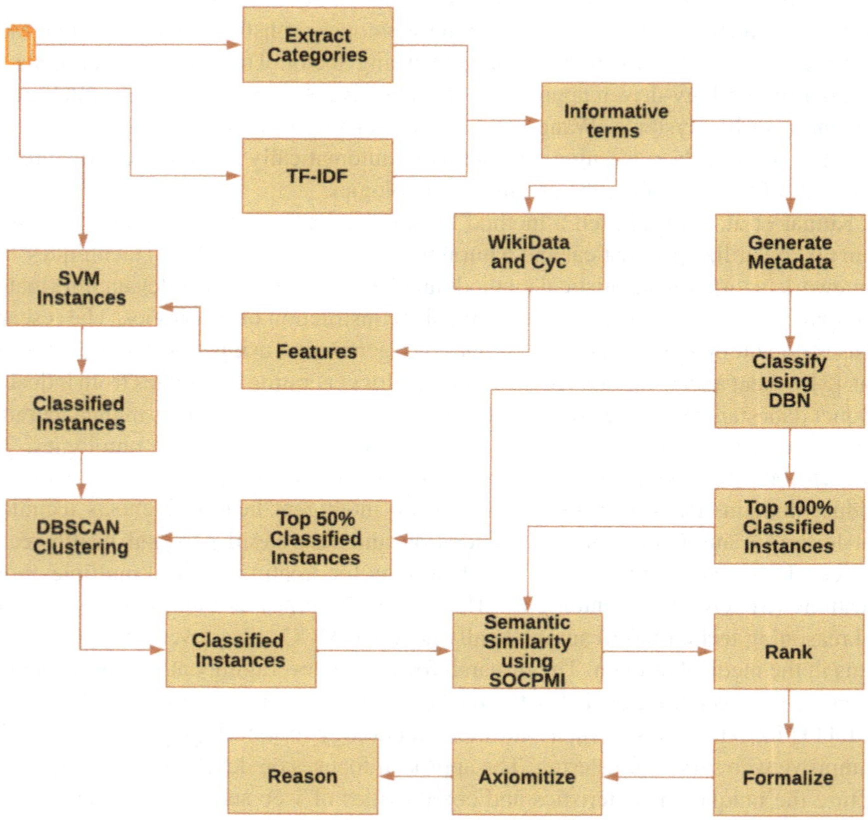

Fig. 1. System architecture for the Ontology Synthesis on Peace and Conflict Studies as the domain of choice

The initial term set undergoes enrichment and contextualization by leveraging the knowledge stores of WikiData and CYC. Access to WikiData is facilitated through an API, while the CYC knowledge base is accessed via SPARQL querying. Entities obtained from WikiData and CYC serve as features in a Support Vector Machine (SVM) classifier, which is responsible for classifying the dataset. The choice of SVM classifier is justified by its lightweight nature, which helps to minimize computational load. Considering the limited nature of features derived from WikiData and CYC, a lightweight classifier like SVM suffices for this classification step, as it is not the final recommendation. The instances classified by the stream classifier are further utilized within the model. Furthermore, the informative terms extracted from the dataset are utilized for metadata generation, which leverages metadata generation tools to achieve exponential growth.

This process of metadata generation enhances the knowledge base and broadens the scope of the framework.

To generate metadata, the framework utilizes two distinct tools: DSpace and MarkEdit. These tools play a crucial role in enhancing the diversity, volume, and heterogeneity of the metadata. The entities derived from the generated metadata undergo classification using Deep Belief Networks (DBN), rather than being directly utilized. Given the large-scale volume, heterogeneity, and interconnected nature of the metadata, a powerful deep learning classifier like DBN is necessary. Despite its computational complexity, the DBN classifier yields intermediate classification accuracy, making it suitable for handling the metadata. The metadata is further transformed into an atomic level and seamlessly integrated into the localized framework. The entities classified by the DBN classifier, specifically the top 50% of instances, are combined with the dataset's classified outcomes from the SVM classifier. These combined instances then undergo DBSCAN clustering, alongside other classified instances from the dataset. This clustering process enables the grouping of instances based on similarity.

To compute semantic similarity among the clustered instances, the entities from the DBN classifier are subjected to the SOC-PMI (second-order co-occurrence point-wise mutual information) measure, employing a threshold of 0.75. The threshold is set stringently due to the substantial number of entities resulting from the DBN classifier. This process yields the initial solution set, which may still be extensive in terms of quantity and volume. To refine and optimize this solution set, the framework employs the Clonal Selection Principle optimization algorithm, a robust meta-heuristic optimization algorithm. The SOC-PMI measure is used as an objective function with the same threshold to further enhance and optimize the initial solution set. The optimized solution set is then ranked based on the increase in the Composite Performance Measure (CPM). Based on this ranking, edges are created and formalized to establish the ontology's structure. The ontology is subsequently axiomatized using concepts such as "is-a-part-of," "has-a-part-of," "is a superclass of," and "is a subclass of," as well as composition, generalization, and specialization relations. For reasoning purposes, a Pellet reasoner can be employed, and further instantiation can be accomplished using any standard knowledge base. In this case, Protégé is used to populate the instances.

DBSCAN (Density-Based Spatial Clustering of Applications with Noise) is an algorithm used for clustering data points based on their density and proximity. It has the unique ability to identify clusters of arbitrary shape without the need to predefine the number of clusters. DBSCAN defines clusters as regions of dense points separated by areas with lower point density, allowing for effective noise and outlier detection. The algorithm begins by selecting a point and checking if there are enough neighboring points within a specified radius. If the point has a enough neighbors, it becomes a core point, and the algorithm expands the cluster by including other nearby points that also have enough neighbors. This process continues until no more points can be added to the cluster. Points within the radius of a core point but with fewer neighbors are classified as border points. Points that do not belong to any cluster and lack a enough neighbors are labeled as noise points. Eq. (1) ($f^{pmi}(t_i, w)$) calculates the PMI value between a target word t_i and a context word w. PMI measures how often the two words appear together compared to their individual occurrences in the entire corpus.

$$f^{pmi}(t_i, w) = \log_2 \frac{f^b(t_i, w) \times m}{f^t(t_i) f^t(w)} \tag{1}$$

For a given word w, we define a set of words (X^w) sorted by their PMI values with 'w'. We select the top β words from this set that have positive PMI values. Equation (2) ($f(w_1, w_2, \beta)$) computes a summation of PMI values between the words in the set X^{w_1} and a target word w_2. The parameter γ is used to adjust the importance of individual PMI values.

$$f(w_1, w_2, \beta) = \sum_{i=1}^{\beta} \left(f^{pmi}(X_i^{w_1}, w_2) \right)^{\gamma} \tag{2}$$

The semantic PMI similarity function ($Sim(w_1, w_2)$) between two words, w_1 and w_2, is defined in Eq. (3). It calculates the semantic similarity by normalizing the values obtained in Eq. 11 with respect to two different sets of top β words.

$$Sim(w_1, w_2) = \frac{f(w_1, w_2, \beta_1)}{\beta_1} + \frac{f(w_2, w_1, \beta_2)}{\beta_2} \tag{3}$$

The semantic word similarity is normalized, so that it provides a similarity score between 0 and 1 inclusively.

4 Results

The evaluation outcomes of the proposed OSPC framework, which is a model for synthesizing ontologies in the field of peace and conflict studies as well as conflict resolution, were validated using various metrics including precision, recall, accuracy, F-measure percentages, and false discovery rate (FDR). Precision, recall, and accuracy metrics were employed to measure the significance of the results, whereas FDR indicated the number of false positives, representing the error rate in the model. These metrics were selected as the preferred evaluation criteria. Based on Table 1, the proposed OSPC framework achieved the highest average precision of 95.33% and the highest average recall of 96.78%. It demonstrated a high accuracy percentage of 96.055% and the highest F measure of 96.04952787%, along with a low FDR of 0.05%, indicating a minimal number of false positives. To evaluate the performance of OSPC, a comparison was made with similar models such as COD, TGOS, and the OMS framework. As per table one, COD showed an average precision of 90.74%, an average recall of 92.73%, an average accuracy of 91.735%, and an average F measure of 91.72420777%, along with an FDR of 0.05%. TGOS achieved an average precision of 91.45%, an average recall of 93.07%, combined with an average accuracy of 92.26% and an average F measure of 92.25288858%, along with an FDR of 0.09%. On the other hand, the OMS framework provided an average precision of 92.09%, an average recall of 94.07%, combined with an average accuracy of 96.055% and an average F measure of 96.04952787%, along with an FDR of 0.08%.

Table 1. Comparison of Performance of the proposed OSPC with other approaches

Model	Average Precision %	Average Recall %	Average Accuracy %	Average F-Measure %	FDR
COD [1]	90.74	92.73	91.735	91.72	0.10
TGOS [2]	91.45	93.07	92.26	92.25	0.09
OMS [3]	92.09	94.07	93.08	93.06	0.08
Proposed OSPC	95.33	96.78	96.055	96.04	0.05

The Ontology Synthesis on Peace and Conflict Studies (OSPC) has achieved impressive precision, F-measure percentages, and a low False Discovery Rate (FDR) due to several key factors. OSPC is a pioneering framework designed specifically for peace and conflict studies, addressing the need for ontologies in this field. Its high precision and F-measure percentages are attributed to the semi-automatic approach it adopts. The model is 95% automated, with only 5% requiring human review to incorporate cognitive input. This combination of automation and human involvement ensures a robust and accurate synthesis process. OSPC leverages powerful techniques like TF-IDF to extract informative terms from the dataset. It enriches the dataset with valuable entities by incorporating knowledge resources such as WikiData and CYC. These features are then used in the stream classifier for dataset classification. The metadata generation process in OSPC employs reliable tools, and the metadata is accurately classified using Deep Belief Networks (DBN), a robust deep learning classifier. This precise classification improves the overall accuracy of the framework. The selective association of classified instances from the DBN with the dataset instances and the clustering process using DBSCAN contribute to establishing a robust and formal model. The application of SOC PMI (second-order co-occurrence point-wise mutual information) under the clonal selection principle ensures the computation of the most relevant entities. The ranking, formalization, and axiomatization of entities lead to ontology generation. Reasoning through semantics, numerical computation, and preferential selection, along with optimization using the clonal selection principle, further enhance the quality and optimality of the results.

Experimentations are conducted on four distinct independent datasets strategically obtained from various sources. The datasets used are as follows: Peace and Conflict Studies 164B - Spring 2007 from UC Berkeley, Peace Negotiations in Civil Conflicts from Harvard Data verse, and Sudan - Data on Conflict Events from HDX. These datasets are strategically formalized by using a customized annotation process to categorize and generate annotations for the data. To enhance the datasets, a customized web crawler is employed to crawl relevant documents from the World Wide Web based on the annotations and degrees present in the data. This process creates a large document repository that is integrated with the original dataset's annotations. As a result, a comprehensive and prioritized integrative dataset for peace and conflict studies, as well as civil conflicts, is established, and the experimentations are conducted using this enriched dataset. The underperformance of the COD model can be attributed to several factors. Firstly, the model lacks the incorporation of learning models, which restricts its ability to adapt and improve over time. Additionally, the absence of external sources hampers the enrichment of the ontology. The collaborative nature of the COD model revolves around providing a formal platform for knowledge conduction, management, and assimilation through the integration of human resources. While this approach fosters collaboration and generates a highly cognitive ontology, it heavily relies on human intervention throughout the entire process. Consequently, the model lacks the desired level of automation, with only a small percentage (5%) being automated. To overcome these limitations, it is necessary to evolve the COD model into a more advanced and automated framework. This evolution should involve incorporating learning models, leveraging external sources for knowledge enrichment, and reducing the reliance on manual human intervention. A potential next step could be exploring the TGOS model.

The TGOS model underperforms when compared to the proposed model for several reasons. Firstly, although TGOS provides a general configuration for ontologies, the process of knowledge conceptualization relies on limited resources and lacks the application of constraint sets to refine the initial resources of the ontologies. As a result, the preferential selection of entities is not achieved effectively. Additionally, the TGOS model lacks the ability to consume knowledge from heterogeneous sources and lacks mechanisms for relevance computation. Moreover, the model lacks automation and does not incorporate learning models, which hinders its performance. Therefore, these shortcomings in the TGOS model lead to its underperformance when compared to the proposed model. The OMS model is a framework that integrates ontologies and object modeling within the software engineering domain. It utilizes data modeling to interact with objects and incorporates ontological concepts into object-oriented methodologies. While this approach is intelligent, it lacks a concrete implementation. The model emphasizes the formalization of ontologies through the establishment of axioms and relationship sets. However, it falls short in adequately addressing unstructured data and lacks a learning mechanism. Consequently, the framework does not generate ontologies of the highest quality and lacks a comprehensive ecosystem for learning and semantic reasoning. Considering the limitations in the baseline models, the proposed model stands out due to its robust knowledge synthesis paradigm using metadata and TF-IDF, as well as its knowledge permeability paradigm facilitated by the DBN classifier and SVM classifier for dataset permeability. The utilization of DBSCAN clustering enables effective initial entity clustering, while strong clustering of associations is achieved. The incorporation of semantic reasoning, specifically SOC-PMI based semantic elements computation, further enhances the framework. Additionally, the application of the clonal selection principle for optimization ensures that the proposed framework excels and establishes itself as a best-in-class model.

The line graph illustrates the distribution of precision values for recommendations between the proposed OSPC and the baseline models. In Fig. 2., the proposed OSPC model achieves the highest position in the hierarchy, while COD occupies the lowest position and OMS is placed second from the top.OSPC strikes a balance between automation and human involvement, with 95% of the synthesis process automated and only 5% requiring human review. This semi-automatic approach ensures efficiency while incorporating cognitive input. In contrast, both COD and TGOS heavily rely on human

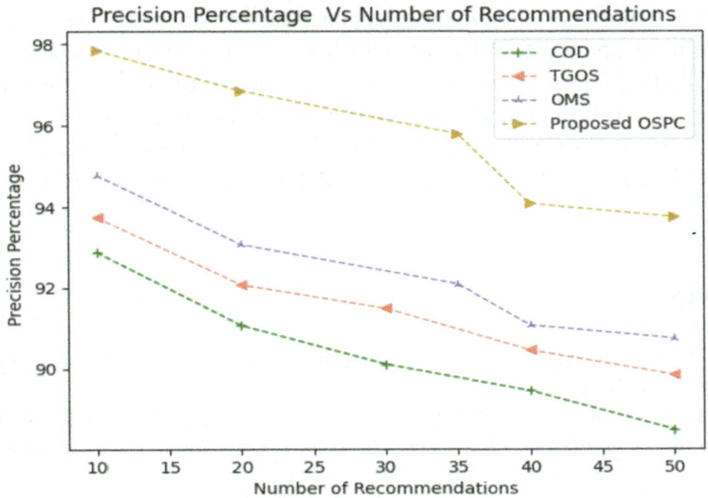

Fig. 2. Precision Vs Number of Recommendations Graph.

intervention throughout the process, limiting automation and scalability. OMS lack a concrete implementation, which further inhibits its performance. OSPC leverages powerful techniques and external knowledge resources to enhance the ontology synthesis process. It incorporates techniques like TF-IDF, deep learning classifiers (e.g., Deep Belief Networks), and clustering algorithms (e.g., DBSCAN). Additionally, OSPC enriches the dataset with valuable entities from resources like WikiData and CYC. These advanced techniques and knowledge resources improve the precision, F-measure percentages, and overall accuracy of OSPC. In contrast, COD lacks learning models and external sources for knowledge enrichment, while TGOS relies on limited resources and lacks mechanisms for relevance computation. OMS lack a comprehensive ecosystem for learning and semantic reasoning.

5 Conclusion

This paper presents a framework for developing ontologies tailored to peace and conflict studies, a distinct field. The framework employs a model that extracts pertinent terms from a dataset and generates knowledge by utilizing resources such as WikiData and CYC. These characteristics are subsequently inputted into a Support Vector Machine (SVM) classifier for dataset classification. Furthermore, the extracted terms are utilized to generate metadata, which is classified using a robust deep learning Deep Belief Network (DBN) classifier. By combining the SVM classifier for the dataset and the DBN classifier for the metadata, the framework effectively manages computational workload and ensures computational feasibility. Additionally, the framework incorporates Density-Based Spatial Clustering of Applications with Noise (DBSCAN) to cluster the initial instances between the classified instances of the dataset and metadata. This clustering procedure establishes an initial strong upper ontology for semantic relevance computation. The SOC-PMI technique is applied to assist in this process. Using an empirically determined threshold and the clonal selection principle, optimization is employed to acquire the most optimal solution set, which is then utilized to formalize the initial ontology. The formalized ontology is further subjected to reasoning, instantiation, and a receive operation. The paper presents an evaluation conducted on standard integrated datasets, achieving an overall precision of 95.33% percent, recall of 96.78%, Accuracy of 96.055%, F-measure of 96.04952787%, and the lowest False Discovery Rate (FDR) of 0.05%.

References

1. Holsapple, C.W., Joshi, K.D.: A collaborative approach to ontology design. Commun. ACM **45**(2), 42–47 (2002)
2. Soininen, T., Tiihonen, J., Männistö, T., Sulonen, R.: Towards a general ontology of configuration. AI EDAM **12**(4), 357–372 (1998). https://doi.org/10.1017/S0890060498124083
3. Siricharoen, W.: Ontology modeling and object modeling in software engineering. Int. J. Softw. Eng. Appl. **3** (2009)
4. Alobaidi, M., Malik, K.M., Sabra, S.: Linked open data-based framework for automatic biomedical ontology generation. BMC Bioinf. **19**(1), 1–13 (2018)
5. Huang, C., Cai, H., Xu, L., Xu, B., Gu, Y., Jiang, L.: Data-driven ontology generation and evolution towards intelligent service in manufacturing systems. Futur. Gener. Comput. Syst. **101**, 197–207 (2019)
6. Wang, Y., et al.: Formal ontology generation by deep machine learning. In: 2017 IEEE 16th International Conference on Cognitive Informatics & Cognitive Computing (ICCI*CC), pp. 6–15. IEEE, July 2017
7. Kumar, N., Kumar, M., Singh, M.: Automated ontology generation from a plain text using statistical and NLP techniques. Int. J. Syst. Assurance Eng. Manage. **7**, 282–293 (2016)
8. Fraga, A.L., Vegetti, M.: Semi-automated ontology generation process from industrial product data standards. In: III Simposio Argentino de Ontologías y sus Aplicaciones (SAOA)-JAIIO 46 (Córdoba, 2017) (2017)
9. Kim, G.W., Lee, D.H.: Intelligent health diagnosis technique exploiting automatic ontology generation and web-based personal health record services. IEEE Access **7**, 9419–9444 (2019)
10. Leo, J., et al.: Ontology-based generation of medical, multi-term MCQs. Int. J. Artif. Intell. Educ. **29**, 145–188 (2019)
11. Rupasingha, R.A., Paik, I., Kumara, B.T.: Specificity-aware ontology generation for improving web service clustering. IEICE Trans. Inf. Syst. **101**(8), 2035–2043 (2018)
12. Surya, D., Palvannan, S., Deepak, G.: Bi-CSem: a semantically inclined bi-classification framework for web service recommendation. In: International Conference on Innovations in Bio-Inspired Computing and Applications, pp. 425–438. Springer, Cham, December 2022
13. Jagan, N.D., Deepak, G., Santhanavijayan, A.: MPTR: a metadata-driven prospective tag recommendation model based on integrative classification. In: Advances in Data and Information Sciences: Proceedings of ICDIS 2022, pp. 291–301 (2022)
14. Deepak, G., Santhanavijayan, A.: OntoBestFit: a best-fit occurrence estimation strategy for RDF driven faceted semantic search. Comput. Commun. **160**, 284–298 (2020)
15. Pushpa, C.N., Deepak, G., Thriveni, J., Venugopal, K.R.: Onto Collab: Strategic review oriented collaborative knowledge modeling using ontologies. In: 2015 Seventh International Conference on Advanced Computing (ICoAC), pp. 1–7, December 2015

16. Deepak, G., Rooban, S., Santhanavijayan, A.: A knowledge centric hybridized approach for crime classification incorporating deep bi-LSTM neural net-work. Multimedia Tools Appl. **80**(18), 28061–28085 (2021)
17. Rithish, H., Deepak, G., Santhanavijayan, A.: Automated assessment of question quality on online community forums. In: International Conference on Digital Technologies and Applications, pp. 791–800. Springer, Cham, January 2021
18. Kumar, N., Deepak, G., Santhanavijayan, A.: A novel semantic approach for intelligent response generation using emotion detection incorporating NPMI measure. Procedia Comput. Sci. **167**, 571–579 (2020)
19. Deepak, G., Gulzar, Z., Leema, A.A.: An intelligent system for modeling and evaluation of domain ontologies for crystallography as a prospective domain with a focus on their retrieval. Comput. Electr. Eng. **96**, 107604 (2021)
20. Kynshi, L.D.L., Deepak, G., Santhanavijayan, A.: MagnetOnto: modelling and evaluation of standardised domain ontologies for magnetic materials as a prospective domain. Int. J. Intell. Enterp. **8**(4), 459–475 (2021)
21. Manoj, N., Deepak, G., Santhanavijayan, A.: OntoINT: a framework for ontology integration based on entity linking from heterogeneous knowledge sources. In: Congress on Intelligent Systems: Proceedings of CIS 2021, vol. 2, pp. 27–35. Springer, Singapore, July 2022

Blockchain's Potential in International Criminal Justice: A Blue Ocean Analysis and Literature Review

Nuno Santos[1(✉)], Joel Curado[1,3], and Joao C. Ferreira[1,2,3]

[1] ISTAR, ISCTE-Instituto Universitário de Lisboa, 1649-026 Lisbon, Portugal
nuno.mbr.santos@gmail.com
[2] Inov Inesc Inovação—Instituto de Novas Tecnologias, 1000-029 Lisbon, Portugal
[3] Casperlabs, Zug, Switzerland

Abstract. Evidence lies at the forefront of any judicial proceeding, and assumes an even more central role when within the context of international criminal justice. These courts have no enforcement abilities, thus require the support of other entities or nations to collect evidence that could [2] build a case against the perpetrators of the crimes under the court's mandate. Evidence management systems, and its repositories are now, more than ever, exposed to threats which could include the exfiltration, tampering or destruction of evidence. Blockchain technology with its characteristics of time-stamping, authentication, immutably and trust among independent parties could be the answer to become anti-fragile against these growing threats. This paper uses the Blue Ocean Strategy analysis to verify the validity of a blockchain solution versus current evidence management solutions within the international criminal justice sector. The paper later analyses the literature available within this context to understand the current state of the art in this area of knowledge.

Keywords: Blockchain · Chain of Custody · Evidence Management · Court · Judicial · International Criminal Justice · Blue Ocean Strategy

1 Introduction

The success of International Criminal Courts, could be perceived based on the fairness of the trials in charging the accused, and this is highly dependent on the ability of the Prosecution, Defence and the Court's Registry to manage evidence properly since collection until disclosure [1].

The advent of blockchain technology has sparked innovative possibilities in various domains, including the realm of international criminal justice. As the custodian of evidence and the guarantor of due process, the criminal justice system demands a meticulous approach to evidence management and chain of custody. However, traditional methods face challenges in maintaining transparency, security, and authenticity.

This paper delves into the intersection of blockchain technology and the field of international criminal justice, seeking to explore the potential of blockchain in revolutionizing the management of evidence and ensuring the integrity of the chain of custody.

V. Sakalauskas et al. (Eds.): IBICA 2023, LNNS 1231, pp. 12–26, 2025.
https://doi.org/10.1007/978-3-031-78946-5_2

Embracing a Blue Ocean Strategy analysis, the research compares existing practices with a potential blockchain based solution as an innovative approach focused on bringing value to the international judicial system.

The primary objectives of this study are twofold. Firstly, employing a Blue Ocean Strategy lens, we aim to understand the potential of blockchain technology in transforming evidence management within the context of international criminal justice. Secondly, through a comprehensive literature review using the PRISMA methodology, the paper evaluates the existing academic discourse and insights related to the utilization of blockchain technology in this domain.

The use of a strategic analysis framework, such as the Blue Ocean Strategy, provides a clear view of the value such solution could bring, which combined with a literature review reinforces the need for further research and expansion of the body of knowledge in this area.

2 Blue Ocean Strategy Applied to Evidence Management in International Criminal Justice

2.1 Identification of Current Body of Knowledge

The Blue Ocean Strategy [2] was invented in 2004 and analyses innovations that enable the creation of uncontested market spaces that make the competition irrelevant. This strategy, in its essence, helps create a roadmap that enables businesses to create or venture into new markets and able to keep costs down and reducing competition [3].

A search for Blue Ocean Strategy applied to judicial system was done in 3 databases (Scopus, ACM Digital Library and IEEE) on 11/10/2023, with the following string:

"Blue Ocean Strategy" AND (Court OR Judicial OR evidence)

Only a total of 22 articles were found in the combined database, and having looked at the title of each one of them, none made reference, direct or indirect, to the judicial sector, as the results only arose from a combination of the first term with the work evidence. Alternatively, using the following string returned only one result [4]:

"Blue Ocean Strategy" AND Blockchain

The search string with only "Blue Ocean Strategy" returned almost 300 results, and after a high level analysis on the titles of the documents, it was clear that this strategy has been researched in a broad range of areas, such as banking industry [4], academia [5], small and medium enterprises [3], logistics [6], and many others.

Transposing this strategy to the area of international criminal justice, as each court has a well-defined mandate and mutually exclusive, the origination of this type of strategy focuses on their evidence management capabilities from a general perspective.

2.2 Strategy Canvas Approach

The authors have several years of combined experience in blockchain, public sector, and national and international Judicial systems within technology areas. The identification of major characteristics of current judicial evidence management systems was done as well as what characteristics they could incorporate to improve and leapfrog the current state.

Using the Four Actions Framework [2], a comparison was done between traditional evidence management systems and a blockchain based one, hereinafter denominated MultitrustBloc.

Factors considered -*Eliminate* - Reliance on controlling party for evidence integrity: it reflects the situation that evidence is at the mercy of the party who controls it. Only when the evidence is disclosed, other parties are able to guarantee the immutability of the evidence and they will have different copies of the same evidence.

MultiTrustBloc resolves this by keeping an immutable record once the evidence is registered through the creation of a digital signature that is stored in the solution and decentralized and replicated among other nodes.

Witness corroboration required: due to the usual long lead time between evidence collection and disclosure and presentation in Court, it is often required at least a witness to validate the veracity of evidence being presented in Court. Additionally, the relative easiness to tamper with the data requires a third party, for example a witness, to validate the evidence authenticity.

MultitrustBloc: by creating a digital signature before the evidence is considered as such, and by replicating it, the court and its judges will have many more guarantees of the authenticity of the evidence, without resorting to other ways of corroboration.

Reduce - Evidence tampering vectors: there are several points when the evidence can be tampered with, or accessed without permission, depending on the security level of the custodian, from collection, to custody, to handing over, there is low visibility and higher chance than desired that evidence could be adulterated.

MultiTrustBloc: by creating a signature of the evidence early in the process, any change to it will be detected as not complying with the original registration of the data, meaning that from that point, the record is immutable.

Multiple copies of the same evidence: for digital evidence, it is very likely that, depending on the mean of storage, the evidence is copied multiple times, leading to increased chances of leakage and tampering.

MultiTrustBloc: allows for having a clear identification of where de evidence is, and if in digital form, to have it directly linked with the evidence record, allowing that the parties with access to access the same file directly without having each one a different place they access it.

Raise - Streamlined Chain of Custody: ensuring that the evidence was always within the possession and control of a trusted custodian is of paramount importance to help guarantee the authenticity and integrity of the evidence.

MultiTrustBloc: Every single change in the evidence is recorded, ensuring that any change in status is well accounted for and should there be any need, all changes easily identified and retrieved.

Scalable evidence reviews and sharing when a legal team receives a piece of evidence from a third party, requires it to register, perform its due diligence to minimize any possibility the evidence is not authentic or invalid in Court. For each piece or batch of evidence received, it requires processing it from the very beginning.

MultiTrustBloc: keeps the log of all changes in the data and allows for accredited organizations to first review the data, and then share it with the legal team for their review. This will allow the latter to have much more information about the evidence, including its origin and if any analysis has been done on it.

Cooperation with interested entities: The IJC relies on a wide range of entities to source evidence that could support the building of a case, as many times it has no physical direct access to persons or areas of interest.

MultiTrustBloc: Opens the network to trusted and non-trusted partners as it allows the former to register possible evidence in the solution, and thus do much of the preparatory work that needs to be done. It also empowers these organizations to become more active as they are able to have a bigger impact in the fulfilling of their objectives. *Create* - Digital evidence and associated record shared synchronously: The evidence and the chain of custody are two different artifacts that attempt to be in as much synchrony as possible, however it is very possible that both are kept in different areas and is linkage could easily break due to human error or outsider's action.

MultiTrustBloc: By creating a strong link between the evidence record, including chain of custody and the evidence itself, the solution can provide a cohesive management of both the evidence and the record.

Realtime chain of custody: As it is built nowadays, the chain of custody is normally done in manual form and each team or entity could have their own protocols.

MultiTrustBloc: establishes a standard regarding the chain of custody. It is also extremely easy and quick to retrieve the whole chain of custody of evidence, by retrieving all the blocks created by the change of status during its lifetime.

2.3 Strategy Canvas

To represent this in a visual form, a Strategy Canvas [2] has been designed to take into consideration the factors mentioned above. Figure 1 depicts how MultiTrustBloc fits into the incumbent evidence management solution.

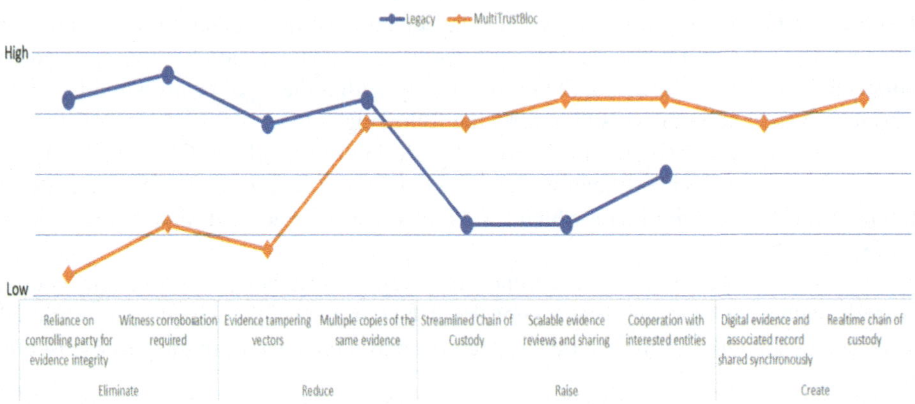

Fig. 1. Strategy Canvas for current Evidence Management solutions and MultiTrustBloc

Validating the Blue Ocean Characteristics
The authors that coined the Blue Ocean strategy [2] established a five-point checklist for understanding if an innovation fits its strategy. This sub-chapter compares the MultiTrustBloc with those same requirements:

Creates uncontested market space?

Yes. While international criminal courts don't have competition, by implementing such innovative solution, their reach will expand by allowing other entities to participate in their activities. It creates a platform where (potential) evidence could be registered, even before the legal team has the approval to investigate a specific situation.

Makes the competition irrelevant?

Not applicable, each court has a unique and very well-defined mandate.

Creates and captures new demand?

Yes. It allows a multitude of organizations to directly participate in its activities and help it pursue the accomplishment of its mandate.

Breaks the value-cost trade-off?

Yes. It becomes a matter of trust and confidence that the data is correct. Teams and Judges can focus on the content rather than on its authenticity.

Aligns the court's activities in pursuit of differentiation and low cost?

Yes. Tasks such as chain of custody verification, or evidence integrity are drastically reduced, since it is done automatically, and easily confirmed. Time and effort are dramatically reduced, also because digital evidence will require less corroboration, meaning less witnesses in the Courtroom testifying, thus less costs in this activity. These can be particularly high.

3 Literature Review

After concluding that a blockchain solution could add value to International Criminal Justice, when compared with the current systems, a systematic literature review was made using PRISMA methodology [7]. The selection of the studies to be included in this review, and within the scope of this research, consisted in all studies that focused on Blockchain and the use case of Chain of Custody within the legal or investigative area. The search string chosen, was therefore, the following:

Blockchain AND "Chain of Custody" AND (Court OR Judicial OR Evidence)

The sources of this identification were four databases (Scopus, Web of Science, ACM Digital Library and IEEE) with which ISCTE has access agreements and were accessed on the 22nd June 2023.

A total of 65 documents were screened, after removing a total of 41 duplicate records from the databases searched. 12 were not able to be retrieved without incurring additional costs. 8 studies with IoT in the title were removed as it deviates from the purpose of this research.

Once the first screening of studies was concluded, 45 articles were assessed for eligibility. 15 of these had an abstract which was not relevant, and for 3 after analyzing the whole article, 2 didn't fit the scope of the research and 1 was in essence a duplication of another relevant article, despite the different title.

In summary, 27 studies qualified for the systematic review, as depicted in Fig. 2.

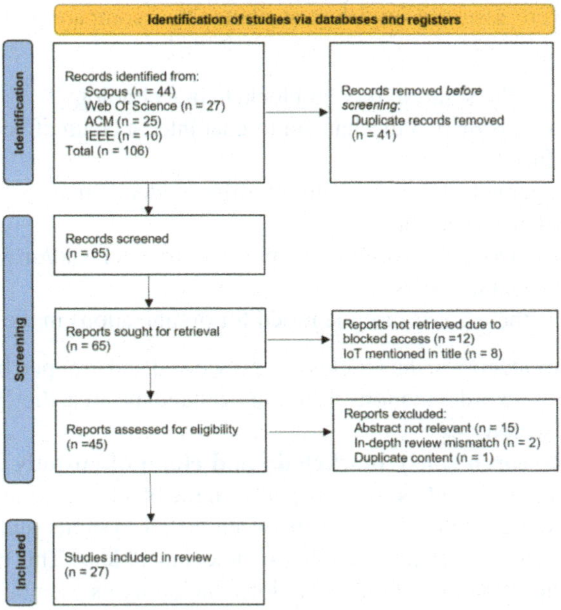

Fig. 2. PRIMA instance of the literature review for this dissertation

3.1 Studies Reviewed

The literature around this topic is still very nascent, with the first related study being published in 2018. Figure 3 shows an increased interest in this topic, despite the trough that occurred in 2022, with only 3 studies published.

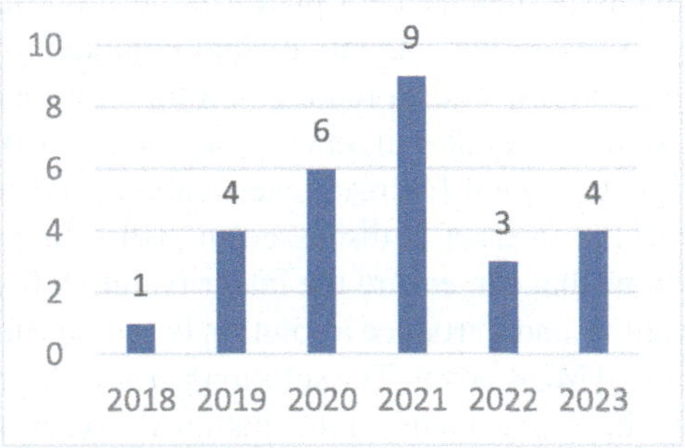

Fig. 3. Number of eligible articles published per year

In this chapter, a variety of articles related to the intersection of blockchain technology and chain of custody are discussed and evaluated, focusing primarily on the applicability of blockchain to improve the integrity, authenticity, and secure handling of digital evidence within a legal and public sector context.

The studies were assessed on 5 key attributes that compare with the area of this research, being:

1. Blockchain: Does the solution rely on blockchain technology?
2. Chain of Custody: Is there a mechanism to guarantee a chain of custody for data, in particular evidence?
3. Evidence Management: Does the solution offer evidence management capabilities, through the evidence lifecycle?
4. Multi-level trust: Does the solution support the transfer and/or access of data by different, independent, parties?
5. Public Sector: Is there a use case identified for organizations in the public sector?

Based on these categories, the articles were reviewed and grouped based on the same matching categories, in order to more clearly understand the area, they focus on and how it relates to this research.

Group A – Research solely blockchain and chain of custody attributes - This group is represented by 4 studies focused solely on the blockchain and chain of custody attributes. Given the complex interplay between social systems and technology, the authors of [8] discuss blockchain through a sociotechnical lens. This perspective offers a more nuanced understanding of the technology, including its societal implications and potential applications. They highlight that blockchain technology could revolutionize the chain of custody process due to its inherent characteristics like transparency, security, and immutability. As with any new nascent technology, they argue that it is very hard to predict the path blockchain will take and propose a learning-by-doing approach to better understand blockchain's capabilities and limitations.

The article by Olukoya [9] provides an analysis on how blockchain technology can be integrated with incident response software to enhance digital investigations. The author suggests that a blockchain ledger can serve as a powerful tool for preserving the integrity of security investigations and associated metadata. Building on models from an open-source incident management platform, he proposes a framework using blockchain technology to ensure the authenticity and immutability of forensic findings and subsequent analysis actions. The stored data on the blockchain, being immutable and verifiable, provides a robust audit trail for every action taken during an investigation.

The topic of multimedia digital forensic investigations and securing the exchange of multimedia data over the internet is discussed in [10]. The authors emphasize the pressing need for solutions that can ensure the integrity, reliability, and trustworthiness of multimedia investigations and introduce a solution based on Blockchain Hyperledger Sawtooth to which they call MF-Ledger. The solution permits these stakeholders to form a private network where they can exchange information and reach consensus on different investigation actions. The authors argue that this addresses some of the key challenges facing the multimedia industry in the era of global digital connectivity and heightened cybersecurity risks.

The article in [11] discusses the proposal of a tamperproof timestamped provenance ledger (TTPL) utilizing a public and trusted blockchain. It offers various levels of integrity verification, and has been designed with a focus on privacy, with scalable characteristics. The authors highlight the feature that allows the data originator to create a permanent proof via a web browser, demonstrating they had access to potentially sensitive data at a specific point in time. This proof can be generated without compromising, revealing, or externally storing any sensitive information. Additionally, this proof is anchored in a timestamped provenance record contained within a public blockchain.

Group B – Research addresses Evidence management with chain of custody on Blockchain, but not multi-level trust nor the public sector - This group includes 7 studies that not only addresses blockchain and chain of custody attributes, but also incorporates evidence management.

The article [12] addresses the current challenges in evidence preservation technologies, which heavily rely on standalone cryptography technology, making them susceptible to tampering. Liu et al. propose a data preservation model that is based on a decentralized blockchain, bolstered by smart contracts. The integration of smart contracts automates the management and verification of this process, significantly reducing the likelihood of human error or tampering and the utilization of blockchain technology in this context provides a decentralized, immutable ledger to record the handling of evidence, thus enhancing its security and integrity.

The authors in [13] underline the vital role of the chain of custody in demonstrating that digital evidence has not been tampered with during any phase of an investigation. Their work is focused on digital image forensics, combining the use of grey hash and blockchain technologies. Grey hash is a form of digital fingerprint for images, and combined with blockchain they present a robust solution for maintaining the integrity of the digital evidence.

Sathyaprakasan and co-authors delve into the fundamental concepts of blockchain and the chain of custody [14]. They bring forward a high-level framework for managing digital forensic evidence utilizing Hyperledger Fabric, a permissioned blockchain infrastructure. They focus on illustrating how blockchain technology can be effectively applied in the domain of digital forensics. The framework they propose aims to integrate the chain of custody (the process of documenting the movements and locations of the evidence from the time of collection) into the Hyperledger Fabric.

Another proposed solution that tackles the problem of evidence management within a resilient chain of custody supported by blockchain is presented in [15]. The authors propose a solution that integrates private and public blockchains, where the public blockchain acts as a guarantor of data integrity and decentralization. By doing so, they aim to create a system that is not only reliable and secure but also allows independent verification of the evidence's chronological journey.

Another work within this group is [16] where the author devises a Blockchain based solution named CustodyBlock to support chain of custody and integrity of digital evidence. It also aims to facilitate transparency and trust between different parties involved in the evidence's life cycle, from its initial collection to its use in legal proceedings.

In [17] the authors propose a system that fuses blockchain technology with machine learning in the form of K-means clustering. The K-means clustering technique is used to determine the storage location of the digital evidence. The primary focus of the proposed system is the security of digital evidence information. The use of machine learning techniques aids in efficient and secure data storage, making the retrieval process more manageable and efficient.

Another blockchain based solution for managing digital evidence was identified in [18] with the characteristic of also tracking the entities accessing this data as a major differentiator from other proposals. This feature is crucial in maintaining a secure chain of custody for digital evidence, assuring that the evidence has not been tampered with during the investigation process. Additionally, Proof of Work (PoW) consensus algorithm is employed to verify the integrity of the blocks in the blockchain.

Group C – Research covering all attributes except 5. Public Sector - This group of 8 studies extends to cover all attributes except public sector area. Zhang et al. [19] focus their research on cloud forensics and identify an approach to validate the provenance of digital forensics from Cloud Service Providers from different jurisdictional areas. It bases the system in a Certificate authority that guarantees the anonymity of the users.

Looking at the topic from the viewpoint of mobile forensics, the authors in [20] identify that while current mobile forensic tools are primarily concerned with the efficiency of data extraction, it is very important to maintain the integrity of such data for its use as evidence in court. They propose a specially designed blockchain system built from the ground up to fulfil the requirements of preserving digital evidence.

The authors in [21] explore the potential of utilizing blockchain technology in the management of chain of custody. They propose the use of a private and permissioned blockchain, showing its applicability in this context. They also evaluate the performance of this system, illustrating the practicality and effectiveness of their solution for the management of digital forensics evidence. They address the essential requirements of a chain of custody (CoC) process which include Integrity, Traceability, Authentication, Verifiability, and Security (being tamper-proof).

In [22] the authors utilize blockchain technology and specifically incorporate a Boneh-Lynn-Shacham signature, known for its short signature size and efficiency. The unique aspect of their proposal is the integration of attribute-based encryption for fine-grained access control. This means that access to the evidence is controlled based on the attributes of the user or the evidence itself, ensuring that only authorized individuals can access specific pieces of evidence. This approach offers an added layer of security and control in the management of digital evidence.

The article [23] introduces a blockchain-based framework for managing digital evidence in a manner that a multitude of stakeholders can accept it. The proposed framework aims to ensure the integrity and immutability of digital evidence. Through this approach, the authors aim to create a real-time, tamper-proof method of managing digital forensic evidence that can effectively meet the demands and challenges of modern forensic investigations. It is constructed in three logical layers:

- The Evidence Layer: This layer supports a trusted storage medium for digital evidence, ensuring its safety and authenticity.
- The Blockchain-based Layer: This layer is where the blockchain technology operates, providing a decentralized and secure system that guards against tampering and allows for the verifiability of the stored evidence.
- The Network Layer: This layer provides peer-to-peer connectivity among all parties that come into contact with the evidence. It facilitates communication and cooperation between different stakeholders.

The article [24] presents a blockchain-based system designed to maintain digital forensics' chain of custody. In this system, the authors introduce a method for managing and exchanging data between different parties, ensuring the integrity and verifiability of the digital evidence. The solution they propose includes several key capabilities such as:

- Evidence Creation: The ability to record and store digital evidence securely.
- Evidence Hash Transfer: The use of cryptographic hash functions to maintain the integrity of digital evidence while it is transferred between different parties.
- Evidence Display: A mechanism for transparently displaying the digital evidence to all relevant stakeholders.

Through these features, the system created by Chopade et al. ensures that the chain of custody in digital forensics is maintained, thus enhancing the credibility and trustworthiness of the digital evidence.

The authors in [25] explore how blockchain technology can be utilized to ensure the authenticity and legality of various processes and procedures in the realm of digital forensics. They recognize the need for a comprehensive view of transactions that affect specific data, extending back to their origin and propose using blockchain technology to create an immutable record of these transactions. To validate their concept, Lone and Mir present a proof of concept using Hyperledger Composer, and finalize the study by measuring the performance of their proposed architecture, demonstrating the potential effectiveness of their blockchain-based approach to managing the chain of custody in digital forensics.

The last article within this group focuses on a practical case study for the creation of smart contracts for managing the chain of custody of digital evidence [26]. It expands on previous work conducted by Bonomi et al. [21]. The authors focus on exploring different aspects of the chain of custody in digital forensics, specifically examining different architectures: centralized, distributed, and multi-blockchain. The authors focus particularly on distributed architecture, which can provide decentralized and tamper-resistant record-keeping, enhancing the integrity and traceability of digital evidence. To showcase the practical implementation of their ideas, the authors develop a prototype system using a distributed architecture combined with smart contracts. These smart contracts incorporate zero-knowledge proof, a cryptographic method by which one party can prove to another that they know a value, without conveying any information apart from knowing the value.

Group D – Research covering all attributes except 4. Multi-level trust - This group of 5 studies covers every attribute except multi-level trust, emphasising public sector applicability. With a focus on log files gathered from a Cloud Service Provider (CSP) [27], the authors identify and analyse the benefits of incorporating blockchain into the process of capturing and maintaining log files in a cloud environment. They also compare this approach to the conventional methods of collecting and preserving forensic evidence.

In the article [28], the authors devise a blockchain based chain of custody solution that introduces the concept of a "digital witness". These digital witnesses are named "Hearsay Digital Witnesses" as they are designed to identify, retrieve, and securely store digital data, functioning as a layer of data replication that further secures the evidence.

Another study fitting within this group is called LEChain [29]. It is designed to oversee the entire judicial process, from evidence collection during police investigations to jury voting during court trials, utilizing blockchain technology to supervise the entire flow of evidence as well as all court-related data.

In [30] a blockchain based evidence management system is discussed as a way to address the challenges of evidence tampering and obstruction of justice prevalent specifically in South Asia. They propose a solution in the form of a hybrid blockchain-based system designed to create a tamper-proof environment for storing and handling evidence. The system is designed to protect the identities of whistleblowers and offers the option for the original uploaders of confidential evidence to disclose it if they believe the involved parties are not acting appropriately.

In [31], Tsai presents an approach to managing evidence within criminal investigations by leveraging the capabilities of blockchain technology, namely the Ethereum [42] platform, supplemented by smart contracts.

Grouping E – Individual studies which had unique coverage of attributes - This group of 3, represents individual studies each with unique coverage of attributes. Sanda et al. [32] focus their contribution within the Cloud Service Providers sector, specifically in Virtual Machines running within those cloud providers. They address the challenges and solutions related to maintaining evidence integrity in cloud forensics. Acknowledging the potential for collusion among stakeholders to tamper with evidence, it emphasizes the crucial role of blockchain in ensuring the integrity of cloud forensic artifacts, such as cloud logs, chain of custody, and file metadata. The authors don't identify how Evidence Management process could be supported.

Akello et al. [33] explore the potential of blockchain technology in addressing gun violence issues, especially in the United States. They propose a blockchain-based framework that enables tracking the chain of custody of firearms as they are transferred between individuals. This study doesn't cover multi-level trust nor evidence management.

The authors of [34] propose a data preservation system called BoCA (Blockchain-of-Custody Application), which is compatible with various blockchains, including Ethereum and BitCash. Their solution allows to store data, encrypted or otherwise, in a transparent and verifiable manner.

3.2 Literature Review Conclusions and Gaps Identified

The literature available within this topic clearly demonstrates the increased attention to the value of blockchain to maintain tamperproof chain of custody and to support the management of evidence in those platforms. Several types of blockchain can be used to implement this and the configuration can be done differently focusing on specific needs from the researchers.

It is clear how blockchain technology can improve the resiliency and integrity of a chain of custody for many areas, including forensic analysis or evidence management. The studies analysed approach this subject in different and novel ways.

During the systematic review no study would encompass all five attributes this research aims to focus on, and as such this is an important addition to the body of knowledge within this area. Table 1 depicts the studies, grouped by attribute set.

Other gap that was identified in the literature reviewed is the absence of use cases and solutions tailored for the international legal public sector, or a sector where a network of providers feed the main entity with evidence.

Table 1. Literature review overview and comparison with this research

Group	Blockchain Framework	Chain of Custody	Evidence Management	Multilevel-trust	Public Sector
A – Research solely block-chain and Chain of Custody at-tributes	Y	Y	N	N	N
B – Research addresses Evidence management with Chain of Custody on Blockchain, but not multilevel trust nor the public sector	Y	Y	Y	N	N
C – Research covering all at-tributes except 5. Public Sector	Y	Y	Y	Y	N
D – Research covering all attributes except 4. Multi-level trust	Y	Y	Y	N	Y
E - [32]	Y	Y	N	Y	Y
E - [33]	Y	Y	N	N	Y
E - [34]	Y	N	Y	N	N

4 Conclusion

The examination of blockchain technology's potential within the domain of international criminal justice, guided by a Blue Ocean Strategy analysis and a comprehensive literature review, has illuminated the transformative capabilities of blockchain in evidence management and chain of custody and a lack of literature in these two domains combined.

Through the lens of the Blue Ocean Strategy, this paper unveils unexplored opportunities for leveraging blockchain technology to enhance the integrity, security, and transparency of evidence management systems within the legal sphere. The decentralized, immutable, and transparent nature of blockchain emerged as a promising solution to mitigate challenges associated with evidence custody and integrity.

The comprehensive literature review, following the PRISMA methodology, provided insights into the evolving landscape of blockchain applications within the judicial context. This area of research could be expanded by the conceptualization of this solution, gathering of detailed requirements and later implementation of a Proof-of-Value to demonstrate the benefits of such solution.

References

1. Cogan, J.K.: The problem of obtaining evidence for international criminal courts. Hum. Rights Q. **22**(2), 404–427 (2000). https://doi.org/10.1353/hrq.2000.0021
2. Kim, W.C., Mauborgne, R.: Blue Ocean Strategy. Harvard Business Review, 01 October 2004, Accessed 15 Jun 2023. https://hbr.org/2004/10/blue-ocean-strategy
3. Awladthani, F.S., Porkodi, S., Saranya, R., Pandurengan, V.: A systematic literature review of the adoption of a blue ocean strategy by small and medium enterprises for sustainable growth. J. Sustain. Sci. Manag. **18**(2), 197–230 (2023). https://doi.org/10.46754/jssm.2023.02.014
4. Jindal, P., Chavan, L.: Role of blockchain technology in creating blue ocean strategy for banking products and services. In: Contemporary Studies of Risks in Emerging Technology: Part B, pp. 169–182 (2023). https://doi.org/10.1108/978-1-80455-566-820231008
5. Nasereddin, A.Y.: Impact of the blue ocean strategy dimensions in achieving competitive advantage from the perspective of faculty members. Inf. Sci. Lett. **12**(6), 2685–2698 (2023). https://doi.org/10.18576/isl/120639
6. Kim, C., Yang, K.H., Kim, J.: A strategy for third-party logistics systems: a case analysis using the blue ocean strategy. Omega **36**(4), 522–534 (2008). https://doi.org/10.1016/j.omega.2006.11.011
7. Page, M.J.: et al.: The PRISMA 2020 statement: an updated guideline for reporting systematic reviews. BMJ n71, March 2021 https://doi.org/10.1136/bmj.n71
8. Ehrenberg, A.J., King, J.L.: Blockchain in Context. Inf. Syst. Front. **22**(1), 29–35 (2020). https://doi.org/10.1007/s10796-019-09946-6
9. Olukoya, O.: Distilling blockchain requirements for digital investigation platforms. J. Inf. Secur. Appl. **62** (2021). https://doi.org/10.1016/j.jisa.2021.102969
10. Khan, A.A., Uddin, M., Shaikh, A.A., Laghari, A.A., Rajput, A.E.: MF-ledger: blockchain hyperledger sawtooth-enabled novel and secure multimedia chain of custody forensic investigation architecture. IEEE Access **9**, 103637–103650 (2021). https://doi.org/10.1109/ACCESS.2021.3099037
11. Jaquet-Chiffelle, D., Casey, E., Bourquenoud, J.: Tamperproof timestamped provenance ledger using blockchain technology. Forensic Sci. Int.-Digit. Investig. **33** (2020). https://doi.org/10.1016/j.fsidi.2020.300977
12. Liu, G., He, J., Xuan, X.: A data preservation method based on blockchain and multidimensional hash for digital forensics. Complexity **2021** (2021). https://doi.org/10.1155/2021/5536326
13. Ali, M., Ismail, A., Elgohary, H., Darwish, S., Mesbah, S.: A procedure for tracing chain of custody in digital image forensics: a paradigm based on grey hash and blockchain. Symmetry **14**(2), (2022). https://doi.org/10.3390/sym14020334

14. Sathyaprakasan, R., Govindan, P., Alvi, S., Sadath, L., Philip, S., Singh, N.: An implementation of blockchain technology in forensic evidence management. In proceedings of IEEE 2021 International Conference on Computational Intelligence and Knowledge Economy (ICCIKE), Naranje V., Singh B., and Velan S., Eds., Institute of Electrical and Electronics Engineers Inc., pp. 208–212 (2021). https://doi.org/10.1109/ICCIKE51210.2021.9410791
15. Burri, X., Casey, E., Bollé, T., Jaquet-Chiffelle, D.-O.: Chronological independently verifiable electronic chain of custody ledger using blockchain technology. Forensic Sci. Int. Digit. Investig. **33** (2020). https://doi.org/10.1016/j.fsidi.2020.300976
16. Alruwaili, F.E.: CustodyBlock: a distributed chain of custody evidence framework. Information **12**(2), 88 (2021). https://doi.org/10.3390/info12020088
17. Kusuma, R.S.: Digital evidence security system design using blockchain technology. Int. J. Saf. Secur. Eng. **13**(1), 159–165 (2023). https://doi.org/10.18280/ijsse.130118
18. Gopalan, S.H., Suba, S.A., Ashmithashree, C., Gayathri, A., Jebin Andrews, V.: Digital forensics using blockchain. Int. J. Recent Technol. Eng. **8**(2) Special Issue 11, 182–184 (2019). https://doi.org/10.35940/ijrte.B1030.0982S1119
19. Zhang, Y., Wu, S., Jin, B., Du, J.: A blockchain-based process provenance for cloud forensics. In: IEEE International Conference on Computer and Communications, ICCC, Institute of Electrical and Electronics Engineers Inc., pp. 2470–2473 (2018). https://doi.org/10.1109/CompComm.2017.8322979
20. Manjre, B.M., Goyal, K.K.: A novel and custom blockchain approach for the integrity assurance of the digital evidences extracted during the extraction and decoding of mobile artifacts from the mobile forensic tools. In: AIP Conference Proceedings, American Institute of Physics Inc., (2023). https://doi.org/10.1063/5.0127910
21. Bonomi, S., Casini, M., Ciccotelli, C.: B-CoC: a blockchain-based chain of custody for evidences management in digital forensics. In: OpenAccess Ser. Informatics. In: Danos, V., Herlihy, M., Potop-Butucaru, M., Prat, J., Tucci-Piergiovanni, S., (eds.), Schloss Dagstuhl-Leibniz-Zentrum fur Informatik GmbH, Dagstuhl Publishing (2020). https://doi.org/10.4230/OASIcs.Tokenomics.2019.12
22. Yan, W., Shen, J., Cao, Z., Dong, X.: Blockchain based digital evidence chain of custody. In: Proceedings of the 2020 the 2nd International Conference on Blockchain Technology, in ICBCT 2020. New York, NY, USA: Association for Computing Machinery, pp. 19–23, May 2020. https://doi.org/10.1145/3390566.3391690
23. Ahmad, L., Khanji, S., Iqbal, F., Kamoun, F.: Blockchain-based chain of custody: towards real-time tamper-proof evidence management. In: ACM International Conference Proceedings Series, Association for Computing Machinery (2020). https://doi.org/10.1145/3407023.3409199
24. Chopade, M., Khan, S., Shaikh, U., Pawar, R.: Digital forensics: maintaining chain of custody using blockchain. In: Proceedings of International Conference on I-SMAC IoT Social, Mobile, Analytics Cloud, I-SMAC, Institute of Electrical and Electronics Engineers Inc., pp. 744–747 (2019). https://doi.org/10.1109/I-SMAC47947.2019.9032693
25. Lone, A.H., Mir, R.N.: Forensic-chain: Blockchain based digital forensics chain of custody with PoC in Hyperledger composer. Digit. Investig. **28**, 44–55 (2019). https://doi.org/10.1016/j.diin.2019.01.002
26. Santamaria, P., Tobarra, L., Pastor-Vargas, R., Robles-Gomez, A.: Smart contracts for managing the chain-of-custody of digital evidence: a practical case of study. Smart Cities **6**(2), 709–727 (2023). https://doi.org/10.3390/smartcities6020034
27. Awuson-David, K., Al-Hadhrami, T., Alazab, M., Shah, N., Shalaginov, A.: BCFL logging: an approach to acquire and preserve admissible digital forensics evidence in cloud ecosystem. Future Gener. Comput. Syst. **122**, 1–13 (2021). https://doi.org/10.1016/j.future.2021.03.001

28. Al-Khateeb, H., Epiphaniou, G., Daly, H.: Blockchain for modern digital forensics: the chain-of-custody as a distributed ledger. In: Jahankhani, H., Kendzierskyj, S., Jamal, A., Epiphaniou, G., Al-Khateeb, H. (eds.) Blockchain and Clinical Trial. Advanced Sciences and Technologies for Security Applications, Springer, Cham (2019). https://doi.org/10.1007/978-3-030-11289-9_7
29. Li, M., Lal, C., Conti, M., Hu, D.: LEChain: a blockchain-based lawful evidence management scheme for digital forensics. Future Gener. Comput. Syst.- Int. J. Escience. **115**, 406–420 (2021). https://doi.org/10.1016/j.future.2020.09.038
30. Shahaab, A., Hewage, C., Khan, I.: Preventing spoliation of evidence with Blockchain: a perspective from South Asia. In: 2021 the 3rd International Conference on Blockchain Technology, in ICBCT 2021, New York, NY, USA, Association for Computing Machinery, pp. 45–52, July 2021. https://doi.org/10.1145/3460537.3460550
31. Tsai, F.-C.: The application of Blockchain of custody in criminal investigation process. In: Knowledge-Based and Intelligent Information & Engineering Systems (kse 2021), Watrobski, J., Salabun, W., Toro, C., Zanni-Merk, C., Howlett, R.J., Jain, L.C. (eds.), Amsterdam: Elsevier Science Bv, pp. 2779–2788 (2021). https://doi.org/10.1016/j.procs.2021.09.048
32. Sanda, P., Pawar, D., Radha, V.: Blockchain-based tamper-proof and transparent investigation model for cloud VMs. J. Supercomput. **78**(16), 17891–17919 (2022). https://doi.org/10.1007/s11227-022-04567-4
33. Akello, P., Vemprala, N., Lang Beebe, N., Raymond Choo, K.-K.: Blockchain use case in ballistics and crime gun tracing and intelligence: toward overcoming gun violence. ACM Trans. Manag. Inf. Syst. **14**(1), 7:1–7:26 (2023). https://doi.org/10.1145/3571290
34. Martin, T., Hammoudeh, M.: Data preservation system using BoCA: Blockchain-of-custody application. In: The 5th International Conference on Future Networks & Distributed Systems, in ICFNDS 2021, New York, NY, USA, Association for Computing Machinery, pp. 70–77, April 2022. https://doi.org/10.1145/3508072.3508084

InCVQA: Incremental Knowledge Derivation Scheme for Visual Question Answering Using Gated Recurrent Units

Anamaya Vyas[1] and Gerard Deepak[2](✉)

[1] Department of Computer Science and Engineering, Vellore Institute of Technology, Vellore, India

[2] Department of Computer Science and Engineering, Manipal Institute of Technology Bengaluru, Manipal Academy of Higher Education, Manipal, India

gerard.deepak.christuni@gmail.com

Abstract. In response to requests for a comprehensive Web 3.0 visual question-answering system, this paper presents a strategic framework. This framework is more coherent than Web 2.0. It converges document and image collections. These datasets build a term pool using category-label extraction or TF-IDF. Latent semantic indexing, metadata generation, and deep learning classification using a resilient GRU model aggregate supplementary knowledge. The YAGO Knowledge Store Repository preprocesses and improves queries. Wikidata API helps enrich intermediate auxiliary knowledge. This strengthens the semantic network and lays the groundwork for semantic computations. The Bose-Einstein index and SoCPMI quantify semantic similarity. Integrating an imperialistic competitive algorithm accelerates convergence to the best solution by computing intermediate optimal solutions. Impressively, the suggested framework achieves 95.81% precision, 96.07% false discovery, and 0.05% false discovery. These results strengthen the framework's position as a top visual question-answering solution, exceeding baseline models.

Keywords: Visual Question Answering · Oceanography · Web Crawler

1 Introduction

Visual Question Answering (VQA) transforms how people interact with visual content. In social networking, e-commerce, and healthcare, visual material like images and videos is vital. However, collecting relevant information from these visual assets is difficult, requiring an intelligent and efficient approach. The innovative InCVQA system uses an Incremental Knowledge Derivation Scheme with Gated Recurrent Units (GRUs) to solve the complex problems of Visual Question Answering (VQA). InCVQA is a powerful and easy-to-use visual data query tool to bridge the gap between visual material and human cognition. Image and video sharing has made a good Visual Question Answering (VQA) system essential. In order to make accurate diagnosis, doctors must analyse medical imagery. Healthcare practitioners can use natural language queries to get fast

V. Sakalauskas et al. (Eds.): IBICA 2023, LNNS 1231, pp. 27–38, 2025.
https://doi.org/10.1007/978-3-031-78946-5_3

and accurate answers from the InCVQA system. Human-like questioning and computer vision have improved medical diagnosis, benefiting patients and doctors. In the age of artificial intelligence and data-driven decision-making, InCVQA is a promising solution that leverages visual data and natural language processing. InCVQA uses the Incremental Knowledge Derivation Scheme to learn from data, improving its performance and understanding of varied visual situations. Gated Recurrent Units (GRUs) enable contextually informed inference and response creation. It becomes a powerful tool with many uses outside medical, including retail, education, and entertainment.This study examines the InCVQA framework's architectural design, training methods, and performance assessment. This study uses specific examples and performance measures to highlight the importance and potential impact of InCVQA as an innovative method to Visual Question Answering across different sectors. InCVQA's fundamentals show that it can revolutionise how humans interact with visual content. Thus, it advances an intelligent and instinctual future.

Motivation: The suggested paradigm blends auxiliary knowledge and argumentation for semantics-driven visual question-answering and is Web 3.0-compliant. Visual question-answering frameworks that combine semantic reasoning and learning are needed as the web moves towards Web 3.0. Existing frameworks frequently use standalone machine learning or deep learning methods without semantics. The hybrid intelligence technique for web video recommendation uses semantics-oriented knowledge-based reasoning and learning.

Contribution: The proposed research uses GRU (Gated Recurrent Units) for metadata classification in a Knowledge Centric approach to learning infrastructure. Topic modelling, semantic indexing, and Wikidata querying add knowledge to two datasets to extract terms and categories. The YAGO Knowledge Store expands knowledge. This metadata-generating augmentation method increases knowledge density exponentially. The Bos-Einstein index and SoCPMI measure aid semantic thinking, whereas a competitive algorithm is a strategic metaheuristic optimisation model that efficiently converges to optimality.

***Organization*:** The manuscript is structured in a well-organized manner, beginning with a comprehensive literature review. It then proceeds to outline the Proposed System Architecture, followed by the execution of the model and a thorough performance analysis of the results achieved.

2 Related Works

Antol, S et al. [1] introduced a solution for the visually impaired called Visual Question Answering (VQA), involving accurate responses to open-ended image queries. This demands image comprehension and reasoning beyond captions. A dataset with images, questions, and answers is provided for VQA evaluation. Baseline methods and human performance are compared. Wang. P et al. [2] proposed Fact-based VQA which tackles the limitations of standard VQA that only deal with straightforward questions. FVQA involves questions demanding external information for more profound reasoning. FVQA augments regular VQA datasets with image-question-answer-supporting fact tuples, represented as structured triplets. Baseline models' performance is assessed on FVQA, and a model employing supporting-facts reasoning is presented. Zheng, S et al. [3] advocated a drug-protein interaction prediction technique for drug development. Similar to visual question answering, they use 2D distance maps for proteins and molecular notations for medications in deep learning. This method improves accuracy over benchmarks. Visualising molecular regions of interest deepens understanding. Marino, K et al. [4] created OK-VQA to alleviate the constraints of current Visual Question Answering (VQA) benchmarks by integrating external knowledge questions. New benchmark OK-VQA has almost 14,000 questions that require external knowledge. This innovative challenge challenges state-of-the-art VQA models, showcasing the dataset's complexity. The benchmark promotes knowledge-based VQA research. K. Chen et al. [5] proposed a new attention-based deep learning architecture for Visual Question Answering (VQA). ABC-CNN employs question-guided attention to appropriate visual regions to provide accurate answers. Attention is determined by convolving picture characteristics with customizable convolutional kernels based on query meaning. Liu, S et al. [6] developed BPI-MVQA, a bi-branched model for medical visual question answering. The model's first branch extracts image sequence and spatial characteristics using a transformer-structured parallel network. A multi-head self-attention system integrates these multi-modal properties. The second branch retrieves image feature similarities from a VGG16 network to label text. Liu, L et al. [7] developed a Fact-based Visual Question Answering (FVQA) system that blends picture comprehension and external knowledge utilising two modules (System 1 and System 2). System 1 represents images and queries with a dual BERT-style model, whereas System 2 uses a graph neural network with fact and semantic networks for dual-level reasoning. Results on FVQA and OK-VQA datasets show its advantages over baseline approaches and capacity to reveal reasoning process insights. Interactive Question Answering (IQA) by Gordon, D et al. [8] involved an autonomous agent interacting with a dynamic visual environment to answer questions. This requires scene perception, object interaction, and action planning. The Hierarchical Interactive Memory Network (HIMN) uses controllers for different temporal abstraction levels. IQUAD V1, a new dataset with 75,000 questions and scene settings in a photo-realistic simulated environment, was used to evaluate HIMN. The inverse problem of Visual Question Answering (iVQA) was established by Liu, F et al. [9] to test visuo-linguistic understanding. iVQA generates questions for picture-answer pairings, requiring more visual knowledge than VQA. Their iVQA approach uses partially generated questions and answers for dynamic inference and attention adjustment. A novel ranking metre and linguistic metrics demonstrate the model's ability to generate different, relevant, and grammatically correct questions and replies. An ASP-based visual question-answering framework called AQuA (ASP-based Question Answering), which uses YOLO for object identification, was proposed by Basu, K. et al. [10]. It converts user inquiries into ASP queries and then extracts semantic relations for more thorough understanding. AQuA outperforms humans in the CLEVR test, achieving 93.7% accuracy and offering justifications without any prior training. This human-like strategy, as opposed to machine learning techniques, ensures correctness and explanation.

Underwater Oceanographic Sciences relies on Visual Question Answering (VQA). The dataset includes US Geological Survey Shallow Reef Track Imaging System photos [11], geotagged Seafloor Images, and position data from Cape Cod Bay, Massachusetts bottom photos [12]. The paper includes the Fish Object Direction Dataset [13]. A customized annotator tool meticulously annotated these datasets. Each annotation is vital to reprioritizing records in datasets based on labels or annotations. The project uses a web crawler to harvest deep-sea and oceanographic photos from the current World Wide Web framework to enrich data. This creates and maintains an image repository with labels. This aggregated dataset of underwater photographs is the basis for our experiments and a carefully managed library for domain study and analysis.

The study carefully collected a dataset from three sources to facilitate Visual Question Answering (VQA). The first source is USGS Coastal and Marine Geology Programme static oceanographic observations at Martha's Vineyard Coastal Observatory, which illuminate oceanic processes [14]. The second source covers northeastern US seas and ground fishing villages from 1997 to 2014 [15]. The NOAA National Marine Fishery Service Vessel Trip Report data [16] provides this extensive resource. In [17–31] several frameworks in support of the literature of the proposed work have been depicted.

3 Proposed Architecture

The suggested system architecture is a framework for visual question answering (VQA) that takes a knowledge-centric and semantically inclined approach to the problem. Questions posed by users act as the key motivator for the development of this architecture. The user's inquiry is first subjected to preprocessing, during which various methods, including tokenization, lemmatization, stopword removal, and named entity recognition (NER), are utilized to isolate relevant keywords. These keywords, which were retrieved from the preprocessing stage, will be handled as separate terms. After then, a query language known as SPARQL is used to interrogate the Yago Knowledge Store using the results. The system obtains pertinent information from the Yago Knowledge Store using SPARQL queries as the primary method of data retrieval. The gathered information is subsequently put to use by the VQA model, which generates responses that are not only correct but also meaningful to the user's inquiries.

A dataset of documents and another dataset of images are both incorporated into the model as separate but complementary lightweight datasets because these datasets contain a substantial number of annotations, classification is not necessary. A lookup agent is used to manage the data rather than classification because of its efficiency. Labels are retrieved from the photos that make up the image collection, and these labels, together with the related categories that serve as annotations, are then used. The term pool is then expanded with the addition of these labels and categories. The technique known as Term Frequency-Inverse Document Frequency (TF-IDF) is the one that is used when analyzing the document dataset. This procedure locates phrases that appear often throughout the document corpus and are judged to be informative. After that, these illuminating terms are added to the pool of potential terms as well. Latent semantic indexing, often known as LSI, is a topic modelling framework that is used to perform further processing on the term pool that was collected from the datasets. By using the structure of the World Wide Web as a reference corpus, LSI is able to aid in the discovery of previously unknown but pertinent subjects. The LSI technique allows for the identification and incorporation of newly discovered entities into already established terms.

After the entities have been combined, the MetaTag-Harvest tool is used to produce metadata about the new entity using the merged data. In this step of the process, we will extract any pertinent information or metadata that is associated with the entities that make up the term pool. The entities' representations within the system are improved as a result of the metadata being generated since it adds more context and descriptive information about the entities. Through the utilization of LSI and MetaTag-Harvest, the model is able to acquire a more in-depth grasp of the concepts and entities that are present in the term pool, hence improving its capacity to offer responses that are correct and informative.

The task of extracting tags from the word pool dataset falls on the shoulders of the MetaTag Harvester tool. These tag phrases, which are retrieved through the harvester, can be rather extensive and difficult to work with if there is no additional processing or classification performed on them. In order to solve this problem, the tags are put through a classification process with the assistance of a GRU classifier, which is an effective deep learning classifier. The large-scale volume, variety, and density of the metadata led to the selection of a deep learning classifier such as GRU as the optimal solution. In addition, GRUs have the benefit of autohandling feature selection, which positions them in a favourable position for the responsibility at hand. The GRU classifier is responsible for assigning the instances to specified categories or classes after they have been classified.

After that, the instances that have been categorised are received from the GRU classifier and are used to construct a semantic network. This is accomplished by the use of Renyi Entropy in the computation of the nodal entropy. The evaluation of the information content and value of each node in the semantic network is aided by the use of the nodal entropy, which makes it possible for the interactions between the entities to be represented in a manner that is more meaningful. The model obtains a structured semantic network through the use of the MetaTag Harvester, GRU classification, and Renyi Entropy. This improves the model's capability to explore and comprehend the interconnection of the metadata.

The terms that were found through LSI and those that were found in the term pool set are merged together to create a bigger merged network, which is then used to formalise the semantic network. The Renyi entropy metric is utilised in the computation of the nodal information measure for this combined network. In addition, each entity that is part of the existing structure of the semantic network is routed through an agent before being added to the WikiData API. In order to uncover and create a sizable enriched semantic network, this technique retrieves a big number of associated entities that are connected to the precise nodes. Following this, the enriched semantic network is placed through a computation that determines its level of semantic similarity to the Yago-enriched query terms. Measures of Socpmi (second-order co-occurrence point-wise mutual information) and Bose Einstein's index are used in this computation. The step deviation is set to 0.25. Both the Bose Einstein index and the Socpmi have been calibrated to have a step deviation of 0.25, while the Socpmi has been calibrated to have a median step deviance of 0.05.

The production of powerful semantics is helped along by the combination of Socpmi's strength and the influence of Bose Einstein's index. However, getting there is not a straightforward procedure at all. Alternately, it is carried out in accordance with the Imperialistic Competitive algorithm. The semantics matrix is constructed and initial solutions are obtained as a first step in the process. In order to maximise the effectiveness of the given initial feasible solution set, the Imperialistic Competitive algorithm is put to use. The Bose-Einstein index and the Socpmi are each treated as their own separate objective function by the metaheuristic optimisation technique that is being discussed here. By repeating the process in an iterative fashion, a solution set that is more practicable can be obtained, which in turn improves the semantic network's quality as a whole.

After the first solutions have been obtained, they are then subjected to another round of optimisation with the help of a metaheuristic optimisation algorithm called the Imperialistic Competitive algorithm. Maintaining the Bose Einstein index and Socpmi with the same thresholds as their respective separate objective functions is the way to accomplish the goal of making the initial feasible solution set into a more feasible solution set. This approach of optimisation entails carrying out alternative iterations in a random order in order to arrive at a solution set that is both optimal and practicable (Fig. 1).

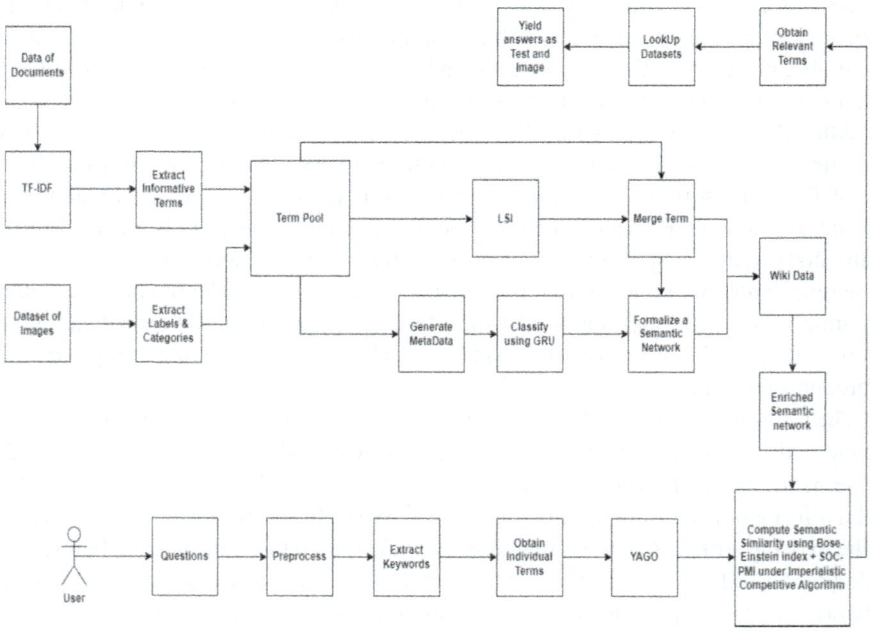

Fig. 1. Proposed system architecture

The optimal solution set that was obtained in this manner is then provided as the initial relevant words, which are ordered in ascending order according to the Socpmi measure. These pertinent terms act as markers that identify the occurrence of the query, in addition to the photos and the dataset. The presence of these pertinent terms is connected, and the answers are extracted from the exact statements included inside the document with the use of a lookup agent that was developed with agent speak. The paper is combed through to determine the response to the inquiry, and then that response, together with any pertinent images linked with it, is shown to the user. The completed product includes the question, a textual response to the inquiry, and any supporting visuals that are pertinent to the subject. After the output has been generated, it is normally transmitted somewhere for inspection and additional analysis.The Shannon entropy provides the foundation for the Rényi entropy, which is a mathematical term that expands the idea of entropy. It evaluates the degree to which a probability distribution is subject to uncertainty or unpredictability. Equation (1) depicts the equation for the Renyi entropy.

$$H_\alpha(X) = \left(\frac{1}{1-\alpha}\right) * \log\left(\sum_{i=0}^{n} P(X_i)^\alpha\right) \tag{1}$$

In this context, the Rényi entropy of a random variable X is denoted by the symbol H(X), and its associated probability distribution is denoted by the symbol P(X). The importance that is placed on the various probabilities in the distribution is determined by the parameter known as. As gets closer to 1, the Rényi entropy will eventually match up with the Shannon entropy. In order to determine the Rényi entropy, we must first compute the sum of the probabilities multiplied by the power of for each of the potential outcomes X_i of the random variable X. This will allow us to determine the entropy. After performing a logarithmic transformation on this sum (with a base of 2), the final expression is multiplied by $\left(\frac{1}{1-\alpha}\right)$, which brings it closer to its final form.

The amount of information or uncertainty included within a probability distribution can be quantified with the use of the Rényi entropy. It is possible to capture a variety of properties of the distribution by modifying the value of the parameter. For example, when is given the value 2, it corresponds to the quadratic Rényi entropy. This concept has applications in a variety of domains, including information theory, data compression, and machine learning, amongst others. The Renyi entropy makes it possible to quantify and evaluate the degree of uncertainty across a variety of probability distributions, which in turn simplifies the process of conducting data and information system analyses.

4 Results

The proposed InCV QA model, an Incremental knowledge derivation scheme for visual question answering employing gated recurrent units, was rigorously evaluated utilising a variety of performance metrics. InCV QA attained exceptional precision, recall, accuracy, and f-measure percentages of 95.81%, 96.33%, 96.07%, and 96.07%, respectively, as shown in Table 1. Moreover, its false discovery rate (FDR) was an impressively low 0.05. This evaluation demonstrates that InCV QA outperforms other methods in terms of precision, recall, accuracy, and f-measure percentages, while reducing the false discovery rate with remarkable efficiency. These results validate the efficacy of the InCV QA model and its viability as a robust solution for visual question-answering tasks.

To evaluate and compare the performance of the proposed InCV QA model, baselines for three distinct models were established: SVSBV QA, LTR, and QRU. The FDR for the SVSBV QA model was 0.12, and its precision, recall, accuracy, and F-measure percentages were 88.72%, 90.8%, 89.4%, and 89.40%, respectively. The average precision of the LTR model was 90.69%, the average recall was 90.18%, the average accuracy was 91.435%, the average F-measure was 91.43%, and the FDR was 0.10. Similarly, the QRU framework obtained average precision, recall, accuracy, and F-measure percentages of 90.69%, 93.07%, 92.03%, and 92.02%, respectively, with an FDR of 0.10. In terms of precision, recall, accuracy, and F-measure percentages, it is evident that the InCV QA model outperforms the other models based on the data presented in Table 1. Additionally, it has the lowest FDR value of all models. The success of the InCV QA model is attributable to its semantically intelligent framework, which incorporates robust auxiliary knowledge via knowledge derivation. In comparison to the baselined models, these results demonstrate the superiority of the InCV QA model and its potential as an effective solution for visual question-answering tasks.

Table 1. Comparison of Performance of the proposed InCVQA with other approaches

Model	Average Precision %	Average Recall %	Average Accuracy %	Average F-Measure %	FDR
SBVQA [17]	88.72	90.08	89.4	89.40	0.12
LTR [18]	90.69	92.18	91.435	91.43	0.10
QRU [19]	90.99	93.07	92.03	92.02	0.10
Proposed InCVQA	95.81	96.33	96.07	96.07	0.05

The InCV QA model's multi-step knowledge generation improves visual question responding. TF-IDF identifies informative keywords in the dataset, prioritising relevant phrases across the dataset. A term pool is created from these useful terms. The term pool is enriched with semantic correlations from image categories and labels. The model uses latent semantic indexing-based topic models to uncover term patterns and associations to grow its knowledge base. Advanced Metadata Generator tools add valuable metadata to the semantic network, generating exponential knowledge. GRU, a sequential data processing deep learning model, simplifies metadata handling. InCV QA formalises the semantic network, organising extracted knowledge and links. The model combines Wikidata, a community-contributed and verified knowledge database, to expand its knowledge base. This complete method creates a powerful and well-structured semantic network, enabling the InCV QA model to excel in visual question answering with a plethora of knowledge.

In the proposed InCV QA approach, formalised Semantic Networks efficiently organise gradually generated information, ensuring a reliable and well-structured basis. User questions are preprocessed to extract keywords and enhanced with Yago knowledge to optimise query processing. The enriched dataset's semantic network is enhanced with a resilient step deviance and semantic reasoning approach to handle difficult questions and leverage semantic similarity. InCV QA uses the Imperialistic competitive algorithm to refine the initial solution set and improve results. These improvements and optimisations boost the model's speed, making it the best visual question answering system solution. The suggested InCV QA model excels at processing visual questions and providing accurate and meaningful responses by combining incremental knowledge acquisition, semantic reasoning, and optimisation strategies. It excels at visual question answering thanks to its well-structured semantic network and excellent optimisation algorithms.

The efficacy of the LTR baseline model is deemed inadequate as a result of various significant variables. The relevance computation in Learning to Rank (LTR) is shown to be weak, despite the integration of Neural module networks and an end-to-end Module network. These networks utilise knowledge-based architectures similar to semantic networks. Consequently, the knowledge produced by the model remains unoptimized and lacks the requisite fine-tuning to offer precise and all-encompassing responses. Furthermore, the model has a deficiency in atomicity, as it does not effectively decompose intricate problems into more manageable and discrete components. The aforementioned constraint hinders the system's capacity to efficiently process complex inquiries. In contrast, the InCV QA model that has been proposed has a higher level of performance in comparison to the LTR model. The InCV QA model produces exact and meaningful outcomes in visual question answering through the utilisation of incremental knowledge expansion, powerful semantic reasoning, and optimisation with the Imperialistic competitive algorithm. The semantic network of this system is well organised and incorporates sophisticated approaches, enabling it to overcome the limitations of the LTR model and position itself as a very effective solution within the field. The subpar performance of the SbQA model can be attributed to its dependence on rudimentary baseline approaches. The model employs Convolutional Neural Networks (CNNs) and utilises Recurrent Neural Networks (RNNs) to extract features at the word level. However, relying solely on word embedding presents a constraint in efficiently training the model. Consequently, the learning process becomes constrained, resulting in underfitted training as a consequence of inadequate knowledge representation. On the other hand, the InCV QA model that has been suggested demonstrates superior performance due to its incorporation of a full methodology for knowledge derivation, semantic reasoning, and optimisation. The InCV QA model demonstrates significant advancements in visual question answering by employing a systematic approach to knowledge expansion, leveraging a meticulously organised semantic network, and including sophisticated algorithms such as the Imperialistic competition algorithm. This model effectively addresses the constraints of the SbQA paradigm and delivers noteworthy outcomes. The InCV QA model demonstrates superior performance compared to the SbQA model and establishes itself as a very effective and advanced solution for visual question answering tasks, through leveraging robust learning paradigms, sophisticated semantic networks, and optimised knowledge representation. QRU's underperformance is due to its question representation update for visual question answering. It uses numerous reasoning layers and intricate visual relations, but not term-level relations. Visual reasoning may yield appealing results, but it ignores labels and term-level information. Despite pretrained CNN and GRU, the QRU model lacks substrate knowledge to interpret and use auxiliary information. CNN handles complicated visual relations effectively, while the GRU's limited term-level relation capture leads in underfitted knowledge representation. Thus, the QRU model's performance lags below other baseline models due to a lack of complete knowledge and term-level linkages.

Incremental knowledge derivation, semantic reasoning, and optimisation using advanced algorithms solve these problems in the proposed InCV QA model. This comprehensive approach makes the model better at visual question answering than the QRU model and other baseline models. The proposed InCV QA model outperforms baseline models due to its complete knowledge system, progressive knowledge production, and robust semantic similarity metrics. The approach uses semantic similarity measurements like Bose-Einstein index and SoC PMI to reason semantically and gain significant insights from data. Wikidata integration enriches the model's semantic network, providing a solid foundation for accurate and informed answers. The InCV QA paradigm is unique in its ability to use a sophisticated knowledge enrichment environment. The Imperialistic competitive algorithm optimises the InCV QA model's solutions, improving performance and efficiency. The Imperialistic competitive algorithm, robust knowledge system, incremental knowledge generation, semantic similarity metrics, and optimisation enable the proposed InCV QA model to outperform all baseline models. Its semantic reasoning and accuracy make it a top visual question answering system (Fig. 2).

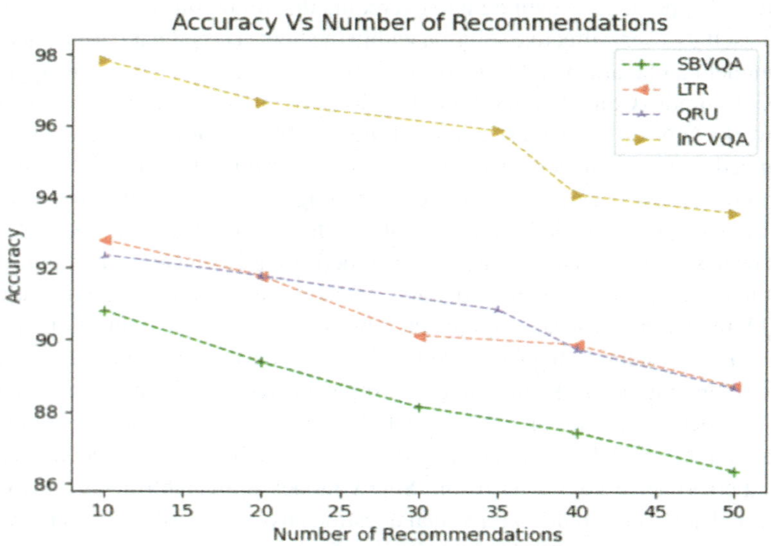

Fig. 2. Precision Vs No. of Recommendations

5 Conclusion

This research offers a strategic inCVQA paradigm for knowledge-centric visual question-answering augmentation. A shared term pool from dataset terms unifies the model despite using two datasets. This term pool generation, metadata creation, and topic model enrichment contextualise knowledge, creating a more dense framework. Semantic network formalisation and Wikidata enrichment enhance knowledge accumulation exponentially. The model's knowledge base is expanded by query word preprocessing and enrichment utilising the YAGO repository, a standard knowledge source. Bose-Einstein Index, SoCPMI with differential thresholds, and step deviance metrics improve semantic understanding. Metaheuristic optimisation technique the utilitarian cooperative algorithm correlates visual inquiries and image-based replies for best results. The suggested algorithm's Precision of 95.81, Recall of 96.33, Accuracy of 96.07, F-Measure of 96.07, and low FDR of 0.05 make it a top visual question-answering paradigm.

References

1. Antol, S., et al.: Vqa: visual question answering. In: Proceedings of the IEEE International Conference on Computer Vision, pp. 2425–2433 (2015)
2. Wang, P., Wu, Q., Shen, C., Dick, A., Van Den Hengel, A.: Fvqa: fact-based visual question answering. IEEE Trans. Pattern Anal. Mach. Intell. **40**(10), 2413–2427 (2017)
3. Zheng, S., Li, Y., Chen, S., Xu, J., Yang, Y.: Predicting drug–protein interaction using quasi-visual question answering system. Nat. Mach. Intell. **2**(2), 134–140 (2020)
4. Marino, K., Rastegari, M., Farhadi, A., Mottaghi, R.: Ok-vqa: a visual question answering benchmark requiring external knowledge. In: Proceedings of the IEEE/CVF Conference on Computer Vision and Pattern Recognition, pp. 3195–3204 (2019)
5. Chen, K., Wang, J., Chen, L.C., Gao, H., Xu, W., Nevatia, R.: ABC-CNN: an attention based convolutional neural network for visual question answering (2015). arXiv preprint arXiv: 1511.05960
6. Liu, S., Zhang, X., Zhou, X., Yang, J.: BPI-MVQA: a bi-branch model for medical visual question answering. BMC Med. Imaging **22**(1), 1–19 (2022)
7. Liu, L., Wang, M., He, X., Qing, L., Chen, H.: Fact-based visual question answering via dual-process system. Knowl.-Based Syst. **237**, 107650 (2022)
8. Gordon, D., Kembhavi, A., Rastegari, M., Redmon, J., Fox, D., Farhadi, A.: Iqa: visual question answering in interactive environments. In: Proceedings of the IEEE Conference on Computer Vision and Pattern Recognition, pp. 4089–4098 (2018)
9. Liu, F., Xiang, T., Hospedales, T.M., Yang, W., Sun, C.: IVQA: inverse visual question answering. In: Proceedings of the IEEE Conference on Computer Vision and Pattern Recognition, pp. 8611–8619 (2018)
10. Basu, K., Shakerin, F., Gupta, G.: Aqua: asp-based visual question answering. In: International Symposium on Practical Aspects of Declarative Languages, pp. 57–72. Springer, Cham, January 2020
11. https://search.dataone.org/view/ef75efb3-9b8c-406c-b519-06b34fed8457
12. https://www.geoplatform.gov/metadata/27e94277-5335-5249-8eca-5102cb8bc67d
13. https://public.roboflow.com/object-detection/fish/1
14. https://search.dataone.org/view/2355f767-0944-4dea-a83b-6f62b087b487
15. https://erddap.bco-dmo.org/erddap/info/bcodmo_dataset_752624/index.html
16. https://exaly.com/journal/15993/deep-sea-research-part-i-oceanographic-research-papers/
17. Zhou, B., Tian, Y., Sukhbaatar, S., Szlam, A., Fergus, R.: Simple baseline for visual question answering (2015). arXiv preprint arXiv:1512.02167
18. Hu, R., Andreas, J., Rohrbach, M., Darrell, T., Saenko, K.: Learning to reason: end-to-end module networks for visual question answering. In: Proceedings of the IEEE International Conference on Computer Vision, pp. 804–813 (2017)
19. Li, R., Jia, J.: Visual question answering with question representation update (qru). In: Advances in Neural Information Processing Systems, vol. 29 (2016)
20. Manoj, N., Deepak, G.: ODFWR: an ontology driven framework for web service recommendation. In: Data Science and Security: Proceedings of IDSCS 2021, pp. 150–158. Springer, Singapore (2021)

21. Kannan, B.D., Deepak, G.: I-DLMI: web image recommendation using deep learning and machine intelligence. In: International Conference on Hybrid Intelligent Systems, pp. 270–280. Springer, Cham, December 2022
22. Yethindra, D.N., Deepak, G.: A semantic approach for fashion recommendation using logistic regression and ontologies. In: 2021 International Conference on Innovative Computing, Intelligent Communication and Smart Electrical Systems (ICSES), pp. 1–6. IEEE, September 2021
23. Agrawal, D., Deepak, G.: HSIL: hybrid semantic infused learning approach for course recommendation. In: International Conference on Digital Technologies and Applications, pp. 417–426. Springer, Cham, January 2022
24. Adithya, V., Deepak, G.: HBlogRec: a hybridized cognitive knowledge scheme for blog recommendation infusing XGBoosting and semantic intelligence. In: 2021 IEEE International conference on electronics, computing and communication technologies (CONECCT), pp. 1–6. IEEE, July 2021
25. Krishnan, N., Deepak, G.: KnowCommerce: a semantic web compliant knowledge-driven paradigm for product recommendation in e-commerce. In: Congress on Intelligent Systems: Proceedings of CIS 2021, vol. 2, pp. 37–45. Springer, Singapore, July 2022
26. Giri, K.S.V., Deepak, G.: A semantic ontology infused deep learning model for disaster tweet classification. Multimed Tools Appl. (2023). https://doi.org/10.1007/s11042-023-16840-6
27. Deepak, G., Trivedi, I.: A hybridized deep learning strategy for course recommendation. Int. J. Adult Educ. Technol. (IJAET) **14**(1), 1–16 (2023)
28. Deepak, G., Santhanavijayan, A.: OntoBestFit: a best-fit occurrence estimation strategy for RDF driven faceted semantic search. Comput. Commun. **160**, 284–298 (2020)
29. Valecha, A., Deepak, G. and Surya, D.A.: KIASOntoRec: a knowledge infused approach for socially aware ontology recommendation. In: International Conference on Innovations in Bio-Inspired Computing and Applications. Springer, Cham (2022)
30. Shrivastava, R.R., Deepak, G.: BiCropRec: a Bi-classifier approach for crop recommendation based on inclusion of semantic intelligence and topic modelling. In: Advances in Data and Information Sciences: Proceedings of ICDIS 2022. Springer, Singapore, pp. 151–160 (2022)
31. Shaw, H., Deepak, G.: MetaBlog: a metadata driven semantics aware approach for blog tagging. In: International Conference on Digital Technologies and Applications, pp. 406–414. Springer, Cham, January 2022

StrategicVideoRec: A Strategic Approach for Scientific Video Recommendation Integrating BERT and Fact Driven Semantics

M. Prathibha[1] and Gerard Deepak[2](✉)

[1] Department of Biomedical Engineering, Sri Sivasubramaniya Nadar College of Engineering, Chennai, India

[2] Department of Computer Science Engineering, Manipal Institute of Technology Bengaluru, Manipal Academy of Higher Education, Manipal, India

gerard.deepak.christuni@gmail.com

Abstract. This paper proposes a semantically oriented strategic framework for web video recommendation. The transition of the web from the conventional, standard Web 2.0 to a much more intelligent and semantic Web 3.0 necessitates the utilization of a video recommendation framework that is compliant with Web 3.0. Metadata generation forms the basis of the proposed strategic video recommendation framework, which expands the auxiliary knowledge exponentially within the framework. BERT classifier is integrated into the model to classify the metadata, while Convolutional Neural Network (CNN) is utilised to categorize the dataset. Normalised Pointwise Mutual Information (NPMI) computation serves as a strategic threshold and Genetic Algorithm helps in semantics-oriented reasoning and optimisation through Metaheuristics. Salient Semantic Analysis (SSA) helps in yielding the most optimal final solution set. The proposed framework, the best-in-class framework for video recommendation on Web 3.0, has attained an overall precision of 95.75% with lowest value of FDR of 0.05 and an accuracy of 96.07%.

Keywords: Web Video Recommendation · Fact Driven Semantics · Machine learning · Artificial Intelligence

1 Introduction

In the digital era, the proliferation of web-based video content has revolutionized the way people consume information, entertainment, and education. The sheer volume of videos available on various online platforms has led to a new challenge: how to efficiently guide users through this vast content landscape to discover videos that align with their preferences and interests. This challenge has given rise to the critical role of web video recommendation systems in shaping users' experiences and interactions with online video platforms. Web video recommendation systems utilize sophisticated algorithms and data analysis techniques to predict users' preferences and interests, thereby suggesting videos that are more likely to resonate with their individual tastes. The overarching goal is to create a curated and tailored video consumption journey that not only

V. Sakalauskas et al. (Eds.): IBICA 2023, LNNS 1231, pp. 39–49, 2025.
https://doi.org/10.1007/978-3-031-78946-5_4

enhances user satisfaction but also prolongs their engagement with the platform. As online video platforms continue to evolve and diversify, the significance of effective video recommendation systems becomes increasingly apparent.

Motivation: The primary motivation behind the proposed web video recommendation stems from the ongoing transformation of the web into a realm of greater intelligence and semantic understanding, commonly referred to as Web 3.0. The existing paradigm for web video recommendation primarily rely on statistical models or only on semantic similarity measures or only on learning paradigms such as machine learning and deep learning or standalone clustering algorithms. A hybrid semantics-oriented learning-reasoning with factual and epistemic reasoning is to be the basis of Web 3.0 which is absent in the existing paradigm. Henceforth, this is the primary motivation for the semantically inclined knowledge-centric hybrid intelligence driven web video recommendation framework.

Contribution: The encompassment of metadata generation and classifying the metadata using a strong BERT classification model contributes in making the metadata more permeable and atomic. The encompassment of CNN which is a strong deep learning classifier used twice independently into the framework helps in proper classification of the dataset which is concrete and dependable. The Normalised Pointwise Mutual Information (NPMI) promotes semantic similarity computation and helps in semantic reasoning. Genetic algorithm helps in generating the most optimal dataset and Salient Semantic Analysis (SSA) accelerates the quality of the relevance of the entities which are suggested as user's facets and subsequently the Web 3.0 videos. In contrast to the other baseline models, the suggested framework has higher precision, recall, accuracy, and F-measure and lower FDR.

Organization: The structure of this paper is as follows – Sect. 2 provides and extensive overview of "Related works", while Sect. 3 outlines the "Proposed system architecture". Section 4 delves into the "Performance evaluation and results" and lastly, Sect. 5 ends with the 'Conclusion' of the paper.

2 Related Works

Mei et al. [1] extend VideoReach, a unique online video recommendation system that reduces users' efforts in discovering the most pertinent videos based on recent viewings without the need for a sufficient number of user profiles. Pu et al. [2] offer a multimodal topic learning technique to produce video topics offline by utilising descriptors, headings, and cover visuals. The topics thus generated function as semantic topic attributes to aid in determining user preferences and generating recommendations. Li et al. [3] introduce a personalized course recommendation approach that utilizes the BERT model, combining deep learning and big data technology within a MOOC system.

Tarus et al. [4] provide an overview of existing research on ontology-driven recommendation systems in the context of online education. The diverse recommendation methods utilized by online-learning recommenders based on ontology are sorted into categories. Wang et al. [5] introduce a novel movie recommendation model that enhances accuracy by combining LSTM and CNN architectures. The model leverages LSTM to capture the sequential patterns in user ratings, ensuring context comprehension. Ferracani et al. [6] introduce a video recommendation system based on content analysis leveraging visual prominence for enhancing video feature representation. Konapure et al. [7] propose a model that utilizes NLP methods to categorize video metadata and match it with suitable advertisements. The second module treats videos as sequences of frames, employing a Convolutional Neural Network trained on an image dataset to classify video topics, with comparison to a pre-trained model. Jia [8] presents a novel algorithm (BTCBMA) for recommendation of web-based educational courses that emphasizes the quality of the learner's educational experience. Kim et al. [9] propose a method that predicts users' future book choices using BERT for processing past and current book usage. The approach recommends popular books from different categories like the predicted category in BERT, adding serendipity. In [10–29] several models and datasets in support of the literature of the proposed scheme has been discussed.

3 Proposed System Architecture

The hybrid semantic intelligence-driven video recommendation framework's proposed system architecture is illustrated in Fig. 1. The primary input is the user query which undergoes a preprocessing stage, that includes tokenization, lemmatization, stop words removal and Named Entity Recognition (NER). For pre-processing, The Word Sense Disambiguation (WSD) is applied to the query words. Independent query words (Qw) are acquired following the completion of the pre-processing stage. The aforementioned query words (Qw) are further enriched by generating the metadata directly by using the structural metadata from Web 3.0.

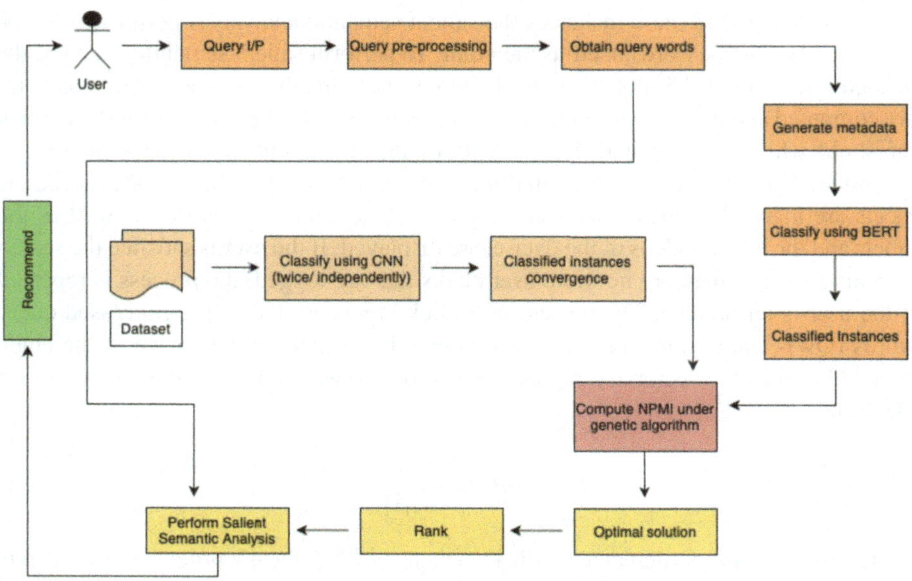

Fig. 1. Proposed System Architecture for StrategicVideoRec

Metadata is generated using two different tools namely - the Dspace and the Open Calais. The reason for using two distinct tools is to increase the amount and density of metadata in terms of volume, scale, heterogeneity and diversity. Since the metadata is generated using two distinct tools, the Dspace and the Open Calais, it cannot be used as it is. Rather, it is required that metadata is atomized and made permeable by subjecting to classification. The classification of metadata is achieved using a strong deep learning BERT classifier. The classifier instances that come out on the perspective of the query words from the BERT classifier is made use of. Subsequently, the dataset of web videos which is categorical in nature and labelled is subjected to classification twice independently using the same CNN classifier but two independent classifiers are made use of. The first classifier classifies the from the standpoint of the video contents itself by using the video features and the second time, CNN classifies the dataset using the labels and the annotations as features and classifies them based on the labels. The classified instances from both the CNN classifiers are subjected to convergence and the most convergent entities are prioritized first. These convergent entities or instances that come out of from the two different CNN classifiers of the dataset are subjected to normalized pointwise mutual information (NPMI) computation in line with the classifier instances acquired from BERT. Since, the instances that come out of the BERT classifier are extensively large, NMPI will compute it with a threshold of 0.75. Also, yields a very large initial feasible solution set which is optimized using the genetic algorithm with NPMI itself as the objective function. This is run subsequently until all the instances which come out of the BERT classifier gets repeated to yield an optimal solution set. The optimal set of solutions is ranked in the ascending sequence of the NPMI measure. NPMI computed under the Genetic Algorithm has a threshold of 0.75 by considering only the positive values between 0 and 1. The instances that come out of the optimal solution set are ranked and to these instances the salient semantic analysis is performed by the pre-processed query words itself as the input. To perform salient semantic analysis, the needed threshold of 0.5 alone is considered because already the most relevant entities are computed and the optimal solution set has been derived. Therefore, 0.5 is the median threshold which is computed and the entities that come out of the SSA pipeline is recommended to the user as the initial facet. Based on the user click on the facets, the facets are mapped to the datasets based on the annotations, categories as well as the labels and the actual videos of the dataset are displayed. If the user is satisfied the search concludes; Until there are no more user clicks that are logged, the process is repeated if the user is unsatisfied. The present user click inputs are fed as pre-processed query words (Qw). The cosine similarity is computed by evaluating the cosine of the angle formed between the vectorized representations of two words, independent of the vector magnitude.

$$cos\theta = \frac{\vec{w}.\vec{x}}{|\vec{w}||\vec{x}|} \tag{1}$$

Cosine similarity is depicted by Eq. (1) where $\vec{w}.\vec{x}$ is the dot product of the vectors 'w' and 'x', $|\vec{w}|$ and $|\vec{x}|$ are the magnitudes of the two vectors 'w' and 'x' and.

$|\vec{w}|\,|\vec{x}|$ is the regular product of the two vectors 'w' and 'x'. Pointwise Mutual Information (PMI), a statistical metric, employed to evaluate the extent of association between two terms in a given corpus of text. It assesses the likelihood of the co-occurrence of the terms in relation to their individual occurrence probabilities. PMI considers the frequency as well as the simultaneous occurrence patterns of terms within the text corpus. PMI is depicted by Eq. (2)

$$pmi(g; h) = log2p(g|h) - log2p(g) = log2p(h|g) - log2p(h) \tag{2}$$

NPMI is an extension of Pointwise Mutual Information (PMI) that addresses the issue of bias towards frequently observed terms in a corpus of text. By normalizing the PMI scores, NPMI takes into account the prevalence of individual terms and their probabilities of simultaneous occurrence. The normalization process scales the values between -1 and + 1, where -1 represents a lack of co-occurrence, 0 indicates independence, and + 1 denotes an absolute simultaneous occurrence.

$$npmi(l; k) = \frac{pmi(l, k)}{h(l, k)} \tag{3}$$

PMI is depicted by Eq. (3) where h(l,k) denotes the joint self-information - $\log_2 p(l,k)$. The principles of natural selection and genetics form the basis of genetic algorithm (GA), a search and optimization technique. First, the data (sentences) are tokenised by dividing them into grams. Grams represent the individual symbols or words. Then, a population is initialised which comprises of weighted vectors of grams. Random weights ranging from 0 to 1 are assigned to these vectors during the initialisation phase. The initial population represents potential solutions which are assessed using fitness function. Each individual is given a fitness value by the fitness function which represents their reproductive fitness. Individuals possessing higher fitness values are more likely to be chosen for reproduction. A parent selection mechanism is utilised to choose solutions which higher fitness. It guarantees that the most favourable solutions are more likely to transmit their genetic material to the succeeding generation. During the next step, mating pool, the fittest individuals, as determined by the fitness function, are brought together in random to form mating pairs. By creating these random pairs, the generation of new combinations of genetic material is promoted, leading to a wider exploration of potential solutions and an increased likelihood of discovering superior solutions to the optimisation problem. Until the concluding criterion is reached, this iterative process continues for multiple generations. The population of weight vectors is returned and the average is computed to obtain a single gram value upon meeting the termination criterion. If the termination criterion is not met, indicating an unsuitable population, the individuals are advanced to the subsequent generation. Salient Semantic Analysis (SSA) is an approach employed in NLP to extract significant semantic characteristics from text corpus. The primary objective of SSA is to identify the salient aspects of the text while disregarding redundant information. SSA relies on co-occurrence statistics to gauge the semantic connection between a word and the salient aspects present in its vicinity. By capturing the co-occurring words within a specific window, SSA constructs semantic profiles that encompass the primary concepts linked to each word. The degree of association between a given text g and a specific concept r_j can be calculated by summing the association strengths of all the words within the text $g = (s_1,\ldots,s_n)$ towards the concept r_j. It is depicted by Eq. (4).

$$v(g,\ rj) = \sum_{s_i \in g} as(s_i, r_j) \tag{4}$$

The degree to which a word s_i and a concept r_j are associated is determined by the pairwise mutual information (PMI) value computed between the word and the concept. It is depicted by Eq. (5).

$$as(s_j, r_j) = P_{ij} \tag{5}$$

4 Performance Evaluation and Results

In order to assess the performance of the StrategicVideoRec, Precision, Recall, Accuracy, F-Measure percentages, and False Discovery Rate (FDR) are utilised as strategic metrics. These metrics gauge the relevance of the outcomes. The model's error rate is measured using FDR, which measures the quantity of false positives. Table 1 reveals that the highest precision percentage of 95.75, the highest Recall percentage of 96.39, the highest accuracy percentage of 96.07, the highest F-measure percentage of 96.07, and the lowest FDR value of 0.05 were achieved by the proposed StrategicVideoRec. StrategicVideoRec's results for Precision, Recall, Accuracy, and F-measure exist in the mean form. To assess, establish a baseline, and make performance comparisons with StrategicVideoRec, it is compared against three different but similar models EmoWare [10], AMCW [11] and UPVR [12] which are also strategic video recommendation frameworks and hence used for comparison. Table 1 shows that EmoWare [10] produces average values for precision, recall, accuracy, and F-measure of 89.14%, 91.47%, 90.31%, and 0.11 respectively. For AMCW [11], the averages include 91.74% precision, 92.14% recall, 91.94% accuracy, 91.93% F-measure, and an FDR value of 0.09. UPVR [12] has yielded averages of 92.29% precision, 93.09% recall, 92.69% accuracy, 92.69% F-measure, and an FDR value of 0.08. The reason behind the proposed StrategicVideoRec achieving the optimum precision, accuracy, F-measure, and the least FDR value lies in its utilization of two separate classifiers namely - BERT, for targeting the generated metadata and CNN, to target the datasets itself. Although, CNN is not just used once but a mirror of CNN is used twice independently to classify the datasets at two different perspectives - one, of the images and the other, of the text contents. Subsequently, the pre-processed query words which are input from the user is subjected to generation of metadata. Metadata generation insists the perspective of the query words intensify and add auxiliary knowledge in the perspective of the query itself. Thereby, enriching the query to a large extent.

BERT classifies the metadata accurately. Using CNN twice independently on the same dataset with two different perspectives - one, being the perspective of the text and the other, being the standpoint of the video contents, ensures a very strict convergence criteria of the instances which means the relevance computation is also high with respect to the classifier itself. Since the convergence is obtained and most importantly, usage of NPMI for semantics measures oriented with genetic algorithm for generating the optimal solution. Salient semantic analysis to furnish the final recommendations ensures that StrategicVideoRec surpasses the performance of the baseline models and serves as the best-in-class model as it has a strong ecosystem of knowledge in terms of better metadata which is atomized and made permeable using BERT. The calculation of strong relevance relies on Salient Semantic Analysis as well as NPMI, with the utilization of Genetic Algorithm to compute the optimal solution set - this ecosystem and arrangement guarantees that StrategicVideoRec serves as the best-in-class framework for recommendation of videos.

The discrepancy in performance between EmoWare [10] and the proposed framework can be attributed to the fact that EmoWare, a personalised video recommendation system with context awareness, employs deep bidirectional recurrent neural networks. This means that its classification is extremely strong and also uses the actual content of the video such as emotional intelligence and the contents of the video itself is used. Auxiliary knowledge from various sources is not incorporated into this model. The knowledge density is quite weak. Apart from that, the knowledge permeability, semantic relevance and reasoning is absent. The explanation for AMCW [11] falling short of expectations in comparison to the proposed framework lies in its emphasis on automatic multimedia content summarizing to recommend videos, video content summarization is the focus and the actual contents are made use of. The text or labels are not given high importance where the context is lost. Apart from this auxiliary knowledge is absent and the reasoning is also not with respect to the labels or context. The rationale behind UPVR [12] not meeting expectations when compared to the proposed framework is that, despite being a personal recommendation dynamic recurrent neural network, its dynamism is primarily driven by neural networks associated with the video's content. Focusing on user interest makes this approach better but the lateral association of topics remains absent. Density and growth of knowledge is absent. Relevance computation is quite weak. Therefore, the intelligence is lost. Reviewing all the lacunae in the baseline models reveals that the proposed framework StrategicVideoRec has a very strong learning infrastructure in terms of BERT and CNN used twice to the same dataset with different perspective in vantage form.

Table 1. Comparison of Performance of the proposed **StrategicVideoRec** with other approaches

Model	Average Precision %	Average Recall %	Average Accuracy %	Average F-Measure %	FDR
EmoWare [10]	89.14	91.47	90.31	90.29	0.11
AMCW [11]	91.74	92.14	91.94	91.93	0.09
UPVR [12]	92.29	93.09	92.69	92.69	0.08
Proposed StrategicVideoRec	**95.75**	**96.39**	96.07	**96.07**	**0.05**

The experimentations are conducted on four distinct datasets, namely the Raphaela et al.'s "Data_Sheet_1_Communicating Science With YouTube Videos: How Nine Factors Relate to and Affect Video Views" dataset [13], Juan-José Boté et al.'s "YouTube

Dataset - Medical papers analyzing YouTube videos" dataset [14], David A. Robinson et al.'s "National Oceanic and Atmospheric Administration (NOAA) Climate Data Record (CDR) of Northern Hemisphere (NH) Snow Cover Extent (SCE), Version 1, 1966-present" dataset [15] and Sandhya Krishnan's "YouTube Data Science Video Trends" dataset [16]. These datasets were not used as it is but rather these datasets were subjected to customized annotations using customized annotator. All the annotations were further used to crawl the Web 3.0 in order to crawl the scientific videos according to the annotations. The space for videos is separately maintained as a specific repository. Subsequently,.csv files are reprioritized based on annotations and integrated into the large dataset comprising of these annotated datasets and equivalent video space comprising of videos and their laterals which matches the categories/annotations of the dataset is made use of in order to conduct the experimentations. Ground truth was gathered from 100 users for performing experiments on 1894 queries. These users ranged from students who were 18 to adults who were 60 to give atleast 5 relevant key words for each query and 20 queries were given to each of the participants and the ground truth was collected which was used to validate the results. Implementation was carried out with Python3 and Google Colaboratory as the preferred IDE. BERT was configured using Tensorflow framework. CNN was also implemented using Tensorflow. Metadata was generated using two different tools namely - the Dspace and the Open Calais. Identical environment and dataset were used for performing experimentations for the the proposed framework and the baseline models. Validation was performed on the top results, and a correlation analysis was conducted to produce outcomes that stood out as the best-in-class. Figure 2 illustrates the distribution curve of precision against the number of recommendations.

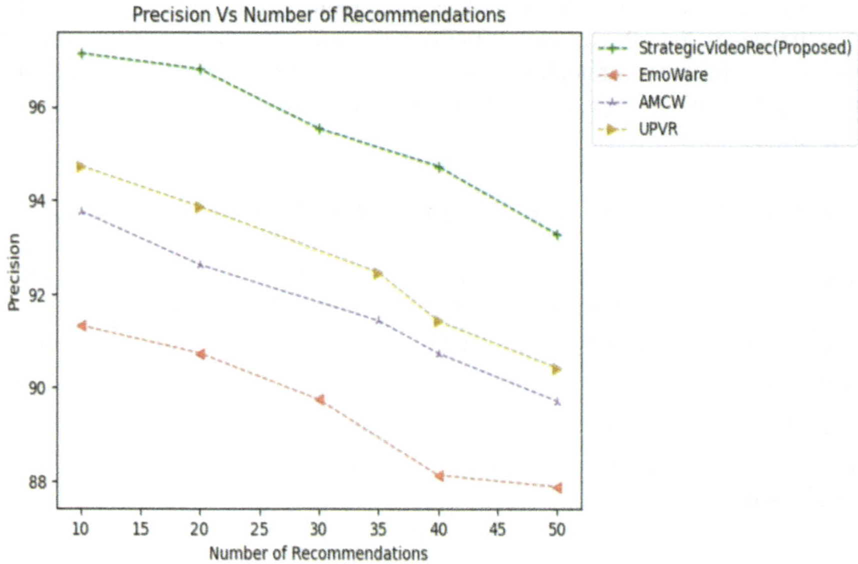

Fig. 2. Precision Vs Number of Recommendations Graph

5 Conclusion

This paper proposes a Web 3.0 compliant framework for recommending videos. It introduces a semantics-oriented learning-reasoning model which encompasses auxiliary knowledge in terms of highly dense metadata which is generated by encompassing query words as specific inputs. BERT driven classification of the metadata helps in making it permeable and thereby increasing the overall intermediate instances for relevance computation and optimal solution generation. The use of CNN twice to classify the same dataset independently helps in proper impromptu classification and the instances conversion yielding from the dataset helps in not only the most relevant intermediate instances which come out of the pipeline but also the instances being quite diverse and heterogenous from the dataset itself are harvested. The NPMI computation, utilizing the NPMI measure, between classified instances of metadata and the dataset, produces an initial solution that stands out as the best-in-class. This initial solution is then refined into an optimal solution using the Genetic Algorithm, a metaheuristic optimization technique within the model. Salient Semantic Analysis (SSA) accelerates the overall relevance of the results. The resultant StrategicVideoRec functions as the best-in-class model for recommending videos. The proposed framework attains the highest precision percentage of 95.75, the highest recall percentage of 96.39, the highest accuracy percentage of 96.07, the highest F-measure percentage of 96.07, and boasts the lowest FDR value of 0.05. These collectively establish it as the best-in-class model for web video recommendation, characterized by semantic orientation and the integration of auxiliary knowledge through metadata for semantics-driven learning-reasoning-based video recommendations.

References

1. Mei, T., Yang, B., Hua, X. S., Yang, L., Yang, S. Q., Li, S.: Videoreach: an online video recommendation system. In: Proceedings of the 30th Annual International ACM SIGIR Conference on Research and Development in Information Retrieval, pp. 767–768, July 2007
2. Pu, S., He, Y., Li, Z., Zheng, M.: Multimodal topic learning for video recommendation (2020). arXiv preprint arXiv:2010.13373
3. Li, B., et al.: A personalized recommendation framework based on MOOC system integrating deep learning and big data. Comput. Electr. Eng. **106**, 108571 (2023)
4. Tarus, J.K., Niu, Z., Mustafa, G.: Knowledge-based recommendation: a review of ontology-based recommender systems for e-learning. Artif. Intell. Rev. **50**, 21–48 (2018)
5. Wang, W., Ye, C., Yang, P., Miao, Z.: Research on movie recommendation model based on LSTM and CNN. In: 2020 5th International Conference on Computational Intelligence and Applications (ICCIA), pp. 28–32. IEEE, June 2020
6. Ferracani, A., Pezzatini, D., Bertini, M., Meucci, S., Del Bimbo, A.: A system for video recommendation using visual saliency, crowdsourced and automatic annotations. In: Proceedings of the 23rd ACM International Conference on Multimedia, pp. 757–758, October 2015
7. Konapure, R.C., Lobo, L.M.R.J.: Video content-based advertisement recommendation system using classification technique of machine learning. In: Journal of Physics: Conference Series, vol. 1854, no. 1, p. 012025. IOP Publishing, April 2021
8. Jia, Y.: BTCBMA online education course recommendation algorithm based on learners' learning quality. Int. J. Inf. Technol. Syst. Approach (IJITSA) **16**(1), 1–17 (2023)

9. Kim, Y., Kim, S.Y., Park, B.: A serendipity recommendation method for book categories using BERT (2023)
10. Tripathi, A., Ashwin, T.S., Guddeti, R.M.R.: EmoWare: a context-aware framework for personalized video recommendation using affective video sequences. IEEE Access **7**, 51185–51200 (2019)
11. Yang, J.C., Huang, Y.T., Tsai, C.C., Chung, C.I., Wu, Y.C.: An automatic multimedia content summarization system for video recommendation. J. Educ. Technol. Soc. **12**(1), 49–61 (2009)
12. Gao, J., Zhang, T., Xu, C.: A unified personalized video recommendation via dynamic recurrent neural networks. In: Proceedings of the 25th ACM international conference on Multimedia, pp. 127–135, October 2017
13. Velho, R.M., Mendes, A.M.F., Azevedo, C.L.N.: Communicating science with YouTube videos: how nine factors relate to and affect video views (2023). https://doi.org/10.3389/fcomm.2020.567606.s001
14. Boté, J.-J.: YouTube dataset - medical papers analising YouTube videos [Dataset] (2019). https://doi.org/10.6084/m9.figshare.7108511.v3
15. Robinson, D.A., Estilow, T.W.: NOAA national centers of environmental information. National Oceanic and Atmospheric Administration (NOAA) Climate Data Record (CDR) of Northern Hemisphere (NH) Snow Cover Extent (SCE), Version 1, 1966-present (2020). https://doi.org/10.7289/V5N014G9
16. SandhyaKrishnan02. YouTube data science video trends (2022). https://www.kaggle.com/datasets/sandhyakrishnan02/youtube-datascience-video-views
17. Manoj, N., Deepak, G.: ODFWR: an ontology driven framework for web service recommendation. In: Data Science and Security: Proceedings of IDSCS 2021, pp. 150–158. Springer, Singapore (2021)
18. Kannan, B.D., Deepak, G.: I-DLMI: Web Image Recommendation Using Deep Learning and Machine Intelligence. In International Conference on Hybrid Intelligent Systems, pp. 270–280. Springer, Cham, December 2022
19. Yethindra, D.N., Deepak, G.: A semantic approach for fashion recommendation using logistic regression and ontologies. In: 2021 International Conference on Innovative Computing, Intelligent Communication and Smart Electrical Systems (ICSES), pp. 1–6. IEEE, September 2021
20. Agrawal, D., Deepak, G.: HSIL: hybrid semantic infused learning approach for course recommendation. In: International Conference on Digital Technologies and Applications, pp. 417–426. Springer, Cham, January 2022
21. Adithya, V., Deepak, G.: HBlogRec: a hybridized cognitive knowledge scheme for blog recommendation infusing XGBoosting and semantic intelligence. In: 2021 IEEE International Conference on Electronics, Computing and Communication Technologies (CONECCT), pp. 1–6. IEEE, July 2021
22. Deepak, M.S.G., P.C N.:CISKGG: a collective intelligence strategy for heterogeneous knowledge graph generation for criminology as a domain of choice. In: 2022 IEEE International Conference for Women in Innovation, Technology & Entrepreneurship (ICWITE), Bangalore, India, pp. 1–10 (2022)
23. Deepak, G., Vibakar, A., Santhanavijayan, A.: OntoInfoG++: a knowledge fusion semantic approach for infographics recommendation. Int. J. Interact. Multimedia Artif. Intell. (527) (2023)
24. Sawarn, S., Deepak, G.: MASSTagger: metadata aware semantic strategy for automatic image tagging. In: International Conference on Digital Technologies and Applications, pp. 429–438. Springer, Cham, January 2022

25. Maddineni, S., Deepak, G., S.P.J.:CDADITagger: an approach towards content based annotations driven image tagging integrating hybrid semantics. In: 2022 IEEE International Conference for Women in Innovation, Technology & Entrepreneurship (ICWITE), Bangalore, India, pp. 1–6 (2022)
26. Deepak, G., Priyadarshini, J.S.: Personalized and enhanced hybridized semantic algorithm for web image retrieval incorporating ontology classification, strategic query expansion, and content-based analysis. Comput. Electr. Eng. **72**, 14–25 (2018)
27. Gulzar, Z., Leema, A.A., Deepak, G.: Pcrs: personalized course recommender system based on hybrid approach. Procedia Computer Science **125**, 518–524 (2018)
28. Deepak, G., Teja, V., Santhanavijayan, A.: A novel firefly driven scheme for resume parsing and matching based on entity linking paradigm. J. Discrete Math. Sci. Cryptography **23**(1), 157–165 (2020)
29. Santhanavijayan, A., Naresh Kumar, D., Deepak, G.: A semantic-aware strategy for automatic speech recognition incorporating deep learning models. In: Intelligent System Design: Proceedings of Intelligent System Design: INDIA 2019, pp. 247–254. Springer, Singapore (2021)

OSMA: Ontology Generation and Synthesis for Sports Medicine and Athletics

Harshada Anavkar[1] and Gerard Deepak[2](✉)

[1] Department of Mechanical Engineering, Indian Institute of Technology Kharagpur, Kharagpur, India

[2] Department of Computer Science and Engineering, Manipal Institute of Technology Bengaluru, Manipal Academy of Higher Education, Manipal, India

gerard.deepak.christuni@gmail.com

Abstract. There is a need for ontology generation and synthesis, for highly important and essential but rare domains like sports medicine and athletics. In order to model a strategic Web 3.0. This paper proposes an ontology synthesis model which is automatic in nature for sports medicine and athletics as a domain of choice in which strategic database terms are extracted through the TFIDF and enhanced using topic modelling, fed through the wiki data pipeline to yield and enrich entities. This is to generate metatags which is subject to classification using the LSTM classifier. Subsequently, the data itself is classified using the generative adverse media networks to yield of the classified instances. The model encompasses hierarchical clustering to cluster the entities that come out of two distinct deep learning classifiers and is formalized in ontology using the normalized pointwise mutual information measure. The ontology is completed through axiomatization, reasoning, instantiation through an agent based cluster. An overall precision of 96.09 percentage with an F-measure of 97.44% and an accuracy of 97.45 percentage has been achieved by the proposed framework for ontology synthesis and the best in class and the first of its kind ontology synthesis model for sports medicine and athletics as a strategic domain.

Keywords: Adaptive Pointwise Mutual Information · NPMI · LSTM · GAN · hierarchical clustering · ontology synthesis · STM

1 Introduction

Ontologies allow a clearer understanding of the complex interplay between numerous elements influencing athlete health, performance, and training by organising domain-specific knowledge into formalised frameworks. This organised representation paves the way for more accurate and effective data integration, resulting in better decisions and personalised treatments. Effective knowledge representation and management have emerged as critical variables in generating breakthroughs and optimising performance in the ever-changing landscape of sports medicine and athletics. Ontologies have emerged as powerful methods for organising and formalising domain-specific information. An

V. Sakalauskas et al. (Eds.): IBICA 2023, LNNS 1231, pp. 50–60, 2025.
https://doi.org/10.1007/978-3-031-78946-5_5

ontology is a structured framework that contains the key concepts, relationships, and attributes of a domain, allowing for seamless information sharing and reasoning. Because of the inherent complexity of the sports medicine and athletics area, a complete approach to knowledge organisation is required. This research digs into ontologies, which offer a richer knowledge of interconnections that can guide informed decision-making by developing explicit relationships between various things such as injuries, training regimens, physiological data, and performance indicators. This work examines the symbiotic link between ontology generation, synthesis, and domain-specific knowledge augmentation via the lens of sports medicine and athletics. Ontology modelling, by offering a formal framework for knowledge representation, offers a viable approach towards unravelling the complexities of sports-related phenomena and optimising athletes' well-being and performance. Automatic ontology synthesis expands on this influence by automating the formerly labour intensive process of ontology construction. It accelerates the evolution of ontologies and maintains them up to date with the fast evolving area of sports medicine and athletics by employing computational tools to collect and consolidate information from multiple sources.

Motivation**:** Motivation for the proposed framework is to model a formal ontology for a domain like sports medicine and athletics which is having high degree of importance in medicine and it is the upcoming branch of medicine. However, experts have to encompass the entities present on the Web 3.0 which is the ultimate strategic hard drive for the cognition and the evolution of the web. So there is a strong need for a formal ontology for sports medicine and athletics. Apart from this, there is a need for automatic strategic deep learning driven frameworks that encompasses semantic intelligence for automatic ontology generation which is required for modelling the Web 3.0 and harvesting entities as ontologies and in turn, evolve it into the knowledge graphs. Owing to this need, the proposition has been made for an automatic ontology modelling framework encompassing semantic artificial intelligence and deep learning models for deriving ontologies.

Contribution**:** The proposition is the first of its kind paradigm for automatic ontology generation for sports medicine and athletics as the domain of choice. The proposal combines two different deep learning strong classifiers—generative adversarial networks and long short-term memory—to classify the data set and the intentionally generated metadata. Knowledge enrichment takes place through primitive term extraction using the TFIDF topic modelling using structural modelling paradigm and depending upon auxiliary knowledge depositories like wiki data. Hierarchical clustering clusters, the intermediate instances and it's normalized into the ontology using the normal point-wise mutual information measure. Agent based planning is also encompassed for instantiation. Apart from this the intermediate steps encompass axiomatization and reasoning in order to yield the best in class ontology precision, recall accuracy, and f-measure and the FDR is decreased.

Organization: This document is structured as follows. Section 2 provides an overview of Related Literature, Sect. 3 discusses the Proposed System Architecture, Sect. 4 presents the Performance Assessment and outcomes, and Sect. 5 concludes with insights.

2 Related Works

D Djurić et al. [1] presents the Ontology Definition Metamodel (ODM), a tool for ontological engineering that makes advantage of Model Driven Architecture (MDA) standards. It examines the distinctions between MOF and OWL concepts and describes ODM using the Meta-Object Facility (MOF). Siricharoen et al. [2] discusses the object model approach to data modeling, comparing it with ontologies models. It draws attention to how object models and ontologies are similar, and it advises modifying current object-oriented software development processes to accommodate the construction of ontologies. Shimizu et al. [3] introduces the Modular Ontology Modelling (MOMo) technique, which solves issues with ontologies that prevent them from being reused, such as different levels of representational granularity, a lack of conceptual clarity, difficulty adhering to sound modelling principles, and a lack of attention on reuse in ontology engineering tooling.

Wang et al. [4] proposes the use of a deep machine learning algorithm called the Algorithm of Formal Ontology Generation (AFOG) to produce quantitative ontologies for knowledge engineering and semantic comprehension. Lonsdale et al. [5] demonstrates the viability of automating ontology generation through ontology reuse by presenting a generic architecture for automated ontology reuse. It experiments with a large generic ontology to generate domain ontologies for natural language web pages. Huang et al. [6] offers a thorough approach to ontology development and evolution for use in applications related to intelligent manufacturing. Data extraction, ontology construction, ontology connection, and ontology evolution are the four steps of the technique. It maps data from different sources to form a unified structure, generates an initial ontology, organizes ontologies based on business elements, and formulates rules for ontology evolution.

Thanh Tho Quan et al. [7] offers the FOGA (fuzzy ontology generation framework) as a method for automatically creating fuzzy ontologies from information including uncertainty. Concept hierarchy generation, fuzzy formal concept analysis, and ontology generation are all included. The system also covers integrating database properties and approximation reasoning for gradual enrichment with fresh data. Thanh Tho Quan et al. [8] discusses the Scholarly Semantic Web and FFCA framework's performance. Fuzzy Formal Concept Analysis (FFCA) is proposed to handle uncertainty in data and interpret concept hierarchy. Fortuna, B et al. [9] provides a semi-automatic ontology editor in a new OntoGen system that integrates text mining and machine learning algorithms for effective user interface, concept recommendation, naming, and visualisation, with positive results in user trails and real-world scenarios. In [10–16] several frameworks in support of the literature of the proposed model have been depicted.

3 Proposed System Architecture

Figure 1 is the proposed system architecture of the proposed ontology synthesis framework for sports medicine and athletics as the domain of choice. The document data set is the primary input to the framework from the data set of documents. The terms are extracted using the TF-IDF (1) or the term frequency inverse document frequency which focuses on yielding the terms which are frequent within the document corpus and rare across the document corpus. In Eqs. 1 and 2, t is term/word, d is document, N is the total number of documents and df is the document frequency.

$$TF - IDF(t, d) = TF(t, d) * \log \frac{N}{df + 1} \quad (1)$$

$$TF(t, d) = \frac{count\ of\ t\ in\ d}{number\ of\ words\ in\ d} \quad (2)$$

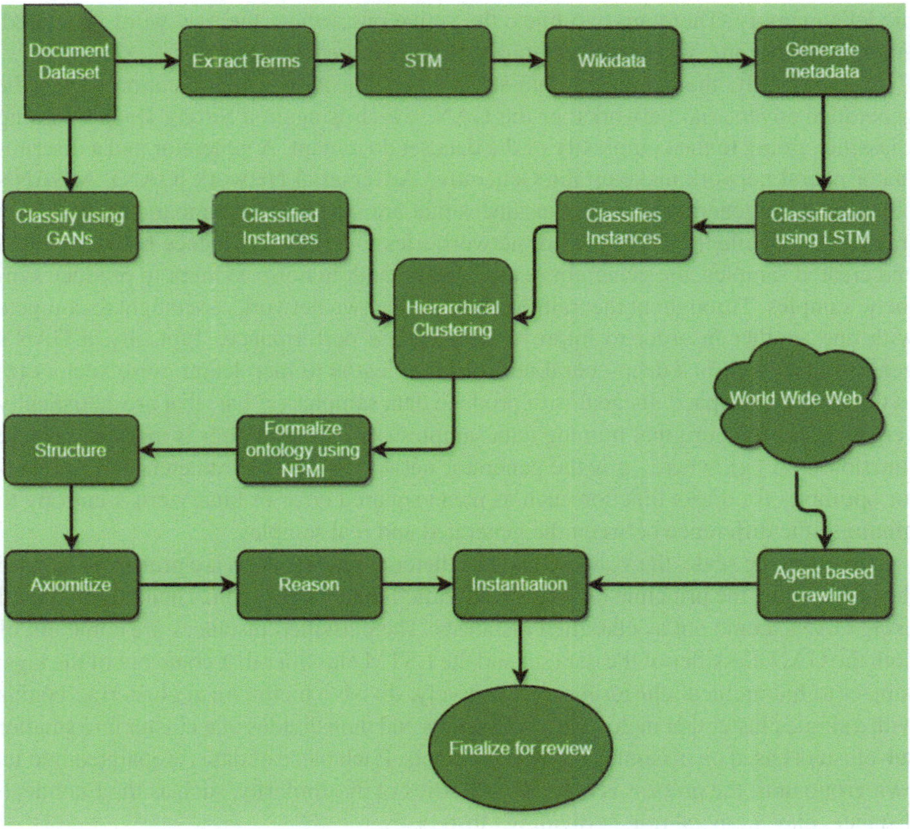

Fig. 1. Architecture Design of OSMA model

The terms extracted using the TFIDF are the informative terms which are subjected to structural topic modelling or the STM in which discovery of entities takes place through topic modelling by keeping the web itself as a reference corpora. Entities that come out of STM pipeline are subjected to wiki data. It's a community contributor, community verified, crowd sourced, knowledge storage, in which hyperlinked, knowledge is housed from the wiki data.

The entities that come out of the Wikilead pipeline are subjected to the community-contributed, community-verified entities. They are subjected to the metatag generation using the Metatag Harvester tool. The Metatag generated through the Metatag Harvester tool is subjected to classification using the LSTM or the long shot of memory, which is a deep learning classifier, because the tags are exponentially large as they are generated from the current structure of the World Wide Web. Owing to the large size and scale of the tags they have to be classified using the LSTM, the Strong Deep Learning Classifier, and LSTM is preferred because of the scale, volume, heterogeneity of the META tags, and also of the efficiency of the LSTM model. The LSTM deep learning classifier is used to address the problems with short-term memory experienced by conventional feed-forward neural networks. LSTM is used to retain previously encountered words in the model's memory. There are two lines: the active one, where the new word is entered, and the passive one, where the previous words are stored in memory.

Subsequently, the document data set is also subjected to classification within the generative adversarial networks, or the GAN, which is again a Strong Deep Learning Classifier, going to the complexity of the data set document. A generator and a discriminator neural network make up the Generative Adversarial Network (GAN). A GAN's objective is to generate new data instances that are statistically comparable to a given training set. While the discriminator network tries to tell the difference between actual and created samples, the generator network uses random noise as input to produce synthetic samples. Throughout the training process, the two networks are taught to compete with one another in order to improve each other's performance. Typically, a GAN's generator network is a deep neural network that learns to map latent noise vectors (z) to the observable space. Its goal is to produce data samples (x_hat) that are statistically very similar to the original training data samples (x). The generator is modelled by the function G(z; _g), where _g is the generator network's set of parameters. The generator optimises for a loss function such as mean squared error or binary cross-entropy to minimise the difference between the generated and real samples.

The size, the scale, the volume, and also heterogeneity GANs are preferred because the operators of the principle of auto –handcrafted feature selection. Then it is classified using GANs, come out as classified instances. The classified instances are come out of both the GAN classifier of the data set and the LSTM classifier that come out of the tags, subject to hierarchical clustering. Alternatively, divisive hierarchical clustering begins with a single cluster that includes all data points and then divides that cluster into smaller sub clusters based on dissimilarity measurements. Each piece of data is separated into its own group until the process is complete. Metrics of dissimilarity, such as the Euclidean distance, play a crucial role in dividing clusters.

A dendrogram, a tree-like structure that graphically displays the clustering process, is generated via hierarchical clustering. In a dendrogram, the leaf nodes represent data points and the interior nodes reflect groups that have combined or broken apart. The similarity or dissimilarity metric used during clustering is reflected in the size of the internal nodes.

When the optimal cluster count is uncertain, hierarchical clustering might be useful because it does not necessitate the upfront determination of the number of clusters. Since it requires computing pairwise distances, it can be computationally costly, especially with large datasets. Clusters might seem very different depending on the distance metric and connection method used. Hierarchical clustering provides researchers and analysts with a versatile and illuminating method for exploring the underlying structure of their data. Hierarchical clustering is achieved to yield entities. These entities are subjected to computation using literal semantic similarity computation using NPMI. PMI can be expressed mathematically as Eq. (3).

$$(PMI)(w1, w2) = \frac{log2(P(w1, w2)}{(P(w1) * P(w2))} \quad (3)$$

In Eq. (1), w1 and w2 are the two words being considered, P(w1, w2) is the probability of them occurring together, and P(w1) and P(w2) are the probabilities of occurrence for each word separately. PMI's strength is in its ability to capture the relationship between words, but its weakness is its sensitivity to frequency of occurrence. This shortcoming is addressed by NPMI, which normalises PMI using the idea of "pointwise mutual information ratio." By dividing PMI by the negative logarithm of the likelihood of co-occurrence, NPMI determines the relative degree of association. The formula for NPMI is depicted in Eq. (4).

$$NPMI(w1, w2) = \frac{PMI(w1, w2)}{-\log(P(w1, w2))} \quad (4)$$

The NPMI scale has a negative endpoint of -1 and a positive endpoint of 1, with 1 indicating the most positive association, 0 indicating no association, and -1 indicating the strongest negative association. NPMI is not set to a threshold, but rather the informative measure in between 0 and 1 is used to correlate the entities which come out of hierarchical clustering and create an edge between the most similar entities and form a ontology, and structure it by formulating an edge and axiomatize it by injecting rules through an axiomatization agent, such as is a part of, has a part of, is a superclass of, is a subclass of, composition, generalization, and specialization relationships are also injected apart from the axiom rule. Once the axiomatization is achieved, the ontological model is subjected to reasoning. This reasoning takes place using a pellet reasoner. The reasoning is achieved to eliminate all the inconsistencies.

Instantiation of instances takes place by using an agent designed using JADE. Through crawling of the World Wide Web, but JADE is set to a very stringent step-deadance of 0.02. Highly stringent because this is the final conception of the instances. A very strong measure like the NPMI is used with positive values and a step-deadance of only 0.02 is considered highly stringent. The instances are added to the modelled ontology and are subjected for review and finalization to domain experts.

4 Performance Review and Results

***Implementation*:** The wiki data was accessed by standard API. Agent was designed using JADE for structural modelling and TF-IDF. Multi-agent coupling was formalized in the model. Agent based crawling was done using JADE. Reasoning was done using the pellet reasoner. The deep learning paradigms using the LSTM and GANs were achieved using the KLS framework. The Meta-tag generation was achieved using the Meta tag harvester. Implementation of the baseline models were done in the exact same model of the exact same data set as the OSMA model and the experimentations were evaluated.

The experiment was run on a single, sizable integrated data set that consists of five different data sets, specifically: The impact factors of Germans of sports medicine and therapy by Exaly 2022 dataset, Statista (2021)-Global sports medicine market, ProPublica-American College of Sports Medicine INC 2020 data sets, the Kerem proposed daily sports activities data sets, ProPublica, Crossroads Tigers Youth Athletics 2020 data sets, these five data sets are integrated together as a single large integrative, comprehensive data set by annotating it using customized annotator and you are using the terms in this data set to crawl documents from the current structure of the Web 3.0 to make the document space and integrating all the records of the data set as a single large dot csv file, and the records are prioritized in a way such that the common annotations take precedence and a single large integrative data set is derived from the check screen. As prospective metrics, precision recall accuracy F-measure percentages in their average form are used to assess the outcomes for the proposed OSMA framework. Because they evaluate the importance of outcomes, precision recall accuracy F-measure are used as the main metrics. The amount of false positives produced by the model is measured by the false discovery rate (FDR), which also serves as an auxiliary metric for measuring model inaccuracy.

Table 1. Comparison of Performance of the proposed OSMA with other approaches

Model	Average Precision %	Average Recall %	Average Accuracy %	Average F-Measure %	FDR
AFO	86.63	88.04	87.335	87.33	0.14
AOD	89.92	90.07	89.995	89.995	0.11
OTAGen	91.17	92.09	91.63	91.63	0.09
Proposed OSMA	96.09	98.82	97.455	97.44	0.04

From Table 1, it is indicated that the proposed OSMA has yielded the highest average precision of 96.09% average recall of 98.82% average accuracy of 97.455% and an average F-measure of 97.44% with the FDR of 0.04. To compare the performance of the proposed OSMA framework, it is base lined on AFO, AOD, OTAGen, which are similar ontology-generational modelling frameworks. However, these frameworks are completely based on different techniques. Since they have a common objective, they are prospective models for comparison. Moreover, they do not operate in the sports and athletics domain ontology traditionally, but they were incorporated as techniques for the exact same data set as the proposed OSMA model and evaluation was carried out. It is clear from Table 1 that AFO produces average precision, recall, accuracy, and F-measure of 86.63%, 88.0%, and 0.14%, respectively. For the AOD, it produced average precision of 89.92%, recall of 90.07%, accuracy of 89.995%, F-measure of 89.995%, and FDR of 0.11 on average. The average precision, average recall, average accuracy, average F-measure, and average FDR for the OTAGen were 91.17%, 92.09%, and 0.09%, respectively. The suggested OSMA has produced the highest average precision, average recall, average accuracy, average F-measure, and the lowest value of FDR of 0.04, according to Table 1.

The OSMA model is first of its kind for generation of ontology in medical and athletics domain as the formal domain of choice. Sports medicine is the upcoming branch of medicine and it is now treated to be independent branch. There will be documents, journal papers, etc. dealing with this new branch of medicine which has to be formally designated, and it is now treated and recognised by information retrieval, summarisation, and systems in the era of Web 3.0 for which a concrete model of ontology is required which is not available as of now. OSMA is the first of its kind. It has a very strong knowledge addition infrastructure to topic modelling using STM, shortlisting of free, informative terms through TF-IDF, Wikidata which houses external open-link data from the World Wide Web, which is community-contributed, community-verified, and crowdsourced. Meta tags, through a tag generator, increases exponentially the number of instances from the World Wide Web into the model which is relevant and thereby reduces the semantic gap of knowledge between the existing web and the localized framework. Furthermore, the usage of LSTM for deep learning classifier and hierarchical clustering to cluster the classified instances that are involved in the metadata and the dataset, and usage of a strong deep learning GAN classifier to classify the document dataset, assures a strong learning infrastructure, a strong clustering in the model, a strong knowledge curation infrastructure, and reasoning through NPMI, normalised-pointwise mutual information in semantic relevance computation, and takes in a semantic measure-based reasoning. Due to the abundant strong semantics, knowledge models, relevant computation mechanisms, and knowledge addition and curation strategies proposed, OSMA in comparison to all the baseline models and serves as the best in class models for ontology synthesis and curation.

When compared to the proposed model, the AFO model does not perform as predicted since, despite the fact that it creates fuzzy ontologies, its primary objective is to create your large-scale ontology. Fuzzy ontologies are not very accurate, not very concrete, but they can depend on formal concept analysis theory and methodology using OWL based representation of concept properties and individuals. Undoubtedly formalize and curate and yield ontology. But the ontology is not diverse enough. It need not be as diverse as the proposed model framework because of the lack of heterogeneous knowledge sources and knowledge regulation mechanisms is definitely weaker in formal when concept analysis is done post paradigm and the learning infrastructure is absent so AFO model lags. Because AOD uses stored knowledge to automatically generate generalisations, it can be thought of as a knowledge base, which is why it lags behind the suggested approach. But meta data is generated but generation is used as it is not curated. That means there will not be much demarcation in the AOD model and thereby it will yield much lesser average precision average recall, average accuracy, and F-measure because the reasoning is weak, the permeability of knowledge is weak, and learning infrastructure is absent henceforth AOD model lags compared to proposed framework. Finally, the OTHAGen model, it is a tuneable generator of ontology that uses agent-based ontology collaboration. Agent no wonder. So agent-based modelling will customize, it will definitely reason out because agents are based on a stare and behaviour model. Reasoning is definitely very high but number of ontological sources available to reason out and curate the knowledge is weak because it only depends upon the substrate data set which does not curate strategic knowledge. Knowledge permeability and knowledge regulatory mechanisms are also absent. Learning model is definitely absent. Therefore, OTHAGen model does not perform as expected.

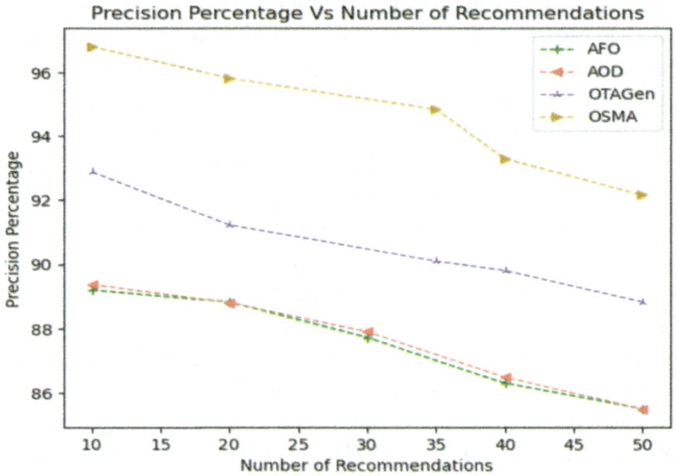

Fig. 2. Recall % vs Number of recommendations of the proposed OSMA and other baseline models

The question was in our recommendation to see which one is better, the proposed OSMA or the baseline models which is depicted in Fig. 2. The proposed OSMA occupies highest in the hierarchy as it utilizes advanced techniques such as STM, TF-IDF, Wikidata, LSTM, hierarchical clustering, and GAN classifier to create a robust and comprehensive knowledge framework. It effectively bridges the semantic gap between the web and localized knowledge, offers strong learning and clustering capabilities, and employs semantic measure-based reasoning. The AFO occupies the lowest in the hierarchy, as it generates fuzzy ontologies rather than large-scale ones. While AFO benefits from formal concept analysis and OWL-based representation, it lacks diverse knowledge sources and regulation mechanisms, resulting in a weaker learning infrastructure. This limitation hampers the AFO model's ability to curate and yield a comprehensive ontology.

The AOD occupies the last but one from the bottom and second from the top. The reason is because it lacks in curation and has weak reasoning capabilities. While AOD automatically generates knowledge, it lacks curated metadata, resulting in less demarcation and lower precision, recall, accuracy, and F-measure. The absence of a learning infrastructure further contributes to the shortcomings of the AOD model compared to the proposed framework. It falls short in terms of the number of ontological sources available for reasoning and curation, relying solely on a substrate dataset without strategic knowledge curation. The absence of knowledge permeability, regulatory mechanisms, and a learning model contributes to the underperformance of the OTHAGen model. Since the proposed framework, the OSMA model has a very strong infrastructure of learning, knowledge curation at different stages from different heterogeneous sources, a strong reasoning mechanism through NPMI-based semantics, and a strong learning infrastructure therefore it delivers the best results.

5 Conclusion

The strategy domain for sports medicine and athletics is proposed in this research as a strategic model for automatic ontology synthesis. Apart from this proposition being the first form of ontology for sports medicine and athletics as a prospective domain, the model is also an automatic ontology generation paradigm in which term extraction is achieved through the TFIDF and topic modelling is achieved in structural topic modelling and the wiki data is used to encompass structurally related, and harvest entities from the world wide web to decrease the cognitive gap between the existing web and the knowledge that permits into the paradigm. Classification of data is achieved using strong deep learning GAN classifier and LSTM is encompassed to classify the intermediately generated meta-tags, hierarchical clustering helps in integrating the similar instances which is formalized using normalized pointwise mutual information measure. Agent based crawling helps in instantiation and addition of instances making the proposed OSMA framework, the best in class, paradigm for automatic ontology generation focused on sports medicine and athletics as a domain of choice, overall, F measure of 97.44 percentage with the recall of 98.82 percentage and the false discovery rate Of 0.04 also has been achieved by the proposed model.

References

1. Djurić, D., Gašević, D., Devedžić, V.: Ontology modeling and MDA. J. Object Technol. **4**(1), 109–128 (2005)
2. Siricharoen, D.W.V.: Ontology modeling and object modeling in software engineering. Int. J. Softw. Eng. Appl. **3**(1), 43–59 (2009)
3. Shimizu, C., Hammar, K., Hitzler, P.: Modular ontology modeling. Semantic Web **14**(3), 459–489 (2023)
4. Wang, Y., et al.: Formal ontology generation by deep machine learning. In: 2017 IEEE 16th International Conference on Cognitive Informatics & Cognitive Computing (ICCI* CC), pp. 6–15. IEEE, July 2017
5. Lonsdale, D., Embley, D.W., Ding, Y., Xu, L., Hepp, M.: Reusing ontologies and language components for ontology generation. Data Knowl. Eng. **69**(4), 318–330 (2010)
6. Huang, C., Cai, H., Xu, L., Xu, B., Gu, Y., Jiang, L.: Data-driven ontology generation and evolution towards intelligent service in manufacturing systems. Futur. Gener. Comput. Syst. **101**, 197–207 (2019)
7. Tho, Q.T., Hui, S.C., Fong, A.C.M., Cao, T.H.: Automatic fuzzy ontology generation for semantic web. IEEE Trans. Knowl. Data Eng. **18**(6), 842–856 (2006)
8. Quan, T.T., Hui, S.C., Fong, A.C.M., Cao, T.H.: Automatic generation of ontology for scholarly semantic web. In: The Semantic Web–ISWC 2004: Third International Semantic Web Conference, Hiroshima, Japan, 7–11 November 2004. Proceedings 3, pp. 726–740. Springer, Berlin, Heidelberg (2004)

9. Fortuna, B., Grobelnik, M., Mladenic, D.: Ontogen: semi-automatic ontology editor. In: Human Interface and the Management of Information. Interacting in Information Environments: Symposium on Human Interface 2007, Held as Part of HCI International 2007, Beijing, China, July 22–27, 2007, Proceedings, Part II, pp. 309–318. Springer, Berlin Heidelberg (2007)
10. Jagan, N.D., Deepak, G., Santhanavijayan, A.: MPTR: a metadata-driven prospective tag recommendation model based on integrative classification. In: Advances in Data and Information Sciences: Proceedings of ICDIS 2022, pp. 291–301. Springer, Singapore (2022)
11. Deepak, G., Santhanavijayan, A.: OntoBestFit: a best-fit occurrence estimation strategy for RDF driven faceted semantic search. Comput. Commun. **160**, 284–298 (2020)
12. Pushpa, C. N., Deepak, G., Thriveni, J., Venugopal, K.R.: Onto Collab: strategic review oriented collaborative knowledge modeling using ontologies. In: 2015 Seventh International Conference on Advanced Computing (ICoAC), pp. 1–7. IEEE, December 2015
13. Deepak, G., Rooban, S., Santhanavijayan, A.: A knowledge centric hybridized approach for crime classification incorporating deep bi-LSTM neural network. Multimedia Tools Appl. **80**(18), 28061–28085 (2021)
14. Rithish, H., Deepak, G., Santhanavijayan, A.: Automated assessment of question quality on online community forums. In: International Conference on Digital Technologies and Applications, pp. 791–800. Springer, Cham, January 2021
15. Kumar, N., Deepak, G., Santhanavijayan, A.: A novel semantic approach for intelligent response generation using emotion detection incorporating NPMI measure. Procedia Comput. Sci. **167**, 571–579 (2020)
16. Anirudh, M., Deepak, G.: ISBRNM: integrative approach for semantically driven blog recommendation using novel measures. In: International Conference on Digital Technologies and Applications, pp. 12–21. Springer, Cham, January 2022

Hospital Remote Care Assistance AI to Reduce Workload

Luís B. Elvas[1,2], Joao C. Ferreira[1,2](✉), and Berit Irene Helgheim[2]

[1] Department of Logistics, Molde University College, 6410 Molde, Norway
luis.m.elvas@himolde.no

[2] Inov Inesc Inovação—Instituto de Novas Tecnologias, 1000-029 Lisbon, Portugal

Abstract. This research aims to create a remote system to reduce the Hospital load for patients with chronic diseases at Home or Recover Centers. A framework of shared Artificial intelligence services is proposed to provide a secure environment for remote patient monitoring. A literature review using the PRISMA methodology provides a knowledge systematization. A new data fusion process with data, image, text, and voice allows for improving remote patient data collected. We apply Design Science Research with CRISP-DM approach to handling the data process. We are developing specific modules to fuse and handle image, voice, text and data. Our model is being validated and applied in a real case from a Portuguese hospital using Alzheimer's patients.

In our research work, we propose a framework for secure information sharing among patients' data and improving health care services. This approach allows the creation of big data systems in a fusion of image, text, voice, and data.

The quick development of AI and associated technologies would assist health-care providers in enhancing patient value and streamlining operational procedures. In this study, we examined how AI technology with data fusion for chronic patients allows professional hospital visits and reduces their workload. Effective management of these opportunities and challenges requires the joint knowledge and tenacity of all healthcare industry players.

Keywords: Blockchain · Identity Management · Supply Chain · Ports · Goods Tracking · Trust · Transparency · Decentralization · Security · Implementation Strategies

1 Introduction

Statistics show that in many European nations, the proportion of elderly individuals in the overall population is continuously rising, and it is anticipated that this trend will continue in the ensuing decades. This will result in a rise in the number of elderly persons who require care [1]. Therefore, the need for home care will increase, since many individuals pick home care above every other type of care, and in an urban setting, the number of persons able to care for dependent family members decreases when the typical large family group is broken up into small family units [2].

V. Sakalauskas et al. (Eds.): IBICA 2023, LNNS 1231, pp. 61–72, 2025.
https://doi.org/10.1007/978-3-031-78946-5_6

Recent years have seen a significant increase in research activity in the field of remote patient monitoring [3]. To identify basic and complicated behaviors including walking, sitting, running, and other activities of everyday life, activity data gathered using a variety of sensors is analyzed [3, 4]. These activities are crucial for giving caregivers and medical rehabilitation staff real-time feedback regarding patients' behaviors, especially for older and special needs patients [5]. Other key uses are in fall detection and postural recognition [6, 7], where older people are at high risk of falling and where understanding what constitutes a true fall might assist to prevent them due to their tendency to have poor health consequences.

Besides de aforementioned, in healthcare settings where the stakes are high, managing worry and stress on an understaffed team is essential to ensure the team can continue to perform. Burnout among nurses at every level of care is a result of a lack of support and anxiety [4]. Nursing staff can have peace of mind knowing their patients are at home safe, thanks to remote patient monitoring [5]. Remote patient monitoring acts as an extra sense taking patients out of the hospital with the same quality [6].

The majority of health issues that older individuals experience is linked to several comorbidities and chronic disorders [1]. The following major aspects have been developed by recent guidelines for health and care workers to aid in the development and implementation of person-centered integrated care for older persons (ICOPE): 1) In order to maximize people's inherent capability and functional ability, healthy ageing should be encouraged; 2) It is important to diagnose and treat intrinsic capability declines as soon as possible; and 3) Capacity losses should be detected at the primary care level and treated appropriately to stop or slow them down. Given the interconnected nature of intrinsic capability reductions, an integrated and person-centered strategy is necessary.

Our research goal is to use Big Data analysis and Artificial Intelligence to remotely collect data of patients at home from different devices and machines such as video cameras, and to build a decision support system. To accomplish this, we do data fusion of structured data with images, text and voice. For the primary stakeholders (patients, doctors, carers, and health authorities), our research offers Decision Support Systems to aid in making more informed choices and reduce the workload of hospital professionals that provide care at home.

Using this collected data, together with structured data from the hospital, more findings and patterns could be discovered in order to provide a more personalised treatment. By developing a focus Big Data platform with the required software programs we are able to process all the heterogeneous data sources. Once processed, this data is stored in a Big Data warehouse in order to have an integrated repository of data to be able to analyse all the generated Big Data in an integrated manner. This is a Home Care Big Data based on pictures from mobile devices, wearables, voice processing, text mining, home devices (e.g. manometers, glucometers or activity devices through smartwatches, etc.). Normally, these data are processed and analysed in an isolated way within the proprietary applications provided by the corresponding vendors.

2 Literature Review

Search Strategy and Inclusion Criteria - In order to assess the work that has been done, we have conducted a literature review. This literature review allows not only to understand the current state of the art but also to identify gaps or problems on the main topic. PRISMA (Preferred Reporting Items for Systematic Reviews and Meta-Analyses) [7] research methodology has been followed. PRISMA methodology is a widely recognized and accepted method for conducting systematic reviews of literature in healthcare research. It provides a structured approach to identifying, selecting, and analyzing relevant studies, ensuring that all relevant studies captured are related to our research question and providing a transparent and reproducible process for conducting systematic reviews, which can help ensure the quality and rigour of our research. Overall, using the PRISMA methodology, we ensure that a rigorous and comprehensive review of the literature on homecare and remote monitoring systems was conducted.

The research question "What is the state of the art on the usage of Internet of Things in Homecare with Artificial Intelligence" was the base for our literature review.

This review was performed on two databases, such as Scopus and Web of Science Core Collection (WoSCC). The filters applied have ensured that all papers were written in English, had to be articles and were published on the last 6 years (2018–2023).

The research strategy used entailed creating a single question from many research perspectives. By consideration of the topic, context, and group under analysis, this made it possible to identify articles that were included in both databases.

Then we searched, in both databases, all the published work with the concept "Artificial Intelligence" or "Machine Learning" or "Data Mining" or "Data Analytics" or "Data Analysis" or "Neural Networks", the target population "Homecare" or "Care Assistance" or "Remote Monitoring System" within a context of "Internet of Things" or "Big Data". Our final query was: "("Artificial Intelligence"OR"Machine Learning"OR"data mining"OR"Data Analytics"OR"Data Analysis"OR"neural networks") AND (Home care OR "Homecare" OR "care assistance" OR "remote monitoring system") AND (IoT OR "Internet of Things" OR "big data")".

The output of the query resulted in 108 documents, but after removing duplicates, we managed to have 77 distinct documents that composed all the papers that should be analyzed. To represent the analysis of all the papers read, a PRISMA process diagram was created. The purpose of this literature review is to understand what exists of homecare, the problems and benefits identified and what can be done to improve the quality of remote monitoring systems to allow patients without serious diseases to go home and be monitored remotely. All the medical staff would benefit from the relief of workload, and patients also could have a better quality of life if they were allowed to be at home. The main topics discussed in all articles are shown in Table 1, with focus on ML and IoT as roles to achieve functional and efficient remote monitoring systems. Goals and Outcomes Analysis - Analyzing all the main goals of this review, a compre hensive analysis of each topic was carried out and shown in Fig. 2. Articles [8–20] discuss the use of Machine Learning (ML) and/or wearable sensors in order to monitor and help self-diagnose some diseases, allowing the patients to be remotely monitored. Systems aim to monitor mainly heart diseases, but there are other examples that monitor.

Metabolic Syndrome, sleep monitoring, diabetes, cerebral palsy, or Parkinson disease. With the same scope but more focused on elderly individuals, [30–33] propose systems for monitoring and management of home care for elderly adults with a focus on elderly needs and acceptance. New solutions are discussed in order to balance the needs of elderly people and the management of their health, but studies show improved qual ity of life for older people when they are at home being monitored. One solution proposed home interactive elderly care two-way video healthcare system that would help the medical staff managing their effort (Fig. 1).

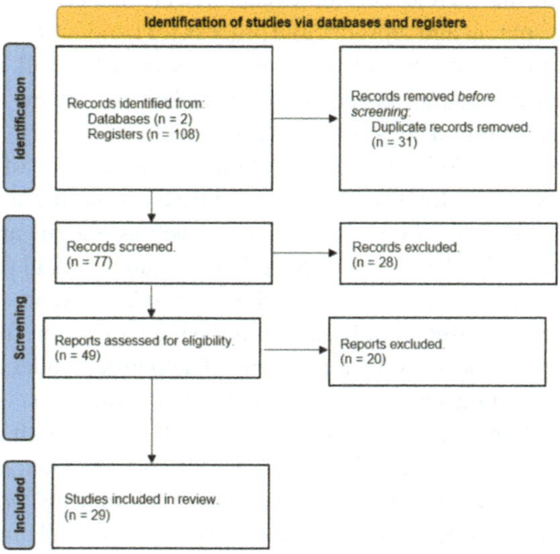

Fig. 1. Results of the application of PRISMA methodology

Tabel 1. Main areas of relevante papers identified

Topics	References	# of papers
IoT and Sensors	[8–29]	22
Machine Learning	[8–20, 22–29]	21
Remote Monitoring Systems	[8–20, 27–34]	21
Elderly	[27–33]	7
Robots	[35, 36]	2

In [21] propose a review of wireless sensor network health monitoring systems and the conclusions are encouraging with the development of more flexible, reliable, secure, real-time and energy efficient systems that improve the quality of remote monitoring systems. In [22–26] discuss and improve the utilization of IoT and sensors on Homecare. They prevent the detection of anomalies on the data stream created with IoT sensors and propose ML models that make effective and timely decisions in the healthcare domain. Overall, sensors are more accurate and efficient with gaps being analyzed and combated.

Another main topic is explored by [27–29], where IoT and ML take a major role on elderly fall prediction and activity recognition. Using sensor or cameras, results appear to be very good.

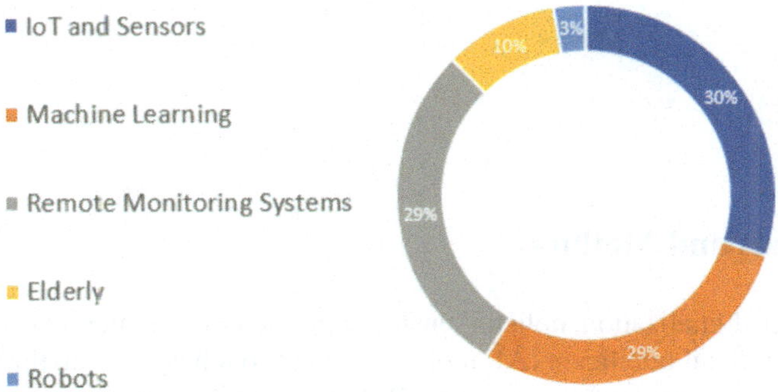

Fig. 2. Comprehensive analysis of each topic, for all relevant papers

From another point of view, [35, 36] suggest the use of robots that would provide valuable tools for in-home care services. In [34] emerge as a caller for the urgent need of home-based remote monitoring systems to avoid the collapse of hospitals first line. A lot of things have been done to improve the performance of remote monitoring systems allowing patients to, if possible, be at home and keep being monitored from possible health problems that could have.

In summary, many studies are focused on the patient side, allowing them to be monitored remotely, but there's a lack of concern on the caregivers and medical staff side. This point is relevant because it places our study on a scientific gap. With these solutions, we can improve the efficiency of medical's staff job by reducing the workload. In a world where Hospitals suffer from overload, it's crucial to give conditions to medical staff and caregivers so they can have a better life quality and increasing their proficiency. In fact, with presented solutions in the literature review we can address different needs from different perspectives (Table 2).

Table 2. Percentage of papers by areas

Topic	Absolut Value	Percentage
IoT and Sensors	22	30%
Machine Learning	21	29%
Remote Monitoring Systems	21	29%
Elderly	7	10%
Robots	2	3%
Total	73	100%

3 Materials and Methods

Care processes digitalization, holistic sensing supported by the Internet of Things (IoT) system and Artificial Intelligent (AI) tools are being actively applied to the health sector giving rise to the smart health paradigm [7]. In this transformative process, the Health Remote Monitoring Systems (HRMS) are recognized as an emerging technology, using sensors and wearable devices to collect patient's data. However, to present clinical value these systems have to be associated with clear clinical processes and therapeutics, so the measurements could be linked with actual patient care.

The implementation of HRMS is based on a health-related IoT system that incorporates, stores, and communicates the information gathered by a set of wearable de-vices and sensors, video and image. The computer senses and records the daily physiological data of the patient by means of a data processing device, data transition, data archive, data analytics, AI [9]. The HRMS is based on four main pillars and aims. Firstly, to develop a system to identify disease development and disease prevention through the use of remote sensors. Secondly, use Big Data processing and analysis, which is used to process multiple heterogeneous data sources in order to process different patient data, to provide high-quality personalized treatment. Thirdly, prediction models built by Artificial Intelligence can be implemented on top of the processed data from Big Data, which would allow the classification of patients, discovering of behavioral patterns. Through this process, alerts can be generated when an abnormality is registered in order to help with disease prevention. Lastly, to create a remote interaction process from hospitals to the patients. Therefore, HRMS enables a data-intensive approach, in which a large amount of health data is generated, stored, and available for data mining, allowing for the generation of useful knowledge.

The device's communication with cloud platforms allows for these sensing data to be stored in the cloud and easily accessed by doctors, allowing a range of remote health monitoring functions. Remote Patient Monitoring (RPM) Technologies require machinery that collects and interprets biometric & physiological data [9]. RPM has many applications such as real-time illnesses detection (through the generation of alerts when an abnormality is registered), continuous monitorization of patients such as those with chronic disease or less severe conditions (reducing patient load at hospitals, as it allows recovering of non-critical patients to happen in their homes) or athlete health monitorization [9].

- Considering the current state of art, this research work proposes AI-based solution us ing data, image, text and voice:
- The ability to take images using standard smartphone technology as a simple sensing tool and the creation of an image data repository.
- Fusion of different data sources (e.g. data, image, text and voice) in a big data for different chronic diseases and we propose to validate in dementia cases.
- Present meaningful and useful data to decision makers to support improved interventions.
- Participatory process involving end-users using hospital and AA patients

It is widely recognized that dementia needs to be seriously considered, and more remote monitor solutions with AI, like the ones proposed in this project, are required in order to reduce the future impact of these diseases in the forthcoming years in the EU and improve the quality of life of these patients. The research work provides a remote monitoring system for health authorities to help reduce the effort and AI and data analytics will help them take better-informed decisions. Another cross-cutting objective that will be specially considered throughout the whole project is the gender aspects in the sense that corresponding developed AI algorithms.

4 Research Plan and Methods

Action Research has been defined to bridge the gap between research and practice. With this aim, it has two clear goals: to generate profits for the customer of the research; and, at the same time, to generate relevant "knowledge of research" [24]. Therefore, Action-Research is an approach highly collaborative between research, re-searchers, and practitioners, focused on both theory and practice and carried out by means of a cyclic process. In addition, Action-Research describes a class of methods, which share the following characteristics [25]: 1) To be oriented to action and change; 2) To be focused on a problem; 3) To have an organic model of process that entails systematic stages and some iterative ones; and 4) To be collaborative between the participants.

In a more formal analysis of the participants, Action-Research, [26] identifies four types of roles that are described in the following along with their related actors.

Considering the above-assigned responsibilities, in order to put the Action-Research methodology into practice, is necessary to establish the activities that will guide the research. We proposed a highly flexible Big Data architecture and the creation of several dashboards for relevant stakeholders following the methodology. This approach [28] elucidates the specific requirements for each stakeholder, and ensures that the implement dashboards that are tailored to different roles and users, being able to derive multiple sets of dashboards that are aligned with the identified in-formation needs.

ICT solutions for healthcare are mostly focused on telehealth platforms designed for adaptive and proactive needs. For these systems to be interoperable, usable, secure, and accurate, strong integration is required. We propose a remote solution that can analyse data, video, text and voice, because this are components of a daily live. Voice and video can be acquired with camera and for privacy issues we do not store them. Our HCS is proposed in Fig. 3.

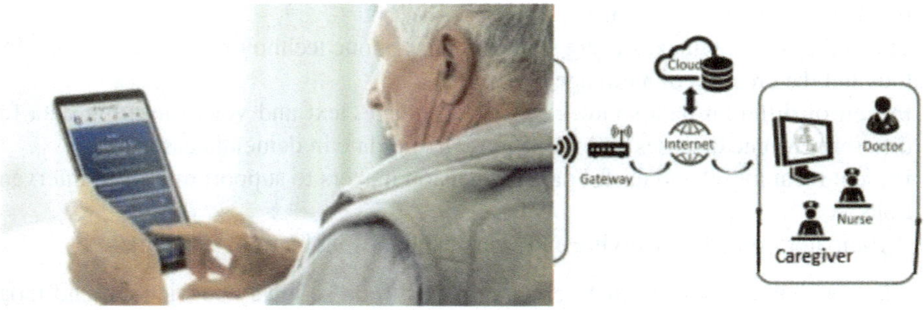

Fig. 3. Proposed System

Based on video data processing from daily/weekly conversations.

1) Face monitor process – Trained algorithm to recognize facial expressions like: pain (try to identify pain levels), happiness, sadness, surprise, anger, fear, disgust, and con- tempt.
2) Head pose detection - We detected three head poses including yaw, pitch, and roll, using the pre-trained OpenFace deep neural network tool
3) Posture recognition - To recognize patient posture, we used a multi-person pose estimation model to localize the anatomical key points of joints and limbs. Then we used the lengths of body limbs and their relative angles as features for recognizing lying in bed, standing, sitting on the bed, and sitting in the chair
4) Extremity movement analysis - We analyzed the data from three accelerometer sensors worn on the patient's wrist, ankle, and arm, and compared the results between delirious and non-delirious patients.
5) Room sound pressure levels and light intensity - The sound pressure levels for delir ious patients' rooms during the night were on average higher than the sound pressure levels of non-delirious patients' rooms (Fig. 4). Average night-time sound pressure levels were significantly different between the delirious and non-delirious patients (p-value < 0.05).

Voice processing, we can detect pain based on voice. A new strategy called pervasive computing focuses on creating intelligent settings where connected devices are inserted and provide continuous, dependable, non-intrusive connectivity with extra benefits. Without a clear understanding of the underlying technological interplay, this system's effect is often an enhancement in human experience and quality of life [15]. As a result, ubiquitous data can also be thought of as information that is always accessible and locations. The whole information required to do an assessment is not always available in critical contexts, such as intensive care units, where choices must be made at the appropriate moment. Despite the fact that not all medical errors may be avoided, the spread of data can be facilitated through ubiquitous data techniques. Reducing redundant information and allowing it to be kept on mobile devices or in strategically placed devices near where choices will be made will improve the quality of healthcare [18]. The system that sent alarms from the monitoring patient system is an example of pervasive data. When patients are outside the parameters of the monitored qualities, these notifications are sent. In general, ubiquitous healthcare aids medical professionals by establishing a setting that enables them to access information at anytime and anyplace.

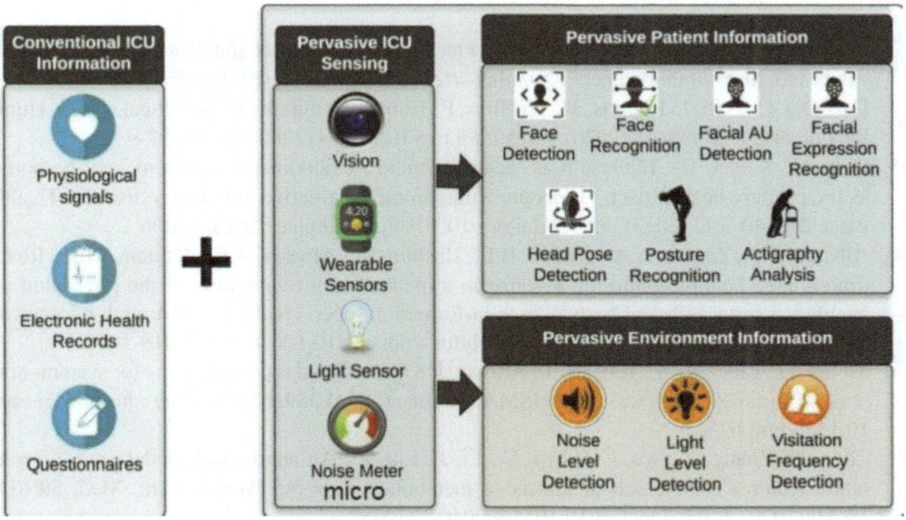

Fig. 4. This is a figure. Schemes follow another format. If there are multiple panels, they should be listed as: (a) Description of what is contained in the first panel; (b) Description of what is contained in the second panel. Figures should be placed in the main text near to the first time they are cited.

5 Conclusions

It is obvious that the quick development of AI and associated technologies would assist healthcare providers in enhancing patient value and streamlining operational procedures. In this study, we looked at how AI technology with data fusion can bring remote chronic patients. This allows them to reduce professional hospital visits and reduce their workload. Effective management of these opportunities and challenges requires the joint knowledge and tenacity of all healthcare industry players.

In conclusion, the implementation of remote home care monitoring systems using AI-powered text mining can significantly reduce the workload of healthcare providers in hospitals. By automating the process of extracting and analyzing relevant information from large amounts of unstructured data, these systems can help healthcare providers to make more informed decisions, improve patient outcomes, and reduce the burden of manual data entry and analysis. As the market for wearable medical devices and remote home care systems continues to grow, it is expected that the use of AI-powered text mining will become increasingly prevalent in the healthcare industry, providing new opportunities for improved patient care and reduced workload for healthcare providers.

References

1. Eurostat, "Europe in figures - Eurostat yearbook 2008," Luxembourg, Office for Official Publications of the European Com-munities (2008). Accessed 6 Sep 2022. https://ec.europa.eu/eurostat/web/products-statistical-books/-/ks-cd-07-001
2. Tarricone, R., Tsouros, A.D.: Home care in Europe: the solid facts. World Health Organ ization. Regional Office for Europe (2008). Accessed 6 Sep 2022. https://apps.who.int/iris/handle/10665/328766
3. Smart homes and home health monitoring technologies for older adults: a systematic review - PubMed. https://pubmed.ncbi.nlm.nih.gov/27185508/, Accessed 02 Nov 2022
4. Dall'Ora, C., Ball, J., Reinius, M., Griffiths, P.: Burnout in nursing: a theoretical review. Hum. Resour. Health **18**(1), 41 (2020). https://doi.org/10.1186/s12960-020-00469-9
5. Gale, N., Sultan, H.: Telehealth as 'peace of mind': embodiment, emotions and the home as the primary health space for people with chronic obstructive pulmonary disorder. Health Place **21**, 140–147 (2013). https://doi.org/10.1016/j.healthplace.2013.01.006
6. Albahri, O.S., Zaidan, A.A., Zaidan, B.B., Hashim, M., Albahri, A.S., Alsalem, M.A.: Real-time remote health-monitoring systems in a medical centre: a review of the pro-vision of healthcare services-based body sensor in-formation, open challenges and meth-odological aspects. J. Med. Syst. **42**(9), 164 (2018). https://doi.org/10.1007/s10916-018-1006-6
7. Moher, D., Liberati, A., Tetzlaff, J., Altman, D.G.: Preferred reporting items for system- atic reviews and meta-analyses: the PRISMA statement. BMJ **339**, b2535 (2009). https://doi.org/10.1136/bmj.b2535
8. Zhou, T., Zhang, Y., Wu, C., Shen, C., Li, J., Liu, Z.: An augmented model with inferred blood features for the self-diagnosis of metabolic syndrome. Methods Inf. Med. **59**(01), 18–30 (2020). https://doi.org/10.1055/s-0040-1710382
9. Miao, F., Wang, X., Yin, L., Li, Y.: A wearable sensor for arterial stiffness monitoring based on machine learning Algorithms. IEEE Sens. J. **19**(4), 1426–1434 (2019). https://doi.org/10.1109/JSEN.2018.2880434
10. Jeng, P.-Y., Wang, L.-C., Hu, C.-J., Wu, D.: A wrist sensor sleep posturemonitoring system: an automatic labeling approach. Sens. Switz. **21**(1), 1–19 (2021). https://doi.org/10.3390/s21010258
11. Hussein, A.F., et al.: An automated remote cloud-based heart rate variability monitoring system. IEEE Access **6**, 77055–77064 (2018). https://doi.org/10.1109/ACCESS.2018.2831209
12. Raj, S.: An efficient IoT-based platform for remote real-time cardiac activity monitoring. IEEE Trans. Consum. Electron. **66**(2), 106–114 (2020). https://doi.org/10.1109/TCE.2020.2981511
13. Premalatha, G., Bai, V.T.: Design and implementation of intelligent patient in-house monitoring system based on efficient XGBoost-CNN approach. Cogn. Neurodyn. **16**(5), 1135–1149 (2022). https://doi.org/10.1007/s11571-021-09754-2
14. Serhani, M.A., El Kassabi, H.T., Ismail, H., Navaz, A.N.: ECG monitoring systems: review, architecture, processes, and key challenges. Sens. Switz. **20**(6), (2020). https://doi.org/10.3390/s20061796
15. Basatneh, R., Najafi, B., Armstrong, D.G.: Health sensors, smart home devices, and the internet of medical things: an opportunity for dramatic improvement in care for the lower extremity complications of diabetes. J. Diabetes Sci. Technol. **12**(3), 577–586 (2018). https://doi.org/10.1177/1932296818768618
16. Conn, N. J., Schwarz, K.Q., Borkholder, D.A.: Nontraditional electrocardiogram and algorithms for inconspicuous in-home monitoring: Comparative study. JMIR MHealth UHealth **6**(5), (2018). https://doi.org/10.2196/mhealth.9604

17. Singh, D.N., Hashmi, M.F., Sharma, S.K.: Predictive analytics & modeling for modern health care system for cerebral palsy patients. Multimed. Tools Appl. **79**(15–16), 10285–10307 (2020). https://doi.org/10.1007/s11042-019-07834-4
18. Shaik, T.,et al.: Remote patient monitoring using artificial intelligence: current state, applications, and challenges. Wiley Interdiscip. Rev. Data Min. Knowl. Discov. (2023). https://doi.org/10.1002/widm.1485
19. Luis-Martínez, R., Monje, M.H.G., Antonini, A., Sánchez-Ferro, Á., Mestre, T. A.: Tech nology-enabled care: integrating multidisciplinary care in parkinson's disease through digital technology. Front. Neurol. **11** (2020). https://doi.org/10.3389/fneur.2020.575975
20. Bolhasani, H., Mohseni, M., Rahmani, A.M.: Deep learning applications for IoT in health care: a systematic review. Inform. Med. Unlocked **23** (2021). https://doi.org/10.1016/j.imu.2021.100550
21. Salunke, G., Singh, V., Jadhav, C.: A review approach to modeling, analysis, and design framework for wireless sensor network health monitoring systems. J. Pharm. Negat. RESULTS **13**, 165–177 (2022). https://doi.org/10.47750/pnr.2022.13.S02.24
22. Kaya, S., Erdem, A., Gunes, A.: A smart data pre-processing approach to effective management of big health data in IoT edge. Smart Homecare Technol. Telehealth **8**, 9–21 (2021). https://doi.org/10.2147/SHTT.S313666
23. Fouad, H., Hassanein, A.S., Soliman, A.M., Al-Feel, H.: Analyzing patient health infor mation based on IoT sensor with AI for improving patient assistance in the future direction. Meas. J. Int. Meas. Confed. **159** (2020) https://doi.org/10.1016/j.measurement.2020.107757
24. Mao, Y., Zhang, L., Wu, X.: Perception analysis and early warning of home-based care health information based on the internet of things. Complexity **2021** (2021). https://doi.org/10.1155/2021/6634575
25. Tiersen, F., et al.: Smart home sensing and monitoring in households with dementia: User-centered design approach. JMIR Aging **4**(3) (2021). https://doi.org/10.2196/27047
26. Grgurić, A., Mošmondor, M., Huljenić, D.: The smarthabits: an intelligent privacy-aware home care assistance system. Sens. Switz. **19**(4) (2019). https://doi.org/10.3390/s19040907
27. Kulurkar, P., Dixit, C.K., Bharathi, V.C., Monikavishnuvarthini, A., Dhakne, A., Preethi, P.: AI based elderly fall prediction system using wearable sensors: a smart home-care technology with IOT. Meas. Sens. **25** (2023). https://doi.org/10.1016/j.measen.2022.100614
28. Vaiyapuri, T., Lydia, E.L., Sikkandar, M.Y., Diaz, V.G., Pustokhina, I.V., Pustokhin, D.A.: Internet of things and deep learning enabled elderly fall detection model for smart homecare. IEEE Access **9**, 113879–113888 (2021). https://doi.org/10.1109/ACCESS.2021.3094243
29. Jansson, J., Hakala, I.: Managing sensor data streams in a smart home application. Int. J. Sens. Netw. **32**(4), 247–258 (2020). https://doi.org/10.1504/IJSNET.2020.106603
30. Zhang, B., et al.: A framework for remote interaction and management of home care El-derly adults. IEEE Sens. J. **22**(11), 11034–11044 (2022). https://doi.org/10.1109/JSEN.2022.3170295
31. Timon, C.M., et al.: Development of an internet of things technology platform (the NEX System) to support older adults to live independently: protocol for a development and usability study. JMIR Res. Protoc. **11**(5) (2022). https://doi.org/10.2196/35277
32. Yi, C., Feng, X.: Home interactive elderly care two-way video healthcare system design. J. Healthc. Eng. **2021** (2021). https://doi.org/10.1155/2021/6693617
33. Miranda, R., et al.: Improved quality of care and reduced healthcare costs at the end-of-life among older people with dementia who received palliative home care: a nationwide propensity score-matched decedent cohort study. Palliat. Med. **35**(9), 1701–1712 (2021). https://doi.org/10.1177/02692163211019321
34. Castelnuovo, G., Pietrabissa, G., Manzoni, G.M., Sicurello, F., Zoppis, I., Molinari, E.: Fighting the COVID-19 pandemic using the technology-based second-line in Italy and Lombardy:

the urgent need of home-based remote monitoring systems to avoid the collapse of the hospital-centred first line. J. Glob. Health **10**(2), 1–2 (2020). https://doi.org/10.7189/jogh.10.020371
35. Yang, G., et al.: Homecare robotic systems for healthcare 4.0: visions and enabling technologies. IEEE J. Biomed. Health Inform. **24**(9), 2535–2549 (2020). https://doi.org/10.1109/JBHI.2020.2990529
36. Zhao, H., et al.: Intelligent indoor metasurface robotics. Natl. Sci. Rev. https://doi.org/10.1093/nsr/nwac266

The Future of News Recommendation: A Blend of User Preferences, Content Analysis, and Social Signals

Sthembiso Mkhwanazi[1] and Absalom E. Ezugwu[2](✉)

[1] School of Mathematics, Statistics, and Computer Science, University of KwaZulu-Natal, King Edward Avenue, Pietermaritzburg Campus, Pietermaritzburg 3201, KwaZulu-Natal, South Africa
49941615@mynwu.ac.za

[2] Unit for Data Science and Computing, North-West University, 11 Hoffman Street, Potchefstroom 2520, South Africa
absalom.ezugwu@nwu.ac.za

Abstract. With the advent of outstanding technological progress in the last decade, there has been a remarkable surge in the number of individuals who rely on online news articles as their primary source of information. This exponential growth has prompted news publishers to transition from traditional newspapers to digital platforms, allowing them to reach a wider audience in real-time. As a result, there is now an abundance of news sources publishing a vast quantity of articles on a daily basis. However, this surge in online content has presented a challenge for news readers, as it has become increasingly difficult to access relevant news in real-time, leading to a state of information overload. To tackle this issue, this study proposes the implementation of a user-personalized news categorization and recommendation system that harnesses the power of advanced Natural Language Processing techniques. The suggested model is built upon hybrid filtering, which combines Collaborative Filtering and Content-Based Filtering methodologies to construct a news model that is tailored to each individual user, taking into account their historical news article preferences. Moreover, the model incorporates social media comments from platforms like Reddit to alleviate challenges associated with these techniques, such as the cold-start problem and data sparsity. To evaluate the performance of the system, we conducted assessments using two key metrics: Mean Average Precision (MAP) and Normalized Discounted Cumulative Gain (nDCG). The results yielded values of 0.67 and 0.70 for MAP and nDCG, respectively, indicating the effectiveness of the proposed system in providing accurate and relevant news recommendations to users.

Keywords: Recommendation system · comment recommendation · natural language processing · NLP · collaborative filtering · content-based filtering · user-personalized news

V. Sakalauskas et al. (Eds.): IBICA 2023, LNNS 1231, pp. 73–87, 2025.
https://doi.org/10.1007/978-3-031-78946-5_7

1 Introduction

The rapid advancement of technology has revolutionized communication and service access, creating a demand for instant solutions in almost every aspect of life. Unsurprisingly, this progress has compelled newspaper publishers to embrace technology by digitizing news articles and delivering real-time news to cater to a vast audience and meet their expectations. Many prominent newspapers have embarked on this journey of digital transformation, resulting in a proliferation of news articles on the internet [1].

The convenience and immediacy of online news services have inspired numerous readers to shift from traditional newspapers to digital news platforms [2]. However, the increasing number of articles available online has made it challenging for readers to access news that aligns with their interests. They are required to browse through a multitude of articles to find the information they seek, leading to time-consuming processes and information overload. These issues significantly impact the news reading experience and frequency of users on the platform. Given the busy lifestyles of people today, they desire to read only those news articles that are of direct interest to them.

Automated techniques that enable efficient news classification and personalized article recommendations are in high demand, as they save users time and effort while reducing information overload. Collaborative Filtering (CF), Content-Based Filtering (CBF), and Hybrid Filtering (HF) are commonly employed techniques in the development of such systems.

- CF leverages user preference databases to evaluate and filter news items based on the opinions of other users. It recommends news articles aligned with a user's interests by analyzing the preferences of similar users [3].
- CBF extracts features from the content of news articles consumed or implicitly shown interest in by the user [4]. It utilizes these features to recommend similar news items that match the user's preferences.
- HF combines the strengths of both CBF and CF approaches. It integrates recommendations from these two methods to generate a final recommendation [5]. By leveraging the advantages of each approach, HF aims to provide users with more accurate and diverse news recommendations.

News providers have adopted information filtering techniques to predict users' preferences based on their past reading habits. This involves utilizing user models and news models to generate personalized recommendations. The user model analyzes users' news consumption history, while the news model focuses on understanding the features and content of news articles [6]. By combining these models, the news platform selects a concise set of potential news articles, ranks them, and recommends the top articles to the user.

To achieve this, advanced Natural Language Processing (NLP) techniques are employed to gain insights into both the user and the news articles on the platform. Machine Learning and Deep Learning approaches are also utilized to facilitate classification and recommendation processes [7]. These techniques enable the system to effectively interpret and analyze textual data, providing accurate and relevant news recommendations tailored to each user's interests.

This study introduces the HF method with a novel approach to further enhance its popularity and application. The integration of another method into HF aims to improve the performance of recommendation systems and contribute to research advancements in the field. The paper highlights the specific contributions of this study.

i. The study introduces a hybrid model that combines CF and CBF using an enriched HF model. This integration leverages the strengths of both approaches, resulting in a more comprehensive and accurate recommendation system.
ii. To address challenges associated with HF, the paper introduces an integrated social-media-enabled technique. This enhancement improves the system's ability to gather and incorporate social media data, leading to improved recommendation accuracy and relevance.
iii. The proposed HF model incorporates advanced NLP techniques to support its implementation. These techniques enable effective analysis and understanding of textual data, contributing to more precise recommendations.
iv. The study designs a tailored data preprocessing pipeline to support machine learning models and enhance the classification process. This pipeline ensures data quality and integrity, resulting in more reliable recommendations.
v. The proposed hybrid model is evaluated using widely accepted benchmark metrics such as Mean Average Precision (MAP) and normalized Discounted Cumulative Gain (nDCG). This evaluation provides valuable insights into the performance and effectiveness of the recommendation system.

In summary, this paper's contributions include the introduction of a hybrid CF-CBF model within the HF framework, the integration of a social-media-enabled technique, the application of advanced NLP techniques, the establishment of a data preprocessing pipeline, and the evaluation of the proposed hybrid model using MAP and nDCG metrics. These contributions collectively aim to improve recommendation system performance and advance research in the domain of personalized news recommendation.

The paper is structured as follows: Sect. 2 provides a comprehensive review of related studies, emphasizing their significance. Section 3 explores the methodology, techniques, and dataset used in developing the new recommendation system, addressing the complexities of news article recommendations. In Sect. 4, the results and findings of the study are presented and analyzed, evaluating the performance and impact on user satisfaction and engagement. Section 5 concludes the paper by summarizing key findings and contributions, while also highlighting potential future extensions and directions for further research and development.

2 Literature Review

Developing perfect personalized news recommendation and classification systems in the dynamic news domain is challenging due to changing article sets, evolving news interests, and cold start problems [9]. Continuous updates and evolving data storage methods of news articles on the internet necessitate ongoing research and techniques to ensure relevance and user satisfaction on online news platforms. This paper provides a comprehensive overview of the state of the art in news recommendation systems, discusses the challenges they face, and emphasizes the importance of addressing these challenges.

Personalized news recommendation systems are categorized into content-based, collaborative, and hybrid filtering techniques. These techniques are widely employed in various online news platforms, news apps, and social networking sites [6]. They differ from non-personalized news recommendations that rely on factors like popularity and geographic location, rather than user interests.

The CBF recommends items based on a user's past preferences, often using implicit feedback like clicks. Lu et al. [10] proposed a CBF approach for news topic recommendation, while Kompan et al. [10] used term frequency (TF) to compute article relevance. Yunanda et al. [11] implemented CBF using Term Frequency-Inverse Document Frequency (TF-IDF) and achieved an 80.77% hit-rate result. However, CBF faces challenges such as the cold-start problem for users with low access to news, computational expenses for TF-IDF calculations, and the inability to recommend newly added news without prior user feedback.

The CF is a successful recommendation system that generates recommendations based on the preferences of similar users. It has subcategories such as neighborhood-based CF, item-based CF, model-based CF, and memory-based CF [4]. Model-based CF has been found to be faster and more effective in generating recommendations compared to item-based CF [12]. Shen et al. [13] achieved an 84.2% recommendation accuracy using CF for news recommendations. Google News and Group Lens are examples of CF-based approaches with reasonable accuracy [14, 15]. However, CF faces challenges such as the inability to recommend newly introduced news articles without prior user ratings, limited effectiveness for users with unique preferences, and scalability issues as the number of news items and users increases. Furthermore, as indicated by Su [3] and Shen et al. [13], the process of searching for potential neighbors becomes more complex as the size of the database expands.

Hybrid filtering combines the strengths of content-based and collaborative filtering approaches to provide personalized news recommendations [16]. It improves accuracy and resolves scalability, data sparsity, and cold-start problems [17]. GroupLens implemented a hybrid collaborative filtering system for Usenet news, utilizing users as content-based filters [15]. To enhance news content understanding, advanced NLP techniques are employed. Systems utilize word2vec embedding with TF-IDF to capture semantic and commonsense information encoded in news articles [6]. Capelle et al. [18], introduced Synset Frequency-Inverse Document Frequency (SF-IDF), a technique using Wordnet synonym sets, to represent news. Neural NLP models, such as autoencoders, are employed to understand the deep representations of news [19]. These methods outperform traditional semantic-based approaches and improve overall performance.

Personalized news recommendation on social media platforms has gained significant popularity in recent years. Users perceive social media as better recommenders than relying solely on journalists and news providers [20]. Social networking sites effectively utilize user interactions such as likes, comments, and shares for personalized news recommendations. Manoharan et al. proposed a fuzzy logic approach using user newsfeeds from Twitter and Facebook for personalization [21]. O'Bannon et al. introduced a content-based method utilizing Twitter to model user interests [22].

Based on the surveyed literature, a hybrid filtering technique is deemed necessary to address the limitations of CBF and CF. In addition, users' Reddit comments are incorporated to build recommender systems. The pretrained Google News word2vec model, developed by Mikolov et al. [23] in 2013, is utilized as a valuable resource for this research. The model combines the Skip-Gram and Continuous Bag Of Words (CBOW) neural network models.

3 Research Methodology

In this section, we outline the methodology employed to develop a hybrid filtering personalized news recommendation system. Through a comprehensive review of the existing literature, we identified the strengths and limitations of previously implemented systems. Leveraging NLP techniques, specifically word2vec tools, we focused on modeling the news text. In the following subsections, we will explore the specific techniques employed to create our hybrid personalized news recommender system. To provide a visual representation of the processes involved, Fig. 1 illustrates the steps we undertook in building the system.

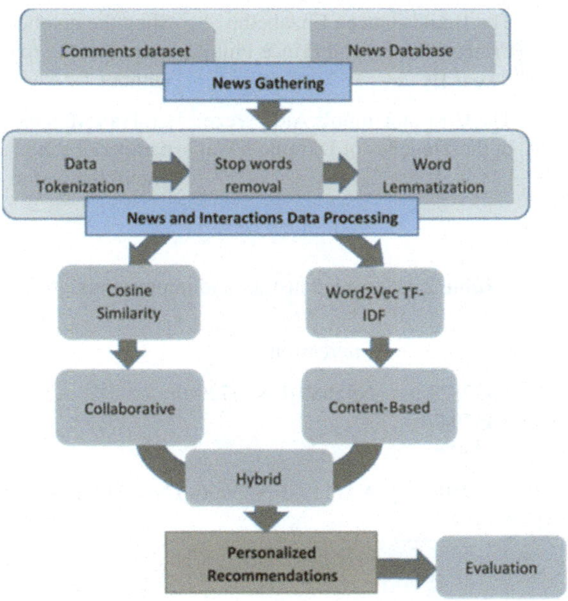

Fig. 1. Methodology summary.

3.1 News and Social Media Dataset

For our system, we utilized the Microsoft News Dataset (MIND), a large-scale dataset specifically designed for news recommendation systems. This dataset comprises approximately 1 million users and includes over 160,000 English articles, each with a category label, title, and abstract. These attributes provide rich contextual information that can be leveraged through NLP techniques to understand the context of each article. The dataset also includes user behaviors towards news, such as click history associated with specific news IDs and distinct user IDs. These user interactions were collected implicitly. The MIND dataset is publicly available for research purposes, and further details can be found in Wu et al.'s paper [24]. Additionally, we incorporated a social media dataset that consists of users' reactions (comments) extracted from various user profiles on Reddit. This dataset includes the specific user's comment, their unique ID, and the category to which the comment belongs. By incorporating this social media dataset, we aimed to enhance the understanding and relevance of user reactions to news articles. Together, the MIND dataset and the social media dataset provided us with a comprehensive foundation for building our personalized news recommendation system, allowing us to leverage both news article attributes and user interactions to create more accurate and context-aware recommendations.

Table 1. A short overview of the news dataset

News ID	Category	Title	Abstract
N55528	lifestyle	The Brands Queen Elizabeth, Prince Charles, and Prince Philip Swear By	Shop the notebooks, jackets, and more that the royals can't live without
N61837	news	The Cost of Trump's Aid Freeze in the Trenches of Ukraine's War	Lt. Ivan Molchany's peeked over a parapet of sand bags at the front line of the war in Ukraine

Table 2. User read history and impressions.

User ID	History	Impressions
U13740	N55189 N42782 N34694 N45794 N18445 N63302	N55689-1 N35729-0
U46596	N47438 N20950 N21317 N5469	N21119-1 N53696-0 N33619-1 N25722-0 N2869-0

In Table 1, it is evident that the dataset does not contain the complete content of the news articles. However, we can utilize the abstract as a substitute because it provides a comprehensive summary of each article, surpassing the title alone. Moving on to Table 2, it presents a summary of users' behavior towards various news articles, displaying only three categories: News ID, History, and Impressions. The News ID serves as a unique identifier for each article, while the user's history of reading news articles determines the implicit rating. Additionally, the Impressions list showcases the news articles presented to the user, with a label of 1 or 0 indicating a click or non-click, respectively.

To construct a collaborative filtering recommendation system, we require users and their ratings for items. Since we are dealing with news articles that are implicitly rated, we assigned a rating of 1 to each article that a user has read and a rating of -1 to the impressions that the user ignored. By utilizing the information provided in Table 2, we obtained the resulting dataset, as illustrated in Table 3. In Table 4 an overview of the Reddit dataset is illustrated.

Just to add some context to this, Reddit is a well-known online platform that facilitates user discussions, content sharing, and participation in topic-specific communities known as "subreddits."

The associated datasets comprise posts, comments, and metadata reflecting user interactions. These datasets encompass various details such as post titles, content, timestamps, user IDs, comment threads, upvotes, downvotes, and more. Researchers and data scientists utilize Reddit datasets for diverse purposes, including sentiment analysis, community detection, recommendation systems, content classification, and the exploration of user behavior and trends.

Table 3. Rating based on user read history and impressions.

User	News Item	News Rating
U13740	N55189	1
U13740	N18445	1
U13740	N35729	−1
U46596	N20950	1
U46596	N21119	1
U46596	N2869	−1

Table 4. A short overview of the Reddit dataset

User ID	History	Impressions
t1_cr92xnl	movies	Well, to each their own I guess. Lotr is another great example. I personally don't mind some cgi, as long as it's good cgi, and it isnt used for every tiny thing. Practical effects do feel more real though
t1_cr9398p	news	It would be her since she fabricated the story. In theory the news should have followed up and investigated the truth of the story, but they don't do that anymore. They just run with it and wait for the riots

3.2 Data Pre-processing

In this subsection, we focus on the cleaning and preprocessing of the dataset to ensure that the raw data is transformed into a structured format suitable for the recommender system.

NLP techniques play a crucial role in this step by enabling machines to comprehend the data effectively. The process involves various techniques such as converting words to lowercase, tokenization, lemmatization, and other methods that aid in organizing the data. By implementing these techniques, we ensure that the data is properly structured and ready for further analysis. Figure 2 illustrates a concise summary of the steps involved in the data preprocessing stage, showcasing the flow of the process.

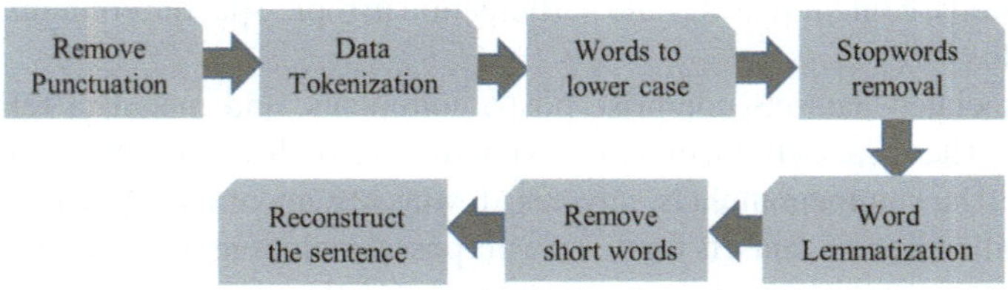

Fig. 2. A pipeline of dataset pre-processing approach.

During the pre-processing steps, we used the NLTK library, which has built-in functions for tasks such as data tokenization, lemmatization, and stopword removal. To further illustrate how these processes operate, consider the following example sentence: "In an effort to combat some of the negative mental impact caused by social media, the company "Instagram" announced it is removing all its filters that give a plastic surgery-like effect."Here is a concise summary of each step:

Remove Punctuation: Punctuation marks were removed to eliminate their influence on news recommendations, reducing execution time. Result: "In an effort to combat some of the negative mental impact caused by social media, the company Instagram announced it is removing all its filters that give a plastic surgery-like effect."

Data Tokenization: The sentence was tokenized into individual words to facilitate processing. Result: "'In', 'an', 'effort', 'to', 'combat', 'some', 'of', 'the', 'negative', 'mental', 'impact', 'caused', 'by', 'social', 'media', 'the', 'company', 'Instagram', 'announced', 'it', 'is', 'removing', 'all', 'its', 'filters', 'that', 'give', 'a', 'plastic', 'surgery', 'like', 'effect'."

Words to Lowercase: All tokenized words were converted to lowercase to ensure consistent representation. Result: "'in', 'an', 'effort', 'to', 'combat', 'some', 'of', 'the', 'negative', 'mental', 'impact', 'caused', 'by', 'social', 'media', 'the', 'company', 'instagram', 'announced', 'it', 'is', 'removing', 'all', 'its', 'filters', 'that', 'give', 'a', 'plastic', 'surgery', 'like', 'effect'."

Remove Stopwords: Commonly appearing words with no significant semantic information, such as pronouns and articles, were removed. Result: "'effort', 'combat', 'negative', 'mental', 'impact', 'caused', 'social', 'media', 'company', 'Instagram', 'announced', 'removing', 'all', 'filters', 'give', 'plastic', 'surgery', 'like', 'effect'."

Word Lemmatization: Words were reduced to their base form or lemma to remove inflectional endings. Result: "'effort', 'combat', 'negative', 'mental', 'impact', 'cause', 'social', 'medium', 'company', 'Instagram', 'announce', 'remove', 'all', 'filter', 'give', 'plastic', 'surgery', 'like', 'effect'."

Remove Short Words: Words that were not removed during previous steps and were not descriptive enough were eliminated.

Reconstruct the Sentence: The pre-processed words were joined to reconstruct the sentence. Result: "effort combat negative mental impact cause social medium company instagram announce remove all filter give plastic surgery like effect."

As part of the pre-processing steps, the news behavior dataset was transformed into a list format, consisting of user-article impressions. If a user clicked on a particular article, it was mapped to 1, while if they only viewed it without clicking, it was mapped to −1. Table 5 provides an overview of the user's behavior towards different articles, highlighting their interactions and engagement levels.

Table 5. Database of users' impressions on the news

User ID	News ID	Verdict
U10149	N39074	−1
U10064	N879	1
U10022	N27448	1
U10045	N20959	−1

To extract the syntactic and semantic information from the comments in the Reddit dataset, we applied the aforementioned techniques for cleaning and preprocessing. Additionally, considering that comments can vary significantly in length, we removed all short comments with a string length of less than 200 characters. This step helped ensure that the remaining comments were more substantial and provided sufficient context for analysis and understanding.

3.3 Content-Based Recommender System

To construct a content-based recommender system, it is essential to comprehend the news articles through their concise and detailed abstracts, as well as user behaviors, in order to create the user model. To capture the similarities between related news articles, we employed word embedding and TF-IDF methods to extract the features, both directly and indirectly.

Word2vec is a widely used tool in NLP models for understanding and vectorizing the contextual meaning of words within a document [6]. Unlike TF-IDF, which relies on word co-occurrence patterns, word2vec models aim to capture syntactic and semantic relationships by considering the similarity of meanings in different contexts. The word2vec model comprises two neural network models: Skip-Gram (SG) and Continuous Bag of Words (CBOW) [23]. Figure 3 provides a visual comparison between these two models, highlighting their differences and characteristics. By utilizing word embedding techniques like word2vec and incorporating TF-IDF, we can effectively capture the essence of news articles and facilitate the recommendation process within the content-based recommender system.

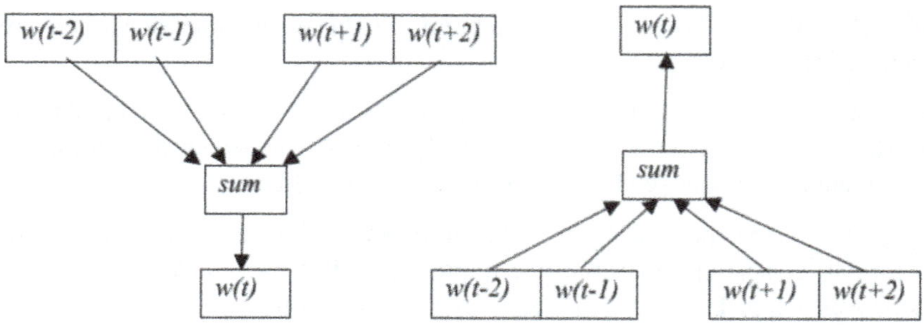

Fig. 3. The visual comparison of CBOW (left) and SG (right)

During the training of the word2vec model, we utilized the abstracts to construct a corpus that served as input for the model. The corpus consisted of approximately 100,000 billion words, and we generated 300-dimensional word embeddings for different words and phrases present in each news article's abstract. To ensure that similar words had comparable vector representations, we utilize the power of TF-IDF. To facilitate this feature extraction process, we imported a pre-trained word2vec model using the Gensim library. We then developed a function to calculate the average word vectors from all the words in the abstract. By mapping the words in the model's vocabulary to their frequency in the corpus, we created an empty array of length 300, which served as a feature vector that was subsequently added to the total.

To determine the weight of each word in the corpus, we employed the TF-IDF method. Equations (1) and (2) represent the calculation formulas for Term Frequency (TF) and Inverse Document Frequency (IDF), respectively. These equations were utilized to compute the TF and IDF values for each word in the corpus, contributing to the overall feature extraction process. By combining the word2vec model and TF-IDF weighting, we obtained meaningful and contextually aware feature representations for the news articles, facilitating the subsequent recommendation process.

$$tf(t,d) = \frac{f(t,\mathrm{d})}{\sum_k (W_k, d)} \tag{1}$$

$$\mathrm{id}f_t = \log\left(\frac{M}{1 + \mathrm{d}f_t}\right) \tag{2}$$

The TF formula mentioned earlier calculates the frequency of a specific term (t) within a document (d), indicating how often the term appears in the document. On the other hand, the inverse document frequency (IDF) measures the commonness of a word in the entire corpus. It uses the total number of documents (M) divided by the number of documents that contain the term (t) as the denominator. By combining TF and IDF, we obtained multiple word vectors by multiplying the TF-IDF values with the accumulated word vectors generated by the word2vec model.

To measure the similarities between two news articles, we employed cosine similarity. Specifically, we considered ten previously read news articles from a user's impression dataset (marked with a value of one) and calculated the cosine similarities with the news dataset. Based on higher similarity scores, we ranked the news articles and selected the top 30 as part of the content-based recommendation system.

For the cold start problem, where a user is new or has not read at least ten news articles, we turned to the social media content-based recommender method. We aggregated additional user comments from Reddit, assuming that what a user comments on reflects their interests and what they find interesting. These comments provided us with insights into the topics a user likes to engage with. To address the cold start and data sparsity challenges, we used these comments as supplementary content. Similar to the user's read history, we leveraged the comments to recommend the top 30 news articles, excluding those that have already been read. More so, by combining both the user's read history and relevant social media comments, we aimed to provide personalized news recommendations even in scenarios where the user's interaction data is limited or unavailable.

3.4 Collaborative Recommender System

For the collaborative recommender system, we utilized a user impression dataset to construct a co-occurrence matrix of users and news article IDs. This matrix allowed us to establish item relationships based on user interactions, specifically focusing on the item-based approach. By calculating cosine similarities between items, we determined their similarity based on user behavior and identified items frequently consumed or viewed together. We ranked the top 50 items with high similarity scores, excluding those already seen by the user, to tailor recommendations to individual preferences. This item-based collaborative recommender system generated personalized recommendations by considering user preferences and behaviors in the dataset.

3.5 Hybrid Recommender System

The hybrid recommender system combines content-based and collaborative approaches to provide comprehensive news recommendations. It merges the two systems based on news article ID, increasing the weighting of items present in both systems. Recommendations are ranked by similarity values, and the top 15 articles are selected for the user. This hybrid approach aims to offer personalized and diverse recommendations by leveraging content characteristics and user behavior.

3.6 Experimental Setup

To implement this system, we utilized Google Collaboratory, an online framework that utilizes various run environments with Python. Due to the large dataset and the need for substantial computational power for preprocessing and training the word2vec model, we leveraged the resources of the Google Cloud Provider. By partitioning virtual machines, we obtained approximately 52 GB of RAM and 200 GB of SSD storage capacity. This infrastructure allowed us to efficiently handle the computational requirements of the system.

4 Results and Discussion

This section presents the results and findings of our proposed system. Evaluating the recommender system offline is challenging, especially for news recommendation without labeled data. Common evaluation metrics like Accuracy, Recall, and Precision work well with labeled datasets and classic machine learning techniques. However, our focus was not solely on accuracy. We aimed to ensure diversity, coverage, serendipity, and novelty in the recommended news.

To measure diversity, we calculated the ratio of different categories recommended compared to past recommendations. Our system maintained at least a 40% diversity ratio, recommending news articles from at least six different categories out of 15 recommendations. Coverage was ensured by recommending every news item to at least one user. To evaluate serendipity and novelty, which are subjective to each user, our hybrid approach incorporating CF facilitated these concepts. The CF recommends news based on what other users find interesting, providing personalized yet novel and exceptional articles.

For evaluation, we used various metrics commonly used in implicit rating recommender systems. We employed the MIND Large Dataset and calculated the overall accuracy using a binary classification of user behaviors. To create the evaluation dataset we mapped the data to 1 or 0 based on being clicked or not, using the user behavior as in Table 5, to create a ranked dataset with clicked news articles at the top. Since our system predicts news based on user history rather than their response to specific articles, we used rank-aware evaluation metrics instead of the mean squared error (MSE). To be specific we used Mean Average Precision (MAP) as in Eq. (3) [25] and Normalized Discounted Cumulative Gain (nDCG) [26] to measure the relevance of the top N recommended news. These evaluation metrics are reliant on the rank values of the news articles in the recommendation list, rather than being based on predicted relevance scores. The dataset used to evaluate the models was based on mapping Our content-based and collaborative-based approaches achieved MAP values of 0.55 and 0.45, respectively, while the hybrid CF-CBF approach achieved around 0.64 for all users in the selection evaluated at k = 10.

$$\text{MAP} = \frac{1}{m} \sum\nolimits_{k=1}^{n} p(k) \times rel(k) \quad (3)$$

where $p(k)$ is the value of precision at position k, $rel(k)$ indicates hit or miss $\{0,1\}$, while the MAP denotes the average precision formula.

To further evaluate and compare with the benchmark set of the dataset, we used the Normalized Discounted Cumulative Gain (nDCG) metric. Similar to MAP, nDCG prioritizes the recommendation of relevant articles at the top. It goes beyond a simple hit-or-miss evaluation and aims to identify the relevance of articles in the corpus. Equation (4) represents the formula used for calculating nDCG.

$$\text{nDCG}_\text{k} = \frac{DCG_k}{IDCG_k} \quad (4)$$

where the discounted cumulative gain is denoted as $DCG_k = \sum_{k=1}^{n} \frac{1}{log_2(i+1)} \times rel(k)$, which represent the graded results in position k, the idealized discounted cumulative gain denoted as I $DCG_k = \sum_{k=1}^{n} \frac{G_i}{log_2(i+1)} \times rel(k)$ to consider the ordering of items based on decreasing relevance [26]. The nDCG scores for CBF, CF, and HF were 0.34, 0.58, and 0.70, respectively at k = 10. These results demonstrate the performance of the recommendation system. In comparison, the MIND dataset has nDCG values of less than 0.5 for both k = 5 and k = 10 on the scoreboard. Table 6 provides the top three benchmark scores from different creators. The scores range from 0 to 1, where higher scores indicate a better and more relevant system for facilitating personalized recommendations for users.

Table 6. Summary of the system results

Approach	MAP@K = 5	MAP@K = 10	nDCG@5	nDCG@10
CBF	0.4000	0.4500	0.3900	0.3400
CF	0.4500	0.5500	0.5000	0.5800
HF	0.6900	0.6700	0.6500	0.7000
UniUM-Fastformer-Pretrain [27, 24]	–	–	0.4180	0.4718
MINER [27, 24]	–	–	0.4102	0.4661
UniUM-Fastformer [27, 24]	–	–	0.4151	0.4684

5 Conclusion and Future Works

In this paper, we propose an HF personalized news recommendation system that combines word2vec, TF-IDF, and social media comments to improve the standard of recommendation systems. We compared the CF and CBF approaches and found that HF outperformed them. This aligns with our expectations, as the hybrid method utilizes the strengths of both approaches to create a robust and accurate personalized recommendation system. Integrating the word2vec model enhanced the accuracy of CBF by capturing syntactic and semantic word relationships in news articles. Social media comments played a crucial role in mitigating cold start issues and data sparsity.

During system development, we identified potential avenues for future research. Integrating social interactions and news interactions from users within a live social network would improve the system's ability to recommend news based on real-time activities and breaking news. Establishing a live test environment would eliminate constraints and biases resulting from limited dataset testing. Additionally, adopting deep learning techniques to extract features from news and user behaviors could further enhance system relevance and personalization.

References

1. Boczkowsk, P.J.: Digitizing the News: Innovation in Online Newspapers (Inside Technology). The MIT Press, New Ed edition (2005)

2. Maier, S.R., Tucker, S.: Online news readers get different news mix than print. Newsp. Res. J. **33**(4), 48–62 (2012). https://doi.org/10.1177/073953291203300405
3. Su, X., Khoshgoftaar, T.M.: A survey of collaborative filtering techniques, vol. 2009, Hindawi (2009). https://doi.org/10.1155/2009/421425
4. Li, M., Wang, L.: A survey on personalized news recommendation technology. IEEE Access **7**, 145861–145879 (2019)
5. Wu, C., Wu, F., Huang, Y., Xie, X.: Personalized news recommendation: methods and challenges. ACM Trans. Inf. Syst. **41**(1), 1–50 (2023)
6. Bellogín, A., Cantador, I., Castells, P., Ortigosa, Á.: Discovering relevant preferences in a personalised recommender system using machine learning techniques. In: Proceedings of the ECML-PKDD 2008 Workshop on Preference Learning (2008)
7. Feng, C., Khan, M., Rahman, A.U., Ahmad, A.: News recommendation systems-accomplishments, challenges & future directions. IEEE Access. **20**(8), 16702–16725 (2020). https://doi.org/10.1109/ACCESS.2020.2967792
8. Lu, Z., Dou, Z., Lian, J., Xie, X., Yang, Q.: Content-based collaborative filtering for news topic recommendation. In: Proceedings of the AAAI Conference on Artificial Intelligence, vol. 29, no. 1, 9 Feb 2015. https://doi.org/10.1609/aaai.v29i1.9183
9. Kompan, M., Bieliková, M.: Content-based news recommendation. In: E-Commerce and Web Technologies: 11th International Conference, EC-Web 2010, Bilbao, Spain, Proceedings 11 2010, pp. 61–72. Springer, Berlin, Heidelberg, 1–3 Sep 2010
10. Yunanda, G., Nurjanah, D., Meliana, S.: Recommendation system from microsoft news data using TF-IDF and cosine similarity methods. Build. Inf. Technol. Sci. (BITS). **4**(1), 277–84 (2022). https://doi.org/10.47065/bits.v4i1.1670
11. Su, X., Khoshgoftaar, T.M.: A survey of collaborative filtering techniques. In: Advances in Artificial Intelligence, vol. 2009 (2009). https://doi.org/10.1155/2009/421425
12. Chen, R., Hua, Q., Chang, Y.S., Wang, B., Zhang, L., Kong, X.: A survey of collaborative filtering-based recommender systems: from traditional methods to hybrid methods based on social networks. IEEE Access. **24**(6), 64301–64320 (2018)
13. Aditya, P.H., Budi, I., Munajat, Q.: A comparative analysis of memory-based and model-based collaborative filtering on the implementation of recommender system for e-commerce in indonesia: a case study PT X. In: 2016 International Conference on Advanced Computer Science and Information Systems (ICACSIS), pp. 303–308. IEEE, 15 Oct 2016. https://doi.org/10.1109/ICACSIS.2016.7872755
14. Shen, L., Huang, S., Mao, X.: A reputation-enhanced model for trust-based collaborative filtering recommender system. In: 2020 International Joint Conference on Neural Networks (IJCNN), pp. 1–7. IEEE (2020). https://doi.org/10.1109/IJCNN48605.2020.9206618
15. Das, A.S., Datar, M., Garg, A., Rajaram, S.: Google news personalization: scalable online collaborative filtering. In: Proceedings of the 16th International Conference on World Wide Web, pp. 271–280, 8 May 2007. https://doi.org/10.1145/1242572.1242610
16. Konstan, J.A., Miller, B.N., Maltz, D., Herlocker, J.L., Gordon, L.R., Riedl, J.: Grouplens: applying collaborative filtering to usenet news. Commun. ACM. **40**(3), 77–87 (1997). https://doi.org/10.1145/245108.245126
17. Li, M., Wang, L.: A survey on personalized news recommendation technology. IEEE Access **7**, 145861–145879 (2019). https://doi.org/10.1109/ACCESS.2019.2944927
18. Liu, J., Dolan, P., Pedersen, E.R.: Personalized news recommendation based on click behavior. In: Proceedings of the 15th International Conference on Intelligent user Interfaces, pp. 31–40, 7 Feb 2010. https://doi.org/10.1145/1719970.1719976
19. Lombo, X., Oyelade, O.N., Ezugwu, A.E.: Crime detection and analysis from social media messages using machine learning and natural language processing technique. In: International Conference on Computational Science and Its Applications, pp. 502–517. Springer, Cham, 4 Jul 2022. https://doi.org/10.1007/978-3-031-10548-7_37

20. Capelle, M., Frasincar, F., Moerland, M., Hogenboom, F.: Semantics-based news recommendation. In: Proceedings of the 2nd International Conference on Web Intelligence, Mining and Semantics, pp. 1–9, 13 Jun 2012. https://doi.org/10.1145/2254129.2254163
21. Okura, S., Tagami, Y., Ono, S., Tajima, A.: Embedding-based news recommendation for millions of users. In: Proceedings of the 23rd ACM SIGKDD International Conference on Knowledge Discovery and Data Mining, pp. 1933–1942, 13 Aug 2017. https://doi.org/10.1145/3097983.3098108
22. Hermida, A., Fletcher, F., Korell, D., Logan, D.: Share, like, recommend: decoding the social media news consumer. J. Stud. **13**(5–6), 815–824 (2012). https://doi.org/10.1080/1461670X.2012.664430
23. Manoharan, S., Senthilkumar, R.: An intelligent fuzzy rule-based personalized news recommendation using social media mining. Comput. Intell. Neurosci. **31**, 2020 (2020). https://doi.org/10.1155/2020/3791541
24. O'Banion, S., Birnbaum, L., Hammond, K.: Social media-driven news personalization. In: Proceedings of the 4th ACM RecSys Workshop on Recommender Systems and the Social Web, pp. 45–52, 9 Sep 2012. https://doi.org/10.1145/2365934.2365943
25. Mikolov, T., Chen, K., Corrado, G., Dean, J.: Efficient estimation of word representations in vector space. arXiv preprint arXiv:1301.3781. (2013)
26. Wu, F., et al.: Mind: a large-scale dataset for news recommendation. In: Proceedings of the 58th Annual Meeting of the Association for Computational Linguistics, pp. 3597–3606, Jul 2020. https://doi.org/10.18653/v1/2020.acl-main.331
27. Microsoft, "MIND: MIcrosoft News Dataset," MSnews (2020). https://msnews.github.io/. Accessed 11 Nov 2022

ACO Bio-Inspired Artificial Intelligence in Automotive Technology: Bridging Natural Systems and Machine Learning for Sustainable Mobility

Bharath Kalyan Ganeshna, G. Maheshwari, Swapnil Nandanwar, and Sumitra Biswal(✉)

Bosch Global Software Technologies (BGSW), Bosch, India
ganeshna.bharathkalyan@bosch.com,
{g.maheshwari,fixed-term.swapnil.vitthalraonandanwar,
Sumitra.Biswal}@in.bosch.com

Abstract. As the automotive industry hurtles toward an era of interconnected, smart vehicles, the need for robust cybersecurity measures becomes increasingly imperative. The advent of modern automobiles equipped with sophisticated digital systems has introduced novel security challenges, particularly in the realm of intrusion detection within vehicular networks. This paper explores the application of Ant Colony Optimization (ACO), a bio-inspired algorithm, as a powerful tool for enhancing automotive network security. By modeling the problem as a path-finding endeavor, ACO excels in detecting sequences of events that could lead to unauthorized actions. This research outlines a comprehensive methodology, including problem modeling, pheromone-based learning, and result visualization, demonstrating the efficacy of ACO in fortifying automotive cybersecurity. However, it is crucial to underscore the necessity of complementing ACO findings with domain knowledge and further analysis.

Keywords: Ant Colony Optimization ACO · Controller Area Network (CAN) · Long Short Term Memory (LSTM) · Intrustion detection system (IDS)

1 Introduction

The automotive sector is in the midst of a dynamic metamorphosis, distinguished by the merging of biological, physical, and digital dimensions. This blurring of boundaries has ignited numerous advancements across various sectors. Artificial Intelligence (AI) plays a pivotal role in propelling innovation and catalyzing change, serving as a driving force in these transformations. Moreover, the emergence of bio-inspired AI drawing inspiration from nature's principles has been pivotal. It offers solutions that are not only efficient but also sustainable, and its influence is felt across a multitude of fields. As the automobile industry progresses

V. Sakalauskas et al. (Eds.): IBICA 2023, LNNS 1231, pp. 88–97, 2025.
https://doi.org/10.1007/978-3-031-78946-5_8

towards autonomous and connected vehicles, the incorporation of advanced digital technologies becomes essential. The realm of automotive cybersecurity has been notably shaped by several bio-inspired algorithms. Among these, the Ant Colony Optimization (ACO) stands out as a foundational algorithm for bio-inspired AI, drawing inspiration from the foraging behavior of ants [1]. These nature-mimicking algorithms, including ACO, have left a marked impact on the cybersecurity framework of modern vehicles. In the realm of automotive cybersecurity, the Vehicle-to-Everything (V2X) communications utilize the ACO, a probabilistic method adept at tackling computational challenges related to optimal pathfinding [2]. V2X is a cornerstone of innovative automotive technology, aiming for a seamless information exchange network [10]. This study investigates ACO in automotive security, focusing on anomaly detection and threat intelligence in Controller Area Network (CAN) protocols. The paper aims to utilize ACO for rapid identification and prediction of security threats in vehicle networks, a novel application in this specific domain. This research is motivated by the need to improve detection and prediction capabilities in automotive cybersecurity. The remainder of this paper is organized as follows: Sect. 2 presents the related work, providing context and background for our study. Section 3, explores the application of ACO and LSTM models in the context of automotive security. Section 4, details the methodology and experimental design of the research. Section 5, presents and analyzes the findings of the experiments. The paper concludes with Sect. 6, where the implications of the research are discussed and directions for future exploration are proposed in this domain.

2 Related Work

The integration of Bio-inspired AI techniques has been instrumental in addressing cybersecurity challenges within the automotive domain. Recent research has explored their application in enhancing security within automotive networks, focusing on aspects like vehicle communication, routing, and autonomous driving systems [5,8]. This section examines the use of bio-inspired AI in automobile security stressing the advantages of ACO.

Due to its capacity to accurately imitate natural occurrences, bio-inspired AI approaches have found wide use in vehicle security. For instance, a joint cluster-based RRM and low-latency framework for NR-V2X networks has been explored [14]. Additionally, the integration of ACO with Random Forest algorithms for feature selection in Intrusion Detection Systems (IDS) has been investigated [15]. The use of evolutionary algorithms in Vehicular Ad-Hoc Networks (VANETs) has also been noted, particularly for optimizing cluster formation, thereby enhancing network performance by drawing inspiration from natural evolutionary processes [4,9]. ACO, in particular, has been utilized for its collective intelligence approach, to improve routing in VANETs and the Internet of Vehicles (IoV) [7,12].

Other ACO-related methods have also been applied to strengthen automobile security. Researchers have proposed an improved multi-objective particle swarm optimization (IMOPSO) method for location privacy protection in road networks

[11], leveraging the social behavior patterns of fish and birds. Additionally, the literature includes studies on ACO-based routing protocols in Wireless Sensor Networks (WSNs), which are crucial in automotive applications for transmitting sensory data to communication and control systems [3,13].

The unique capabilities of ACO, particularly in identifying optimal paths and event sequences, make it highly suitable for detecting unauthorized activities in vehicular networks [7]. It benefits from the ability to represent issues as path finding tasks, that are led by pheromone-based learning in situations where it is necessary to identify sequences of occurrences, such as in the case of unauthorized door opening [12]. Furthermore, the dynamic nature of vehicular networks and the flexibility of ACO to dynamic situations are complementary.

This section provides a comprehensive overview of current research in ACO, bio-inspired AI, automotive technology, and sustainable mobility. It highlights the growing influence of bio-inspired AI methods in vehicle security, detailing their various applications and mechanisms, with a focus on ACO. Given its unique problem-solving abilities, especially in identifying event sequences leading to unauthorized actions, ACO emerges as a promising tool for enhancing vehicle security. Continued research in this area is anticipated to boost the security of connected and autonomous vehicles, contributing to safer and more resilient transportation networks amidst the automotive industry's ongoing digital transformation.

3 Interpreting Different Models from an Automotive Security Use Case Perspective

Bio-inspired AI seeks to emulate the natural world's adaptability and efficiency. This research work explores its application in addressing some critical automotive cybersecurity challenges. The subsequent sub-sections present a detailed automotive cybersecurity use case, identifying the critical issues in this area. The discussion then shifts to the introduction of the ACO and LSTM algorithms, briefly outlining their capabilities to address the identified problem. The principles of both algorithms in the context of the problem statement are elucidated, accompanied by experimentation mechanism diagrams and pseudocodes for both ACO and LSTM. However, relevant equations are provided exclusively for ACO, underscoring the paper's focus on highlighting the potential of ACO in automotive cybersecurity. This detailed exploration lays the groundwork for in-depth experimentation towards resolving the problem statement in the subsequent sections of this research work.

3.1 Unauthorized Vehicle Door Unlocking Use Case

Considering recent automotive cybersecurity breaches, such as the Cherokee Jeep hack where cyber attackers remotely took over essential vehicle controls [6], there's a pressing need to bolster our defenses in the realm of automotive

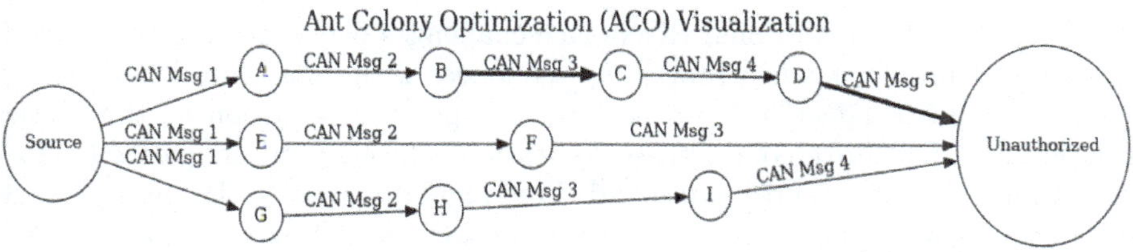

Fig. 1. ACO Visualization

cybersecurity. The act of unauthorized door unlocking is particularly concerning. Gaining illegitimate entry into a vehicle can set the stage for a range of malevolent acts, from theft to vandalism. This research focuses on detecting such unauthorized activities by analyzing the sequences of V2X CAN messages, pivotal for in-vehicle communication. Investigating CAN bus log files allows us to simulate real-life hacking attempts, thereby enhancing our detection methods with the help of ACO (see Fig. 1).

3.2 Introduction to ACO and LSTM

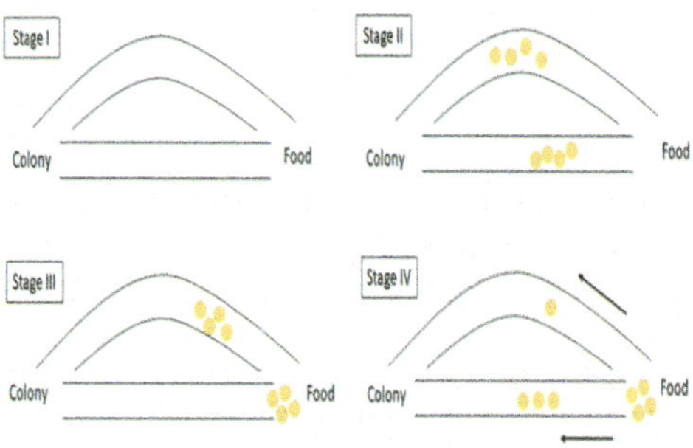

(a) ACO

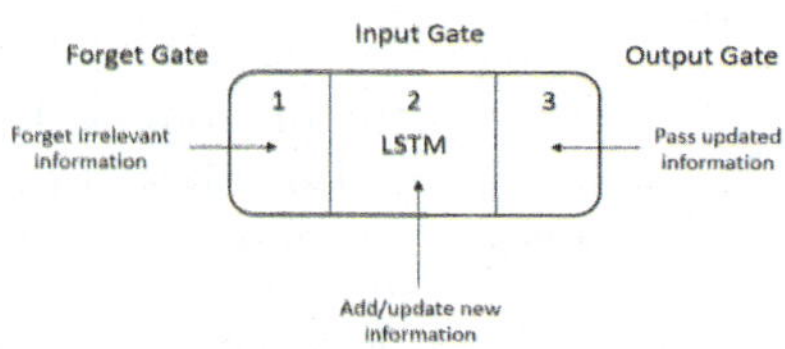

(b) LSTM

Fig. 2. Overview of ACO and LSTM Algorithms.

ACO, drawing inspiration from the natural world, mimics the intelligence of ants in their quest for optimal food paths using pheromones as guidance. A typical ACO mechanism is illustrated (see Fig. 2). In this illustration, Stage I shows paths devoid of pheromones between the food source and ant colony. In Stage II, ants split evenly between two paths, with half on the longer route and half on the shorter. By Stage III, the shorter route accumulates more pheromones due to its earlier completion, attracting more ants. By Stage IV, the increased pheromone concentration on the shorter path, coupled with the faster evaporation on the less-traveled longer path, results in a clear preference for the shorter route among the colony. LSTM is a specialized form of Recurrent Neural Network (RNN) architecture, specifically designed to handle long-term dependencies in sequential data. Traditional RNNs face challenges like vanishing and exploding gradient problems when processing sequences with long-term dependencies, making them less effective for tasks such as speech recognition or time series forecasting. LSTM overcomes these limitations through its unique cell structure, which consists of a cell state and three regulatory gates: the input, forget, and output gates. These gates collectively decide the information to be stored, discarded, or outputted, enabling LSTM to preserve important information over long sequences and making it a popular choice for tasks involving sequential data.

3.3 Why ACO and LSTM?

ACO and LSTM are chosen for this research due to their unique capabilities in handling sequential data and finding optimal paths in a given search space. ACO's nature-inspired approach and LSTM's memory cells make them suitable for detecting unauthorized activities in the complex network of automotive communication systems.

Given a set of states (combinations of `CAN_ID` and `Payload`), the transition probability from state i to state j is given by:

$$P(i,\ j) = \frac{\tau_{ij}^{\alpha}}{\sum_{k \in N_i} \tau_{ij}^{\alpha}} \tag{1}$$

where:

- τ_{ij} represents the pheromone level on the path from state i to state j.
- N_i is the set of all possible states reachable from state i.
- α is a parameter emphasizing the importance of the pheromone.

After each ant constructs a path, the pheromones on the path are updated to reflect the quality of the path. If the path leads to an unauthorized event, the pheromone levels are intensified:

$$\tau_{ij} = \tau_{ij} + \Delta\tau \tag{2}$$

where $\Delta\tau$ is the intensification value.

To simulate the natural evaporation of pheromones over time, an evaporation mechanism is applied:

$$\tau_{ij} = (1 - \rho) \times \tau_{ij} \tag{3}$$

where ρ is the evaporation rate.

3.4 Experimentation Mechanism and Algorithmic Overview

This sub-section comprises a block diagram (see Fig. 3) detailing our experimental setup and the pseudocodes of the employed algorithms, ACO and LSTM (see Algorithm 1 and Algorithm 2). The diagram illustrates the extraction of a .log file using the ICSim tool, which is curated and preprocessed to serve as a dataset for training and testing. This comprehensive approach ensures a robust analysis of the results from both algorithms, offering clear insights into the workflow and the strategic solutions adopted to address automotive cybersecurity challenges. Detailed steps and analyses are elaborated in the experimentation section.

Algorithm 1: ACO Algorithm for Unauthorized Door Unlocking Detection

---

Input: train_data, test_data, alpha, beta, evaporation_rate, intensification, num_ants, num_iterations, max_path_length
Output: Top paths leading to unauthorized events based on pheromone levels
BEGIN
1: Load train_data and test_data;
2: Convert labels to binary (0 for authorized, 1 for unauthorized);
3: Convert `CAN_ID` and `Payload` to numerical format;
4: Initialize pheromones for each state transition to 1;
5: **while** *iteration from 1 to num_iterations* **do**
    while *ant from 1 to num_ants* **do**
        current_state ← random state from states;
        path ← [current_state];
        while *i from 1 to max_path_length* **do**
            probabilities ← pheromones[current_state]$^{\alpha}$;
            Normalize probabilities;
            chosen_index ← choose index based on probabilities;
            next_state ← states[chosen_index];
            Add next_state to path;
            if *next_state leads to unauthorized event* **then**
                break;
            current_state ← next_state;
        while *each transition in path* **do**
            Update pheromones for the transition by adding intensification;
    Multiply all pheromones by (1 − evaporation_rate);
6: Initialize test_paths as empty list;
7: **while** *each transition in pheromones* **do**
    if *pheromone level > intensification* **then**
        Add transition to test_paths;
8: Sort test_paths based on pheromone levels in descending order;
Visualization;
9: Visualize the top paths in test_paths;
END

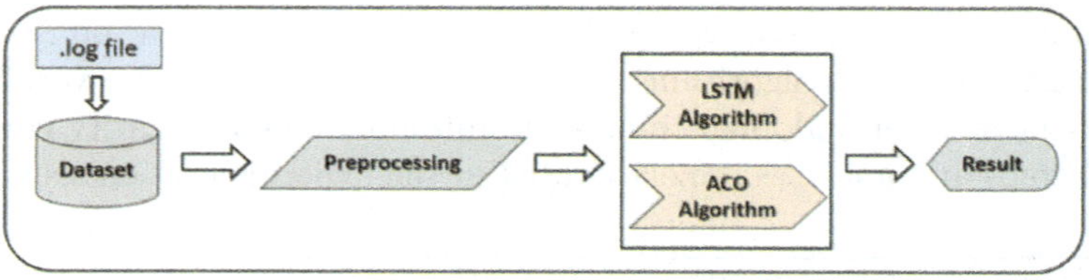

Fig. 3. Block diagram of experiment

Algorithm 2: LSTM for Unauthorized Door Unlocking Detection

Input: Training_data.csv, Testing_data.csv
Output: Visualization of High Risk Paths Identified by LSTM
BEGIN
1: Load training and testing data from CSV files;
2: Pre-process the data by converting `"Status"` to binary values (0 for authorized and 1 for unauthorized);
3: Tokenize and pad the sequences of `CAN_ID` and `Payload` in both training and testing data;
4: Define an LSTM model with appropriate layers;
5: Compile and train the model on the training data;
6: Predict on the testing data and set a threshold for high-risk paths;
7: Identify high-risk paths based on predictions;
8: Initialize a directed graph G;
9: **while** *each path in high-risk_paths* **do**
  while *i from 0 to length of path - 2* **do**
    if *the current element and the next element are not padding* **then**
      Add a directed edge from the current element to the next element in G;

10: Visualize the identified high-risk paths using the graph G;
END

4 Experimental Setup Overview

Our study aims to identify unauthorized door unlocking through CAN bus log analysis, sourced from the ICSim tool and processed on a virtual CAN network. We curate a dataset of about 90,000 log entries, combining both authorized and unauthorized door unlocking logs. After Pre-processing the logs which includes `Timestamp, CAN_ID, Payload,` and a binary `"Status"` (0 for authorized, 1 for unauthorized), we split it into 60% training and 40% testing. We adopt ACO and LSTM for our experimentation.

ACO. CAN messages are mapped to discrete states based on `CAN_ID` and `Payload`. A uniformly initialized pheromone matrix guides ants across state transitions. ACO parameters included $\alpha = 1$ (pheromone influence), $\beta = 1$ (heuristic influence, unused here), an evaporation rate of 0.1, and an intensification factor of 2. Over 50 iterations, 10 ants per iteration explored the state space, with each ant limited to 10 states. Transition probabilities are driven by pheromone levels. Encountering an unauthorized event halts an ant's exploration. Paths are intensified after exploration, and pheromones undergo decay.

LSTM. We employ an LSTM neural network for detecting unauthorized CAN events. Data are tokenized into sequences, mapping unique `CAN_ID` and `Payload` combinations to integers. This is structured with an initial embedding layer to convert these integer tokens into dense 128-dimensional vectors. This is followed by an LSTM layer with 128 units to capture the temporal dependencies in the sequence data. The final dense layer with two output units utilize a softmax activation function to classify each sequence as either authorized or unauthorized. The model is compiled using the Adam optimizer and categorical cross-entropy as the loss function. It is trained over 10 epochs with a batch size of 32. Post-training, predictions on the test data are made, with sequences having a probability greater than 0.8 for the unauthorized class being flagged as high-risk paths.

5 Observations and Result Interpretation

The experimental results (see Fig. 4 and Table 1) highlight the varied performances of ACO and LSTM models. Both models demonstrate high precision, but their differing recall rates, especially the lower rate of LSTM, signal undetected unauthorized events, suggesting a need for further tuning. ACO's results, marked by pheromone levels, offer a clear and actionable insight for cybersecurity experts. It efficiently underscores significant paths, aiding in anomaly detection and proactive defense, with enhanced scalability and focus on sequence transitions. Conversely, LSTM's lower accuracy points to a need for enhancement. Finally, ACO stands out for its clear visualization and results, proving crucial for emphasizing critical paths and potential vulnerabilities in automotive cybersecurity. However, it must be noted that in Fig. 4, due to the intricate nature of the network graph, individual nodes are not distinctly discernible. This is a recognized limitation in the graphical representation of complex data structures. We plan to address this in future advanced experimentation's, ensuring clearer visualizations that further enhance the analytical values of the models.

Table 1. Comparison between ACO and LSTM models

Metrics	ACO	LSTM
Accuracy	0.500	0.333
Precision	1.000	1.000
Recall	0.500	0.333
F1 Score	0.666	0.500

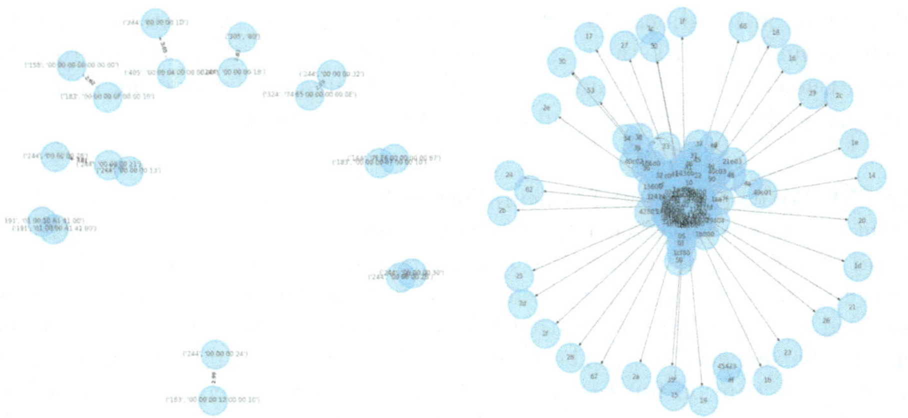

(a) Top Paths Leading to Unauthorized Events on Test Data (b) High Risk Paths Identified by LSTM

Fig. 4. Top paths leading to unauthorized events in Test data. (a) ACO, (b) LSTM

6 Conclusion and Future Work

In the evolving landscape of Advanced Driver-Assistance Systems (ADAS) and Software-Defined Vehicles (SDVs), ensuring robust cybersecurity is paramount. This research introduces a novel approach by applying ACO, traditionally used in different contexts, to the specific challenge of automotive cybersecurity. The innovative use of ACO, in conjunction with LSTM for benchmarking, demonstrates its potential in effectively identifying unauthorized vehicular access, marking a significant stride towards enhancing automotive network security and highlighting the novelty of our approach. The motivation behind this research is to explore and adapt such existing algorithms in new, uncharted domains of automotive cybersecurity. Moving forward, addressing the identified limitations of ACO will be paramount. Automated parameter tuning will be employed to optimize ACO's performance, and strategies like employing parallel algorithms or hierarchical ACO will be explored to enhance scalability. Adaptability in dynamic environments will be investigated by integrating periodic pheromone resets or sliding window mechanisms. The exploration of a hybrid model, combining ACO's heuristic strengths with deep learning capabilities and other potential algorithms, is envisioned to enrich our experimental evaluation in future. These concerted efforts will contribute substantially to fortifying the defenses against unauthorized access, ensuring the security and integrity of sophisticated automotive networks in ADAS and SDVs.

References

1. Correia, S.L.O., Celestino, J., Cherkaoui, O.: Mobility-aware ant colony optimization routing for vehicular ad hoc networks. In: IEEE Wireless Communications and Networking Conference, pp. 1125–1130. IEEE (2011)
2. Liu, B., El Kamel, A.: V2X-based decentralized cooperative adaptive cruise control in the vicinity of intersections. IEEE Trans. Intell. Transp. Syst. **17**(3), 644–658 (2015)
3. Liu, X.: Routing protocols based on ant colony optimization in wireless sensor networks: a survey. IEEE Access **5**, 26303–26317 (2017)
4. Fahad, M., Aadil, F., Ejaz, S., Ali, A.: Implementation of evolutionary algorithms in vehicular ad-hoc network for cluster optimization. In: Intelligent Systems Conference (IntelliSys), pp. 137–141. IEEE (2017)
5. Li, M., Wu, X., He, X., Yu, G., Wang, Y.: An eco-driving system for electric vehicles with signal control under V2X environment. Transp. Res. Part C Emerg. Technol. **93**, 335–350 (2018)
6. Miller, C.: Lessons learned from hacking a car. IEEE Des. Test **36**(6), 7–9 (2019)
7. Nguyen, T. H., Jung, J.J.: ACO-based approach on dynamic MSMD routing in IOV environment. In: 16th International Conference on Intelligent Environments (IE), pp. 68–73. IEEE (2020)
8. Srivastava, A., Prakash, A., Tripathi, R.: An adaptive intersection selection mechanism using ant colony optimization for efficient data dissemination in urban VANET. Peer-to-Peer Network. Appl. **13**(5), 1375–1393 (2020). https://doi.org/10.1007/s12083-020-00892-8
9. Ma, H., Li, S., Zhang, E., Lv, Z., Hu, J., Wei, X.: Cooperative autonomous driving oriented MEC-aided 5G–V2X: prototype system design, field tests and AI-based optimization tools. IEEE Access **8**, 54288–54302 (2020)
10. Sun, Y., Wang, P., Liu, F.: Design and implementation of cooperative adaptive cruise control based on V2X. In: IEEE MTT-S International Wireless Symposium (IWS), pp. 1–3. IEEE (2020)
11. Tian, C., Xu, H., Lu, T., Jiang, R., Kuang, Y.: Semantic and trade-off aware location privacy protection in road networks via improved multi-objective particle swarm optimization. IEEE Access **9**, 54264–54275 (2021)
12. Ghawy, M.Z., Amran, G.A., AlSalman, H., Ghaleb, E., Khan, J., Al-Bakhrani, A.A., et al.: An Effective wireless sensor network routing protocol based on particle swarm optimization algorithm. Wirel. Commun. Mob. Comp. (2022)
13. Upeksha, R.G.C., Pemarathne, W.P.J.: Ant Colony optimization algorithms for routing in wireless sensor networks: a review. In: Recent Advances in Electrical and Electronic Engineering and Computer Science: Selected articles from ICCEE, Malaysia, pp. 47–57 (2022)
14. Waqas, S.M., Tang, Y., Yu, L., Abbas, F.: A joint cluster-based RRM and Low-latency framework using the full-duplex mechanism for NR-V2X networks. Comput. Commun. **209**, 513–525 (2023)
15. Lifandali, O., Abghour, N., Chiba, Z.: Feature selection using a combination of ant colony optimization and random forest algorithms applied to isolation forest based intrusion detection system. Procedia Comput. Sci. **220**, 796–805 (2023)

Improving Evaluation Measures Using Ensemble Technique in Diabetes Dataset

Rabindra Bista[1](✉), Binay Sharma[1], Santosh Khanal[1], Sanjog Sigdel[1], and Joao C. Ferreira[2,3,4]

[1] Department of Computer Science and Engineering, Kathmandu University, Dhulikhel, Kavre 45200, Nepal
{rbista,skhanal}@ku.edu.np

[2] Inov Inesc Inovação—Instituto de Novas Tecnologias, 1000-029 Lisbon, Portugal

[3] Instituto Universitário de Lisboa (ISCTE-IUL), ISTAR, 1649-026 Lisbon, Portugal

[4] Faculty of Logistics, Molde University College, NO-6410 Molde, Norway

Abstract. Early Prediction done on the right class for a certain disease in the medical field is very critical and the effects of misclassification of the disease could be very risky which may lead to the mistreatment of the patient. The important classification performance measurements in medical fields are error rate, recall, specificity and accuracy. This research aims to focus on these four different measurements and improvement on the misjudgments while classifying a person to a disease that will provide him/her from getting the correct treatment. Thus, the accuracy in classifying such medical data should be at the highest with lowest error. Many attempts are made in order to identify this problem with the objective of high precision and better accuracy with different machine learning algorithms. Where these machine learning help to identify the disease, prediction of the disease risk, decision making with treatment, disease measurement. This research aims to investigate the performance of the different machine learning algorithms. Diabetes dataset with a total 2000 rows and 8 columns have been analyzed. After all the framework and model being tested with ensemble classifiers and profound with the highest accuracy from proposed weighted ensemble method i.e. 97.5 with minimum error rate of 0.025. Which is greater than neural network and majority ensemble technique applied by previous researcher.

Keywords: Diabetes dataset · Ensemble Technique · Boosting · Stacking · Machine Learning · Supervised Learning Algorithm

1 Introduction

Diabetes is a major public health problem in developing countries, but today more children are being diagnosed with the disorder, probably due to the rise in childhood obesity [1]. There's no cure for type 2 diabetes, but maintaining weight, eating habits and proper exercise can help manage the disease [2]. If diet and exercise aren't enough to manage your blood sugar well, you may also need diabetes medications or insulin therapy [3]. According to the National Diabetes, 2015, around 9.4% of the population in the U.S. had diabetes, including 30.2 million adults aged 18 years and over [4].

V. Sakalauskas et al. (Eds.): IBICA 2023, LNNS 1231, pp. 98–108, 2025.
https://doi.org/10.1007/978-3-031-78946-5_9

Machine learning is the science of getting computers to act without being explicitly programmed. Machine learning is closely related to (and often overlaps with) computational statistics, which also focuses on prediction-making through the use of computers. It has strong ties to mathematical optimization, which delivers methods, theory and application domains to the field. Machine learning is sometimes conflated with data mining [5].

Diabetes prediction and diagnosis in the traditional method have some limitations in terms of the error of having a positive diagnosis for the related disease [6]. The main objective of this research is proposed a mechanism that works on diabetes dataset which minimizes the error of getting false treatment. The proposed approach in research consists of major steps called ensemble technique applied to the method after the feature selection approach is applied to get a much better result [7]. Initial steps applied to the diabetes dataset applied to different machine learning algorithms to predict the outcomes that are accurate and further selective features were used to get much more desired results with highest accuracy. In the modern times accuracy won't be the better option to select the appropriate model while trained and test the dataset and other evaluation metrics need to be calculated [8]. It is done after the ensemble method reduces false positive and false negative values in the confusion matrix. This results in better accuracy and better evaluation matrices like specificity, recall/sensitivity and error rate that helps to minimize the person getting the wrong treatment [9]. Proposed weighted ensemble methods in extreme gradient boosting and bagging with parameter tuning helps to get better accuracy with highest specificity, recall/sensitivity with lowest error rate than another ensemble method [10].

The main objective of this research is to survey the current trends and algorithms in the diabetes diagnosis system and propose a model that performs efficient and reliable prediction for disease identification.

2 Theoretical Background

2.1 Objective

The main objective of the experiment in this research is to compare different machine learning algorithms and optimize the performance matrices using ensemble techniques in the diabetes dataset. Another objective of this experiment is to understand how the models/algorithms are created and implemented in this scenario, understand about data pre-processing and how it is done, know about different evaluation matrices like accuracy, error rate, specificity and sensitivity and finally, understanding how the results can be reported in confusion matrix.

2.2 Motivation

Usually real-world data contains some kind of noise. So, preprocessing of data is done to reduce the noise. Each attribute of the data may have a very different range of values [11]. The main aim to preprocess is to bring the values of different attributes in the same scale and while in the preprocessing either median, mean. There are various ensembling used such as bagging, boosting, ada-boosting, and stacking [12]. Instead of majority voting, a weighted classifier is applied in ensemble technique to obtain better accuracy, recall/sensitivity. Specificity with less error rate and weighted ensemble number of boosted tree to fit called and estimator is applied with numeric value, maximum tree depth for based learner with numeric value is applied similarly, learning rate and sub sample ratio of the training instance value is applied with these weighted values output is determined to be better among the previous work done on ensemble technique [13]. The previous method uses a majority voting ensemble to obtain certain evaluation measures but this research focuses on using weighted ensembles which not only increase the performance of machine learning models but also increase the other evaluation measures having less error in the machine learning models.

2.3 Environment

All data pre-processing, data analysis and building different machine learning models were done using python. Some of the tools that has been used for the experiment of this research are Python programming language and the platform use was Anaconda along with the Spyder as scripting with the built-in libraries like SciPy, NumPy, Pandas, matplotlib, scikit-learn, keras, tensorflow.

2.4 Algorithms

Classification is a technique where we categorize data into a given number of classes. The main goal of a classification problem is to identify the category/class to which new data will fall under [15]. In classification technique the whole dataset is divided into a set of predefined classes. That's why classification is also called supervised learning. This technique divides the whole process into two phases. In the first phase a model is built with the help of training data and in the next phase this model is tested with test tuples and its accuracy is determined. Following are the different classification algorithms used in Machine Learning and algorithms like Logistic Regression, Decision Tree, KNN, Support Vector Machine [14], Adaboost and Extreme gradient boosting [15, 16].

3 Related Works

Ömer D and Kose U [17] proposed a diabetes prediction using RNN and Diabetes disease contributes to heart health as well as kidney failure, blindness, damage to nerves and blood vessels. For this reason, the diagnosis of diabetes is very important, and many studies have been done for the complete diagnosis of diabetes and the work being done is usually a classification exercise, and it is mostly a matter of increasing the performance of the classification. Author study a classification study for diabetes mellitus was conducted using the Pima Indian diabetes dataset. Author used a multilayer feed forward neural network structure trained by the Bayesian regularization algorithm and the mean square error function. The artificial neural network was retrained 10 times and this study was repeated 20 times. The obtained lowest all classification accuracy is 93.9% and the obtained highest all classification accuracy is 95.5% finally the results of the study were compared with the results of previous which focuses on the diagnosis of diabetes studies using the same UCI machine learning dataset and the obtained results higher than the previously mentioned classification methods and evaluations show that the proposed method is very efficient and increases the classification success.

Kema A, Sen B [18] proposed a study which successfully distinguishes diabetes or normal persons contains two major steps. In the very first step, the feature selection or weighting methods are analyzed to find the most effective attributes for this disease and in the further step, the performances of AdaBoost, Gradient Boosted Trees and Random Forest ensemble learning algorithms are evaluated. Then experimental results, the prediction accuracy of the combination of Stability Selection method and AdaBoost learning algorithm is a little better than other algorithms with the classification accuracy by 73.88% at the first phrase, the dataset is divided 3 times randomly, parts of 60%, %70 and 80% of dataset for learning step and parts of 40%, 30% and 20% of dataset which are remaining parts for the testing step. Then after, feature selection algorithms are employed on these datasets in order to identify important attributes.

Prema N, Varshita V, Yogeswar J [19] proposed various machine learning algorithms are applied to predict diabetes, based on specific attributes. The performances of the algorithms are compared in terms of accuracy, voting based ensemble techniques are applied for the normalized pima diabetes data for which a highest accuracy is achieved. Prediction of diabetes is done using ensemble voting classifiers for pima Indian diabetes dataset, in comparison with different classification algorithms, the highest accuracy of 80% and 81% is achieved for data set by using 10-fold cross validation and by spitting data into 30% testing and 70% training.

4 Methodology

A novel approach for improving the evaluation measures of the diabetes datasets are described in this section. Figure 1 illustrates the research framework built to perform the research process from dataset input, preprocessing till its evaluation using ensemble techniques.

Existing approaches for the machine learning models have the working mechanism based on the traditional approaches like data mining, hybrid approach. The deep learning neural network which does have certain disadvantages while training the machine learning models and also requires large amounts of datasets. This can be removed by having a smaller dataset that can also predict the diabetes disease to get early treatment.

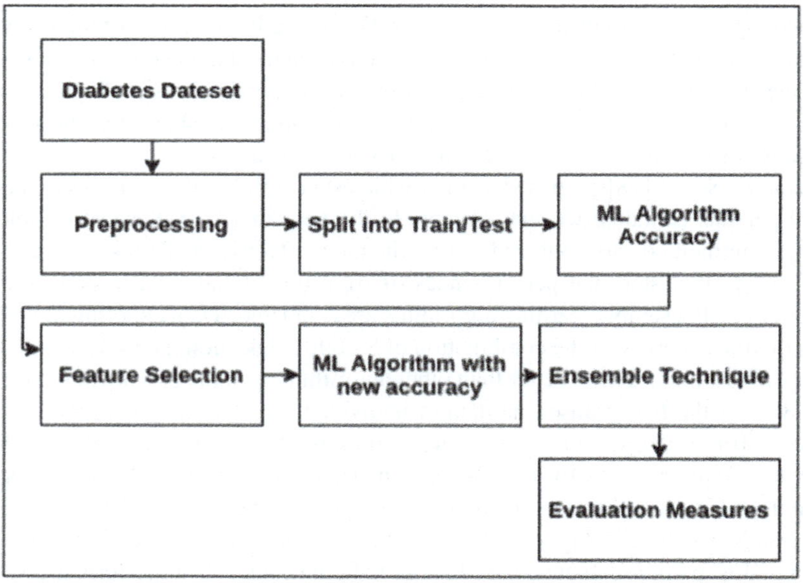

Fig. 1. Computational Research Framework

5 Results

In this section, results of the weighted ensemble techniques are explained in detail. The main objective of this ensemble experiment is to find out the better evaluation measures for the effective machine learning model selection and this section shows the visualization as well as the numeric values that determine that the proposed ensemble method outperforms the previous work done.

Most values in the dataset will be close to mid values also these attributes have zero values in the datasheet. The distribution is roughly symmetric so after applying the median method all the zero values will be deleted shown in Fig. 2.

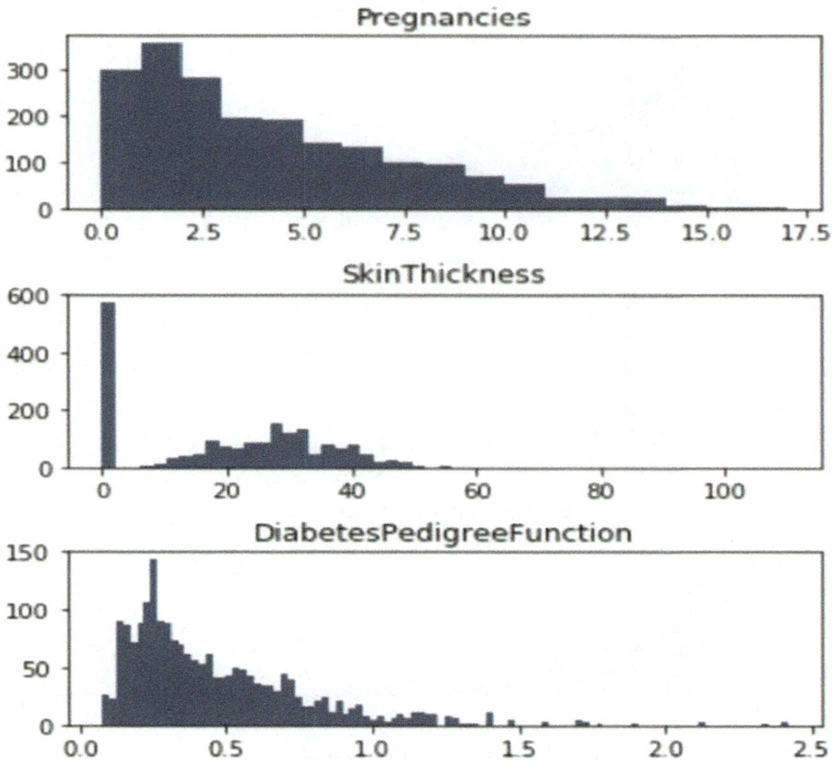

Fig. 2. Histogram of original datasets

Feature selection, the process of finding and selecting the most useful features in a dataset, is a crucial step of the machine learning pipeline. Here in Fig. 3 represents the important feature of the datasheet, having the highest value represent the highest priority and this datasheet represent the highest value of Glucose attribute followed by Body Mass Index, Pregnancies and Age factor (Fig. 4).

Accuracy measured where Table 1 represents the different classification algorithms with their accuracy. Where accuracy represents the accuracy of train and testing datasheet and new accuracy represents after applying feature selection also called priority by the highest feature.

Applying K-Fold cross validation, it splits the entire data randomly into k folds (value of k shouldn't be too small or too high, ideally, we choose 5 to 10 depending on the data size). The higher value of K leads to a less biased model (but large variance might lead to overfit), whereas the lower value of K is like the train-test split approach and here the dataset has been folded into 10 k-fold (Table 2).

After applying Ensemble technique both bagging and boosting methods were analyzed and these ensemble methods are Weighted Ensemble that include Number of estimators = 300, Maximum depth = 8, learning rate = 0.1 and Sub sample = 0.5. Above table shows the different weighted ensemble techniques used to calculate the

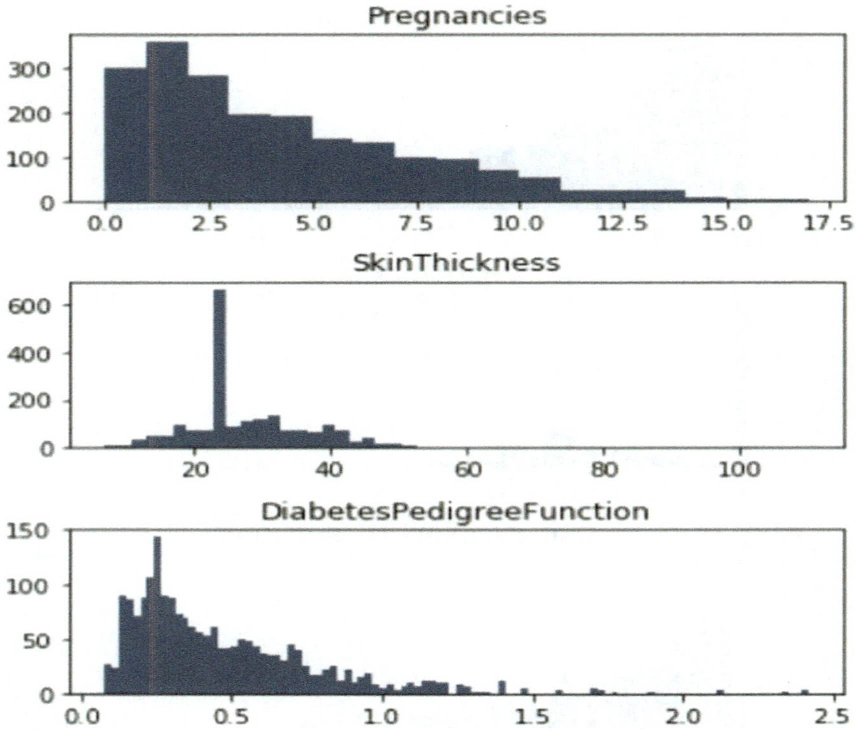

Fig. 3. Histogram after preprocessing

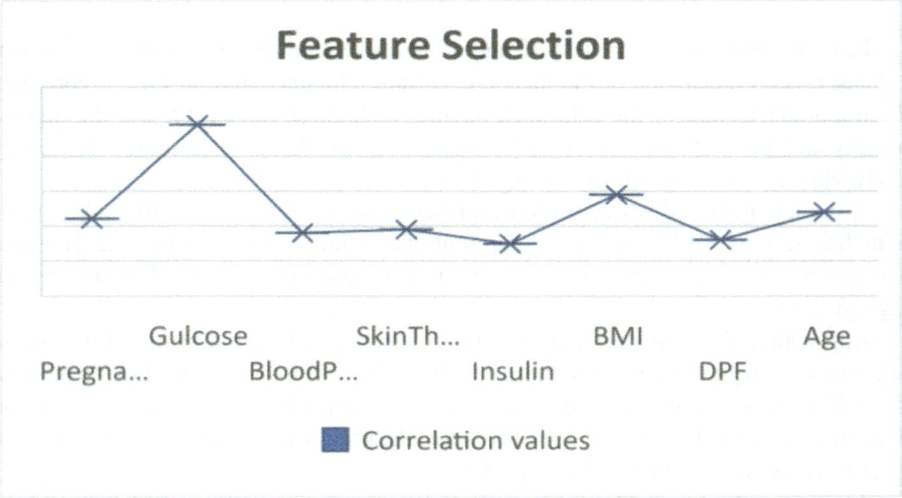

Fig. 4. Feature Selection of the datasets

Table 1. Accuracy of different classification algorithm

Algorithms	Accuracy	New Accuracy after Feature Selection
Linear SVM	77.0	78.0
Radial SVM	96.6	80.5
LR	78.0	77.1
KNN	86.1	85.5
Decision Tree	97.1	97.3
Random Forest	78.8	80.25
Neural Network	77.2	80.5
Gradient Boost	80.0	80.3
Ada Boost	86.0	86.3

Table 2. Cross validation of Calculated Accuracy

Algorithms	CV Mean (in %)
Linear Svm	78
Radial Svm	80.5
Logistic Regression	77.1
KNN	85.5
ANN	80.5
Decision Tree	97.3
XGB	80.3
AdaBoost	86.3

performance metrics like accuracy, recall/sensitivity, specificity, precision and error rate for the higher prediction rate.

Table 3. Comparison of Different Evaluation Measures

Algorithm	Accuracy	Sensitivity	Specificity	Precision	Error rate
Voting Classifier (DTC + KNN)	0.975	0.979	0.966	0.986	0.025
Ensemble XGB classifier (Base Learner	0.87	0.86	0.83	0.8	0.20
Ensemble XGB Classifier (Weighted)	0.975	0.979	0.966	0.986	0.025

The graphical representations of the Table 3 result are shown in the Fig. 5 and ROC for the highest accuracy achieved using ensemble methods were shown in the Fig. 6.

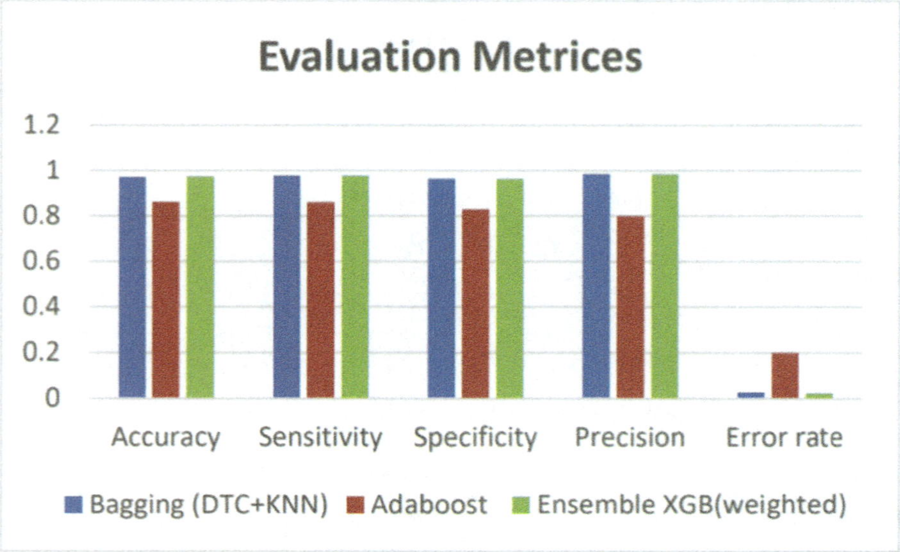

Fig. 5. Evaluation Measure Histogram

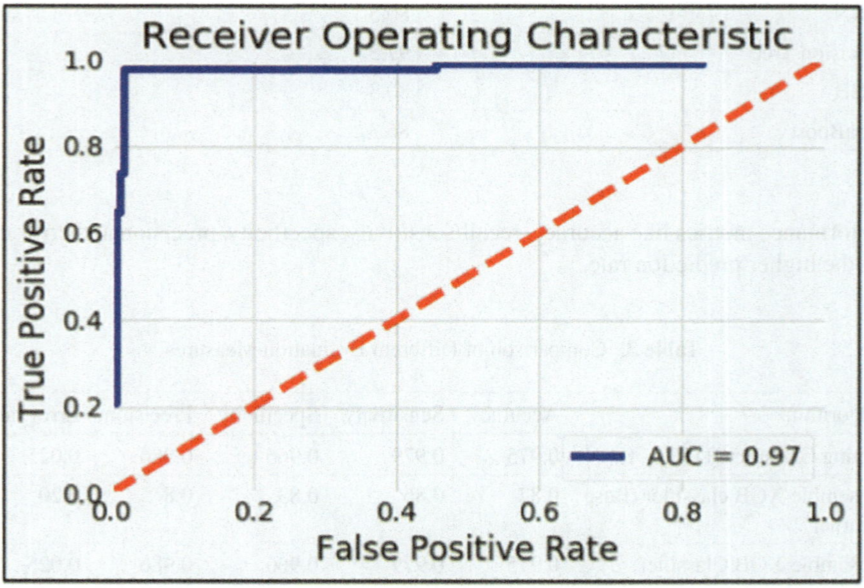

Fig. 6. ROC Curve of Highest Accuracy Achieved

6 Conclusion

The objective of this thesis work is to provide an insight of different techniques and experiments done in this research dealing with the diabetes dataset using popular Machine Learning algorithms. Diabetes is one of the major causes of death from AIDS and Cancer worldwide and the early prediction and nutrition coaching is very important which can increase the survival rate of the patient. The objective of this thesis work is to provide an insight of different techniques that can be employed in automated diabetes diagnosis systems minimizing the error rate. Different classification and regression algorithms are compared, and it has been observed that using ensemble techniques boost the performance of the model not only on the accuracy but also in the specificity, sensitivity with reducing the error rate up to 0.005% with 99.5% accurate on the patient's diabetes dataset.

These results suggest that proposed ensemble technique has the potential to significantly improve the conventional ensemble methods used in the study. It's also shown that the most important attributes for diabetes diagnosis are glucose, pregnancy, body mass index. The proposed system greatly reduces the cost of treatment and improves the quality of life by predicting diabetes and helps to provide nutrition coaching for the one suffering from diabetes. There are certain limitations of this research that the dataset collected from the UCI repository is less to test with deep learning. Also, while testing datasets were divided in a ratio of 70:30 that can be addressed getting better results and all these limitations can be reduced in the future for better machine learning algorithms.

As a part of future work the data size can be increased for the models to perform with high accuracy. Next, various tuning mechanisms could be applied to increase the accuracy of the current model and also with the increasing datasets deep neural network can be used for the real implementation of the model which can be done in future by developing a system so that it would be easier to work on a GUI or application based diabetes diagnosis and individual results can be obtained and early identification of the patient.

Acknowledgements. This work was supported by the project "BLOCKCHAIN.PT(RE-C05-i01.01– Agendas/Alianças Mobilizadoras para a Reindustrialização, Plano de Recuperação e Resiliência de Portugal na sua componente 5 – Capitalização e Inovação Empresarial e com o Regulamento do Sistema de Incentivos "Agendas para a Inovação Empresarial", aprovado pela Portaria N.° 43-A/2022 de 19.01.2022). We would like to express our sincere gratitude to Nepalese Education in E-health Masters (NEEM) project under GA101083048 at Health Informatics Lab in Kathmandu University for supporting this research.

References

1. MayoClinic, (2022, Dec 15), Overview on Type 2 Diabetes. Accessed 14 Jan 2023. www.mayoclinic.org/diseases-conditions/type-2/symptoms-causes
2. Kadam, W.D.: Preprocessing for machine learning algorithms. Int. J. Comput. Sci. (2004)
3. Hu, F.B., Diet, J.A.E.: Lifestyle, and the Risk of Type 2 Diabetes Mellitus in Women, 2001
4. Kotsiantis, S., et al.: Data preprocessing for supervised learning. Int. J. Comput. Sci. (2007)
5. Tyler, C.D.: An approach to graphs of linear forms, IEEE, 2004

6. Kang, S., Ko, P., et al.: An efficient and effective ensemble of support vector machines for anti-diabetic drug failure prediction. Expert Syst. Appl. (2015)
7. Kreifeldt, J.G.: An analysis of surface-detected EMG as an amplitude- modulated noise, presented at the 1989 Int. Conf. Medicine and Biological Engineering, Chicago, IL
8. Buhmann, M.: Radial Basis Functions: Theory and Implementations. Cambridge University Press, Cambridge (2003)
9. Omer, A.: A rule induction algorithm for knowledge discovery and classification. Turk. J. Elec. Eng. Comp. Sci. (2013)
10. Durgabai, R.: Feature selection using ReliefF algorithm. Int. J. Adv. Res. Comput. Commun. Eng. (2014)
11. Dietterich, G.: Ensemble methods in machine learning. In: Kittler, J., Roli, F. (eds.), First International Workshop on Multiple Classifier Systems, LeNCS, Springer - Verlag, New York, 2001
12. Breiman, L.: Bagging Prediction. Machine Learning, 2000
13. Li, L.: Diagnosis of diabetes using a weight-adjusted voting approach. In: IEEE International Conference on Bioinformatics and Bioengineering, 2014
14. Prema, G., Pushpalatha, M.P.: Prediction of gestational diabetes mellitus (GDM) using classification. In: IEEE International Conference on Science, Technology, Engineering and Management (ICSTEM), 2017
15. Sitar-Taut, Y.V.A., et al.: Using machine learning algorithms in cardiovascular disease risk evaluation. J. Appl. Comput. Sci. Math. (2009)
16. Makhtar, M., Khalid, M., et al.: Optimizing sensitivity and specificity of ensemble classifiers for diabetic patient. J. Theor. Appl. Inf. Technol. (2015)
17. Ömer, D.U.K.: Diabetes Determination Using Retraining Neural Network. IEEE, 2018
18. Kema, A., Baha, S.: Diabetes mellitus data classification by cascading of feature selection methods and ensemble learning algorithm. IJ Mod. Educ. Comput. Sci. (2018)
19. Prema, N., Vrashith, V., et al.: Prediction of diabetes using ensemble technique, Int. J. Recent Technol. Eng. (2019)

Gait Analysis for Early Detection of Cardiovascular Diseases Using MPU-6050 Sensor: An Analytical Framework with Data Augmentation Algorithm

Rabindra Bista[1(✉)], Sanjog Sigdel[1], and Joao C. Ferreira[2,3,4]

[1] Department of Computer Science and Engineering, Kathmandu University, Dhulikhel-4, Kavre, Nepal
rbista@ku.edu.np
[2] Inov Inesc Inovação—Instituto de Novas Tecnologias, 1000-029 Lisbon, Portugal
[3] Instituto Universitário de Lisboa (ISCTE-IUL), ISTAR, 1649-026 Lisbon, Portugal
[4] Faculty of Logistics, Molde University College, NO-6410 Molde, Norway

Abstract. Walking pattern analysis known as Gait analysis, is one of the key indicators among various clinical parameters, such as cardiovascular disease, in identifying symptoms of diseases. With sensors, the walking patterns of people can be retrieved to identify the differences between patients and healthy controls. Collection of health data is a crucial subject as it requires approval from different individuals/authorities. This paper presents an analysis of Gait datasets with nine Kinematic gait parameters by using classification models, a new data augmentation algorithm. Classification analysis of the Gait dataset has been performed using Neural Network which has 98.65% accuracy. We develop a new GDAA (Gait Data Augmentation Algorithm) for augmentation of Gait data and its time complexity is O(nf). The result of GDAA falls between the minimum range and maximum range of the original dataset. To evaluate optimum classification results, we performed data analysis of the augmented datasets of varying sizes. After a rigorous analysis of augmented data, we found the accuracy of the Neural Network was 97.1%.

Keywords: Gait Analysis · Cardiovascular Diseases · Explainable Artificial Intelligence · Sensors · Digital Health

1 Introduction

Digital health is the convergence of digital technologies with health, healthcare, living, and society to enhance efficiency of healthcare service delivery and make medicine more personalized and precise [1]. Walking is one of the most common activities humans do every day. With the help of sensors, we can analyze the walking pattern of a person which provides useful health information [2]. Gait analysis is a systematic study, which involves analysis of measurement, description, and assessment of qualities that characterize human locomotion [3]. Gait analysis cannot definitively diagnose or predict a

V. Sakalauskas et al. (Eds.): IBICA 2023, LNNS 1231, pp. 109–118, 2025.
https://doi.org/10.1007/978-3-031-78946-5_10

particular disease. However, the analysis results can be a valuable addition for medical professionals to understand patients' walking behavior. MPU-6050 sensors provide a non-invasive, cost-effective, and easily accessible method for identifying individuals at risk. This ultimately leads to better outcomes and reduced healthcare costs [5].

2 Theoretical Background

2.1 Objective

Getting abundant healthcare data for research needs to go through different processes and approvals. The primary objective of this research is to augment gait data from the original dataset. This enables us to perform gait research using the augmented data in any volume we want.

In shortage of gait data, the original dataset can be used to augment large volumes of data appropriate for the research. Additionally, this research performs machine learning analysis of the augmented data. We chose the dataset which has data on healthy controls and cardiovascular patients. Once an algorithm is developed we can use it to tweak the parameters and develop huge iterations of augmented data for other areas. This helps researchers in overcoming the problem or shortage of research data.

2.2 Motivation

Gait analysis provides insights into an individual's physical function and mobility. It is useful in the diagnosis and management of various health conditions, including neurological disorders, musculoskeletal conditions, and cardiovascular diseases. By analyzing Gait data collected from sensors, such as the MPU-6050, machine learning techniques can be applied to develop prediction model/s for identifying individuals at risk of developing certain health conditions or for monitoring the progression of an existing condition [6].

Research of healthcare data is a crucial subject where the data collection process is complex. To overcome this challenge, an augmentation algorithm is developed using the primary data [7]. We aimed to compare the augmented data and analyze the results among healthy controls and cardiovascular patients.

2.3 Environment

Using the primary data from research [7] we developed a data augmentation algorithm which is presented in Sect. 2.4. Data pre-processing, data analysis, data augmentation algorithm, and analysis of data using machine learning, all were done using the python programming language. The analysis is done using different libraries such as Sci-py, Scikit-Learn, and Pandas.

2.4 Algorithms

A Gait Data Augmentation Algorithm (GDAA) was developed with the assumption of minimum and maximum values for each parameter to be that from the previous dataset. A literature study presented that researchers used datasets available in different data portals and implemented them in their research as a secondary source of data [8, 9]. Augment requires the same proportion of Healthy Controls and Patients. The Gait parameters should be within the range of values in the original dataset. The data augmentation algorithm is presented below:

Algorithm 1. Gait Data Augmentation Algorithm (GDAA)

Input: **Dataframe:** Gait Dataset file
List: Empty list of all features from Gait Dataset
Integer: k
Output: **File:** Augmented Gait Dataset

Start
Convert Input file as a DataFrame.
Independent[] ← Independent gait Features
Dependent[] ← Dependent gait Features
Calculate Minimum and Maximum of the Independent[] gait parameters
For values in Range(0, k):
    Append the subject title to the empty list
    Function Calls: *GenerateStrideTime(); GenerateStrideCount(); GenerateSwingTime(); GenerateStanceTime();*
        Generate StrideTime, Generate StrideCount, Generate SwingTime & Generate Stance Time
        Append StrideTime, StrideCount, SwingTime & StanceTime to the list
End Function
Function Calls: *CalculateStrideLength(); CalculateStrideFrequency(); CalculateWalkingTime(); CalculateStrideFrequency(); CalculateStrideCadence();*
    Append StrideLength, StrideFrequency, WalkingTime, StrideFrequency, StrideCadence to the list.
End Function
End For
Zip the lists which have generated data
Create a dataframe from the zipped list
Store final value as CSV to the directory
End

Initially, the dependent and independent gait features are highlighted in their respective lists, minimum and maximum of independent parameters are calculated. For a range defined by the integer k, for all the independent parameters data is generated using a random function generator. The minimum and maximum values of each parameter define the range of random function generators. Those minimum and maximum values are calculated from the input file's data range. Generation of independent Gait data using a random function generator takes place in steps 8–11.

For all four independent parameters, random data is generated and appended to the list respectively. The implementation of this random data generation involves four functions that call random functions separately using their min and max data range. As four functions do the same task repeatedly steps 8–11 in the algorithm are condensed.

Generation of dependent Gait data takes place in steps 12–15. For all the dependent gait parameters, data is calculated using their respective formula. The dependent gait parameters formula includes some independent gait features generated at that instant of for loop steps 8–11. Calculation of dependent Gait data functions in steps 12–15 includes five functions for each dependent gait parameter. Since the calculation of five different parameters is a similar task, the steps are condensed in the algorithm. Each Gait data is appended in their respective lists.

At the end of a single iteration, all parameters will have a value, and those values are stored in the respective lists which are initially empty. As the loop ends, all lists are grouped i.e. zipped into a single list. That list is then converted into a new data frame which is then exported as a CSV file in the desired directory.

Machine learning analysis was done using k-Nearest Neighbors, Classification and Regression Tree, Support Vector Machine, Naive Bayes, Logistic Regression models and Neural Network [10, 11]. Zheng and team performed gait research using machine learning analysis using three supervised learning algorithms: Support Vector Machine, KStar, and Random Forest [12] which performed accuracy results of 86.9%, 84.7%, and 84.9%. Researchers used a 3D-accelerometer and a 3D-gyroscope to identify the spatial gait parameters and used Deep Convolutional Neural Network to perform the machine learning analysis [13]. An accuracy of 90% was performed from the dataset in comparison to the benchmark dataset results.

Recent work on gait monitoring and measurements has been carried out for Parkinson's Disease using Artificial Intelligence-based wearable gait monitoring [14]. Research focuses on monitoring and assessment of gait in patients with Parkindon's disease and also proposes a wearable physiograph for qualitative and quantitative gait assessment by performing bilateral tracking of the foot biomechanics and unilateral tracking of the arm balance. AI-based decision support is built using gait parameters in Convolutional Neural Network (CNN) with 95% accuracy, 90% sensitivity, and 95% precision.

3 Related Works

One of the symptoms of heart-failure in old age people was identified as gait speed. It was identified as one of the parameters of the research which consisted of 331 subjects 70 years old or above [15]. Research by Barry and team performs temporal gait analysis to evaluate adaptive gyroscope-based algorithms [16]. Algorithms used in the gait research adaptively calculated thresholds to determine Initial Contact and Terminal Contact. Authors have published methods to detect the quasi-real-time gait event using uniaxial shank-attached gyroscopes [17].

Research work by Sina and the team introduced a new approach to achieving a more reliable method to assess the physical performance status (PPS) among different groups of cancer patients undergoing chemotherapy [18]. With the use of body-mounted inertial sensors in 23 body segments, a six-minute walk test (6MWT) was performed. It was observed that a higher gait speed with a greater range of body motion is related to an appropriate PPS. Slower gait speed resulted in a smaller range of physical motions which caused a limitation in a variety of clusters along with two or more consecutive sequences.

Research on gait disorder and balance is performed on patients with Alzheimer's disease. Stride and gait cycle decomposition were retrieved from the sensors for analysis. Research performed an ANOVA test which reflected a significant measurable difference in gait parameters [19]. Gait analysis has played an important role in analyzing

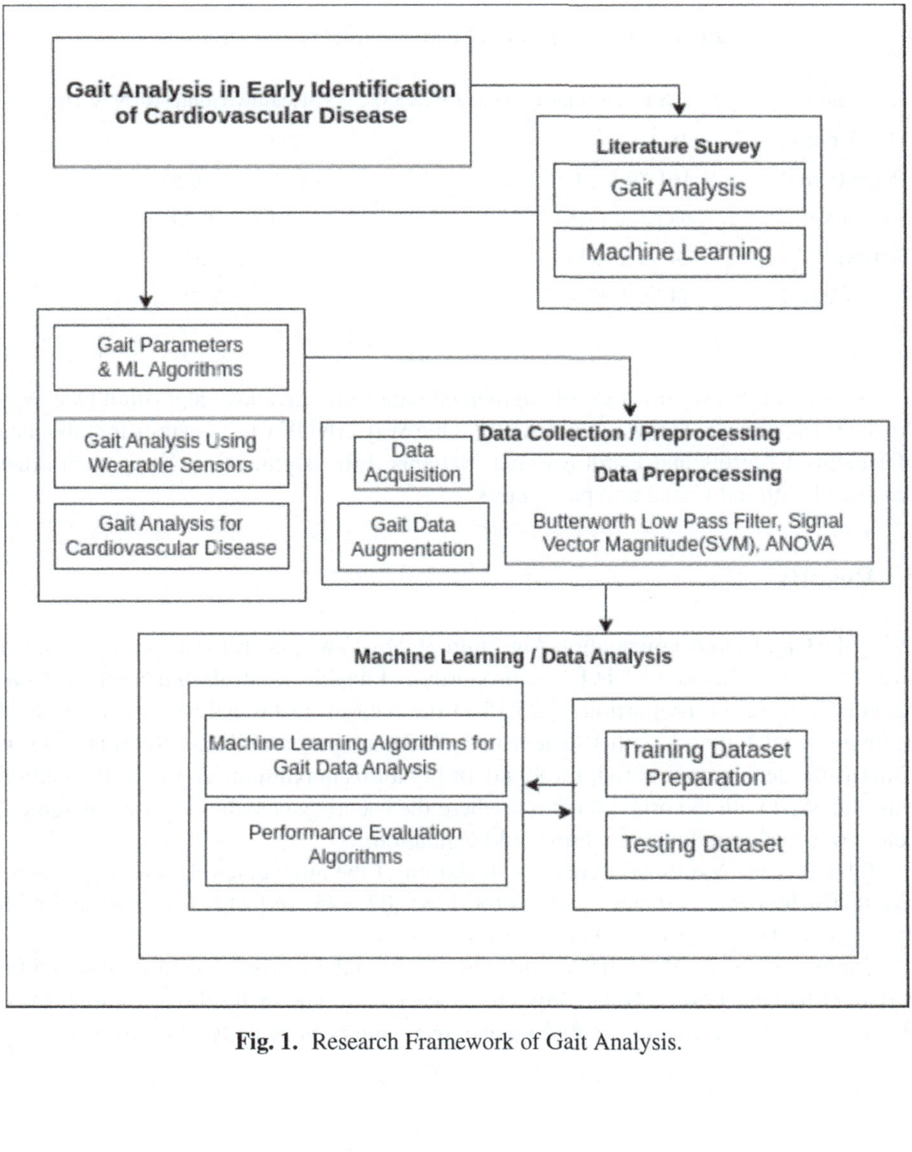

Fig. 1. Research Framework of Gait Analysis.

the step-by-step spatiotemporal parameters of the elderly suffering from hemiparetic, parkinsonian, and choreic gait (Fig. 1).

4 Methodology

A detailed research methodology was implemented in performing the gait analysis result. We analyzed the original dataset and compared it to the augmented data. However, this paper scopes its findings on the augmentation algorithm, data augmentation and analysis of augmented data using various machine learning models.

The original dataset consists of 16 healthy controls and 53 cardiovascular patients and their characteristics are mentioned below (Table 1):

Table 1. Characteristics of Participants of Gait Analysis

Parameters	Cardiovascular Patients (N = 53)	Healthy Controls (N = 16)
Men/Women(n)	28/25	5/11
Height(cm)	161.79 ± 11.53	162.88 ± 10.85
Weight(kg)	63.57 ± 11.52	62.19 ± 10.23
BMI(kg/m2)	24.30 ± 4.24	23.49 ± 3.64
Age(years)	44.58 ± 17.01	32.44 ± 6.20

To evaluate the significance of augmented data from the above algorithm One Way Analysis of Variance test was chosen [20]. One Way ANOVA test ensures that the data of Healthy Controls and Cardiovascular patients differ significantly from each other compared with individual gait parameters.

5 Results

Using the Gait Data Augmentation Algorithm (GDAA) we generated augmented data of sizes 69, 138, 207, 345, and 414. The proportion of healthy controls and cardiovascular patients were in the proportion of 25:75 in the dataset. Data analysis was conducted using a Neural Network with 100 neurons in hidden layers and ReLu (Rectified Linear Unit) as the activation function for 80 no. of iterations maximum. The experimentation was done first with the original dataset where the accuracy calculation listed in Table 2 below was performed with ten cross-fold validation.

Table 3 is the results of accuracy calculation of the augmented dataset of the same size that is 69 rows. A dataset of sizes 69, 138, 207, 345, and 414 was generated with the augmentation algorithm where the proportion is 75:25.

Comparing the result of the Augmented dataset and the original dataset of size 69 it was identified that besides Naive Bayes there wasn't any gain in the classification results. However, the results are above 90% on average, which can be considered for accepting

Table 2. Accuracy calculation of Classification Models (A)

Model	CA	F1	Precision	Recall
k-Nearest Neighbors	0.93	0.931	0.936	0.93
Classification & R Tree	0.95	0.95	0.951	0.95
Support Vector Machine	0.973	0.973	0.975	0.973
Naive Bayes	0.882	0.888	0.915	0.882
Logistic Regression	0.955	0.954	0.955	0.955
Neural Network	0.986	0.986	0.986	0.986

Table 3. Accuracy calculation of Classification Models (B)

Model	CA	F1	Precision	Recall
K-Nearest Neighbors	0.928	0.931	0.945	0.928
Classification & R Tree	0.942	0.943	0.946	0.942
Support Vector Machine	0.928	0.93	0.936	0.928
Neural Network	0.971	0.972	0.974	0.971
Naive Bayes	0.899	0.904	0.929	0.899
Logistic Regression	0.884	0.886	0.89	0.884

augmented data. Also, these results have improved with the increase of the augmented dataset in the section below.

The classification results of the augmented dataset were near to that of the real dataset. Classification Analysis of Augmented Data for all dataset sizes was performed. The result favored optimum results for an augmented dataset of size 276(212P-64H). The result is shown in Table 4.

Table 4. Classification Analysis of Augmented Data. Data Size: 276(212P-64H)

Model	CA	F1	Precision	Recall
K-Nearest Neighbors	0.953	0.954	0.957	0.953
Classification & R Tree	0.978	0.978	0.979	0.978
Support Vector Machine	0.975	0.975	0.977	0.975
Neural Network	0.982	0.982	0.983	0.982
Naive Bayes	0.96	0.961	0.966	0.96
Logistic Regression	0.942	0.942	0.943	0.942

The gain in Classification accuracy, precision, recall, and F1 score were reflected in models like k-Nearest Neighbors, Classification and Regression Tree, and Naive Bayes. The last gain was observed with the model Support Vector Machine. The optimum results for the augmentation data were observed for the data size 276(212P-64H). The optimum result for neural network analysis was observed after experimenting with the analysis in several iterations shown in Table 5.

Table 5. Neural network analysis with varying iterations

No. of Iterations	CA	F1	Precision	Recall
80	0.975	0.975	0.977	0.975
90	0.975	0.975	0.977	0.975
100	0.975	0.975	0.977	0.975
110	0.978	0.979	0.98	0.978
120	0.978	0.979	0.98	0.978
130	0.978	0.979	0.98	0.978
140	0.982	0.982	0.983	0.982
150	0.982	0.982	0.983	0.982
160	0.982	0.982	0.983	0.982

Increasing the number of iterations to 140 iterations the optimum classification results were 98.2% accuracy, 98.2% F1 Score, 98.3% precision, and 98.2% recall.

6 Conclusion

The objective of this experiment was to generate augmented data and perform analysis using machine learning algorithms. We were able to generate augmented data of various sizes and compare it with machine learning algorithms. From the results highlighted in Table [1–5], it is clear that the augmented data performs accurate results similar to the one generated from the original dataset. The accuracy results of the machine learning algorithms we used in our research for the augmented dataset are above 95%. This also ensures the augmented data and the data augmentation algorithms can be further expanded introducing additional features.

As a future work we can conduct machine learning analysis using a huge augmented dataset generated from the GDAA. We can enhance the algorithm by adding additional gait parameters and their rules for generating augmented data. Furthermore we can experiment with the model of generating the algorithm in selecting a separate data domain. In terms of research domain we can explore gait feature ranking and perform analysis on dependent and independent gait features. During the data selection we can collect and categorize different types of cardiovascular diseases. This enables a new interpretation on what gait analysis compared to patients with different cardiovascular patients presents. Also gait parameters and its impact to each category of cardiovascular disease can open different question for future research in the field of gait analysis.

Acknowledgements. This work was supported by the project "BLOCKCHAIN.PT(RE-C05-i01.01– Agendas/Alianças Mobilizadoras para a Reindustrialização, Plano de Recuperação e Resiliência de Portugal na sua componente 5 – Capitalização e Inovação Empresarial e com o Regulamento do Sistema de Incentivos "Agendas para a Inovação Empresarial", aprovado pela Portaria N.° 43-A/2022 de 19.01.2022). We would like to express our sincere gratitude to Nepalese Education in E-health Masters (NEEM) project under GA101083048 at Health Informatics Lab in Kathmandu University for supporting this research.

References

1. Bhavnani, S.P., Narula, J., Sengupta, P.P.: Mobile technology and the digitization of healthcare. Eur. Heart J. **37**(18), 1428–1438 (2016). https://doi.org/10.1093/eurheartj/ehv770
2. Nilpanapan, T., Kerdcharoen, T.: Social data shoes for gait monitoring of elderly people in smart homes. In: The 2016 Biomedical Engineering International Conference (BMEiCON-2016), 2016
3. Ghoussayni, S., Stevens, C., Durham, S., Ewins, D.: Assessment, and validation of a simple automated method for the detection of gait events and intervals. Gait Posture **20**, 266–272 (2004)
4. Lai, D.T.H., Begg, R.K., Palaniswami, M.: Computational intelligence in gait research: a perspective on current applications and future challenges. IEEE Trans. Inf. Technol. Biomed. **13**(5), 687–702 (2009). https://doi.org/10.1109/titb.2009.2022913
5. Tanwar, R., Nandal, N., Zamani, M., Manaf, A.A.: Pathway of trends and technologies in fall detection: a systematic review. Healthcare **10**, 172 (2022). https://doi.org/10.3390/healthcare10010172
6. Aminian, K., Najafi, B., Büla, C., Leyvraz, P., Robert, P.: Spatiotemporal parameters of gait measured by an ambulatory system using miniature gyroscopes. J. Biomech. **35**(5), 689–699 (2002). https://doi.org/10.1016/s0021-9290(02)00008-8
7. Joshi, S., Khanal, S., Bista, R.: Spatiotemporal gait analysis for cardiovascular disease. In: 2022 14th International Conference on Software, Knowledge, Information Management and Applications (SKIMA), Phnom Penh, Cambodia, 2022, pp. 181–186, 2022. https://doi.org/10.1109/SKIMA57145.2022.10029510
8. Nandy, A., Chakraborty, P.: A new paradigm of human Gait Analysis with Kinect. In: 2015 Eighth International Conference on Contemporary Computing (IC3), pp. 443–448, 2015. https://doi.org/10.1109/IC3.2015.7346722
9. Clark, R.A., Bower, K.J., Mentiplay, B.F., Paterson, K., Pua, Y.-H.: Concurrent validity of the Microsoft Kinect for assessment of spatiotemporal gait variables. J. Biomech. **46**(15), 2722–2725 (2013). https://doi.org/10.1016/j.jbiomech.2013.08.011
10. Zhang, H., Guo, Y., Zanotto, D.: Accurate ambulatory gait analysis in walking and running using machine learning models. IEEE Trans. Neural Syst. Rehabil. Eng. (2019)
11. Allseits, E., Lučarević, J., Gailey, R., Agrawal, V., Gaunaurd, I., Bennett, C.: The development and concurrent validity of real-time algorithm for temporal Gait Analysis using inertial measurement units. J. Biomech. **55**, 27–33 (2017). https://doi.org/10.1016/j.jbiomech.2017.02.016
12. Zheng, H., Yang, M., Wang, H., McClean, S.: Machine learning statistical approaches to support discrimination of neuro-degenerative diseases based on gait analysis. Intell. Patient Manag. 57–70 (2009). https://doi.org/10.1007/978-3-642-00179-6_4

13. Hannink, J., Kautz, T., Pasluosta, C., Gasmann, K., Klucken, J., Eskofier, B.M.: Sensor-based gait parameter extraction with deep convolutional neural networks. IEEE J. Biomed. Health Inform. **21**(1), 85–93 (2017). https://doi.org/10.1109/jbhi.2016.2636456
14. Ilesan, R.R., et al.: Proof of concept in artificial-intelligence-based wearable gait monitoring for Parkinson's disease management optimization. Biosensors **12**, 189 (2022). https://doi.org/10.3390/bios12040189
15. Pulignano, G., et al.: Incremental Value of Gait Speed in Predicting Prognosis of Older Adults With Heart Failure. Insights from the IMAGE-HF study, pp. 289–298 (2016)
16. Barry, G.R., McGrath, D., O'Neill, R., O'Donovan, K.J., Burns, A., Caulfield, B.: An adaptive gyroscope-based algorithm for temporal gait analysis. Med. Biol. Eng. Comput. **48**(12), 1251–1260 (2010). https://doi.org/10.1007/s11517-010-0692-0
17. Lee, J.K., Park, E.J.: Quasi-real-time gait event detection using shank-attached gyroscopes. Med. Biol. Eng. Comput. **49**(6), 707–712 (2011)
18. Sina, A., Fazel, N., David, S., Golshah, N., Morteza, A.: Objective clinical gait analysis using inertial sensors and six minute walking test. Pattern Recognit. (2016). https://doi.org/10.1016/j.patcog.2016.08.002
19. Zhang, Y., Yang, H.: A prediction model for early identification of cardiovascular disease using Gait Analysis with the MPU-6050 sensor. J. Med. Syst. **44**(7), 151 (2020)
20. Mileti, I., et al.: Accuracy evaluation and clinical application of an optimized solution for measuring spatio-temporal gait parameters. In: 2020 IEEE International Symposium on Medical Measurements and Applications, 2020

Wearable Devices for Long-Term Care – Survey and Opportunities

Sudip Phuyal[1], Luís B. Elvas[1,2], João C. Ferreira[1,2,4](✉), and Rabindra Bista[3]

[1] Instituto Universitário de Lisboa (ISCTE-IUL), ISTAR, 1649-026 Lisbon, Portugal
sudip.phuyal@iscte-iul.pt, joam@himolde.no
[2] Inov Inesc Inovação—Instituto de Novas Tecnologias, 1000-029 Lisbon, Portugal
[3] Department of Computer Science and Engineering, Kathmandu University, Dhulikhel-4, Kavre, Nepal
[4] Faculty of Logistics, Molde University College, NO-6410 Molde, Norway

Abstract. The incorporation of wearable devices within long-term care represents a groundbreaking advancement in healthcare. These devices, leveraging sensor and intelligent technologies, possess the potential to revolutionize the supervision and handling of chronic illnesses, enriching the lives of individuals reliant on extended care. This paper conducts an exhaustive examination of wearable devices in the long-term care context, spotlighting their wide-ranging applications. It encompasses devices tailored for monitoring vital signs, mobility, medication adherence, cognitive health, and other critical facets of long-term care. Through an in-depth exploration of their capabilities and limitations, it illuminates their capacity to ease the strain on healthcare systems and enhance patient well-being. Moreover, this paper investigates the hurdles related to wearable device adoption in long-term care, addressing concerns like data security, user acceptance, and interoperability. By tackling these challenges, it endeavors to chart a path for the seamless integration of wearable technologies. Drawing insights from current research and industry trends, it identifies emerging prospects such as personalized care plans, remote monitoring, and data-driven healthcare interventions, fostering collaboration among healthcare stakeholders. By harnessing the full potential of wearable technologies, the quality of life of individuals in long-term care can be improved and also, it can contribute to achieve more efficient and sustainable healthcare systems.

Keywords: Wearable Devices · Long-Term Care · Healthcare Monitoring · Chronic Illness Management · Remote Monitoring · Patient-Centric Care · Healthcare Technology

1 Introduction

In 2022, the global wearable medical devices had a market value of US$ 59.12 billion. The publication projects that the market will grow at a compound annual growth rate of 28.6% from 2023 to 2030, bringing the total value to US$ 428.92 billion [1]. Activity trackers and smartwatches are examples of medical equipment that fall within the

V. Sakalauskas et al. (Eds.): IBICA 2023, LNNS 1231, pp. 119–128, 2025.
https://doi.org/10.1007/978-3-031-78946-5_11

category of wearables in the healthcare industry. They are made expressly to track and gather data in real time on an individual's fitness and well-being. About 40% of American citizens use applications connected to healthcare, and 35% use wearable healthcare devices, according to a 2023 article released by mHealth Intelligence [2]. The demand for people to improve their health and the recent technical developments in wearables have spurred a number of industry participants to produce additional smart devices, including biosensors, hearing aids, and Fitbits. In addition, one of the main drivers expected to drive wearable medical device sales during the forecast period is the rising prevalence of chronic illnesses and disorders associated to lifestyle choices and also, most of the patients support incorporating wearable device data in healthcare, which is being valuable for the healthcare professionals to identify the nature and cause of the health problems [2, 3].

The wearable devices market for long-term care has been growing rapidly in recent years and is expected to continue to grow in the coming years. The growth of this market can be attributed to several factors [4], including: 1) Increasing demand for remote patient monitoring and telemedicine services, particularly in the wake of the COVID-19 pandemic; 2) Advances in wearable technology, including the miniaturization of sensors and the development of new applications for wearables in healthcare; 3) Growing awareness of the benefits of wearable devices for long-term care, including improved patient outcomes, increased patient engagement, and reduced healthcare costs; 4) Increasing investment and support from governments and private sector organizations in the development of wearable technology for healthcare [4, 5].

Furthermore, the google updates API also played the significant role for encouraging the for patient COVID-19 pandemic also encouraged the huge population to use the android applications which can collect the vaccination and tests results of the patients and helped to suggest the necessary actions based on their health data [6]. Overall, the wearable devices market for long-term care is expected to continue to grow as the demand for these devices increases and technology advances. Businesses looking to enter or expand in this market can take advantage of this growth potential and capitalize on the increasing demand for wearable devices in long-term care. Here have been several successful use cases of wearable devices for long-term care that demonstrate the potential benefits of these devices for both patients and healthcare providers. Some examples include:

- Remote patient monitoring: Wearable devices equipped with sensors and communication capabilities can allow healthcare providers to monitor patient health data remotely, such as vital signs, physical activity, and medication adherence. This can improve patient outcomes, reduce healthcare costs, and provide patients with more convenient and accessible care [7].
- Fall detection: Wearables that are equipped with a variety of sensors and sophisticated accelerometer technology can identify falls and provide notifications to healthcare practitioners in real time. This proactive feature is an essential intervention mechanism that allows quick actions to stop accidents and improve patient safety in healthcare settings [8].

- Medication reminders: Personalized schedules are programmed into wearable medicine reminders, which alert users with alerts or messages at predetermined intervals [9]. These notifications, which are frequently programmable and connected to companion applications, provide feedback choices and integration with health data to assist users in adhering to their drug schedule. Wearables' user-friendly interfaces and connection characteristics help to enhance healthcare management by enabling remote monitoring, greater adherence, and timely drug ingestion [10].
- Chronic disease management: Wearable devices can provide continuous monitoring and tracking of health data for patients with chronic diseases, such as diabetes, heart disease, and respiratory conditions. This can improve disease management and prevent complications, leading to improved patient outcomes and reduced healthcare costs [9, 11].

Remote healthcare, combined with advanced wearable technology, can revolutionize the way healthcare is delivered. By enabling patients to obtain remote medical assistance, physicians to perform procedures remotely, and both to have access to vital health information, remote healthcare can improve access to care and increase efficiency. In addition, wearable devices, equipped with sensors and other technologies, can provide real-time health monitoring, enabling healthcare providers to identify potential health issues and respond quickly. This can be particularly beneficial for individuals with chronic conditions requiring frequent monitoring and support and those living in remote areas with limited healthcare services. As a result, remote healthcare and wearable technology have the potential to improve outcomes, reduce costs, and increase patient satisfaction. The reports from different sources like ReportsnReports [12], MarketsandMarkets [13], ResearchAndMarkets [14] also provided the current market projection of the wearable devices market for long-term care, including market size, growth. They also provide insights into key drivers, challenges, and trends affecting the market, as well as an analysis of market share and competitive landscape.

Traditional methods of training medical teams with new techniques and tools can be disrupted in the face of fast-moving medical threats, such as pandemics, natural disasters, and other emergencies [7]. In these situations, remote medicine can play a critical role in closing the gap and ensuring that medical teams have access to the latest information and training. With remote medicine, healthcare providers can participate in training sessions and access educational resources from anywhere, without the need for travel or in-person interaction. This can help ensure that medical teams have the knowledge and skills they need to provide effective care, even in fast-moving and unpredictable circumstances. Remote medicine can also improve access to care by enabling healthcare providers to reach and treat patients who are unable to travel or are located in remote areas. This can be particularly important in the context of fast-moving medical threats, where access to care can be limited.

The COVID-19 pandemic has made remote medicine more attractive and necessary in many ways. Many patients have become worried about seeking in-person medical care due to the risk of exposure to the virus and the need to avoid crowded medical centers. At the same time, safety rules and guidelines have prevented physicians from physically approaching patients, making it difficult to provide effective care in a traditional setting. In response, remote medicine has become an increasingly important solution, providing patients with medical care and support while reducing their risk of exposure to the virus. Remote medicine has also allowed healthcare providers to continue delivering patient care, despite the pandemic's challenges. This has been critical in ensuring that individuals with chronic conditions and other health needs continue to receive the care they require, even in the context of a global health crisis.

The demand for wearable technologies is growing for a number of reasons, including the need to reduce healthcare costs, address the needs of aging populations, and manage the growing burden of chronic disease. Wearable technologies have the potential to improve health outcomes, increase patient engagement, and reduce the cost of care, making them an attractive solution for many individuals and healthcare organizations. By providing real-time monitoring of vital signs, tracking of medication adherence, and facilitating communication between patients and healthcare providers, wearable technologies can help individuals manage chronic conditions more effectively and proactively. Additionally, the growth of remote medicine has further increased the demand for wearable technologies, as these devices can provide valuable data and insights that can be used to support remote care and improve health outcomes.

The integration of AI, ML, and big data analytics in wearable technology has significantly enhanced the usefulness of sensor data in many ways. By processing and analyzing large amounts of data generated by wearable sensors, AI and ML algorithms can identify patterns and make predictions that were not previously possible [15]. This can help healthcare providers and individuals better understand their health status and make informed decisions about their care. For example, AI algorithms can be used to detect early signs of chronic conditions and alert patients and healthcare providers before the conditions become serious. By utilizing machine learning algorithms and big data analytics, wearable technology can make sense of complex data streams and provide actionable insights that can improve health outcomes, reduce costs, and increase patient engagement. The power-efficient nature of AI and ML algorithms, combined with advances in cost-efficient and power-efficient electronics, has made wearable technology increasingly accessible to individuals and healthcare organizations [16, 17].

Wearable technology has become an essential part of home care systems, providing a variety of functions to improve patient outcomes. These multipurpose gadgets, which are fitted with a variety of sensors, monitor vital indicators such as blood pressure, body temperature, and heart rate continuously, giving instantaneous information on a person's health. Beyond monitoring, wearables are excellent in medication management. They accurately ensure that patients take their medications as prescribed and diligently remind them to do so. Furthermore, in the case of an accident, their fall detection capabilities provide prompt warnings to emergency contacts or caregivers, promoting speedy response procedures. These gadgets are also excellent at monitoring activity and mobility, providing thorough insights into a person's physical well-being. Moreover, the smooth remote connection between patients and their caregivers or healthcare providers is made possible by their integration with speech and video features. Wearables that are enhanced with emergency buttons can quickly notify responders or selected contacts in an emergency. Furthermore, these devices are essential for controlling chronic disorders such as diabetes since they continually measure activity levels, check glucose levels, and provide regular injection reminders. All things considered, wearable technology is a valuable asset to the home care industry, combining various features to support chronic disease management, patient care, and safety.

This comprehensive article carefully examines the revolutionary role wearable technology has taken on in the healthcare industry. It highlights their diverse contributions by exploring the ways in which these devices facilitate the detection, tracking, and treatment of a range of health-related problems. The article emphasizes the importance of wearable technology and looks in-depth at how it may empower medical personnel by providing real-time insights into patients' vital signs, activity levels, and medication adherence. Wearable technology is equipped with advanced sensors and data analytics. Moreover, it offers a perceptive examination of the growing market demand for wearable technology, clarifying the elements propelling its uptake and its multifarious uses in healthcare environments. This scrutiny encompasses the devices' usability, interoperability, and their potential to revolutionize diagnostics, preventive care, chronic disease management for long-term care.

2 Literature Review

Wearable devices can play a significant role in managing chronic conditions by providing real-time monitoring and tracking of key health metrics, such as heart rate, sleep quality, activity levels, and medication adherence. Additionally, some wearables can be equipped with sensors that can detect early warning signs of chronic conditions, such as fluctuations in heart rate or blood pressure. Wearables can also be integrated with digital health platforms and linked to electronic health records (EHRs), enabling healthcare providers to access up-to-date patient information and monitor progress over time [18, 19]. This information can be used to inform treatment decisions and adjust medications as needed. Wearables can also be used to motivate patients to adopt healthy behaviors, such as exercise and healthy eating, and to provide reminders for self-care activities, such as taking medications and monitoring blood sugar levels.

Several studies, such as those carried out by A. Dutta et al. [20], M. Farid et al. [21], M. S. Borycki et al. [22], K. S. Chiang et al. [23] and K. R. Mann [24], explore deeply into the use of wearable technology in tracking chronic illnesses like diabetes, heart disease, and COPD. Together, they highlight the potential benefits of wearable technology in the management of chronic illnesses, pointing to lower healthcare costs, better patient outcomes, and more patient participation.

Furthermore, these research agree on the obstacles that need to be overcome for implementation to be successful. For wearable technology to be seamlessly integrated into the management of chronic diseases, a number of critical issues need to be addressed. These issues include data privacy, security concerns, and the pressing need for standardized procedures in data gathering and analysis. Their complementary results highlight the need of resolving these issues in order to fully realize the potential advantages that wearable technology may provide for the management of long-term medical disorders.

Wearable devices can play a critical role in medication management by providing reminders and alerts to help individuals keep track of when they need to take their medications. Some wearables are equipped with sensors that can detect when a medication has been taken, helping to ensure adherence to medication schedules. Wearables can also be integrated with electronic health records (EHRs), allowing healthcare providers to monitor medication schedules and to provide real-time feedback on adherence. This information can be used to make adjustments to medication regimens and to provide patients with educational materials to help them understand the importance of taking their medications as prescribed. In addition, wearable devices can provide individuals with personalized feedback and insights into their health and wellness, helping to motivate them to prioritize their self-care and to make healthier lifestyle choices.

J. N. Arslan et al. [25], Y. Kim et al. [26], L. Marengo et al. [10] and J. M. et al. [27] discussed the applications of wearable devices for reminding patients to take their medication and tracking medication adherence. The paper discusses the potential benefits of wearable devices for medication adherence, including improved patient outcomes, increased patient engagement, and reduced healthcare costs. The authors also highlighted the challenges that need to be addressed for effective implementation of wearable technology for medication adherence, including data privacy, data security, and the need for standardization of data collection and analysis methods.

Wearable devices equipped with sensors can be used for monitoring vital signs such as heart rate, blood pressure, blood oxygen levels, and temperature. These devices can include smartwatches, fitness trackers, and specialized wearable devices designed specifically for health monitoring. The data collected by wearable devices can be used for a variety of purposes, including tracking physical activity, monitoring changes in health status, and providing alerts for potential health issues. Wearable devices for vital sign monitoring can potentially improve the quality of life for individuals with chronic conditions and those at risk for health problems. However, there are also challenges associated with using wearable devices for vital sign monitoring, including the need for accurate and reliable sensors, privacy and security concerns, and the need for reliable algorithms to interpret and make use of the data collected by the devices. Despite these challenges, wearable technology for vital sign monitoring is an active area of research and development and has the potential to significantly impact the field of healthcare.

A number of papers [28, 30–32] explore the use of wearable technology for monitoring vital signs such as blood pressure, oxygen saturation, and heart rate. This includes smartwatches, sensors, and similar devices. Every research highlights the potential benefits associated with these technologies, including better patient outcomes, more patient engagement, and decreased healthcare costs. They all accept the challenges that must be overcome in order to successfully use wearable technology in continuous vital sign monitoring, including issues with data privacy, security, and the requirement for standardized data gathering and processing procedures.

The work presented in [33] offers a comprehensive synopsis, including algorithms, assessment techniques, and common issues in this field. Similarly, a thorough examination is provided in [34], which focuses on different algorithms, feature extraction approaches, classification methods, and performance assessment measures. In addition to these evaluations, the work in [35] presents a complete system that makes use of wearable sensors such as gyroscopes and accelerometers to categorize a variety of activities in addition to detecting falls. Further exploring the use of wearable sensors in smart home settings. Moreover, the results in [36] highlights behavioral modeling for fall detection and clarifies the difficulties in implementing such systems. Together, these works cover a wide range of research topics, including system construction, algorithms, and difficulties in using wearable sensors for fall detection. As a result, they greatly advance this important field of study.

3 Opportunities

High-speed communication developments have completely changed the healthcare industry by making it possible to deploy remote technology like wearables for remote monitoring, video conferencing, and smart home health products widely. These advancements lower the risk of infection and save time and money by enabling remote access to patient data, real-time contact with healthcare practitioners, and fewer in-person visits. They also open up more patient access to experts, improving accessibility to treatment overall and launching telemedicine into a new age.

Because cloud computing has made powerful computer resources more accessible, smaller medical institutes may now handle large amounts of medical data and create cutting-edge models without having to make significant infrastructure expenditures. Investments made in healthcare research by the government and insurance companies stimulate innovation and lead to the creation of cutting-edge sensors and remote medical infrastructure. These expenditures guarantee a progressive change in medical technology by considerably improving patient outcomes and lowering healthcare expenses.

In the meanwhile, cutting-edge machine learning algorithms simplify the delivery of healthcare by helping doctors make decisions, gaining insights from the analysis of massive amounts of medical data, and improving resource efficiency. Connectivity advancements and sensor downsizing keep medical technology more accessible and affordable while also enabling more effective device communication and increasing access to healthcare, particularly in underprivileged areas. In the end, these developments offer better health results and more accessibility to healthcare for anyone.

Several factors pose as inhibitors to the widespread adoption and successful implementation of new medical technology. First, the substantial costs involved in both the development and deployment phases can act as a significant barrier, especially for smaller healthcare facilities. Additionally, the necessity for transparent and explainable machine learning algorithms in the medical realm is crucial to instill trust and ensure decisions align with patient welfare. Adjusting healthcare insurance models and infrastructure is vital to support access to these technologies. Stringent regulatory requirements, including compliance with HIPAA and other testing standards, contribute to time-consuming and costly processes. Protecting sensitive medical data, adhering to privacy regulations like HIPAA, remains a paramount concern. Moreover, the challenge of parts obsolescence and lifecycle limitations in medical devices poses hurdles in maintenance and upgrades, driving up costs. Lastly, the complexity of processing vast amounts of medical data necessitates new technologies that can effectively handle these demands.

4 Conclusion

The synthesis of learnings from a range of papers exploring wearable technology's application in long-term healthcare presents an enthralling picture of its revolutionary potential. In this context, wearable technology combined with telemedicine emerges as a disruptive force that makes it easier to monitor patients remotely and collect real-time health data. This integration improves the effectiveness of treatments and accuracy by bridging geographical gaps and creating a complete repository of health information. Navigating this upward trajectory is not without its challenges, though. The most important issues are privacy and security concerns around patient data, which demand strong structures and procedures to support broad implementation. In addition, there are many obstacles to overcome, including the need to secure universal support from stakeholders, establish consistent regulatory frameworks, invest in necessary infrastructures, validate data accuracy for well-informed clinical decisions, and ensure interoperability among various devices and systems. However, wearables present a strong value proposition for long-term care facilities. These devices indicate the ultimate level of continuous monitoring capabilities, early health anomaly identification, real-time data access for increased patient involvement, streamlined provider collaboration, decreased readmissions to hospitals, and the exciting potential for long-term cost savings. Wearable technologies have the potential to completely transform the way that extended healthcare is delivered. However, there are also many possibilities as well as complicated obstacles that must be overcome before wearables become widely adopted and successfully integrated.

Acknowledgements. This work was supported by the project "BLOCKCHAIN.PT (RE-C05-i01.01 – Agendas/Alianças Mobilizadoras para a Reindustrialização, Plano de Recuperação e Resiliência de Portugal na sua componente 5 – Capitalização e Inovação Empresarial e com o Regulamento do Sistema de Incentivos "Agendas para a Inovação Empresarial", aprovado pela Portaria N.° 43-A/2022 de 19.01.2022).

This work was also under the Eramus+ NEEM project.

References

1. Wearable Medical Devices Market Size | Research Report [2030]. https://www.fortunebusinessinsights.com/industry-reports/wearable-medical-devices-market-101070. Accessed 15 Nov 2023
2. Over a Third of Adults Use Health Apps, Wearables in 2023, Up From 2018. https://mhealthintelligence.com/news/over-a-third-of-adults-use-health-apps-wearables-in-2023-up-from-2018. Accessed 15 Nov 2023
3. Most Patients Support Incorporating Wearable Device Data in Healthcare. https://mhealthintelligence.com/news/most-patients-support-incorporating-wearable-device-data-in-healthcare. Accessed 15 Nov 2023
4. Escobar-Linero, E., Muñoz-Saavedra, L.: Sensors FL-P-, 2023 undefined Wearable Health Devices for Diagnosis Support: Evolution and Future Tendencies. mdpi.com
5. Chen, J., Li, T., You, H., et al.: Behavioral interpretation of willingness to use wearable health devices in community residents: a cross-sectional study. mdpi.com
6. Google Updates API for Patient Access to Personal COVID-19 Data. https://ehrintelligence.com/news/google-updates-api-for-patient-access-to-personal-covid-19-data. Accessed 15 Nov 2023
7. Malasinghe, L.P., Ramzan, N., Dahal, K.: Remote patient monitoring: a comprehensive study. J. Ambient. Intell. Humaniz. Comput. **10**, 57–76 (2019). https://doi.org/10.1007/S12652-017-0598-X

8. Chang, M., Wu, Y., Niu, H., et al.: Cross-Platform Gait Analysis and Fall Detection Wearable Device. mdpi.com
9. Atienza, A.: LB-2023 II, 2023 undefined Towards Personalized Medication Management: A Novel Approach to Medication Reminder Using Wearable Sensors and IoT. ieeexplore.ieee.org
10. Marengo, L.L., Barberato-Filho, S.: Involvement of human volunteers in the development and evaluation of wearable devices designed to improve medication adherence: a scoping review. Sensors **23**, 3597 (2023). https://doi.org/10.3390/S23073597
11. Guo, Y., Liu, X., Peng, S., et al.: A review of wearable and unobtrusive sensing technologies for chronic disease management. Elsevier
12. ReportsnReports (2021) Wearable Medical Devices Market by Product (Smartwatch, Smart Glasses, Fitness Tracker, Sleep Tracker, and Insulin Pump), Application (Remote Patient Monitoring, Fall Detection, Medication Reminders, and Chronic Disease Management), End User - Global Forecast to 2025
13. MarketsandMarkets (2021) Wearable Medical Devices Market by Product (Diagnostic & Monitoring, Therapeu-tic), Application (Remote Patient Monitoring, Fitness & Wellness, Medical Training & Rehabilitation), Component (Sensors, Display), End User - Global Forecast to 2022. [Report]. https://www.marketsandmarkets.com/report-search-page.asp?rpt=wearable-medical-devices-market. Accessed 11 Sep 2023
14. Research and Markets (2021) Global Wearable Medical Devices Market By Product, By Application, By Region, Competition Forecast & Opportunities, 2012–2022. https://www.researchandmarkets.com/reports/Global-Wearable-Medical-Devices-Market-By-Product-By-Application-By-Region-Competition-Forecast-Opportunities-2012-2022-Report.html. Accessed 13 Sep 2023
15. Nahavandi, D., Alizadehsani, R.: AK-CM and, 2022 undefined application of artificial intelligence in wearable devices: opportunities and challenges. Elsevier
16. Miao, F., Wu, D., Liu, Z., et al.: Wearable sensing, big data technology for cardiovascular healthcare: current status and future prospective. mednexus.org **136**, 1015–1025 (2022). https://doi.org/10.1097/CM9.0000000000002117
17. Witte, H., Blatter, T.U., Nagabhushana, P., et al.: Statistical learning and big data applications. J. Lab. Med. (2023).https://doi.org/10.1515/LABMED-2023-0037/HTML
18. Epstein, C.: Obstetric TM-J of, Neonatal G&, 2023 undefined Linking Electronic Health Records With Wearable Technology From the All of Us Research Program. Elsevier
19. Assenza, G., Fioravanti, C.: SG- on M for, 2020 undefined new perspectives on wearable devices and electronic health record systems. ieeexplore.ieee.org
20. Dutta, A., Patel, R.B., Sorber, M.J.: Wearable technology for chronic disease management: a systematic review. J. Med. Syst. (2020)
21. Farid, M., Hasan, A.: Wearable technology and chronic disease management: a review. J. Med. Syst.
22. Borycki, M.S., Kaushal, M.C., Jensen, R.H.: Wearable devices for chronic disease management: a review of the literature. J. Med. Internet Res.
23. Chiang, K.S., Chao, M.F.: Wearable technology and chronic disease management: the future is now. J. Med. Syst.
24. Mann, K.R., Lee, K.J.: Wearable technologies for chronic disease management: a review of the literature
25. Arslan, N., Pathak, P.N., Gupta, V.K.: Medication adherence and wearable devices: a review. J. Med. Syst.
26. Kim, Y.J.K., Kim, Y., Kim, J., Kim, J.: Wearable technology for medication adherence monitoring: a systematic review. J. Med. Syst.
27. Babic, J.M., Banavar, S.: Wearable technology for medication adherence: a systematic review. J. Med. Syst.

28. Dutta, A., Patel, R.B., Sorber, M.J.: Wearable vital sign monitoring devices: a systematic review. J. Med. Syst.
29. Wang, X., Liu, L., Li, Y.: Wearable devices for continuous vital sign monitoring: a review of the literature. J. Med. Internet Res.
30. Arslan, J.N., Pathak, P.N., Gupta, V.K.: Wearable technology for continuous vital sign monitoring: a systematic review. J. Med. Syst.
31. Kim, J.S., Kim, J., Kim, J.: Wearable devices for vital sign monitoring: a review of the literature. J. Med. Internet Res.
32. Farid, M., Hasan, A.: Wearable technology for vital sign monitoring: a review. J. Med. Syst.
33. Bet, P., Castro, P.: Informatics MP-I journal of medical, 2019 undefined Fall detection and fall risk assessment in older person using wearable sensors: a systematic review. Elsevier
34. Ojetola, O., Gaura, E.: International JB-2011 S, 2011 undefined Fall detection with wearable sensors--safe (Smart Fall Detection). ieeexplore.ieee.org
35. Maglogiannis, I.: CI-IC, 2016 undefined Fall detection and activity identification using wearable and hand-held devices (2018). content.iospress.com. https://doi.org/10.3233/ICA-150509
36. Forbes, G., Massie, S., Craw, S.: Fall prediction using behavioural modelling from sensor data in smart homes. Artif. Intell. Rev. **53**, 1071–1091 (2020). https://doi.org/10.1007/S10462-019-09687-7

Gamification System for Eco-Driving: Enhancing Driver Motivation and Fuel Savings Through Game Mechanics

Adriana Franco[1], Daniel Cale[1,2], and Joao C. Ferreira[1,2,3,4(✉)]

[1] Iscte - University Institute of Lisbon, Lisbon, Portugal
{adriana_luis_franco,ddces}@iscte-iul.pt, jcafa@iscte.pt
[2] ISTAR-IUL, Information Sciences and Technologies and Architecture Research Center, Lisbon, Portugal
[3] Inov Inesc Inovação—Instituto de Novas Tecnologias, Lisbon, Portugal
[4] Faculty of Logistics, Molde University College, NO-6410 Molde, Norway

Abstract. Implementing a gamification system aimed at motivating drivers to adopt eco-friendly driving behaviors and subsequently achieve fuel savings. The growing concern for environmental sustainability and the imperative to reduce carbon emissions from vehicular activities have led to an increased interest in eco-driving practices. Gamification, leveraging elements from game design in non-game contexts, offers a promising avenue for fostering behavior change. This research presents a comprehensive framework for developing a gamification system that integrates game mechanics to encourage and incentivize drivers to engage in fuel-efficient driving practices.

The study employs a mixed-methods approach, combining data collection from vehicle sensors and user feedback through user testing. The gamification system is meticulously designed, incorporating diverse game mechanics such as challenges, rewards, leaderboards, and progress tracking. By analyzing data on driving habits and fuel consumption, the effectiveness of the system is evaluated in terms of fuel savings and driving behavior modifications. The feedback from participants provides insights into user engagement, motivational factors, and perceptions of the gamified eco-driving experience.

The findings indicate that the gamification system significantly influences driver behavior, resulting in reduced fuel consumption and heightened motivation to practice eco-driving techniques. The integration of real-time feedback and competitive elements resonates positively with users, stimulating sustained interest and participation. The study demonstrates the potential of gamification as an impactful tool for fostering eco-friendly behaviors and contributing to the broader goals of sustainable transportation and environmental preservation.

Through this research, this investigation contributes to the fields of gamification, eco-driving, and sustainable transportation by offering a tangible solution for encouraging positive behavior changes among drivers. The insights gained from the study provide a foundation for future endeavors in designing effective gamification systems that address environmental concerns and promote sustainable practices in various domains.

V. Sakalauskas et al. (Eds.): IBICA 2023, LNNS 1231, pp. 129–137, 2025.
https://doi.org/10.1007/978-3-031-78946-5_12

Keywords: Eco-driving · gamification · sustainable transportation · behavior change · fuel savings · game mechanics · motivation · user engagement

1 Introduction

The advancement of technology has created the necessity to improve driver efficiency. Jointly with the company Santos e Vale, it detains its "own distribution network with more than 500 vehicles and more than 1,200 employees spread across 21 operational platforms, which coordinate more than 8,500 deliveries daily" [1] They demonstrate keen interest in the matter, so they create an environment where they clarify the challenges at hand and delineate areas where our assistance can be of value.

The importance of eco-driving in reducing carbon emissions is one of the topics that are mostly approached in the daily conversation nowadays [2] and is "relatively low-cost and immediate, and the improvement in fuel efficiency can be up to 45%" [3]. "The transport sector consumes about 20% of global energy and is responsible for nearly 25% of global energy-related CO2 emissions, 75% of which are emitted by road transport." [3].

With the intervention of the European Union, they aim to achieve a reduction of 55% in domestic greenhouse gas emissions by 2030 compared to 1990. Furthermore, their goal is to achieve climate neutrality by 2050.[4] The central focus of this project is the impact of the transportation sector, as exemplified in Fig. 1.

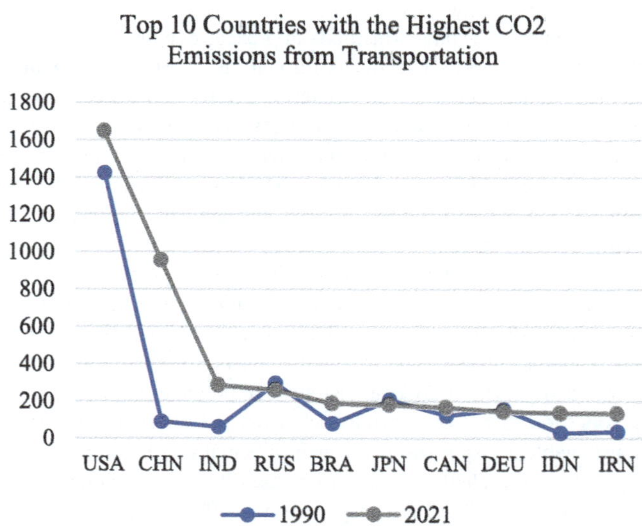

Fig. 1. The Ranking of the Transportation Sector CO2 Emissions produced annually by Country's Top 10. Units: Values in are expressed in Mt CO2/yr (megatons of carbon dioxide per year) [4].

"Driving behavior alone can have a large beneficial impact on fuel consumption." According to Lárusdóttir and Ulfarsson, the Clean Urban Transport for Europe (CUTE)

project believes that there is a connection, between fuel usage and the average speed of vehicles [5].

Other characteristics and situations that can also give rise to the expense are "driving behavior (or driving style), vehicle weight, pavement structure, pavement temperature, hybridization, and regenerative braking" [5].

Based on an analysis of data extracted from 67 diverse models spanning the years 1999 to 2002, the weight has an effect, in this study. The results suggest that all vehicles experience a body weight decrease of 16% in fuel consumption. This finding emphasizes the importance of weight, as a factor influencing fuel efficiency within the context of this analysis [5].

According to the data and analysis mentioned earlier, the company has identified that vehicles G7 and G8 have the fuel consumption.

The consumption of fuel, that led to the spending of CO2 emission originates with the "driving behavior but can also be influenced indirectly by several local measures, e.g., phased traffic lights." [6].

Truck drivers have a very demanding occupation with constant unrelenting pressure, with tight schedules dictated by their superiors. These schedules normally are made with short deadlines. The pressure and responsibility that this profession requires are high and can be amplified when the drivers are transporting dangerous goods [7].

According to Google Trends, the term "Gamification" has become increasingly significant and popular across industries since 2010 and it remains relevant even today, as explained in Fig. 2.

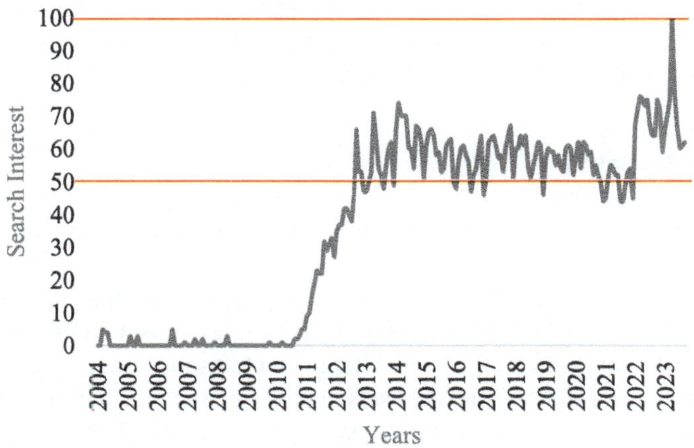

Fig. 2. Search Interest at "Gamification". Source: Google Trends. Legend: "Numbers represent search interest relative to the highest point on the chart for the given region and time. A value of 100 is the peak popularity for the term. A value of 50 means that the term is half as popular. A score of 0 means there was not enough data for this term" [11].

According to Magana and Monuz-Organero [8], drivers should follow the advice that the "driver must not drive at high speed, avoid braking or accelerating sharply, avoid braking or accelerating unnecessarily, the user should drive at steady speed and engine speed should be low." These recommendations should lead to a decrease in CO2 emissions.

Also, they added that a gamification system starts with a level of conformity with suggested driving advices, so the application will indicate what the users are as drivers and should do with actual driving style to improve, this would be applied with a score [8].

Cooperatively with the company they communicate that the problem is that drivers don't have access to information, and they want to give them access to their consumption, braking, sudden acceleration, "ralenti" is when the car is stationary but switched on, and any other relevant aspects that may have an impact on the vehicle.

In collaboration with the company, it is conveyed that the prevailing issue pertains to the lack of information accessibility among drivers. The objective is to provide drivers with access to pertinent data such as fuel consumption, braking patterns, instances of sudden acceleration, as well as the state of "relantim" (i.e. when the vehicle is stationary, but the engine remains operational). These aspects are significant as they can influence the vehicle's overall performance.

The primary objectives encompass enhancing road safety, constructing a segment of an interactive and user-friendly mobile application in collaboration with the company, devising a plan to mitigate carbon emissions, harnessing the potential of data analytics, and implementing machine learning algorithms to acquire a deeper understanding of driving patterns.

These insightful observations contribute to the identification of strategies for enhancing efficiency and safety through data-driven decision-making.

Moreover, a crucial aim is to establish and sustain a transparent mode of communication between drivers and stakeholders. This can be accomplished by employing a proficient gamification system for eco-driving.

2 Literature Review

Eco-driving covers a range of factors and areas that can be improved to achieve efficiency results.

Based on the thesis made by Minh Céline [9], she created an application that "collects and processes data regarding the user's travels and energy consumption" and developed "a smart decision support tool for house energy and transport usage" [9].

The main goal was to motivate the users by highlighting how their lifestyle changes can impact their health, carbon footprint, and energy consumption. This was achieved by assessing the energy usage related to the user's home and vehicle [9].

Initially, a workshop addressed several important topics, including the UI design and attractive visual of the application's interface. In which the goal was to "create awareness about energy consumption" such as carbon footprint reduction. Furthermore, was investigated how to approach the rewards system where the physical reward is a must, and to users who accomplish substantial decreases in their emissions, for example, the "discount on an electric car, a free bus ticket or a lock for a bike." [9] This project was made to operate in both real-time and historical data [9].

About data and carbon emissions, her objectives were that the app had to display information regarding the "approximation of the user's emissions and expenses due to the energy use at home and transportation during a day", "the number of km traveled by the user during a day" and "sensitize the user about the effect of gas emissions." The language of her choosing was Python, for the backend calculations [9].

This approach to carbon emissions introduces a new perspective into the challenge of the gamification of eco-diving and adds a considerable layer of knowledge.

Another thesis that approaches gamification, however related to public transport, so Olofsson [10], analyzed and set out to investigate whether gamification can be successfully applied to the GoOn public transport app to increase user engagement and usage. This initiative resulted in a drop in the use of public transportation during the pandemic.

It begins by addressing whether results can be achieved by implementing the application with the addition of game elements, such as achievements, leaderboards, and other graphic components that can make the experience of using the application more engaging and fun [10].

Gamification framework can be described as [10], the initials GAME: "Gather - Gather information by asking questions. What are you gamifying? Who are your users? Why are you doing it? How do you measure success?; Act - Act on the information you have. Design the best solution for your goals and the engagement/experience of your users. Then test it with the users.; Measure - Measure user activity and goal outcomes. Get feedback and iterate the improvements.; Enrich - Enrich your system over time, people change as do their goals. Keep up with them or one step ahead [10]."

The author states that there are two main types of users: intrinsic and extrinsic. The former include socializers (motivated by social interaction), free spirits (looking for opportunities and authenticity), achievers (overcoming personal challenges), and philanthropists (helping without expecting rewards), while the second type of users includes authors (contributing for rewards), consumers (looking for rewards for existing actions), connectors (looking for useful contacts) and exploiters (finding loopholes in the system for personal gain) [10].

The gamification elements and system structure for developing the application involve a proof of concept, including levels and achievements, with the objectives for the different levels and the unlocking of achievements based on users' actions. The author used tools such as ReactJS, which is a JavaScript library for developing the user interface, this part was destined for the front-end; Amazon S3 was also used, which is an Amazon cloud storage system and finally; and Linux, to configure and run where the app would be hosted [10].

3 Developing Framework

To encourage the approach to achieving the benefits of eco-driving, and reducing the environmental impact of vehicular activities, by developing a gamification system which appears to be a promising alternative for behavioral transformation.

The purpose of the theoretical framework is to delve into the theoretical foundation of your eco-driving gamification system, directly connected to transportation. These objectives are resourceful and multifaceted, the work of Minh Céline [9], says that she developed an application that the main is to motivate users and improve their lifestyle, this point of view can connect with the angle of the company transmitted by helping drivers gain motivation through rewards and as they make better choices as drivers (Fig. 3).

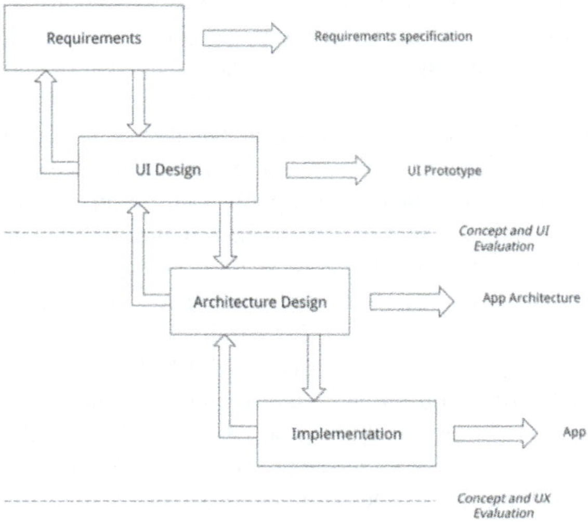

Fig. 3. Development process [9].

Also, Minh Céline [9] emphasizes the connotation that the historical data and in real-time can create impact by allowing the users to understand their emissions and expenses, this is a certainly worthy initiative that can help the drivers of the company gain motivation so can they improve daily. With this approach the users can increase by having a more fostering eco-conscious behavior.

The mechanics of the gamification system is a crucial part for the future change of behavior, the important elements are points, badges, leaderboards, challenges, and rewards are integrated on the system. As Olofsson, highlights on this work by finding information about the power of the elements. His research with conquers and classifications led to the increase of the user involvement, this can make the experience of the drivers more engaging and evolutionary for themselves.

The design principles that make the system whole are fundamental like creating an interface and functionalities that can relate to the users and can make them to adopt more eco-driving habits. Olofsson [10] introduced the structure "GAME" that highlights the importance of collecting information, act on it, track the activity of the user and continuously enrich the system, so they can align with the objectives of the user. This analysis of the users illuminates the possibilities of motivation that them can be involved on the elements of the gamification since social interaction and personal fulfillment. This can lead to the different variety of the users that drive the vehicles and how we can understand them on a deeper level.

This chapter acts as a bridge between the theoretical and practical, to highlight the details that lead to a critical initiative on the part of the components so that they can be motivated to adopt more ecological behaviors as drivers and that this results in a reduction in fuel consumption and its respective environmental impact.

The development of a gamification system for eco-friendly defensive driving hinges on various critical technical aspects. These parameters are pivotal for its successful design and execution. User data tracking is the cornerstone, involving real-time monitoring of driving habits, fuel consumption, and emissions data. This data forms the basis for personalized challenges and rewards aimed at encouraging eco-conscious driving. A point system is pivotal, offering rewards for eco-friendly actions such as fuel efficiency, smooth acceleration, and adherence to speed limits. These points motivate users and can unlock rewards or enhance their leaderboard standings.

The ranking system recognizes and rewards users for reaching milestones and maintaining eco-friendly driving habits, fostering a sense of accomplishment and recognition. Leaderboards foster competition and improvement by displaying user rankings, and motivating users to enhance their driving habits to climb the leaderboard. Challenges play a central role, engaging users with tailored eco-, driving challenges. Completing these challenges earns users additional rewards and sustains their motivation. An effective rewards system offers tangible incentives like fuel discounts, car maintenance services, and eco-friendly products, motivating users to maintain eco-conscious behavior. Data analysis and insights provide users with information on their driving habits, emissions, and expenses, motivating them to make eco-conscious choices.

These technical parameters collectively create a gamification system that effectively motivates users to adopt eco-friendly driving habits, reducing fuel consumption and environmental impact. In the forthcoming application, an exemplary dashboard has been constructed to showcase pertinent data pertaining to working hours, performance scores, and key indicators for enhancement. It has been determined that these aspects hold the utmost significance in fostering the growth and progress of drivers, as elucidated in Fig. 4.

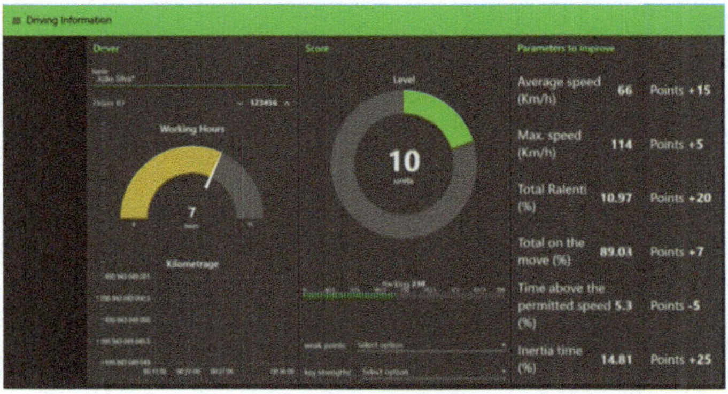

Fig. 4. Example mockup of the application to be developed.

4 Conclusion

In conclusion, the implementation of a gamification system designed to incentivize drivers to embrace environmentally conscious driving habits constitutes a noteworthy advancement in tackling the urgent issues surrounding ecological sustainability and curtailing carbon emissions resulting from automotive operations. The results of this study have shed light on the potential of gamification, which involves incorporating game design elements into non-game settings, as a potent driver of behavior modification.

By employing a rigorous mixed-methods methodology, which incorporates collecting data from vehicle sensors and acquiring valuable user feedback through user testing, we will formulate a comprehensive framework for creating a proficient gamification system. The system will incorporate a wide range of game mechanics, such as challenges, rewards, leaderboards, and progress tracking, to motivate and incentivize drivers to actively participate in fuel-efficient driving practices.

The discernible effect of implementing this gamification system on driver behavior is apparent, as there is a noticeable decrease in fuel consumption and an increased inclination to adopt environmentally conscious driving techniques. The integration of real-time feedback and competitive elements has been well received by users, resulting in the cultivation of enduring interest and active engagement.

This study not only highlights the potential of gamification as a dynamic tool for promoting eco-friendly behaviors but also aligns with the broader goals of sustainable transportation and environmental conservation. This intervention provides a tangible resolution to promote favorable modifications in driver behavior, making a substantial contribution to the domains of gamification, eco-driving, and sustainable transportation.

The implementation of gamification in the context of eco-driving practices holds significant potential for fostering a sustainable and environmentally aware future within the realm of transportation.

References

1. Santos e Vale, "SANTOSEVALE - Logística, Distribuição e Transporte," 2021
2. Harvey, J., Thorpe, N., Fairchild, R.: Attitudes towards and perceptions of eco-driving and the role of feedback systems. Ergonomics **56**(3), 507–521 (2013). https://doi.org/10.1080/00140139.2012.751460
3. Huang, Y., Ng, E.C.Y., Zhou, J.L., Surawski, N.C., Chan, E.F.C., Hong, G.: Eco-driving technology for sustainable road transport: a review. Renew. Sustain. Energy Rev. **93**, 596–609 (2018). https://doi.org/10.1016/j.rser.2018.05.030
4. Crippa, M., et al.: CO2 emissions of all world countries. JRC Science for Policy Report, European Commission, EUR, vol. 31182, 2022
5. Lárusdóttir, E.B., Ulfarsson, G.F.: Effect of driving behavior and vehicle characteristics on energy consumption of road vehicles running on alternative energy sources. Int. J. Sustain. Transp. **9**(8), 592–601 (2015). https://doi.org/10.1080/15568318.2013.843737
6. Van Mierlo, J., Maggetto, G., Van De Burgwal, E., Gense, R.: Driving style and traffic measures - Influence on vehicle emissions and fuel consumption. Proc. Inst. Mech. Eng. Part D: J. Automob. Eng. **218**(1), 43–50 (2004). https://doi.org/10.1243/095440704322829155

7. Pritchard, E., et al.: Impact of work and coping factors on mental health: Australian truck drivers' perspective. BMC Public Health **23**(1), 1090 (2023). https://doi.org/10.1186/s12889-023-15877-4
8. Magana, V.C., Munoz-Organero, M.: GAFU: using a gamification tool to save fuel. IEEE Intell. Transp. Syst. Mag. **7**(2), 58–70 (2015)
9. Minh, C.: A pervasive app for a better management of energy and transportation-Gamification to link activity, carbon footprint, energy consumption and commuting patterns to achieve positive lifestyle changes, Norwegian University of Science and Technology (NTNU), 2017
10. Olofsson, H.: Gamification for public transportation-Improving attractiveness with interactive game elements. 2021
11. Google, "Google Trends: Gamification," 2023. Accessed 22 Sep 2023. https://trends.google.com/trends/explore?date=all&q=Gamification&hl=en

Optimizing Healthcare Analytics: A Zero-Inflated Poisson Approach to Pediatric Emergency Room Visits

Nadashree Bose and Hemlata Joshi(✉)

Department of Statistics and Data Science, CHRIST (Deemed to be University), Bengaluru, India
hemlata.joshi28@gmail.com

Abstract. In various fields, the modeling of count data holds significant importance. The Poisson regression model is a commonly utilized tool for this purpose. However, this model assumes that the data has uniform dispersion, a condition often not met in real-world observations. The nature of overdispersion can vary depending on the specific context. When the overdispersion is primarily due to an excessive number of zero counts, the Zero-inflated Poisson regression model becomes a more suitable choice for modeling count data. The paper commences by offering a summary of the theoretical foundations of both Poisson regression and Zero-inflated Poisson regression. To evaluate their performance, use the Mean-Squared error (MSE) as a comparative metric. Next, apply these models to analyze the frequency of hospital emergency room visits by children between 10–18 years of age. The overdispersion of the visit count in our dataset is mostly caused by the excessive occurrence of zero counts. The findings demonstrate that the Zero-inflated Poisson regression model outperforms the standard Poisson regression model in terms of MSE.

Keywords: Poisson model · Zero-inflated Poisson model · Count data · Excess of zeroes · Mean-Squared error · Emergency room visit

1 Introduction

A generalized linear regression model (or GLM) formulated by [1] is made up of three components:

1. A random component, that defines the conditional distribution of the dependent variable Y_i, within the exponential family, such as the Binomial, Gamma, Poisson, Gaussian, or inverse-Gaussian families of distributions, given the values of the independent variables in the model.
2. A linear predictor - which is a linear function of explanatory variables,

$$\psi_i = \alpha + \beta_1 X_{i1} + \beta_2 X_{i2} + \ldots + \beta_j X_{ij} \tag{1}$$

V. Sakalauskas et al. (Eds.): IBICA 2023, LNNS 1231, pp. 138–148, 2025.
https://doi.org/10.1007/978-3-031-78946-5_13

3. Link function that can be inverted $g(\cdot)$, which establishes a connection between the mean of the response variable, $\theta_i = E(Y_i)$, to the linear predictor:

$$g(\theta_i) = \psi_i = \alpha + \beta_1 X_{i1} + \beta_2 X_{i2} + \ldots + \beta_j X_{ij} \tag{2}$$

The assumptions of a generalized linear model are as follows:

1. The response variable and the explanatory variables are not assumed to have a linear relationship in a generalized linear model. Instead, it presumes a linear relationship between the transformed expected response variable, as specified by the link function, and the independent variables.
2. There should be no multicollinearity among the explanatory variables.
3. Auto-correlation among the residuals should be absent.

Analyzing count data is a central concern in public health research, typically addressed through Poisson regression models. Poisson regression model is a type of generalised linear model where the response variable represents counts and it follows poisson distribution. These models assume that the mean and variance are equal, but this assumption is frequently violated in practical situations due to the turning up of numerous zeros or overdispersion. Overdispersion mainly occurs when the variance of the response variable exceeds its mean. When using the standard Poisson regression model in such cases, the variance of estimated coefficients tends to be underestimated.

To tackle this issue, this study adopts the ZIP model as the preferred count model. The ZIP model by [2] proves more suitable than the standard Poisson regression when dealing with count data exhibiting overdispersion or when there's an unexplained excess of zero counts. This model has the advantage of categorizing the population into two groups: one with zero outcomes and the other with non-zero outcomes. In the context of ZIP regression, the zero outcomes are modeled by including a proportion (π) of additional zeroes and a proportion $(1 - \pi)$ of zeroes generated from a Poisson distribution.

This paper applies the Poisson regression model and the ZIP regression model to a dataset in the public health sector with an emphasis on the applications of these models in that field. Coefficient estimates in both scenarios is calculated and then the outcomes are compared using the MSE criterion. Our primary focus is to shed light on how these two methods differ in handling count data with an abundance of zero values.

The rest of the paper is organized as follows: In Sect. 2, the methodology is outlined, defining the Poisson Regression Model and the ZIP Regression Model, along with their respective maximum likelihood estimation procedures. We also introduce the MSE function in this section. Section 3 provides a practical application of these models in the context of public health, including a detailed description of the dataset, data analysis, and the presentation of results and significant findings. In Sect. 4, the paper is concluded, with an emphasis on the significance of applying such regression models in the specified domain. Finally, in Sect. 5, potential future research directions are explored, and suggestions for further investigation are offered.

2 Methodology

A generalized linear model (GLM) in statistics is an adaptable generalization of conventional linear regression. The GLM generalizes linear regression by enabling each measurement's variance to be a function of its predicted value and by allowing the linear model to be connected to the response variable via a link function.

In order to combine many other statistical models, such as linear regression, logistic regression, and Poisson regression, [1] created generalized linear models. They suggested an iteratively reweighted least squares approach for maximum likelihood estimation (MLE) of the model parameters.

Each result of the dependent variables Y is considered in a GLM to be produced from a specific distribution in the exponential family that includes the normal, binomial, Poisson, and gamma distributions. The independent variables, X, influence the distribution's mean, θ , in the following ways:

$$\mathrm{E}(Y|X) = \theta = g^{-1}(X\beta) \tag{3}$$

where $\mathrm{E}(Y|X)$ is the expected value of Y conditional on X; $X\beta$ is the linear predictor, a linear combination of unknown parameters β; g is the link function.

This approach assumes that the variance V is normally a function of the mean given by:

$$\mathrm{Var}(Y|X) = \mathrm{V}(g^{-1}(X\beta)) \tag{4}$$

Usually, Bayesian, maximum quasi-likelihood, or maximum likelihood estimation techniques are used to estimate the unknown parameters, β.

2.1 Poisson Regression

Poisson regression is a common method for modeling data for counts. It operates ψ the assumption that the response variable follows a Poisson distribution, and the expected value's logarithm can be expressed as a linear combination of unknown parameters. Let y_i represent the response variable. It is assumed that y_i adheres to a Poisson distribution with a mean θ_i, determined by a function of covariates x_i. Consequently, the Poisson regression takes the form:

$$P(y_i) = \frac{e^{-\theta_i}\theta_i^{y_i}}{y_i!}; y_i = 0, 1, 2, \ldots, \theta_i > 0 \tag{5}$$

with mean and variance, $E(y_i) = V(y_i) = \theta_i$. Here, θ_i is the conditional mean, defined as $\theta_i = E(y_i|x_i) = e^{x_i^T\gamma}$. The vector $x_i = (xi_1, xi_2, \ldots, xi_k)$ comprises the covariates, while $\gamma = (\gamma_1, \gamma_2, \ldots, \gamma_k)$ represents the vector of parameters which are unknown. The value of k specifies the dimension of the covariate vector within the model.

In order to estimate the Poisson regression's parameters, maximum likelihood approaches can be employed. Assuming that the observations $(y_i|x_i)$ are independent, the log-likelihood function is used to estimate the parameter vector γ is formulated as:

$$\ln L(\gamma) = \sum_{i=1}^{n} \left[y_i x_i^T \gamma - e^{x_i^T \gamma} - \ln(y_i!) \right] \tag{6}$$

By Fisher Scoring method the Poisson maximum likelihood estimator of γ is obtained as:

$$\hat{\gamma}_{PMLE} = (X'\hat{W}X)^{-1} X'\hat{W}\hat{z} \tag{7}$$

where, $\hat{W} = diag[\hat{\theta}_i]$ and $\hat{z}$ is a vector, with the ith element given by $\hat{z}_i = \log(\hat{\theta}_i) + \frac{y_i - \hat{\theta}_i}{\theta_i}$.

An essential characteristic of the Poisson distribution is that the conditional variance of y_i equals λ_i. This property is known as equidispersion. When the variance of y_i exceeds λ_i, it indicates the presence of overdispersion.

2.2 Zero-Inflated Poisson Regression

For modeling zero-inflation in count data, the zero-inflated Poisson (ZIP) model was proposed by [2]. For count data containing extra zeros, the ZIP model can be thought of as a mixing model. There are two categories of zeros that can be identified in the count data for the ZIP model. The first one, known as structural zeros, belongs to a non-susceptible group. The structural zero happens with a π chance. Random zeros are the second type of zeros in the count data for the ZIP model that come from a vulnerable group. The random zero has a Poisson distribution and occurs with probability $(1 - \pi)$. ZIP regression serves as a suitable model for count data marked by overdispersion and an abundance of zero values. This distribution blends characteristics of both the Poisson and logit distributions. The permissible values for the variable Y encompass non-negative integers, namely 0, 1, 2, 3, and so forth.

Each observation has two probable instances. In scenario 1, there is no count. However, if instance 2 occurs, a Poisson model is used to create counts (including zeros). Let's say that case 1 happens with probability π and case 2 happens with probability $(1-\pi)$. Consequently, the following is a textual representation of the probability distribution of the Zero-inflated Poisson (ZIP) random variable y_i,

$$P(y_i = j) = \begin{cases} \pi_i + (1 - \pi_i)e^{-\theta_i} & \text{if } j = 0 \\ (1 - \pi_i)\frac{\theta_i^{y_i} e^{-\theta_i}}{y_i!} & \text{if } j > 0 \end{cases} \tag{8}$$

where π_i is described as the logistic link function below.

The Poisson component can include the exposure duration t, and the k regressor variables (the x's). These quantities are expressed as follows:

$$\theta_i = \exp(\ln(t_i) + \gamma_1 x_{1i} + \gamma_2 x_{2i} + \ldots + \gamma_k x_{ki}) \tag{9}$$

In many cases, $x_1 \equiv 1$, in which case γ_1 is referred to as the intercept. The unknown parameters that make up the regression coefficients $\gamma_1, \gamma_2, \ldots, \gamma_k$ are calculated from a set of data. Their estimates are represented by $b_1, b_2, \ldots, b_k$. This logistic link function π_i is provided by-

$$\pi_i = \frac{\lambda_i}{1 + \lambda_i} \tag{10}$$

where,

$$\lambda_i = \exp\left(\ln(t_i) + \tau_1 z_{1_i} + \tau_2 z_{2_i} + \ldots + \tau_m z_{m_i}\right)$$

The logistic component consists of a collection of m regressor variables (the z's) and an exposure duration t. The most common method of estimating the coefficients of ZIPRM is to use the maximum likelihood method.

The likelihood function's logarithm is -

$$\mathcal{L}(\gamma; \tau) = \mathcal{L}_1 + \mathcal{L}_2 - \mathcal{L}_3 \tag{11}$$

where,

$$\mathcal{L}_1 = \sum_{i:y_i=0} \ln\left[\lambda_i + e^{-\theta_i}\right]$$

$$\mathcal{L}_2 = \sum_{i:y_i>0} (y_i \ln(\theta_i) - \theta_i - \ln(y_i!))$$

$$\mathcal{L}_3 = \sum_{i=1}^{n} \ln(1 + \lambda_i)$$

The slope of $\mathcal{L}$ is given by:

$$\frac{\partial \mathcal{L}}{\partial \gamma_r} = \sum_{i:y_i=0} \left(\frac{-x_{ir}\theta_i}{\lambda_i \exp(\theta_i) + 1}\right) + \sum_{i:y_i>0} (y_i - \theta_i)\, x_{ir}, \text{ for } r = 1, 2, \ldots, k \tag{12}$$

$$\frac{\partial \mathcal{L}}{\partial \tau_r} = \sum_{i:y_i=0} \left(\frac{z_{ir}\lambda_i \exp(\theta_i)}{\lambda_i \exp(\theta_i) + 1}\right) - \sum_{i=1}^{n} \frac{\lambda_i}{1 + \lambda_i} z_{ir}, \text{ for } r = 1, 2, \ldots, m \tag{13}$$

The second derivatives are:

$$\frac{\partial^2 \mathcal{L}}{\partial \gamma_r \partial \gamma_s} = \sum_{i:y_i=0} \left(\frac{x_{ir}x_{is}\theta_i((\theta_i - 1)\lambda_i \exp(\theta_i) - 1)}{(\lambda_i \exp(\theta_i) + 1)^2}\right) - \sum_{i:y_i>0} (\theta_i x_{ir} x_{is}), \text{ for } r, s = 1, 2, \ldots, k \tag{14}$$

$$\frac{\partial^2 \mathcal{L}}{\partial \tau_r \partial \tau_s} = \sum_{i:y_i=0} \left(\frac{z_{ir} z_{is} \lambda_i \exp(\theta_i)}{(\lambda_i \exp(\theta_i) + 1)^2} \right) - \sum_{i=1}^{n} \frac{z_{ir} z_{is} \lambda_i}{(\lambda_i + 1)^2}, \text{ for } r, s = 1, 2, \ldots, m \quad (15)$$

$$\frac{\partial^2 \mathcal{L}}{\partial \gamma_r \partial \tau_s} = \sum_{i:y_i=0} \frac{x_{ir} z_{is} \lambda_i \theta_i \exp(\theta_i)}{(\lambda_i \exp(\theta_i) + 1)^2}, \text{ for } r = 1, 2, \ldots, k; s = 1, 2, \ldots, m \quad (16)$$

Using the simplex approach, this complex likelihood function is estimated. The final step of the estimated coefficients is defined as -

$$\hat{\gamma}_{ZIPMLE} = (X^T \hat{W} X)^{-1} X^T \hat{W} \hat{v} \quad (17)$$

where, W = diag$[\hat{\theta}_i]$ and $\hat{v}$ is a vector where the ith element equals to $\hat{v_i} = ln(\hat{\theta}_i) + \frac{y_i - \hat{\theta}_i}{\hat{\theta}_i}$.

The general form of Mean-Squared error(MSE) for the Poisson MLE and ZIP MLE is defined as:

$$MSE(\hat{\gamma}_{PMLE}) = MSE(\hat{\gamma}_{ZIPMLE}) = tr[(X'\hat{W}X)^{-1}] = \sum_{k=1}^{p} \frac{1}{\sigma_k} \quad (18)$$

where, σ_k is the eigen value of the $(X^T \hat{W} X)$ matrix.

3 Empirical Application

With the aid of actual data, both models' applications are demonstrated in this section. The drawbacks of the Poisson Regression Model when the regressand variable has a large number of zero counts are then discussed.

The Emergency Room Dataset sourced from the NHIS 2022 (U.S.) database is examined in this study, with the focal point being the count of emergency room visits made by children between the age of 10–18 years within the preceding 12 months. This dataset encompasses a total of 13 variables, with 12 covariates alongside the target variable, all of which are comprehensively detailed in Table 1. The data consists of 495 samples of children from the age group of 10–18 years.

The pertinent question here is how frequently a child has visited a hospital emergency room for concerns about their health over the course of the previous 12 months. 61% of them are male, and 39% are female. The minimum and maximum heights are determined to be 43 and 96 in., respectively. The range of weights is 50 pounds at the minimum and 996 pounds at the maximum. Between the ages of 10 and 18, hospital emergency room visits by children have a mean of 0.3575 and a standard deviation of 0.9548. It is evident that the variance is significantly larger than the mean. This shows that the data is overdispersed. Also, in Fig. 1 the distribution of emergency room visits is shown, where the excess of zeros is clearly visible. A Poisson regression model is typically not the best model to fit the data in these scenarios.

Table 1. Variable description for the Emergency Room Dataset

Variable Name	Variable Type	Variable Description
Visits	Discrete	Emergency Room Visits by Children below age 18 (past 12 months)
Sex	Binary	Gender of the Child (1 if Male, 2 if Female)
Age	Discrete	Age of the Child (between 10–18 years)
PHstat	Categorical	General Health Status (1 if Excellent, 2 if Very Good, 3 if Good, 4 if Fair, 5 if Poor)
Height	Discrete	Height of the Child (in Inches)
Weight	Discrete	Weight of the Child (in Pounds and without shoes)
BMI	Categorical	Body Mass Index of the Child (1 if Underweight, 2 if Healthy Weight, 3 if Overweight, 4 if Obese, 9 if Unknown)
Medication	Categorical	Child needed Prescription Medication but did not get it due to cost (past 12 months) (1 if Yes, 2 if No, 8 if Not Ascertained)
Traffic	Categorical	Traffic causes Safety Issues in the Neighborhood (1 if Yes, 2 if No, 8 if Not Ascertained, 9 if Don't Know)
Behaviour	Categorical	Difficulty controlling Behaviour of the Child (1 if No Difficulty, 2 if Some Difficulty, 3 if a lot of Difficulty, 4 if Can't do it at all)
Dental	Categorical	Time since last Dental Exam/Cleaning (0 if Never, 1 if Within the past year, 2 if Within the last 2 years, 3 if Within the last 3 years, 4 if Within the last 5 years, 5 if Within the last 10 years, 8 if Not Ascertained, 9 if Don't Know)
Asthma	Binary	Asthma Emergency Room Visit by the Child (past 12 months)
Diabetes	Categorical	Child ever had Diabetes or not (1 if Yes, 2 if No, 9 if Don't Know)

The 12 independent variables are encoded and the regression is first performed with the new 39 variables. The Shapiro-Wilk test is performed and the p-value is obtained, which is nearly equal to zero. This shows that the residuals are not normally distributed. Hence, a GLM is a better fit here. Among the 39 variables, 9 variables are chosen based on their significance. Those variables are, "medication.1", "medication.2", "dental.1", "dental.2", "dental.3", "dental.4", "dental.8", "asthma.1" and "diabetes.1". The significance of these variables are assessed with the help of Lagrange Multiplier Test or the Likelihood Ratio Test, which gives a p-value of 0.9975 (>0.05). The same variables are taken and fitted into both Poisson regression model and ZIP regression model and the summary is shown in Table 2 and 3. All the variables for the Poisson regression model and

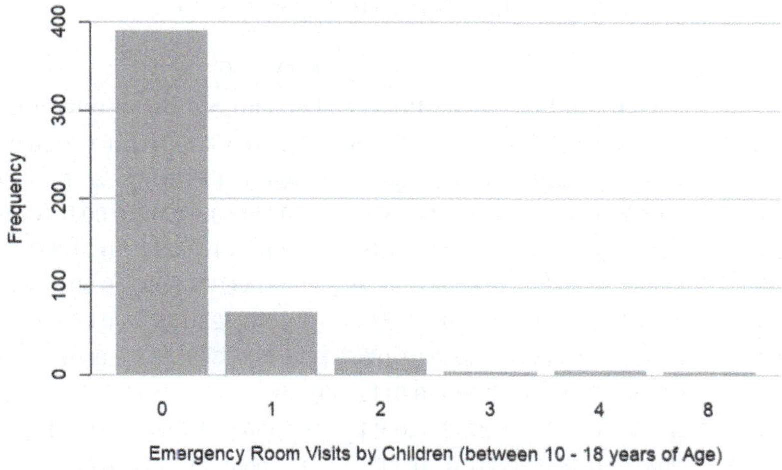

Fig. 1. Frequency Distribution of Emergency Room Visits by Children(between 10–18 years of age)

the count part of the ZIP regression model are significant based on the p-value, as it is less than 0.05.

Table 2. Estimation of Poisson Regression Model

	Estimate	Std.Err.	z value	Pr(>|z|)
(Intercept)	3.539637	0.621023	5.699685	<2e-16
medication.1	–3.32571	0.528658	–6.29087	<2e-16
medication.2	–3.53964	0.368335	–9.60983	<2e-16
dental.1	–1.4602	0.510558	–2.86	0.004236
dental.2	–1.89712	0.645497	–2.93901	0.003293
dental.3	–1.86835	0.766793	–2.43658	0.014827
dental.4	–2.203	0.878014	–2.50907	0.012105
dental.8	–1.4602	0.65371	–2.23371	0.025502
asthma.1	1.86142	0.205768	9.04621	<2e-16
diabetes.1	2.376486	0.458989	5.177655	<2e-16

The performance of the models are evaluated in terms of Mean-Squared Error(MSE) and Akaike Information Criteria(AIC) value. The results are shown in Table 4.

The findings presented in Table 4 unequivocally demonstrate that the ZIP Regression Model outperforms the Poisson Regression Model, as evidenced by its lower MSE. Furthermore, the AIC for the ZIP Regression Model is also notably

Table 3. Estimation of ZIP Regression Model

	Pr(Y > 0)				Pr(Y = 0)			
	Estimate	Std.Err.	z-value	Pr(> $\|z\|$)	Estimate	Std.Err.	z-value	Pr(>$\|z\|$)
(Intercept)	3.8255	0.6567	5.8254	<2e-16	-24.851	301404.2	-0.0001	0.9999
medication.1	-2.8801	0.5229	-5.5079	<2e-16	-1.6684	626101.5	<2e-16	0.9999
medication.2	-2.4592	0.3996	-6.1534	<2e-16	25.9229	301404.2	0.0001	0.9999
dental.1	-1.746	0.5534	-3.1551	0.0016	-0.381	1.1923	-0.3196	0.7493
dental.2	-3.2634	0.662	-4.9299	<2e-16	-16.8143	3714.999	-0.0045	0.9964
dental.3	-2.5735	1.1341	-2.2692	0.0233	-1.4576	2.9148	-0.5001	0.617
dental.4	-2.1969	0.9143	-2.4029	0.0163	43.6785	548883.5	0.0001	0.9999
dental.8	-1.746	0.6877	-2.539	0.0111	0.5307	320948.8	<2e-16	0.9999
asthma.1	0.8306	0.2537	3.2742	0.0011	-60.4554	548886.5	-0.0001	0.9999
diabetes.1	1.296	0.4845	2.6751	0.0075	-17.1966	2714.247	-0.0063	0.9949

Table 4. Performance Metrics of Poisson and ZIP regression Models

	MSE	*AIC*
Poisson	5.592646	677.3876
ZIP	0.387683	661.0843

lower, underlining its superior fit to the data. Hence, in the presence of overdispersion in the count dependent variable, a ZIP regression model better explains the overabundance of zeros than the general Poisson regression model.

The count part of the zero inflated poisson regression model is given by:

$$\begin{aligned}\text{Poisson with log link } \ln(\theta) =& log(3.8255) - 2.8801(medication.1) - 2.4592(medication.2)\\ &- 1.7460(dental.1) - 3.2634(dental.2) - 2.5735(dental.3)\\ &- 2.1969(dental.4) - 1.7460(dental.8) + 0.8306(asthma.1)\\ &+ 1.2960(diabetes.1)\end{aligned} \tag{19}$$

Asthma.1 and diabetes.1 have positive signs, indicating that a positive change in these factors induces an increase in the number of visits to the emergency room by children of age 10–18 years. The number of positive signs the same with the Poisson regression model. However, there are differences in the percentage changes. The percentage change in medication.2 for the Poisson regression model is –97%, and for the ZIP regression model is –91%, indicating those who could afford medication did not have to visit the emergency room as often than those who haven't. Similarly, in asthma.1 and diabetes.1 the percentage of change for Poisson regression model and ZIP regression model are 543%, 976%, and 129%, 265%, respectively. It indicates those who had visited the emergency room of a hospital due to asthma in the past 12 months have higher chance of getting admitted than those who did not go, and those who ever had diabetes had a greater chance of visiting the emergency room against those who did not ever have diabetes. Odds ratio can be regarded while interpreting the parameters of the zero outcomes model. The odds ration in dental.2 is almost 100%, indicating those who did not have any dental clinic visit within last 2 years, have 100% higher probability of visiting the emergency room than the others.

By means of all these findings, it is simple to conclude that the ZIP regression model surpasses the poisson regression model in terms of accurately predicting the quantity of visitors to a hospital's emergency room. This analysis underscores a key insight, in discrete data scenarios where an overabundance of zero counts is observed in the dependent variable, the ZIP Regression Model consistently outperforms the Poisson Regression Model.

4 Conclusions

In conclusion, this research highlights the critical significance of count data modeling in various fields, particularly in the realm of public health. While the Poisson regression model serves as a common tool for this purpose, it often assumes uniform dispersion, a limitation frequently encountered in real-world datasets. This study demonstrates that the ZIP regression model emerges as a superior choice when dealing with overdispersion caused by an excess of zero counts. Through a meticulous analysis of hospital emergency room visit data for children between 10–18 years of age, this article have substantiated that the ZIP Regression model consistently outperforms the traditional Poisson Regression model, as evidenced by the lower MSE. These findings underscore the necessity of adopting ZIP Regression models in public health research, where datasets with excess zero counts are prevalent. Adopting this model can result in more precise and informative studies, which will ultimately improve healthcare decision-making and resource allocation in the public health sector.

5 Future Scope

The article's findings suggest several compelling future research directions in count data modeling, especially in the context of public health. These include applying the ZIP regression model to more complex healthcare datasets, refining the model itself, exploring predictive modeling for healthcare service utilization, analyzing spatial and temporal patterns, and investigating clinical applications and health policy evaluations. Additionally, integrating machine learning techniques into count data modeling offers exciting prospects for enhancing our understanding of healthcare utilization and improving evidence-based decision-making in public health. These diverse avenues promise to expand the field's horizons and address critical healthcare challenges.

References

1. Nelder, J.A., Wedderburn, R.: Generalized linear models. J. Roy. Stat. Soc. **135**(3), 370–384 (1972)
2. Lambert, D.: Zero-inflated poisson regression, with an application to defects in manufacturing. Technometrics **34**(1), 1–14 (1992)
3. Nanjundan, G., Naika, T.R.: Asymptotic comparison of method of moments estimators (MME) and maximum likelihood estimators (MLE) of parameters in zero-inflated poisson model. Appl. Math. **3**(6), 610–616 (2012)
4. Mouatassim, Y., Ezzahid, E.H.: Poisson regression and Zero-inflated Poisson regression: application to private health insurance data. Eur. Actuar. J. **2**(2), 187–204 (2012)
5. Eggers, J.: On statistical methods for zero-inflated models. Behav. Res. Methods **49**(5), 1805–1825 (2017)
6. Agresti, A.: An Introduction to Categorical Data Analysis. Wiley, New York (1996)
7. Centers for Disease Control and Prevention. National Health Interview Survey. U.S. Department of Health & Human Services (2023). https://www.cdc.gov/nchs/nhis/2022nhis.htm. Accessed 22 Sep 2023

Enhancing Education Policy Estimation: A Novel Ridge Fuzzy Regression Approach for Handling Multicollinearity with Fuzzy Input Data

Tanmoy Das(✉) and Hemlata Joshi(✉)

Department of Statistics and Data Science, Christ (Deemed to be University), Bengaluru, India
das.tanmoy0811@gmail.com, hemlata.joshi28@gmail.com

Abstract. Multicollinearity often complicates regression analysis, both in classical and fuzzy input setup. This research introduces a new approach that combines ridge regression with fuzzy regression to tackle correlated covariate's impact, with a specific focus on improving education policy systems. Our method utilizes the α-level estimation algorithm and a dataset where Grade Point Average (GPA) serves as a fuzzy input, while input variables remain crisp. We assess our estimator's performance using RMSE and MAPE. This applied research showcases the potential of our method in enhancing education policies through more accurate data-driven decision-making.

Keywords: Education Policy · Multicollinearity · Ridge Regression Model · Ridge Fuzzy Regression Model · Multiple Fuzzy Linear Regression

1 Introduction

Fuzziness often emerges in the domain of education policy formulation and curriculum design within universities due to the multifaceted nature of educational systems. Education policies must contend with diverse student populations, varying learning styles, and evolving societal needs, all of which introduce inherent uncertainty. Curriculum settings, likewise, need to accommodate a wide range of disciplines, teaching methods, and evolving educational paradigms. These complexities make it challenging to precisely quantify the impact of policy changes or curriculum modifications, giving rise to fuzzy elements in decision-making processes. Consequently, incorporating fuzzy logic and fuzzy regression models can provide a more nuanced and adaptive approach to address the intricacies of education policy and curriculum planning in the ever-evolving landscape of higher education.

Among some commonly applied statistical techniques, Linear and Ridge regression models are well-established for handling precise datasets effectively.

V. Sakalauskas et al. (Eds.): IBICA 2023, LNNS 1231, pp. 149–160, 2025.
https://doi.org/10.1007/978-3-031-78946-5_14

Nevertheless, real-world data often deviates from precision, presenting us with non-precise data, including linguistic or imprecise data. Such data structures fall outside the scope of classical statistical theory. To address this challenge, researchers have introduced the fuzzy regression model.

The inception of the fuzzy regression model is attributed to Tanaka et al. (1982) [1], and since the beginning, it has found utility in the analysis of unconventional datasets. Analogous to traditional statistical theory's multiple regression models, fuzzy multiple regression models also contend with the challenge of multicollinearity. To address this concern, we introduce the ridge fuzzy regression model, a cutting-edge method that combines ridge regression and the fuzzy regression model. This fusion's main goal is to lessen multicollinearity's negative effects when they occur. We propose the α-level estimation approach, which is based on the α-level ridge loss function, to build the ridge fuzzy regression model.

The format of this paper goes as, Sect. 2, begins by presenting the methodology encompassing the fuzzy ridge regression model, also fuzzy numbers and the alpha-level estimate process are also explained. Additionally, we outline the criteria for evaluating performance. Moving on to Sect. 3, We explore the use of the fuzzy multiple regression model in practice, applying the α-level estimation method to a dataset pertaining to student performance. Here, we discuss the fuzzy coefficients alongside their respective alpha-levels, concluding with a performance comparison against a simpler fuzzy regression model. Section 4 marks the culmination of our work, where we emphasize the significance of these estimation techniques within the realm of education. Lastly, in Sect. 5, we explore potential avenues for future research in this domain.

2 Methodology

The primary model of multiple linear regression is defined as,

$$Y = X\beta + \epsilon \tag{1}$$

with the least square MSE for the regression coefficient is

$$\hat{\beta_{OLS}} = (X'X)^{-1} X'Y \tag{2}$$

in the presence of multicollinearity or when $p > n$, model experiences non-orthogonality hence makes the estimators biased. Ridge regssion model in handles these situations well.

2.1 Ridge Regression Model

Hoerl and Kennard (1976) [4] introduced the ridge regression model to resolve the issue of non-orthongonality by incorportaing a positive constant with $X'X$ and got the estimate of β as,

$$\hat{\beta}_{RIDGE} = (X'X + kI)^{-1} X'Y \tag{3}$$

, where $k(> 0)$ is the postive tuning parameter.

The formula for the MSE of the Ridge Estimator is given as,

$$MSE(\hat{\beta}_{RIDGE}) = \sigma^2 \left((X'X + kI)^{-1}X'X(X'X + kI)^{-1}\right) + (k^2\beta'(X'X + kI)^{-2}\beta) \tag{4}$$

The bias-variance trade-off idea serves as another justification for using the ridge regression model. As the regularization parameter (k) increases, it forces the regression coefficients closer to zero, effectively shrinking the penalty term. This process reduces the model's flexibility, leading to decreased variance but increased bias. Consequently, by substantially curbing variance while accepting a moderate increase in bias, ridge regression can yield improved outcomes compared to the least squares estimation method.

2.2 Fuzzy Regression Model

However, in practical scenarios, we frequently encounter linguistic or imprecise data, exemplified by terms like 'young,' 'tall,' or 'high.' To address such data types, Zadeh's introduction of fuzzy sets [5] has proven invaluable. Fuzzy regression models, harnessing the power of fuzzy data represented through fuzzy sets, have garnered attention and demonstrated their utility across various domains. Analogous to traditional multiple regression techniques, multicollinearity issues often arise within fuzzy multiple regression models. This creates a requirement for the ridge fuzzy regression model, which smoothly combines the ridge regression principles with fuzzy data in an effort to overcome the difficulties presented by multicollinearity.

2.3 Fuzzy Numbers

A fuzzy set is an ordered pair set $A = \{(x, \mu_A(x)) : x\epsilon X\}$, where $\mu_A(X) : X \to [0, 1]$ represents the membership function of x in a set A.

A fuzzy set's support is determined by $H(A) = \{x\epsilon R | \mu_A(x) > 0\}$. A set $A_\alpha = \{x\epsilon R^1 : \mu_A(x) \geq \alpha\}$ contains all the elements in X for an α-level set, where A is crisp set with range of α as $[0, 1]$, while A_α membership value in A is greater than or equal to α. An α-level set of fuzzy set A can be represented as,

$$A(\alpha) = [l_A(\alpha), r_A(\alpha)] \tag{5}$$

where A is normal, i.e. $\mu_A(x_0) = 1$ for some $x_0\epsilon\mathbb{R}$ and $\mu_A(x)$ is convex.

A fuzzy number A is considered to be an LR-fuzzy number as a special case if the following membership function describes it:

$$\mu_A(x) = \begin{cases} \mathcal{L}_A\left(\frac{n-x}{l_n}\right) & \text{for } 0 \leq n - x \leq l_n \\ \mathcal{R}_A\left(\frac{x-n}{r_n}\right) & \text{for } 0 \leq x - n \leq r_n \\ 0 & \text{otherwise} \end{cases} \tag{6}$$

where, $'n'$ is the mode, and $'l'_n$, $'r'_n$ are called the left and right spreads respectively and $'r'$ and $'l' > 0$. These spreads, which can be symmetric or asymmetric, illustrate how fuzzy a number is. $\mathcal{L}_A$ and $\mathcal{R}_A$ are shape functions of A and They are constant, left-continuous, and non-increasing functions that confirm the characteristics of the class of fuzzy sets in order to ensure that:

$$\mathcal{L}_A(0) = \mathcal{R}_A(0) = 1 \tag{7}$$

and,

$$\mathcal{L}_A(x) = \mathcal{R}_A(x) = 0, \quad x \in \mathbb{R}[0, 1) \tag{8}$$

In particular, if

$$\mathcal{L}_A(x) = \mathcal{R}_A(x) = 1 - x \tag{9}$$

Afterwards, A is known as a triangular fuzzy number and is represented by,

$$A = (n, l_n, r_n) \tag{10}$$

A fuzzy set can be represented by either its membership function or its α-level sets, according to Zadeh's (1975) resolution identity theorem [7]. Assume that A is a fuzzy number with a membership function of $\mu_A(x)$ and a set of $A(\alpha)$ levels. This theorem states that it can be written as follows:

$$\mu_A(x) = \sup_{\alpha \in [0,1]} \{\alpha \cdot I_A(\alpha)(x)\} \tag{11}$$

where $I_A(\alpha)(x)$ displays the defining characteristic function for the α-level set of A.

2.4 Fuzzy Regression Model

Consider the multiple fuzzy linear regression model:

$$Y_i = A_0 + A_1 X_{i1} + \cdots + A_p X_{ip} + E_i, i = 1, \ldots, n \tag{12}$$

The classical regression model is a specific case of the fuzzy regression model, just as crisp data is a special example of fuzzy data.

2.5 α-Level Estimation Algorithm

To estimate the parameters of our ridge fuzzy regression model, we use the α-level ridge loss function. When used with the traditional ridge regression model, it serves to reduce the conventional ridge loss function. The following is a description of the α-level estimation [2] algorithm, which relies on the α-level ridge loss function:

Step 1: Make the dependent variable Y's $\alpha - level$. The set of a "alpha-level" is given by for any positive integer j. $A = \{\alpha_l : \alpha_l \epsilon (0, 1), l = 1, 2, ..., s\} \cup \{0, 1\}$.

Step 2: Find the estimator $l\hat{A_j}(1)$ using the ridge estimation method. $r\hat{A_j}(1)$ of $l_{A_j}(1)$ and $r_{A_j}(1)$ by reducing the modified ridge loss with the α-level functions:

$$\sum_{i=1}^{n}\left(l_{Y_i}(1)-\sum_{j=1}^{p}l_{A_j}(1)l_{X_{ij}}(1)\right)^2+k\sum_{j=1}^{p}l_{A_j}(1)^2 \tag{13}$$

and

$$\sum_{i=1}^{n}\left(r_{Y_i}(1)-\sum_{j=1}^{p}r_{A_j}(1)r_{X_{ij}}(1)\right)^2+k\sum_{j=1}^{p}r_{A_j}(1)^2 \tag{14}$$

Step 3: In Step 1, we make the assumption that the set A contains the elements $\alpha^* = \max_{x_j} A$. The intermediate estimators $l\bar{A_j}(\alpha^*)$ and $r\bar{A_j}(\alpha^*)$ are then discovered by minimizing the following:

$$\sum_{i=1}^{n}\left(l_{Y_i}(\alpha^*)-\sum_{j=1}^{p}l_{A_j}(\alpha^*)l_{X_{ij}}(\alpha^*)\right)^2+k\sum_{j=1}^{p}l_{A_j}(\alpha^*)^2 \tag{15}$$

and

$$\sum_{i=1}^{n}\left(r_{Y_i}(\alpha^*)-\sum_{j=1}^{p}r_{A_j}(\alpha^*)r_{X_{ij}}(\alpha^*)\right)^2+k\sum_{j=1}^{p}r_{A_j}(\alpha^*)^2 \tag{16}$$

respectively.
Thus, we obtain the estimators $l\hat{A_j}(\alpha^*)$ and $r\hat{A_j}(\alpha^*)$ through altering the intermediate estimators $l\bar{A_j}(\alpha^*)$ and $r\bar{A_j}(\alpha^*)$ so that The estimated coefficients obtained from Step 3 are used to create a membership function with a predetermined shape. For the same, we employ the following min and max operators, which are:

$$l\hat{A_j}(\alpha^*) = \min\left\{l\bar{A_j}(\alpha^*), l\hat{A_j}(1)\right\} \tag{17}$$

and

$$r\hat{A_j}(\alpha^*) = \max\left\{r\bar{A_j}(\alpha^*), r\hat{A_j}(1)\right\} \tag{18}$$

Step 4: Moving forward, for any $\alpha\epsilon(0,1)$, we obtain the intermediate estimators $l\bar{A_j}(\alpha_l)$ and $r\bar{A_j}(\alpha_l)$ of $l_{A_j}(\alpha_l)$ and $r_{A_j}(\alpha_l)$, employing the altered ridge loss functions created in Step 3. We also discover the estimators. $l\hat{A_j}(\alpha_l)$ and $r\hat{A_j}(\alpha_l)$ of $l_{A_j}(\alpha_l)$ and $r_{A_j}(\alpha_l)$ by altering the intermediate estimators so that they adopt the form of the pre-defined form's membership function.

$$l\hat{A_j}(\alpha_l) = \begin{cases}\max\{l\hat{A_j}(\alpha^*), \min\{l\bar{A_j}(\alpha_l), l\hat{A_j}(1)\}\} & \text{if } \alpha_l \le \alpha^* \\ \min\{l\bar{A_j}(\alpha_l), l\hat{A_j}(\alpha^*)\} & \text{if } \alpha_l > \alpha^*.\end{cases} \tag{19}$$

and

$$r\hat{A_j}(\alpha_l) = \begin{cases}\min\{r\hat{A_j}(\alpha^*), \max\{r\bar{A_j}(\alpha_l), r\hat{A_j}(1)\}\} & \text{if } \alpha_l \le \alpha^* \\ \max\{r\bar{A_j}(\alpha_l), r\hat{A_j}(\alpha^*)\} & \text{if } \alpha_l > \alpha^*\end{cases} \tag{20}$$

Step 5: Carry out the process repeatedly to discover $l\hat{A}_j(0)$ and $r\hat{A}_j(0)$ of $l_{A_j}(0)$ and $r_{A_j}(0)$.
Step 6: Finally, using linear regression on the estimated $\alpha - level$ sets, find the membership functions $\mu_{A_j}(x)$ for the fuzzy regression coefficients $A_j\ (j = 0, 1, \ldots, p)$.

$\hat{A}_j(\alpha_l) = \left[l\hat{A}_j(\alpha_l), r\hat{A}_j(\alpha_l)\right]\ (l = 1, 2, \ldots, p), \hat{A}_j(0), \text{and} \hat{A}_j(1)$ Adding a constraint such as: $\mu_{\hat{A}_j}\left(l\hat{A}_j(1)\right) = \mu_{\hat{A}_j}\left(r\hat{A}_j(1)\right) = 1$ at the top of the predefined memebership function satisfying the condition of α level 1.

3 Empirical Application

In this section the performance the ridge fuzzy estimator is illustrated through an example of fuzzy input and crisp output. Where, the mid-point of the response variable (Y_i) is assumed to be centered, and the covariates standarized.

In particular, the Diamond distance is employed as the distance metric, and the measures of RMSE (Root Mean Square Error) and MAPE (Mean Absolute Percentage Error) are used to evaluate the performance.

$$RMSE_{Fuzzy} = \frac{1}{n}\sum_{i=1}^{n} d^2\left(Y_i, \hat{Y}_i\right) = \sqrt{\frac{1}{n}\sum_{i=1}^{n} \left(y_{il} - \hat{y_{il}}\right)^2 + \left(y_{im} - \hat{y_{im}}\right)^2 + \left(y_{ir} - \hat{y_{ir}}\right)^2} \tag{21}$$

and

$$MAPE_{Fuzzy} = \frac{100\%}{n}\sum_{i=1}^{n}\left(\left|\frac{y_{il} - \hat{y_{il}}}{y_{il}}\right| + \left|\frac{y_{im} - \hat{y_{im}}}{y_{im}}\right| + \left|\frac{y_{ir} - \hat{y_{ir}}}{y_{ir}}\right|\right) \tag{22}$$

where, the observed values are as $Y_i = (y_{il}, y_{im}, y_{ir})$ and the fitted values as $\hat{Y}_i = (\hat{y_{il}}, \hat{y_{im}}, \hat{y_{ir}})$.

3.1 Example

A survey was conducted among 25 students from an estmeed University. Consider the response below in a form of a dataset in the Table 1:

Table 1. Student Performance Dataset

Slno.	(y, s)	x_1	x_2	x_3	x_4
1	3.7 (0.25)	1370	3.6	18	52000
2	3.9 (0.3)	1490	3.8	20	61000
3	3.6 (0.35)	1320	3.5	22	48000
4	3.8 (0.2)	1540	3.9	16	69000
5	3.8 (0.25)	1385	3.7	19	54000
6	3.9 (0.3)	1460	3.8	18	59000
7	3.5 (0.35)	1290	3.4	24	45000
8	4.0 (0.15)	1570	4	15	72000
9	3.4 (0.4)	1250	3.3	25	41000
10	3.9 (0.3)	1505	3.9	17	63000
11	3.7 (0.3)	1400	3.7	21	56000
12	3.9 (0.25)	1510	3.9	18	64000
13	3.6 (0.4)	1310	3.5	23	47000
14	4.0 (0.2)	1560	4	16	71000
15	3.7 (0.25)	1360	3.6	20	51000
16	3.8 (0.3)	1485	3.8	19	60000
17	3.5 (0.35)	1280	3.4	22	44000
18	4.0 (0.15)	1580	4	14	73000
19	3.4 (0.4)	1265	3.3	24	42000
20	3.9 (0.3)	1520	3.9	17	65000
21	3.7 (0.3)	1420	3.7	20	57000
22	3.9 (0.25)	1535	3.9	18	68000
23	3.6 (0.4)	1330	3.5	21	46000
24	4.0 (0.2)	1555	4	15	70000
25	3.8 (0.25)	1390	3.7	19	55000

where, the variables are as follows:

Variable Index	Variable Name	Description
y	GPA	Student's Current GPA (Fuzzy Input)
s	Spread	The fuzzy spread of the GPA variable
x_1	SAT Score	Student's High School SAT Score
x_2	High School GPA	Student's GPA in high school
x_3	Study Hours Per Week	Hours a student study per week now
x_4	Family Income	Student's current family income (Annual in Rs.)

In the analysis, two diagnostic tools: Variance Inflation Factors (VIF) and the condition number (CN) are used, to thoroughly assess the multicollinearity within our dataset. The VIF values are summarized in Table 2, which initially hinted the multicollinearity concerns, a deeper examination via the CN metric revealed a stark reality. The formula for the Condition Number is given as,

$$CN = \sqrt{\frac{\lambda_{max}}{\lambda_{min}}} \quad (23)$$

, where λ_{max} and λ_{min} are respectively the matrix's $(X'X)$ largest and smallest eigen values. With a calculated CN of 1483060 (>30)-the dataset exhibited severe multicollinearity. This convergence of evidence from both VIF values and the condition number underscores the unequivocal presence of multicollinearity. Consequently, we deemed Ridge Regression as the appropriate approach to address this challenge, given its capacity to mitigate multicollinearity's adverse effects, enhancing our dataset's predictive accuracy and model stability.

Table 2. VIF Values of the Covariates

SAT Score	High School GPA	Study Hours Per Week	Family Income
92.5265	47.7367	8.6340	74.6281

The recently suggested ridge fuzzy regression in this investigation to assess its performance on the response variable, GPA Score. The outcomes, as illustrated in Table 3, compare the fitted values obtained by the multiple fuzzy linear model to those produced by the ridge fuzzy regression model. The results unequivocally demonstrate that the ridge fuzzy regression method offers a notably superior description of the original data when comparable to the method of multiple fuzzy linear regression. This result highlights the ridge fuzzy regression model's improved accuracy and effectiveness in capturing the underlying patterns and subtleties within the dataset, highlighting its potential as a useful tool in educational data analysis.

Table 3. Fitted Values of GPA

Sl no.	Fitted Values		
	$Y = (y, s)$	$\hat{Y}_{OLS}$	$\hat{Y}_{RIDGE}$
1	(3.7, 0.25)	(3.690, 2.566)	(3.705, 0.459)
2	(3.9, 0.3)	(3.820, 2.799)	(3.810, 0.490)
3	(3.6, 0.35)	(3.588, 2.557)	(3.591, 0.510)
4	(3.8, 0.2)	(3.878, 2.808)	(3.933, 0.438)
5	(3.8, 0.25)	(3.785, 2.609)	(3.736, 0.473)
6	(3.9, 0.3)	(3.856, 2.714)	(3.822, 0.462)
7	(3.5, 0.35)	(3.485, 2.544)	(3.513, 0.536)
8	(4.0, 0.15)	(3.974, 2.839)	(3.998, 0.426)
9	(3.4, 0.4)	(3.401, 2.497)	(3.442, 0.548)
10	(3.9, 0.3)	(3.940, 2.769)	(3.894, 0.450)
11	(3.7, 0.3)	(3.750, 2.670)	(3.720, 0.500)
12	(3.9, 0.25)	(3.923, 2.795)	(3.885, 0.464)
13	(3.6, 0.4)	(3.593, 2.559)	(3.572, 0.523)
14	(4.0, 0.2)	(3.980, 2.841)	(3.980, 0.439)
15	(3.7, 0.25)	(3.689, 2.586)	(3.673, 0.485)
16	(3.8, 0.3)	(3.837, 2.773)	(3.819, 0.476)
17	(3.5, 0.35)	(3.510, 2.492)	(3.533, 0.509)
18	(4.0, 0.15)	(3.969, 2.837)	(4.017, 0.413)
19	(3.4, 0.4)	(3.395, 2.503)	(3.462, 0.535)
20	(3.9, 0.3)	(3.917, 2.794)	(3.904, 0.451)
21	(3.7, 0.3)	(3.744, 2.684)	(3.742, 0.487)
22	(3.9, 0.25)	(3.877, 2.836)	(3.903, 0.465)
23	(3.6, 0.4)	(3.615, 2.555)	(3.603, 0.497)
24	(4.0, 0.2)	(3.997, 2.815)	(3.989, 0.425)
25	(3.8, 0.25)	(3.774, 2.617)	(3.740, 0.473)

Finally, Table 4 shows the peformance metrics of the ridge fuzzy regression against the multiple fuzzy linear regression model.

Table 4. Performance Metrics

	RMSE_Fuzzy	MAPE_Fuzzy (%)
Ridge Fuzzy Regression	0.2893	110
Multiple Fuzzy Regression	3.4003	128

When compared to the Multiple Fuzzy Regression Model in the context of GPA analysis, it was shown that the Ridge Fuzzy Regression model performed better. This superiority was evident in the form of lower Mean Squared Error (MSE) and Mean Absolute Percentage Error (MAPE) values (shown in Table 4).

One crucial factor contributing to this performance gap was the presence of multicollinearity within our dataset. The Ridge Fuzzy Regression model outperformed the Multiple Fuzzy Regression Model in terms of predictive accuracy because it successfully addressed and mitigated the negative impacts of multicollinearity.

The ridge fuzzy regression model is illustrated visually in the plot showing observed versus fitted values (see Fig. 1). The fitted values from the ridge fuzzy multiple regression are indicated by red triangles in this image, while the observed values are shown by black triangles.

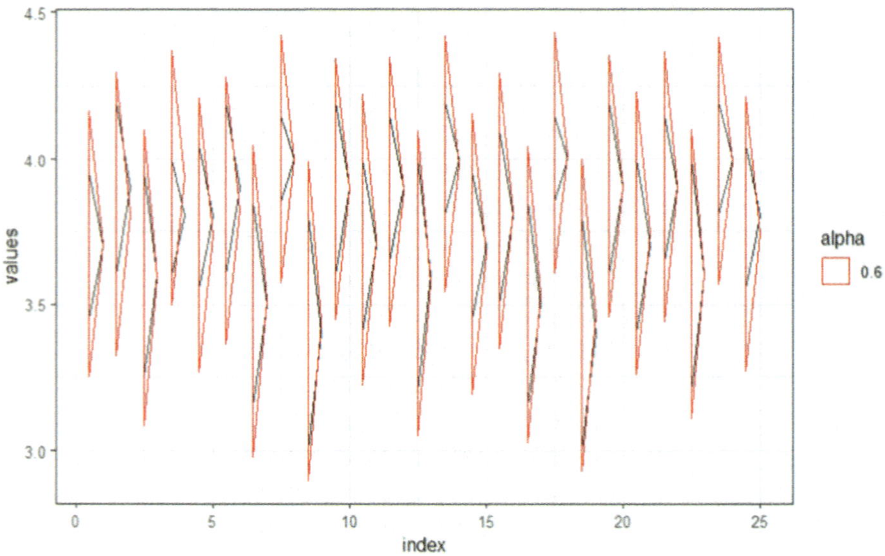

Fig. 1. The Observed versus the Estimated values for the Ridge Fuzzy Regression Model

In summary, the findings underscore the robustness of the Ridge Fuzzy Regression model, emphasizing its capability to deliver more accurate predictions when dealing with multicollinear data, making it a valuable tool for modeling complex relationships in such scenarios.

4 Conclusion

In summary, this investigation paves the way for academic institutions, educational governing bodies, ministries, online learning platforms, and diverse institutions to formulate more effective education policies tailored to the needs of students, considering their academic performance and various attributes. Notably, the study demonstrates the superior efficacy of the ridge fuzzy regression model over its classical fuzzy linear regression counterpart in the presence of multicollinearity. This research contributes significantly to the comprehension of handling linguistic data and addressing their inherent fuzziness through key statistical methodologies, elucidating valuable insights attainable through regression analysis techniques.

Furthermore, the practical application of the fuzzy regression model remains largely untapped on a broader scale in real-world scenarios, where human ambiguities prevail across diverse fields. The current research successfully illuminates a noteworthy application within the pivotal realm of education policy, shedding light on the model's potential to navigate linguistic uncertainties effectively. The implications extend beyond the confines of this study, urging a broader exploration of fuzzy regression applications in addressing complex, real-world challenges where ambiguity is a recurrent facet.

In conclusion, this research serves as a significant contribution to understanding and harnessing the potential of fuzzy regression models, particularly in the critical domain of education policy. The findings underscore the relevance of these models in navigating linguistic nuances and offer a promising avenue for future research and practical implementations in diverse sectors encountering inherent uncertainties.

5 Future Scope

The research described here lays the foundation for a wide range of prospective future projects involving the use of the Ridge Fuzzy Regression Model and educational data analysis. Firstly, there is a pressing need to refine and expand the repertoire of fuzzy regression models, exploring hybrid variants and their suitability for complex, multidimensional educational datasets. Secondly, future studies should extend their focus beyond data analysis and delve into the practical implementation and evaluation of education policies informed by the Ridge Fuzzy Regression Model. Investigating the real-world impact of such policies is paramount. Furthermore, the integration of machine learning algorithms with fuzzy regression models holds considerable potential, offering opportunities for more adaptable and robust models. Beyond education, cross-disciplinary applications of the Ridge Fuzzy Regression Model deserve exploration. Lastly, optimizing educational resource allocation based on the model's insights and enhancing interpretability and visualization techniques are vital aspects of future research. These endeavors collectively aim to advance our understanding of educational data, refine modeling techniques, and foster tangible improvements in education policies and practices across diverse contexts.

References

1. Tanaka, H., Uejima, S., Asai, K.: Linear regression analysis with fuzzy model. IEEE Trans. Syst. Man Cybernet. **12**(6), 903–907 (1982)
2. Choi, S.H., Jung, H.Y., Kim, H.: Ridge fuzzy regression model. Int. J. Fuzzy Syst. **21**(7), 2077–2090 (2019). https://doi.org/10.1007/s40815-019-00692-0
3. Kim, H., Jung, H.Y.: Ridge fuzzy regression modelling for solving multicollinearity. Mathematics. **8**, 1572 (2020). https://doi.org/10.3390/math8091572
4. Hoerl, A.E., Kennard, R.W.: Ridge regression: biased estimation for nonorthogonal problems. Technometrics **12**(1), 55–67 (1970)
5. Zadeh, L.A.: Fuzzy sets. Inf. Control **8**, 338–353 (1965)
6. Zadeh, L.A.: Probability measures of fuzzy events. J. Math. Anal. Appl. **23**, 421–427 (1968)
7. Zadeh, L.A.: The concept of a linguistic variable and its application to approximate reasoning-III. Inf. Sci. **9**, 43–80 (1975)

A Comprehensive Review on Integration of Blockchain and IoT

P. L. Divya[1(✉)], B. Soniya[2], and J. S. Jayasudha[3]

[1] Department of Computer Engineering, Sree Chitra Thirunal Colllege of Engineering, Kerala Technological University, Trivandrum, Kerala, India
divyapl@sctce.ac.in

[2] Department of Computer Science and Engineering, SCT College of Engineering, Trivandrum, India

[3] Department of Computer Science, Central University of Kerala, Kasaragod, India

Abstract. Blockchain is a new technology that is recently been used in many different applications, including Bitcoin, supply chain management, smart grid, health care, and smart farming. Numerous studies have been conducted in this area recently. IoT is critical to the advancement of various fields. IoT networks are expanding in terms of the number of devices linked and the system's size and complexity. There are numerous challenges in IoT, including security, availability, dependability, and authentication. Blockchain technology has the ability to solve challenges in the Internet of Things ecosystem. Blockchain's main characteristics are transparency, security, immutability, traceability, and so on. A thorough examination of blockchain characteristics and applications was conducted by many researchers. We organise and summarise the work that has been done in this area. An architectural design for integrating blockchain and IoT is also presented, ensuring distributed data storage, authentication, and processing and an algorithm for registering IoT devices to the blockchain is also proposed.

Keywords: IoT · Blockchain · security · data storage · authentication · secure transactions

1 Introduction

The Internet of Things (IoT) describes a dynamic environment of interconnected computing devices that use a variety of interoperable components to enable continuous connectivity and data transfer. The IoT market is predicted to increase from more than Fifteen billion devices in 2015 to more than Seventy-Five billion in 2025, resulting in at least 25 individual IoT devices for each person on the planet. IoT systems include all the parts and pieces necessary to put together such a system, present fresh threats to users' personal information and security [1]. Devices connected to the Internet of Things share data with potentially billions of others. Since they allow for anonymous, widespread communication, they attract criminals. Website attacks frequently involve the use of IoT devices. These devices are distinguished by their self-configuration and their dynamic behavior.

V. Sakalauskas et al. (Eds.): IBICA 2023, LNNS 1231, pp. 161–170, 2025.
https://doi.org/10.1007/978-3-031-78946-5_15

Since the Internet of Things can link together anything from tiny wearables to massive industrial systems, it has enabled a wide range of uses. Each of these programs relies on a unique set of frameworks. The success of their implementation depends on their security mechanisms [2]. IoT networks with a centralised architecture may struggle to meet the trust requirement. Therefore, reliable information is crucial to the proper operation of IoT networks. Blockchain is a slashing technology that provides an environment for IoT networks that is secure and decentralised, which is crucial if data processing is carried out by centralised servers with high information processing for the huge volumes of data generated by IoT.

Blockchain technology integrates a variety of cryptographic and computational techniques into a distributed data structure to enable the establishment of trust in an otherwise untrustworthy setting.Some of the difficulties of implementing IoT have been mitigated by combining blockchain technology with it. The potential of blockchain technology needs to be investigated because it is relatively new. There is now considerable academic interest in exploring the potential of an IoT built on blockchain technology. With the help of blockchains, IoT could have an ecosystem where services could be monetised fairly and transparently. Microtransactions for Internet of Things services and data can be conducted securely in the blockchain network environment.

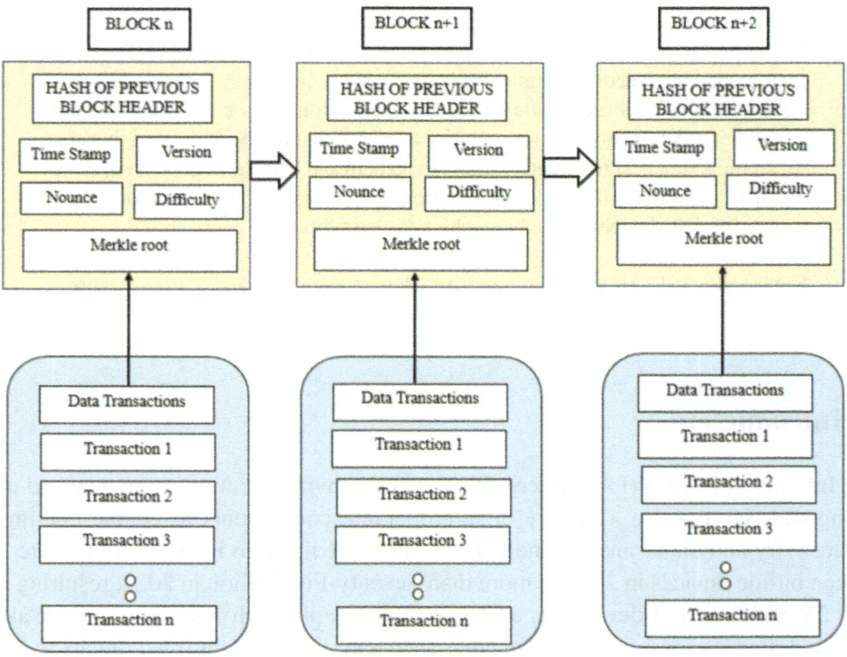

Fig. 1. Basic structure of blockchain

Blockchain is a distributed ledger that relies on the efforts of thousands of interconnected nodes to keep its records updated. The network nodes then verify the trade. After the transaction has been validated, it will be added to a new block along with other transactions that have also been validated. Next, the cryptographic principles are applied to lock down this block. A block will be generated by each node in the network. The network then applies its own set of criteria to this single block to determine its final choice. It is then added to the Blockchain network to verify its security and immutability. Figure 1 depicts the basic layout of a blockchain network. The header and body make up a Block. Time, version, nounce (a random number), difficulty, and merkle root all make up the header. The meat of the document is made up of actual data transfers. Having a hash of the prior block included in every new block ensures that the chain can never be altered. Blockchains employ the Merkle tree data structure. It's also known by the name "hash tree." Fig. 2 depicts the basic layout of a merkle tree. It serves the dual purpose of keeping everything in sync and checking the accuracy of the data. This node's hash is the hash of its children, which are nodes that are not leaves. It safeguards the reliability of the system's data.

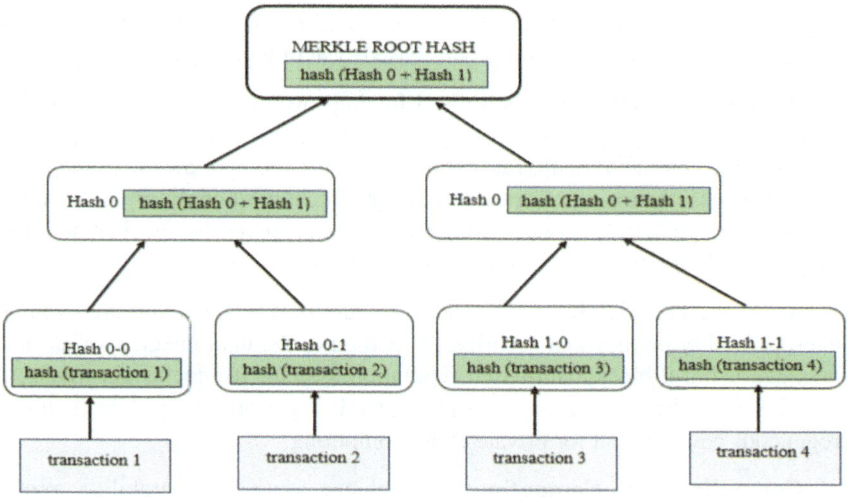

Fig. 2. Merkle tree Structure

2 Blockchain for IoT Applications

IoT network uses blockchain for security and privacy purposes since it is a permanent, transparent and public ledger for recording all data transactions. The blockchain can be applied to many domains in IoT. In Table 1, a comparison of different research in the blockchain and IoT is given. Different applications that use blockchain are:

Internet of Things-Based Healthcare: Blockchain is used in healthcare to provide security and privacy for data sharing. [3] proposes a blockchain-based privacy scheme for IoT healthcare protection. A swarm exchange paradigm is used to improve EHR facilitation in which no central authority is required. Connecting hospitals, patients, health bureaus, and healthcare communities create a consortium blockchain for data sharing, care audit, and medical record review.

Smart Farming: IoT-based smart farming has increased crop yield, efficiency, and decreased farming costs for farmers. An investigation by the "Food and Agriculture Organization of the United Nations" in [4] explains the expanding influence of blockchain in agriculture.

Bitcoin: it is a peer-to-peer network that is based on a method called "blockchain." It's made up of three parts: transactions, a consensus protocol, and a network for communicating. In [5], a system called Cecoin that uses a distributed blockchain instead of Bitcoin was proposed. This takes out the need for a 3rd party in the system. In [6], adding a homomorphic Paillier encryption system to handle plain text in transactions is suggested as a way to make a framework.

Smart Grid: It is an electric network that uses 2-way digital communication to send and receive electricity. It improves efficiency, cuts energy use and costs, and makes the energy supply chain more reliable and clearer. In a transactional energy management system for IoT-enabled smart homes was built on blockchain. Simulations and experiments were used to test how well and if the blockchain-based system will work.

Supply Chain Management: It is utilised for improved quality assurance, effective information exchange, and stringent handling requirements. Domains in supply chain management include the agricultural products industry, the gem industry, and the consumer electronics industry, among others. Using blockchain's homomorphic encryption, implements supply chain data privacy protection. The properties of blockchain improve the supply chain's traceability, reliability, and transparency.

IoT Device Security and Privacy: Blockchain can supplement IoT systems by improving interoperability privacy and security. [7] computes the data integrity of an incoming message using a blockchain-based reputation system. In blockchain transactions [8], updates to embedded devices are propagated. [9] proposes a blockchain-based heterogeneous environment for private cloud computing.

Edge Computing/Fog Computing: To achieve data recoverability property, blockchain is applied to edge computing. [10] discusses artificial intelligence-enabled edge Internet of things. This scheme is safe and secure, and it offers trading and knowledge management. [11] proposes a contract signing protocol for fog computing. It employs three schemes: cryptographic currency, knowledge coin, and proof reading. [12] proposed a blockchain-based mobile edge computing architecture. In a decentralised environment, the proposed architecture components video broadcasting and transcoding service.

Public Sector: Blockchain is used in different government sectors. It ensures that the activities of government officials and citizens are transparent. [13] discusses key aspects of decentralised e-governance. [14] discusses the benefits of blockchain in e-governance and analyses blockchain architectures. [15] proposed using blockchain to build a smart city. The security framework for blockchain and smart device communication platforms was discussed.

Digital Forensics: Blockchain's decentralised nature lends itself well to evidence collection and data integrity in digital forensics. [16] proposed a Digital Forensics investigation structure in the IoT and social system that protects privacy. [17] proposed a model for differential investigation using blockchain in a cloud environment.

Table 1. Comparison of different applications that use blockchain

Author/Research Paper	year	Application	Contribution
P. P. Ray et al.	2021	e-health	The swarm exchange paradigm and consortium blockchain are constructed
A. Vangala et al.	2021	Smart faming	Authentication Scheme and hybrid blockchain
H. Huang et al.	2018	Bitcoin	Fair payment protocol and commitment-based sampling scheme are used
Q. Yang et al.	2021	Energy Management	Developed blockchain and the smart contract to support the holistic TEM system
J. Wan et al.	2019	Smart grid	Blockchain architecture for distributed networks for IioT
S. Wang et al.	2019	Smart grid	Operational model of CES and a two-phase operation algorithm
Z. Li et al.	2017	Smart grid	consortium blockchain for a secure energy trading system
V. Hassija et al.	2021	Supply chain management	Blockchain offers solutions for addressing vulnerabilities within the current supply chain infrastructure
M. Du et al.	2020	Supply chain management	Homomorphic encryption is used in the blockchain for data privacy protection
H. -N. Dai et al.	2019	IoT Privacy and Security	Proposal of BCoT architecture
Y. Liu et al.	2019	Mobile Edge Computing	A blockchain-based MEC architecture proposed
S. Li et al.	2019	Digital forensics	Block-enabled forensics framework for IoT

2.1 Research Challenges for Integration of IoT and Blockchain

The IoT paradigm has been salvaged by the merging of IoT with blockchain by creating beneficial prospects and resolving many of its issues. There are several limitations. The integration procedure itself purportedly raises additional obstacles to innovative research ideas. Most of the literature at the moment concentrates on blockchain characteristics that would enhance IoT design and greatly expand its applicability. Issues like security, connectivity, integrity, scalability, and trust will all be enhanced when blockchain is included in IoT systems. Before the integration's potential can be realised, researchers should address and resolve the problems that result from the integration.

Resources constraints for IOT: IoT devices typically possess constrained resource availability. These devices typically have poorer computing abilities, weaker network connections, less storage, and shorter battery life. Blockchains, on the other hand, have unique requirements of their own. The consensus algorithm, for starters, requires a lot of computing power, which uses energy, making it impractical for low-power IoT devices. Second, storing blockchain data is impossible due to its large size.

2.2 Option of Payment Mechanism

Blocks in blockchain networks are mined using a variety of incentives. Some people use Proof of Work (POW), while others use Proof of Stake (PoS. In blockchains, there are two categories of incentive mechanisms. The IoT paradigm has been salvaged by the merging of IoT with blockchain by creating beneficial prospects and resolving many of its issues. The majority of the research at the moment focuses on the blockchain characteristics that would considerably enhance IoT design.

2.3 Awareness to Security Risk

Due to their scalability and practicality, wireless networks are being used by more and more industries. However, there are numerous security flaws in the wireless medium, including passive eavesdropping, spoofing, denial of service, and others.

2.4 Standardization of Blockchain

IoT developers view blockchain standardization as a crucial issue that will determine the future of their integration because it is anticipated to provide both developers and customers with the necessary direction. It is important to note that when developing blockchain standards, consideration should be given to current relevant industry standards, particularly those that relate to IoT. To boost market confidence, many European nations established standards for financial blockchain transactions.

3 Overview of the Proposed Architectural Design

An architecture for blockchain-IoT- IoT integration is proposed in Fig. 3. It mainly consists of 4 layers the IoT device layer, Blockchain Fog Layer, Cloud Layer, and Data consumer layer. This model preserves privacy and security in an IoT environment. This also increases the authentication efficiency and increase the data storage and processing capabilities of the system.

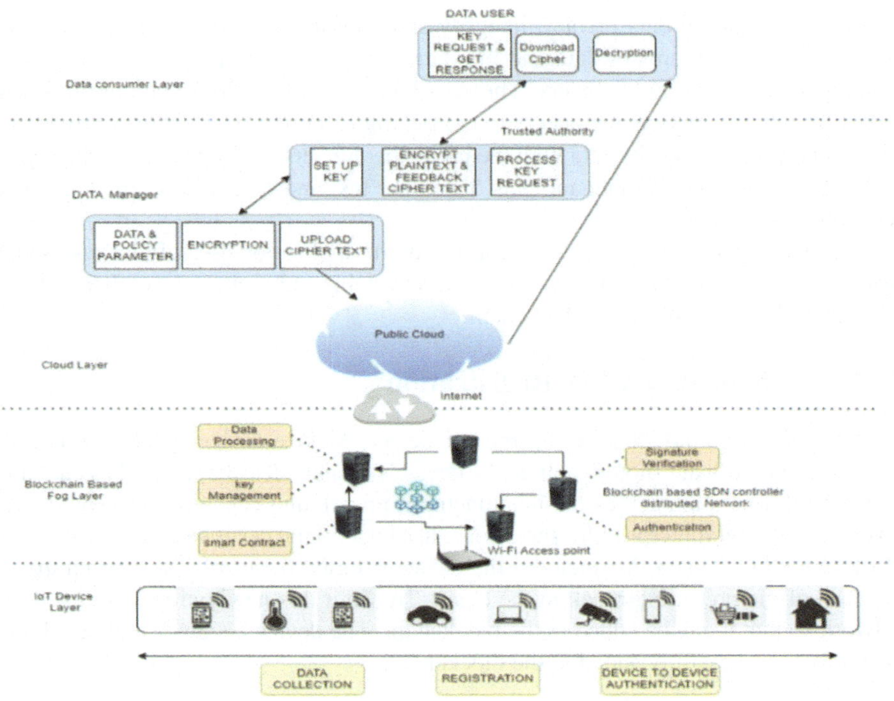

Fig. 3. Architecture of distributed Blockchain and IoT integration

Features of the four layers are given below:

IoT Device Layer (Sensing Layer): This layer consists of different IoT devices having sensing capabilities which are used for monitoring environmental conditions or other parameters depending on the applications. The applications specified are supply chain management, smart agriculture, smart grid, Healthcare systems etc. These devices send filtered data to the Fog layer. This layer performs requested services. This layer is responsible for the delivery of services to the end users or providing data for storing it in the servers/repositories. First registration of device Data collection is done in this layer. Device to Device authentication is required to unauthorized device access.

Fog Layer: The Iot devices transmit the sensed data to this Layer. The fog layer consists of controllers that are connected in a distributed fashion. This layer consisted of controllers which were connected in a distributed fashion. This equipment was capable of performing complex tasks that can process large amounts of data. Processing could be done in a distributed manner. This layer provides services closer to end devices This layer authenticates the devices in the device layer. Smart contracts are responsible for providing authentication mechanisms in the system. The internet allows a fog node to connect to the distributed cloud. It can deploy application services and computing availability flexibly. If fog nodes lack the resources to process the local stream of data, they can send workloads to the cloud.

Cloud Layer: This layer consists of powerful devices for processing, analysis, and storage in addition to communication. These devices are configured as blockchain nodes which are capable of data privacy, integrity and security. This layer is responsible for storing massive amounts of data and big data processing. The processed data will be communicated to the lower layers. The distributed nature of blockchain ensures scalability, reliability, on-demand access to computing resources and fast data processing. The data owner uploads its data with the policy parameters and sends this to the central authority. It encrypts the data and gets the cipher text and feedback to the data owner who uploads the data.

Data Consumer Layer: This layer consists of the application users. If data users want to access device data, they must prove the access rights and request the cipher text. It is transformed by keys from the blockchain.

4 Blockchain-Based IoT Registration

Create a one-of-a-kind identifier for the IoT device. Make a device-specific digital certificate. The certificate should include the device's unique identifier as well as any other relevant information, such as the manufacturer, model, and capabilities of the device. To create a digital fingerprint for the device, hash the digital certificate and create a new block in the blockchain that includes the device's digital fingerprint, a timestamp, and the hash of the preceding block in the chain. Transmit this newly created block to the network so that it can be added to the blockchain. The device becomes registered once the block is successfully added to the blockchain.

Algorithm: Registration of IoT Device

Input: Parking area ID PA_j Device ID Idi
Begin
for each device Idi of parking area P A_j, j=1 to num do
Ki(S) ← PU F (I D_i)
Ki (Pu, Pv) ← ECC (Ki(S))
Block ← Create block { Pu, has(Pu), P A_j, I Di, T}
$Block_s$ig ← Signature (Block)
BlockTrans ← {Block, $Block_s$ig}
if P Aj nor in Blockchain **then**
    Block ← Add {Block + BlockTrans}
    Deploy SmartContract update in cloud
end if
End

4.1 Consensus Algorithm

The Consensus algorithm in blockchain is important for the fog layer and cloud layer. It provides secure and timely access and quality computing services. The Partial Byzantine Fault Tolerance (PBFT) is used in both layers. PBFT offers a distributed network to reach sufficient and valid consensus even if there are malicious nodes in the network. PBFT guarantees reduce the impact of rogue nodes in order to safeguard the system from catastrophic failures [21]. Our decision to use the PBFT algorithm in our architecture is supported by many benefits, the key ones of which are as follows:

1. Rapid execution of the transaction: Multiple confirmations are not needed to validate and finalize a transaction
2. Energy Efficient: PBFT does not necessitate a high level of energy consumption. Computation requires less energy needs.
3. Low reward solution: One of the central concerns with blockchain and IoT connectivity is miner incentives. This problem is resolved by PBFT because it calls for group decision-making through voting on documents by signing messages. Every node in the PBFT network is capable of being rewarded

5 Conclusions

The article conducts a comprehensive review of the integration of blockchain and IoT with a main focus on current research efforts in this area. The main characteristics of blockchain, which is a distributed immutable digital ledger system implemented in a distributed fashion are discussed. We have conducted in-depth survey research on IoT applications that use blockchain. A distributed blockchain architecture model for managing large amounts of data from IoT devices was introduced in this paper. The raw data streams are obtained from the IoT devices. The suggested architecture is based on trending technologies such as blockchain, SDN, and fog computing. This system ensures low latency, real-time data transfer, great scalability, privacy, and availability. The end-to-end delay between IoT devices can be decreased using this architecture. In the Fog layer and Cloud layer partial Byzantine Fault Tolerance can be used as the consensus algorithm in the system. An IoT device registration algorithm is also proposed.

References

1. Butun, I., Österberg, P., Song, H.: Security of the Internet of things: vulnerabilities, attacks and countermeasures. IEEE Commun. Surv. Tutor. (2020)
2. Sengupta, J., Ruj, S., Bit, S.D.: A comprehensive survey on attacks, security issues and blockchain solutions for IoT and IIoT, J. Netw. Comput. Appl. **149**, 102481 (2020)
3. Ray, P.P., Chowhan, B., Kumar, N., Almogren, A.: BIoTHR: electronic health record servicing scheme in IoT-blockchain ecosystem. IEEE Internet Things J. **8**(13), 10857–10872 (2021). https://doi.org/10.1109/JIOT.2021.3050703
4. e-Agriculture in Action: Blockchain for Agriculture - Opportunities and Challenges, Food and Agriculture Organization of the United Nations and International Telecommunication Union, Bangkok, 2019. http://www.fao.org/3/CA2906EN/ca2906en.pdf
5. Qin, B.: Cecoin: a decentralized PKI mitigating MitM attacks. Futur. Gener. Comput. Syst. (2017)
6. Huang, H., Chen, X., Wu, Q., Huang, X., Shen, J.: Bitcoin-based fair payments for outsourcing computations of fog devices. Futur. Gener. Comput. Syst. **78**, 850–858 (2018)
7. Dai, H.-N., Zheng, Z., Zhang, Y.: Blockchain for internet of things: a survey. IEEE Internet Things J. **6**(5), 8076–8094 (2019). https://doi.org/10.1109/JIOT.2019.2920987
8. Liu, B., Yu, X.L., Chen, S., Xu, X., Zhu, L.: Blockchain-based data integrity service framework for IoT data. In: IEEE International Conference on Web Services (ICWS), pp. 468–475 (2017)
9. Yang, K. Zheng, K., Yang, K., V.C. Leung, V.C.: A blockchain-based reputation system for data credibility assessment in vehicular networks. In: Personal, Indoor, and Mobile Radio Communications, pp. 1–5. IEEE (2017)

10. Lin, X., Li, J., Wu, J., Liang, H., Yang, W.: Making knowledge tradable inedge-aienablediot: a consortium blockchain-based efficient and incentive approach. IEEE Trans. Ind. Inf. **15**(12), 6367–6378 (2019)
11. Huang, H., Li, K.-C., Chen, X.: Blockchain-based fair three-party contract signing protocol for fog computing. Concurr. Comput. Pract. Exp. **31**(22), e4469 (2019)
12. Liu, Y., Yu, F.R., Li, X., Ji, H., Leung, V.C.: Decentralized resource allocation for video transcoding and delivery in a blockchain-based system with mobile edge computing. IEEE Trans. Veh. Technol. **68**(11), 11169–11185 (2019)
13. Atzori, M.: Blockchain technology and decentralized governance: is the state still necessary? (2015)
14. Ølnes, S., Ubacht, J., Janssen, M.: Blockchain in government: benefits and implications of distributed ledger technology for information sharing (2017)
15. Biswas, K., Muthukkumarasamy, V.: Securing smart cities using blockchain technology. In: High-Performance Computing and Communications. IEEE, pp. 1392–1393 (2016)
16. Marsal-Llacuna, M. -L.: Future living framework: is blockchain the next enabling network? Technol. Forecast. Soc. Change (2017)
17. Lei, A., Cruickshank, H., Cao, Y., Asuquo, P., Ogah, C.P.A., Sun, Z.: Blockchain-based dynamic key management for heterogeneous intelligent transportation systems. IEEE Internet Things J. **4**(6), 1832–1843 (2017)

Predicting Public Perceptions of Electric and Hybrid Vehicles with ML Algorithms in Response to Oil Price Changes

Yaser Issam Hamodi[1(✉)], Ammar E. Abdelkareem[2], and Ahmed M. Fakhrudeen[3]

[1] Computer Engineering Department, College of Electrical and Electronic Techniques, Iraqi Minister of Higher Education and Scientific Research, Minister Office, Baghdad, Iraq
yaserissam.hamodi@gmail.com

[2] Departement of Computer Networks Engineering, College of Information Engineering, Al-Nahrain University, Baghdad, Iraq

[3] Software Department, College of Computer Science and Information Technology, University of Kirkuk, Kirkuk, Iraq

Abstract. The global automotive industry is witnessing a significant shift towards electric and hybrid vehicles, driven by environmental concerns and recent increases in oil prices. This study explores the collective viewpoints and outlooks of the public regarding electricity and hybrid technology vehicles against the backdrop of increasing oil prices. Policymakers, manufacturers, and researchers can gain valuable real-time insights into public sentiment and opinions regarding sustainable transportation alternatives by analyzing social media discussions. A mixed-methods approach was employed to investigate people's attitudes toward electric and hybrid cars, incorporating quantitative and qualitative data. We collected and analyzed the data from Twitter using natural language processing techniques. The Twitter dataset consisted of public posts and conversations about electric and hybrid vehicles, focusing on discussions surrounding oil prices, sustainability, and driving experiences. Preliminary findings indicate that Twitter users actively discuss electric and hybrid cars in response to rising oil prices. The analysis revealed a range of viewpoints and opinions. Some users express enthusiasm and optimism, highlighting the potential benefits of electric and hybrid vehicles, such as reduced dependence on fossil fuels and lower fuel costs. Others express skepticism and concerns, raising charging infrastructure, battery technology, and affordability issues.

Keywords: Electric and hybrid vehicles · public opinion · social media data · Twitter analysis · machine learning

1 Introduction

Electric and hybrid vehicles have gained significant attention as viable alternatives to traditional internal combustion engine vehicles, particularly in light of environmental concerns and recent increases in oil prices [1, 2]. Rising oil prices have prompted individuals to reevaluate their transportation choices and explore more sustainable options [3].

V. Sakalauskas et al. (Eds.): IBICA 2023, LNNS 1231, pp. 171–181, 2025.
https://doi.org/10.1007/978-3-031-78946-5_16

Understanding people's views and perspectives on electric and hybrid cars in response to these factors is crucial for policymakers, manufacturers, and researchers. This study aims to investigate public sentiment and attitudes towards electric and hybrid vehicles after the rise in oil prices, utilizing data collected from the social media platform Twitter and employing machine learning (ML) algorithms for analysis. Social media platforms have become a rich source of real-time information and public opinions on various topics. With its vast user base and diverse content, In the context of increasing oil prices, Twitter offers valuable insights into the public sentiment towards electric and hybrid technology cars [4]. At the same time, ML models can be trained to classify sentiments expressed in tweets, such as positive, negative, or neutral [6]. This automated analysis provides a comprehensive overview of public perceptions, identifying prevalent sentiments and key themes concerning electric and hybrid vehicles following the surge in oil prices [7].

By integrating Twitter data and ML algorithms, this study aims to address several research questions. Firstly, it seeks to identify the dominant sentiments expressed by Twitter users towards electric and hybrid vehicles, examining whether the rise in oil prices has influenced public attitudes positively or negatively. Secondly, the objective is to delve into the primary subjects and subjects linked with conversations regarding electrical and hybrid technology cars within the framework of escalating oil prices. Therefore, this research aims to bridge the gap between public sentiment, emerging trends in social media discussions, and the implications for sustainable transportation planning and policy-making.

The structure of this paper is organized as follows. Section 2 introduces the reader to some background and reviews related works. Section 3 presents the methodology of our work. Section 4 lists and explains the conducted steps to evaluate our proposals. The results are presented and discussed in Sect. 5. Finally, Sect. 6 concludes this research and identifies some future directions.

2 Theoretical Background and Literature Review

2.1 Intelligent City

Through information and communication technologies (ICT), smart cities manage their public resources to improve the quality of services, enhance comfort, facilitate maintenance and achieve sustainability in the first place [8]. Electric Vehicles (EVs), such as electric cars, buses, and neighborhood vehicles, will be the main force in the transportation industry in the near future. In addition, the shift forward will include electrifying the entire transport system as part of a citywide effort to reduce hydrocarbon emissions [9, 10]. Nevertheless, based on new energy distribution and traffic management issues, the efficiency of these new transportation systems cannot be guaranteed in the context of a typical city. As a result, a smart city can help achieve this national objective [11]. EVs It is indispensable in many smart city programmers'; consequently, smart cities and EVs are closely related [12]. One of the main objectives of various intelligent city efforts is to decrease emissions and significantly enhance urban air quality. With their reduced maintenance needs and lower operational expenses than conventional vehicles, EVs play a key role in achieving this goal. Moreover, smart city infrastructure like advanced traffic management systems and charging stations can facilitate EVs' seamless integration and uptake [13].

To ensure that the population's needs are met, careful planning and effective monitoring of the charging infrastructure for these vehicles is required. All stakeholders, including government, businesses and individuals, must coordinate and cooperate to ensure EVs are successfully and effectively integrated into the smart city infrastructure. [11, 14]. Due to the development of smart cities, urban areas may become more efficient, sustainable, and inhabitable [13]. Nevertheless, municipalities need to solve implementation challenges to ensure efficiency and fairness. Cities can achieve the success and benefits of their smart city initiatives by investing in technology and data management, developing strong privacy and cybersecurity policies and protocols, collaborating efficiently with stakeholders, and formulating inclusive initiatives accessible to all [15, 16].

2.2 Related Works

The rise in oil prices and increasing concerns about environmental sustainability have led to a heightened interest in electrical and hybrid technology vehicles as alternatives for conventional cars that run on petroleum fuel (gasoline) [1]. With the emergence of social media platforms, researchers have begun to explore these platforms as valuable sources of data to understand public opinions and viewpoints on electric and hybrid cars. ML algorithms, including Logistic Regression (LR) [17], Naive Bayes (NB) [18], and Random Forest (RF) [19], have been employed to analyze Twitter data and uncover insights into people's perspectives on electrical and hybrid technology vehicles in the context of increasing oil prices.

Several studies have utilized LR to analyze Twitter data and investigate public sentiments toward electric and hybrid vehicles. For instance, the authors in [4] employed LR to classify tweets related to electric vehicles and oil prices into positive, negative, or neutral sentiments. Their findings indicated that positive sentiments were associated with environmental friendliness, potential cost savings, and technological advancements. In contrast, negative sentiments often concern charging infrastructure and limited driving range. Similarly, NB classifiers have been widely used for sentiment analysis of Twitter data concerning electric and hybrid vehicles. In a study [20], NB was applied to classify tweets and identify sentiments toward electric vehicles after the surge in oil prices. The analysis revealed positive sentiments encompassing benefits such as reduced dependence on fossil fuels, lower emissions, and potential long-term cost savings. Negative sentiments were commonly linked to concerns about limited charging infrastructure and the relatively high initial cost of electric and hybrid vehicles. Lastly, the RF algorithm has also been employed to analyze Twitter data and examine public perceptions towards electric and hybrid cars. For instance, the effort in [21] utilized RF to classify tweets and detect sentiment polarity towards electric vehicles amid rising oil prices. The study found positive sentiments often associated with environmental sustainability, technological advancements, and potential economic benefits. On the other hand, negative sentiments typically centered around concerns regarding charging infrastructure and range anxiety.

While these studies have made significant contributions, there are still opportunities for further research. Future investigations could explore ensemble methods that combine multiple ML algorithms to enhance sentiment classification accuracy. Additionally, incorporating user demographic information, such as age, gender, and location, into the analysis could provide deeper insights into the variations in public viewpoints and opinions.

3 Methods and Modelling

This Section is dedicated to presenting our methods. As shown in Fig. 1, the methods of our methodology are as follows:

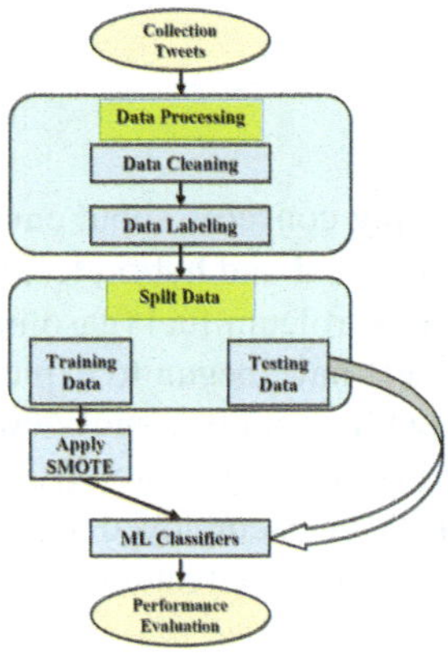

Fig. 1. The methodology of the proposal.

3.1 Data Collection

The methodology begins with collecting a large dataset of tweets related to electric and hybrid cars, oil prices, and associated keywords using the Twitter API. The timeframe for data collection should be defined to capture relevant discussions after the rise in oil prices. Various data collection methods were proposed for social media data analysis. Twitter and other social media microblogs make user messages public. To investigate public perception and data classification of EVs, we crafted a script to capture relevant Twitter data. Instead of the official Twitter API, Python was used to design and develop this application. For data capture, we utilized the most relevant hashtags on Twitter and Google Trends keywords. Table 1 lists the utilized hashtags via the dataset. After data collection, additional elements were extracted from the collated tweets. Examples include hashtags, language, tweet date, and ID. For analysis, these were saved as values separated by commas. In the experiment, we collected 310,218 tweets between the beginning of January 2022 and the end of December 2022 in response to the global oil price increase to measure its effects during the crisis. After filtration, 152,112 tweets were left as a final dataset.

Table 1. The related used hashtags.

#Evoil	#EVroadtrip	#getthepower
#Electriccars	#evcharging	#HybridElectricCars
#electriccars,	#ElectricVehicles	#HybridCars

3.2 Data Preprocessing

3.2.1 Data Cleaning

In the context of Twitter data for NLP, the following are considered as noise: Hashtags, URLs and whitespaces, non-standard text and symbols, names of users and Duplicate tweets. After removing the noise, the following preprocessing steps were performed: *Text Cleaning*: The collected tweets undergo preprocessing steps, such as removing special characters, URLs, and user mentions. Tokenization is performed to break the text into individual words or tokens. *Stop word Removal*: Common stop words (e.g., "and," "the," "is") that do not contribute to sentiment analysis are eliminated. *Stemming or Lemmatization*: Words are normalized using stemming or lemmatization techniques to reduce variations (e.g., "cars," "car's" reduced to "car").

3.2.2 Data Labeling

Data labeling is crucial in supervised ML, providing the necessary ground truth for training algorithms. The quality and accuracy of data labeling directly impact the performance and generalization ability of the trained model. We used **Vader sentiment analysis** to label the final dataset because an automated data labeling tool was developed to assist and improve the efficiency of the process. If a single tweet were calculated as more significant than 0, it would be marked as positive. Nevertheless, if the tweet were less than 0, it would be marked as a negative tweet. Lastly, we have two binary classes (0 and 1) 0s represent negative tweets, and 1s represent positive tweets. Table 2 shows the labeled dataset and class tweet counts [31].

3.3 Synthetic Minority Oversampling Technique (SMOTE)

Table 2 presents the results of data labeling, revealing 123,458 positively labeled tweets and 28,654 negatively labeled tweets. To conduct our experiment, we adopted an 80:20 split ratio to create training and testing datasets. However, the table also highlights that our dataset is imbalanced, which can lead to biased predictive results favoring one class over the other when using selected models. To address this issue, we employed SMOTE. It works by analyzing examples from the minority class and using the k-nearest neighbor algorithm to identify a random nearest neighbor. Subsequently, this analysis generates a synthetic instance in the feature space [22]. This technique helps balance the dataset, enabling our models to make more accurate predictions and avoid favoring any specific class unfairly [30]. So, Table 3 lists the training dataset after applying the SMOTE (to avoid biases towards any class).

Table 2. Details of the collected dataset.

No. of Tweets	Negative	028,654
	positive	123,458
	Total	152,112

Table 3. Before and after applying SMOTE on the training dataset.

Number of tweets	After	Before
Negative	99,223	24,713
Positive	99,223	99,223

4 Performance Evaluation Steps

4.1 Feature Extraction

1) **Bag-of-Words Representation** [23] represents each tweet as a numerical feature vector. This representation captures the frequency of words or n-grams (contiguous sequences of words) in the tweet.
2) **TF-IDF) Transformation** [24] is applied to weigh the importance of each tweet based on its frequency in the tweet and rarity across the entire dataset.

4.2 ML Classifier Models

1) **Logistic Regression (LR):** The LR model identified key features associated with positive sentiments, including "environmentally friendly," "cost savings," and "advancements in technology." Negative sentiments were commonly linked to concerns about "charging infrastructure" and "limited driving range."
2) **Naive Bayes (NB):** The NB algorithm revealed that positive sentiments were often associated with "reduced dependence on fossil fuels," "lower emissions," and "potential cost savings." Negative sentiments were frequently expressed regarding "charging infrastructure" and "initial cost."
3) **Random Forest (RF):** The RF model highlighted positive sentiments related to "environmental sustainability," "technological advancements," and "potential economic benefits." Negative sentiments often concern "charging infrastructure" and "range anxiety."

4.3 Evaluation Metrics

Multiple criteria are used to assess the efficacy of the various classifiers. The correct predictions are represented by the word "True" as True Positive (***TP***) and True Negative (***TN***). The incorrect predictions are referred to by the word "False" as False Positive (***FP***) and False Negative (***FP***) [25]. Common evaluation metrics, such as accuracy, precision, recall, F1-score, and Matthews correlation coefficient (MCC), are computed to measure the performance of the ML models in sentiment classification [29].

4.4 Cross-Validation

To ensure the robustness of the models, cross-validation techniques like k-fold cross-validation can be employed, where the dataset is divided into k subsets and the models are trained and tested on different combinations of the subsets [26].

5 Results

Here, the analysis is conducted regarding *Sentiment Distribution* and Feature Importance. The sentiment distribution across the collected tweets is analyzed to determine the overall sentiment trends toward electric and hybrid cars after the rise in oil prices [27]. On the other hand, the importance of features (words or n-grams) in predicting sentiment is assessed using techniques such as feature importance scores or coefficients derived from the trained models. This provides insights into the key factors influencing public viewpoints [28].

Here, we evaluated our models regarding accuracy, precision, recall, F1-score, and MCC in sentiment classification. Figure 2 shows the confusion matrix of the utilized algorithms. Suppose the prediction is made well in each of the confusion matrix's four areas (TN, TP, FN, and FP) proportionally to the size of negative and positive elements in the dataset. In that case, MCC can produce a high score. To tackle this problem after implementing ML algorithms to the data set, we preferred to choose the MCC metric.

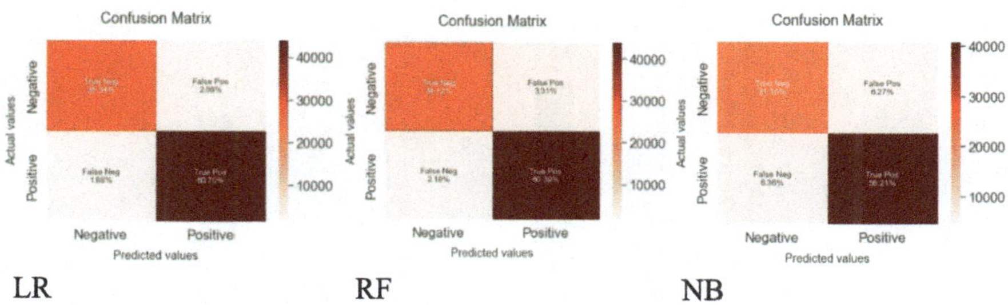

Fig. 2. The confusion matrix of our suggested ML algorithms.

As listed in Table 4, LR achieved better accuracy about (96%). Although the RF algorithm is so close, NB produced the lowest accuracy among the models (87%). Similarly, but in MCC, LR (0.915) outperformed the other models (RF and NB). Additionally, the model achieved the highest MCC with 0.915 scores, whereas RF and NB reached 0.87 and 0.731, respectively. Using the confusion matrix, an in-depth evaluation of the optimal experiment was conducted. Once again, in recall, precision, and F1-score, LR outperformed RF and NB in both classes. Accordingly, the LR classifier is better than RF and NB in all performance evaluation metrics.

Accordingly, the comparison of the performance of the algorithms in sentiment classification demonstrated that the LR algorithm outperformed RF and NB models (see Fig. 3) in terms of MCC and accuracy.

Table 4. Results of applied ML models.

Model	Class	Accuracy	Recall	Precision	F1-score	MCC
LR	0	0.96	0.94	0.95	0.95	0.915
	1		0.97	0.97	0.97	
RF	0	0.95	0.91	0.94	0.93	0.882
	1		0.97	0.95	0.96	
NB	0	0.87	0.83	0.83	0.83	0.731
	1		0.90	0.90	0.90	

Nevertheless, most Twitter users expressed positive sentiments towards electric and hybrid cars after the rise in oil prices, emphasizing the perceived environmental benefits and potential cost savings. Furthermore, concerns about charging infrastructure and limited driving range were prevalent among negative sentiments, indicating the need to address infrastructure gaps and enhance vehicle range capabilities. Moreover, key features identified by the models provide insights into the factors influencing public opinions and can effectively guide policymakers and manufacturers in promoting electric and hybrid vehicles.

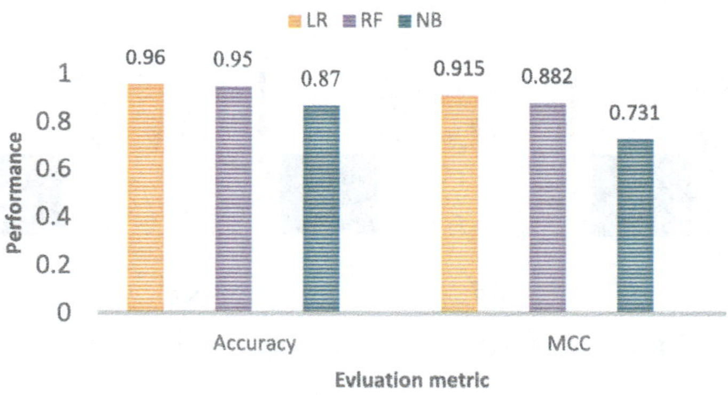

Fig. 3. Comparison of our ML algorithms.

Finally, for ethical consideration, the study ensured Twitter users' data privacy and anonymity throughout the data collection and analysis. Additionally, measures were taken to mitigate biases in the collected data and address potential biases in the models to ensure fair and unbiased sentiment analysis. Therefore, the Twitter data analysis using ML algorithms, including LR, NB, and RF, revealed that most public sentiments towards electric and hybrid cars after the rise in oil prices were positive.

6 Discussion

Twitter data analysis revealed that most sentiments expressed towards electric and hybrid cars after the rise in oil prices were positive. Users highlighted the perceived environmental benefits, cost savings, and technological advancements associated with these vehicles. However, concerns regarding charging infrastructure and limited driving range were commonly expressed as negative sentiments. ML algorithms, including LR, NB, and RF, demonstrated varying performance in sentiment classification. LR achieved an accuracy of 96%, NB achieved an accuracy of 87%, and RF achieved an accuracy of 95%. These models provided insights into the key features influencing sentiments, such as environmental sustainability, cost savings, charging infrastructure, and range anxiety.

The findings have several implications for policymakers and manufacturers in promoting electric and hybrid cars. The positive sentiments expressed by Twitter users indicate a growing acceptance and interest in these vehicles as sustainable alternatives to traditional gasoline-powered cars. Policymakers can leverage this positive sentiment to implement supportive policies, such as increasing charging infrastructure availability and providing incentives for electric and hybrid vehicle adoption. However, addressing the concerns expressed in the negative sentiments is crucial. Efforts should improve charging infrastructure, extend vehicle range capabilities, and reduce upfront costs.

7 Conclusion and Future Directions

Twitter data analysis using ML algorithms has revealed positive sentiments toward electric and hybrid cars after the rise in oil prices. The findings contribute to the ongoing efforts to address environmental concerns and transition toward a more sustainable future in the automotive industry. One of our primary focuses in the future is examining the influence of demographic factors, geographic location, and cultural contexts on public sentiments.

References

1. Tran, D.D., Vafaeipour, M., El Baghdadi, M., Barrero, R., Van Mierlo, J., Hegazy, O.: Thorough state-of-the-art analysis of electric and hybrid vehicle powertrains: topologies and integrated energy management strategies. Renew. Sustain. Energy Rev. **119**, 109596 (2020)
2. Balali, Y., Stegen, S.: Review of energy storage systems for vehicles based on technology, environmental impacts, and costs. Renew. Sustain. Energy Rev. **135**, 110185 (2021)
3. Shaaban, K.: Why don't people ride bicycles in high-income developing countries, and can bike-sharing be the solution? Case Qatar. Sustain. **12**(4), 1693 (2020)
4. Balla, S.N., Pani, A., Sahu, P.K., González-Feliu, J.: Examining shifts in public discourse on electric mobility adoption through Twitter data. Transp. Res. Part D: Transp. Environ. **121**, 103843 (2023)
5. Singh, D.K.S., et al.: Social network analysis for precise friend suggestion for Twitter by associating multiple networks using ml. Int. J. Inf. Technol. Web Eng. (IJITWE) **17**(1), 1–11 (2022)
6. Ali, I., Mughal, N., Khand, Z.H., Ahmed, J., Mujtaba, G.: Resume classification system using natural language processing and machine learning techniques. Mehran Univ. Res. J. Eng. Technol. **41**(1), 65–79 (2022)

7. Agresti, A.: Categorical Data Analysis. John Wiley & Sons, Hoboken (2013)
8. Yeh, H.: The effects of successful ICT-based smart city services: from citizens' perspectives. Gov. Inf. Q. **34**(3), 556–565 (2017)
9. Lai, C.S., et al.: A review of technical standards for smart cities. Clean Technol. **2**(3), 290–310 (2020)
10. Gandy, O.H., Jr., Nemorin, S.: Toward a political economy of nudge: smart city variations. Inf. Commun. Soc. **22**(14), 2112–2126 (2019)
11. Camero, A., Alba, E.: Smart city and information technology: a review. Cities **93**, 84–94 (2019)
12. Wang, T., Luo, H., Zeng, X., Yu, Z., Liu, A., Sangaiah, A.K.: Mobility based trust evaluation for heterogeneous electric vehicles network in smart cities. IEEE Trans. Intell. Transp. Syst. **22**(3), 1797–1806 (2020)
13. Canizes, B., et al.: Electric vehicles' user charging behaviour simulator for a smart city. Energies **12**(8), 1470 (2019)
14. Ismagilova, E., Hughes, L., Dwivedi, Y.K., Raman, K.R.: Smart cities: advances in research—an information systems perspective. Int. J. Inf. Manag. **47**, 88–100 (2019)
15. Daina, N., Sivakumar, A., Polak, J.W.: Electric vehicle charging choices: modelling and implications for smart charging services. Transp. Res. Part C: Emerg. Technol. **81**, 36–56 (2017)
16. Lv, Z., Qiao, L., Cai, K., Wang, Q.: Big data analysis technology for electric vehicle networks in smart cities. IEEE Trans. Intell. Transp. Syst. **22**(3), 1807–1816 (2020)
17. Menard, S.: Applied logistic regression analysis (No. 106). Sage (2002)
18. Rish, I.: An empirical study of the naive Bayes classifier In: IJCAI 2001 Workshop on Empirical Methods in Artificial Intelligence, vol. 3 IBM. New York, pp. 41–46 (2001)
19. Breiman, L.: Random forests. Mach. Learn. **45**, 5–32 (2001)
20. Gautam, J., Atrey, M., Malsa, N., Balyan, A., Shaw, R.N., Ghosh, A.: Twitter data sentiment analysis using Naive Bayes classifier and generation of heat map for analyzing intensity geographically. Advances in Applications of Data-Driven Computing, pp. 129–139 (2021)
21. Novalita, N., Herdiani, A., Lukmana, I., Puspandari, D.: Cyberbullying identification on Twitter using random forest classifier. J. Phys. Conf. Ser. **1192**(1), 012029. IOP Publishing (2019)
22. Khan, M.R.H., Afroz, U.S., Masum, A.K.M., Abujar, S., Hossain, S.A.: Sentiment analysis from Bengali depression dataset using machine learning. In: 2020 11th International Conference on Computing, communication and networking technologies (ICCCNT), pp. 1–5. IEEE (2020, July)
23. Zhang, Y., Jin, R., Zhou, Z.H.: Understanding bag-of-words model: a statistical framework. Int. J. Mach. Learn. Cybern. **1**, 43–52 (2010)
24. Lavin, M.: Analyzing documents with TF-IDF (2019)
25. Jordan, M.I., Mitchell, T.M.: Machine learning: trends, perspectives, and prospects. Science **349**(6245), 255–260 (2015)
26. Baker, M.R., Taher, Y.N., Jihad, K.H.: Prediction of people sentiments on Twitter using machine learning classifiers during Russian aggression in Ukraine. Jordanian J. Comput. Inf. Technol. **9**(03) (2023)
27. Baker, M.R., Akcayol, M.: Priority queue-based estimation of the importance of web pages for web crawlers. Int. J. Electr. Comput. Eng. **9**(1), 330–342 (2017)
28. Jihad, K.H., Baker, M.R., Farhat, M., Frikha, M.: Machine learning-based social media text analysis: impact of the rising fuel prices on electric vehicles. In: Abraham, A., Hong, T.P., Kotecha, K., Ma, K., Manghirmalani Mishra, P., Gandhi, N. (eds.) Hybrid Intelligent Systems. HIS 2022. LNNS, vol. 647. Springer (2023)
29. Abbas, Z., Al-Ani, A.: Detection of anomalous events based on deep learning- BILSTM. Iraqi J. Inf. Commun. Technol. (IJICT) **5**(3) (2022)

30. Hasan, I., Awad, O.: An optimized fuzzy logic controller for wireless network control system using PSO. Iraqi J. Inf. Commun. Technol. (IJICT) **5**(1) (2022)
31. Muhson, M., Al-Ani, A.: Blind restoration using convolution neutral network. Iraqi Journal of Information and Communications Technology (IJICT) Conference Series: The 1st Conference of Applied Researches in Information Engineering (2021, ARIE)

Enhancing Fault Tolerance Level in E-Health Monitoring System Using Proactive Approach

Praynita Karki(✉)

Department of Computer Science and Engineering, Kathmandu University, Kavre, Nepal
praynita.karki@gmail.com

Abstract. E-Health system should provide effective, efficient service delivery to citizens anywhere at any time with the help of information and communication technologies such as the internet, local area network, mobiles, etc. The major things for a better health service system are system availability, reliability, per formability, and security. This paper proposes a proactive fault handling approach to enhance the fault tolerance level of an E-health monitoring system with a machine learning-based model to predict the upcoming fault in the system. It trains 8 different models using 8 different Machine Learning Classifiers and compares those models using 6 evaluation matrices: Accuracy, Mean Absolute Error, Precision, Recall, F-Measure and False Positive Rate. The model with better performance among all is recommended for developing Proactive Fault Handling based E-health Monitoring System. Proposed Systems with Proactive Fault Handling (PFH) and Systems without using Proactive Fault Handling are modeled using a Continuous Time Markov Chain (CTMC), and the Availability, Downtime, and Reliability of the modeled systems are computed and compared.

Keywords: e-health · machine learning · fault tolerance

1 Introduction

In today's context, the world is changing and revolutions are taking place rapidly and the adoption of technologies is also improving. Paper based record in the health sector has adopted the electronic health record system. With the demand of the users, technology has taken a very important place in everyone's life. According to paper [1], "a system failure occurs when the delivered service no longer complies with the specifications, the latter being an agreed description of the system's expected function and/or service". Failure of one component can lead to one or more faults in another, where faults can be in both hardware and software that cause errors. To have secure and reliable system we have to minimize the fault. But, the systems are often under attack. Despite all attacks and different errors and fault conditions the system must work properly and provide its services continuously for its user. Hence handling the fault is essential for achieving the goal of availability and reliability of e-health system. For this, the e-health system must be perceived without any kind of failure, fault, and error. Fault tolerance aims to operate the system correctly even in the occurrence of some faults. The systems have to tolerate the fault to maintain its efficiency. System still should manage to give the output even if there occurs an error in the system.

V. Sakalauskas et al. (Eds.): IBICA 2023, LNNS 1231, pp. 182–191, 2025.
https://doi.org/10.1007/978-3-031-78946-5_17

In health monitoring system, different entities are involved such as doctor, community health workers, nurses, stakeholders, administrators, patients. Health plays very important role in the development of the country. For healthy community there should be proper monitoring system. We should have very strong reliable, proactive system for the seamless delivery of the information and services.

2 Related Works

There are various researches done in the field of e-health monitoring system. Mostly done to identify and monitor diseases such as diabetes, chronic diseases etc. using machine learning.

The paper [2], proposes the methodology for the e-health monitoring system, this system is considered to attain the demands of the system, patient and medical staff in any system, be it physical, network or software with the presence of fault. Authors have suggested achieving a system that can tolerate faults because there is high chance a fault can occur at some point.

In paper [3], Hidden Markov Model based baum-welch algorithm and Support Vector Machine was proposed for the improvement of the reliability of the healthcare system and to predict the abnormality caused by the medical sensors.

Doctors cannot be there every time to regularly monitor the patient that is why e-health monitoring system is suggested which can provide qualitative and security services. Machine learning algorithm such as Support Vector Machine, Decision tree, K-NN and regression algorithm were recommended and Support Vector Machine was selected among all on the diabetes dataset [4].

In the work [5], authors have explained the need of fault tolerant system in e-government system for serving its user at any time and any location. They explained that the system encounters different unfavorable incidents/situations during its execution despite of those unfavorable conditions or situations system should have the capability to tolerate and provide its entire services continuously to its user. To increase the dealing capacity of the system with the unexpected interruptions and faults and achieve the goal of availability and reliability of the system, researchers have designed and proposed architecture. The proposed architecture uses fault tolerant approach in e-government system with the concept of availability and reliability. They have used design diversity and software rejuvenation together for the purpose of increasing the system fault tolerance capability. The architecture proposed by them contains two systems; both of them run together and serve the same purpose. Problems and abnormalities in both systems are monitored by watch dog, if detected then the information is passed immediately to the manager. The defective system is further examined and repaired by the manager with the approach of software rejuvenation. The analysis performed by considering two states: One system with manager and design diversity and another without manager and design diversity illustrated that the proposed architecture has the maximum capability to run and sustain the e-government system with incorporating the features such as availability, survivability and reliability.

Very few researches have been done in the area of the fault tolerance for e-health monitoring system. To increase security, availability, reliability, capability to tolerate fault etc., different fault tolerance techniques such as Check pointing, Job Migration, Replication, Self –Healing, Safety-bag checks, S-Guard, Retry etc. has been used but none of the study has been done to find the fault tolerance level of Proactive Fault Handling based e-health monitoring system till date.

3 Methodology

After going through different literature on fault tolerance systems, came up with a research idea on improving the fault handling system by proactively predicting the possibility of the fault. For carrying out this study, a hypothesis "Availability and Reliability of the E-Health system can be increased whereas the Downtime of the system can be decreased by using Machine Learning based Proactive Fault Handling approach" was created.1999 instances were randomly generated and two classes:

Class **Yes** represents the instances where those results fault in the system and

Class **No** represents the instances that don't results fault.

Among the 1999 dataset, 906 belong to the fault class whereas 1093 belong to the normal class. For carrying out experiments on the proposed framework a random set of 1999 datasets were generated. The dataset thus prepared was used to train and test the model in the experiment phase. The accuracy of all models was computed and compared in the Model selection phase; only the classifier with high accuracy was taken and modeled using CTMC. Availability, Downtime, and Reliability of the selected model was calculated and comparison between a system with PFH and a system without PFH were made.

4 Conceptual Framework

Proposed architecture is based on the principle of design diversity. That is a number of system with different design performs a same task. Each system is equipped with the lm-sensors. The lm-sensor monitors the system affecting parameters. Some of the parameters that are captured by lm-sensors are CPU utilization, Memory consumption, Bandwidth Allocation, Response time etc.

Values sensed by lm-sensor are used to maintain the profile by the System monitoring unit of Proactive Fault handling unit. The profile thus maintained is used by the Fault prediction system of Proactive Fault handling unit to make a prediction. If the prediction is fault then the coordinator takes an action to handle the fault. The proposed conceptual Framework of Proactive Fault Handling based E-Health Monitoring System is shown in Fig. 1.

Coordinator takes an action based on the policy designed in order to tackle the fault situation. Action of coordinator is guided by proactive fault tolerance policy. Policies are designed in a way that it reduces the impact of possible failure. Major policies that will be implemented are:

1. Set the alarm to inform the administrator the possibility of the fault
2. Detect additional node (system)
3. Check if the new request result in fault

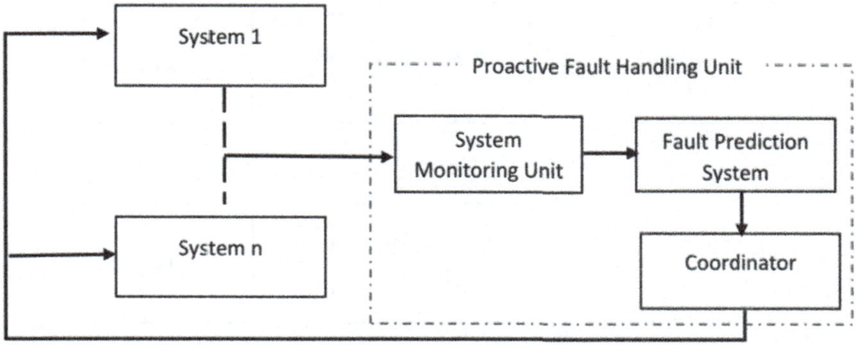

Fig. 1. Conceptual Framework of E-Health Monitoring System with Proactive Fault Handling

4. If yes go to step 2. If no go to step 5
5. Process the request on the newly allocated system
6. Take corrective action on the repair unhealthy node
7. Maintain the log of the action

If any fault arises on the coordinator and the coordinator goes down then a new coordinator is selected by the process of voting.

Working of Proactive Fault handling based system has been divided into two different phases:

4.1 Model Training Phase

Before the actual implementation of the fault detection system, a model is needed to be trained. To train the model features are extracted from the dataset, which are then supplied to the machine learning algorithm along with their respective labels. The machine learning classifier learns from the input provided and produces a model. The training of the model is done in a supervised way. This research trained 8 different models using 8 different classifiers such as Artificial Neural Network, Decision Tree, Hidden Markov Model, Multinomial Naïve Bayes, Naïve Bayes, Random Forest, Support Vector Machine, and ZeroR. Those trained module were compared using 6 different parameters: Accuracy, Mean Absolute Error, Precision, Recall, F-Measure, and False Positive Rate. Features from the dataset are extracted. The extracted features and labels are supplied to the classifier. The classifier learns from the input provided and produces a model.

4.2 Model Implementation Phase

Once the model is trained, it is then used for prediction. Different components that are active after the model is implemented are shown in Fig. 2.

System Monitoring Unit. The System Monitoring Unit utilizes the values sensed by lm-sensors to maintain the profile. The profile generated by the System Monitoring Unit in conjunction with our machine learning model is used to predict fault.

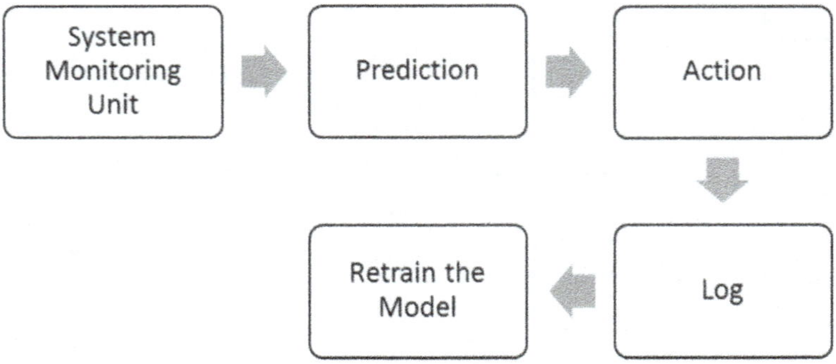

Fig. 2. Model in action

Prediction. Based on the features extracted from the profile generated by the System Monitoring unit the model trained in phase one makes a prediction. Different types of predictions that are made by the model are: TP, TN, FP, and FN. If the prediction made is a fault then a warning is generated.

Action. Based upon the prediction made different actions those are taken. As summarized in the Table 1 preventive actions are taken in case of TP and FP whereas no action is taken in case of TN and FN.

Log. This component is responsible for maintaining log of each incident. Log thus maintained is used for analysis purpose. As well as it is used to retrain the model.

Retrain. In the long go of the model it is required to continuous upgrading of the model for addressing new challenges and unsolved issues with the current version of the model. Model with the new set of data is retrained to enhance its predicting capability.

Table 1. Prediction and Respective Action

Prediction	Action
TP	Preventive Action
TN	No Action
FP	Preventive Action
FN	No action

5 Experiment Setup

A machine with Windows 8 Operating System, 8 GB RAM, Inter Core i5 processor, was set up for the experiment. It had Weka version 3.7 installed on it which was used for training and testing model. SHARPE was used for modeling the system and computing the availability, downtime and reliability of the system. Using the 10-fold cross validation it was made sure that all of the statements were used for testing purpose. Each fold contains 1799 instances as training set and 200 instances as test set.

5.1 Model Selection

In this experiment using Weka, 8 different models were trained using 8 different machine learning classifiers: Decision Tree, Hidden Markov Model, Multinomial Naïve Bayes, Naïve Bayes, Artificial Neural Network, Random Forest, Support Vector Machine and ZeroR. It is observed that model trained using Artificial Neural Network has the highest accuracy (99.13), highest Precision (0.989), highest Recall (0.989), highest F-Measure (0.989), lowest Mean Absolute Error (0.0128) and lowest False Positive Rate (0.011). Thus it was decided to use Artificial Neural Network for training the model. The result of the experiment is summarized in Table 2:

For computing the availability, reliability and downtime of the system values of precision, recall, false positive rate were taken from the model selected i.e. Artificial Neural Network and the values of MTTF = 999 h, MTTR = 1 h, lead-time $\Delta l = 1$ min, PTP = 0.4, PFP = 0.1, PTN = 0.01, k = 2 were taken randomly [6].

Table 2. Performance Summary of models

Classifiers	Accuracy (%)	Mean Absolute Error	Precision	Recall	F-Measure	False positive rate
Artificial Neural Network	99.13	0.0128	0.989	0.989	0.989	0.011
Decision Tree	96.47	0.0395	0.965	0.965	0.965	0.035
Hidden Markov Model	54.68	0.5	0.299	0.547	0.387	0.547
Multinomial Naïve Bayes	72.16	0.2818	0.727	0.722	0.723	0.274
Naïve Bayes	88.05	0.2651	0.884	0.882	0.882	0.115
Random Forest	97.49	0.0656	0.978	0.977	0.977	0.022
Support Vector Machine	81.49	0.1826	0.817	0.817	0.817	0.188
ZeroR	54.68	0.4956	0.299	0.547	0.387	0.547

5.2 Availability, Downtime and Reliability

To calculate the availability, system were modeled using Continuous Time Markov Chain (CTMC) and the Availability of the system with Proactive Fault handling and system without PFH is calculated manually as well as using SHARPE. Availability of the system with PFH was 0.99978 whereas availability of the system without PFH was 0.99900. Figure 3 illustrates the relation between availability of the system with different Repair Rate from 1 time a day to 48 times a day.

Downtime of the system with Repair rate 1 time in a day was 115.322 min with the availability 0.99978 in system with proactive fault handling whereas Downtime of the system with Repair rate 1 time in a day was 525.0749 min in system without proactive fault handling. The relation between downtime of the system with different Repair Rate from 1 time a day to 48 times a day is illustrated in Fig. 4.

System were modeled using Continuous Time Markov Chain (CTMC) to calculate the Reliability and the calculation was divided into two phases i.e. Reliability of Proactive Fault handling system and Reliability of system without Proactive Fault Handling.The system are always fully operational when given time (t) equals to zero, that's why calculation was done in various given time for both system, the system with PFH and system without PFH. This Graph in Fig. 5 gives the clear view of Reliability of PFH based system versus Reliability of a system without PFH for different time (t) in hour.

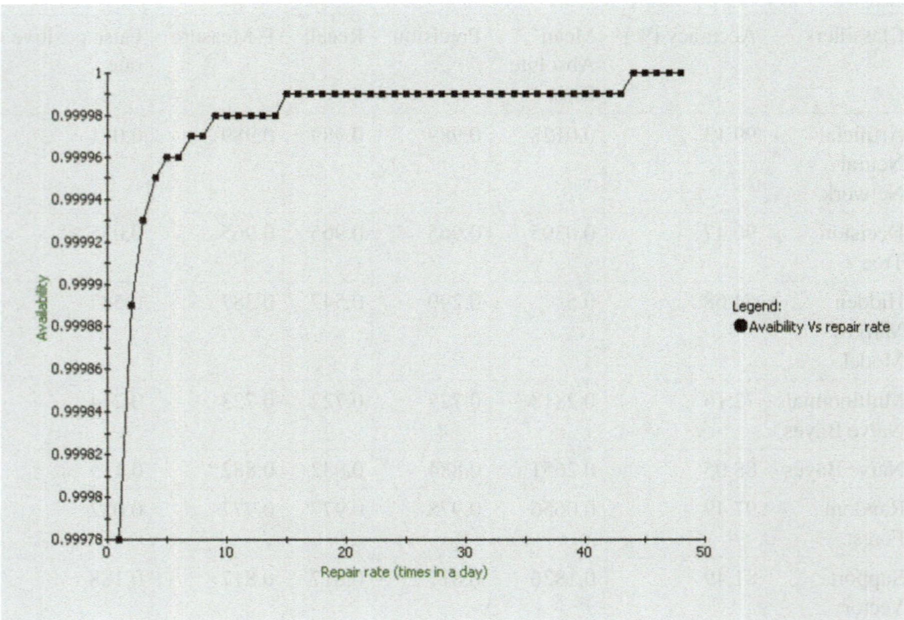

Fig. 3. Repair Rate versus Availability

From the experiment it was found that Reliability of Proactive Fault Handling System is higher than that of system without Proactive Fault Handling.

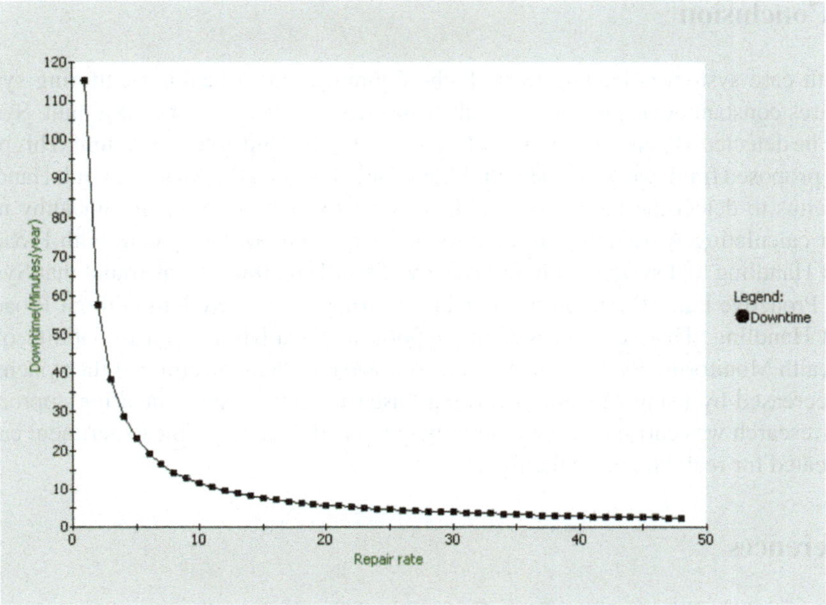

Fig. 4. Repair Rate versus Downtime

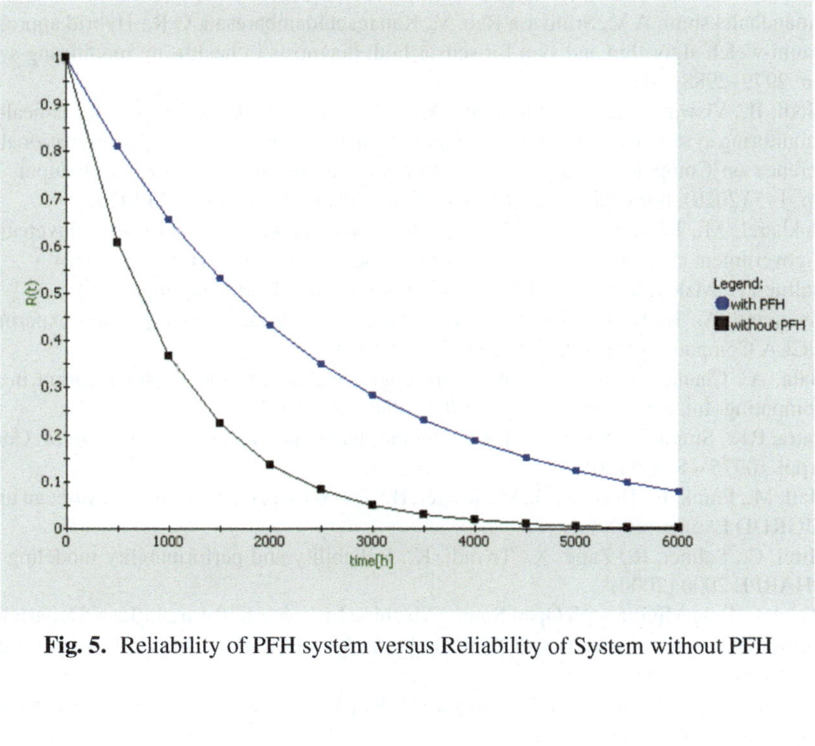

Fig. 5. Reliability of PFH system versus Reliability of System without PFH

6 Conclusion

Health care system is leading towards the digitalization. E-health monitoring system requires constant details of the hospital. At some point, there can occur a fault. System must be detected, diagnosed, recovered from faults to be fault-tolerant system. This paper have proposed framework of E-Health Monitoring System with Proactive Fault Handling that aims to detect the fault early and take corrective action to repair unhealthy node. After calculating Availability, Reliability and Downtime of the system with Proactive Fault Handling and system without Proactive Fault Handling, it was found that System with Proactive Fault Handling has higher performance over System without Proactive Fault Handling. This paper proved the hypothesis "Availability and Reliability of the E-Health Monitoring System can be increased whereas the Downtime of the system can be decreased by using Machine Learning based Proactive Fault Handling approach". This research was carried out on randomly generated data; the same experiment can be replicated for real data set in the future.

References

1. Laprie, J.C.: Dependability: Basic Concepts and Terminology. Springer, Wein, New York (1992)
2. Alahmadi, A., Soh, B.: A proposed methodology for an E-health monitoring system based on a fault-tolerant smart mobile (2014). https://doi.org/10.1007/978-3-319-01766-2_84
3. Anandhalekshmi, A.V., Srinivasa Rao, V., Kanagachidambaresan, G.R.: Hybrid approach of baum-welch algorithm and svm for sensor fault diagnosis in healthcare monitoring system, pp. 2979–2988 (2022)
4. Godi, B., Viswanadham, S., Muttipati, A.S., Samantray, O.P., Gadiraju, S.R.: E-healthcare monitoring system using IoT with machine learning approaches. In: 2020 International Conference on Computer Science, Engineering and Applications (ICCSEA), Gunupur, India, pp. 1–5 (2020). https://doi.org/10.1109/ICCSEA49143.2020.9132937.(2020)
5. Pokharel, M., Park, J.S.: Increasing system fault tolerance with software rejuvenation in e-government system. IJCSNS Int. J. Comput. Sci. Netw. Secur. 160–166 (2010)
6. Salfner, F., Malek, M.: Reliability modeling of proactive fault handling (2006)
7. Avizienis, A., Kelly, J.P.: Fault tolerance by design diversity: concepts and experiments. UCLA Computer Science Department. IEEE (1984)
8. Bala, A., Chana, I.: Fault tolerance- challenges, techniques and implementation in cloud computing. Int. J. Comput. Sci. Issues **9**(1), 288–293 (2012)
9. Patra, P.K., Singh, H., Singh, G.: Fault tolerance techniques and comparative. Int. J. Comput. Appl. (0975 – 8887) (2013)
10. Hall, M., Frank, E., Holmes, G., Pfahringer, B.: The WEKA data mining software: an update. SIGKDD Explor. (2009)
11. Hirel, C., Sahner, R., Zang, X., Trivedi, K.: Reliability and performability modeling using SHARPE 2000 (2000)
12. Weka 3 - Data Mining with Open Source Machine Learning Software in Java. Department of Computer Science, University of Waikato. http://www.cs.waikato.ac.nz/ml/weka/. Accessed 14 Jan 2016
13. Nepal Demographic and Health Survey 2022: Key Indicators Report. Ministry of Health and Population, Kathmandu, Nepal (2022)

14. Balbastro, F., Capozucca, A., Guelfi, N.: On the integration of mobility in a fault-tolerant e-health web information system. In: 21st International Conference on Advanced Information Networking and Applications Workshops (AINAW 2007). IEEE, Niagara Falls (2007)
15. Balbastro, F., Capozucca, A., Guelfi, N.: Analysis and framework-based design of a fault-tolerant web information system for m-health. Serv. Oriented Comput. Appl. **2008**, 111–144 (2008)
16. Gia, T.N., Rahmani, A.-M., Westerlund, T., Liljeberg, P., Tenhunen, H.: Fault tolerant and scalable IoT-based architecture for health monitoring. In: 2015 IEEE Sensors Applications Symposium (SAS). IEEE, Zadar (2015)
17. Nepal, Government of. Digital Health (n.d.). https://digitalhealth.mohp.gov.np/public/. Accessed 5 Sept 2023

Application of Blockchain Technology in the Context of a Security Operations Centre (SOC) Log Management

Carlos Tomaz[1], Inês Marques[1], Lina Galheto[1], Joel Curado[1], and Joao C. Ferreira[1,2,3(✉)]

[1] ISTAR, ISCTE-Instituto Universitário de Lisboa, 1649-026 Lisbon, Portugal
jcafa@iscte.pt
[2] Inov Inesc Inovação—Instituto de Novas Tecnologias, 1000-029 Lisbon, Portugal
[3] Faculty of Logistics, Molde University College, NO-6410 Molde, Norway

Abstract. In the era of digital transformation and the ever-evolving threat landscape, Security Operations Centres (SOCs) play a pivotal role in safeguarding organizations against cyberattacks and ensuring data integrity. The management and security of logs generated by various systems and devices are critical aspects of SOC operations. Traditional log management systems, however, face challenges related to data tampering, trust, and transparency. This work aims to evaluate the application of Blockchain technology in one of the operational processes (Security Operations) of companies that provide services in information security. We intend to understand the advantages and challenges of adopting this technology and how it can be a solution to take into account in order to optimize the services provided by these companies, or even give rise to new services.

Keywords: Blockchain · Data Protection · Data Security · SOC · Security Operations · Logs · Events

1 Introduction

In the business context, access control to information, including authentication and authorization, is one of the fundamental security mechanisms to be implemented in almost all information technology systems [1]. Although there are different ways of implementing access control, most systems rely on a centralized part, which could compromise the integrity and privacy of company data due to the poor security offered in this structure [2].

In this sense, the introduction of Blockchain technology has emerged as a way of solving this problem, eliminating the need for intermediaries, and guaranteeing security in the exchange and sharing of information by means of transaction records that are difficult to manipulate [3].

Blockchain is characterized by decentralization, immutability, transparency, and traceability. In the company that is the target of this work, we tried to identify where one

V. Sakalauskas et al. (Eds.): IBICA 2023, LNNS 1231, pp. 192–196, 2025.
https://doi.org/10.1007/978-3-031-78946-5_18

or more of these characteristics would be applicable, and its applicability to preserving the immutability and traceability of the systems that generate alerts and security incidents is evident. This application stems from the increasingly pressing need to prove with as much guarantee as possible that certain occurrences and actions have been recorded chronologically and without tampering so that we can determine who, what and when.

This work therefore assesses the applicability of Blockchain technology in the event, alert, and incident management process of a SOCaaS (Security Operation Centre as a Service) service.

2 Literature Review

As the authors state in [4], the concept of Blockchain first appeared in Satoshi Nakamoto's article, "Bitcoin: a peer-to-peer electronic cash system", in 2008, and was initially applied to Bitcoin and virtual currency trading systems, offering a distributed and secure system for recording transactions and storing data. Since then, and according to the article [5], the continuous innovation provided by Blockchain has promoted the development of Industry 4.0, with the adoption of this technology being a tool for companies to improve their competitiveness in the context of digital transformation.

Blockchain is defined by the authors in [6] as a decentralized, immutable, and distributed record of transactions that is maintained by a network of nodes. This technology works by creating blocks of transactions that are validated and added to a chain of blocks through consensus between the nodes of the network. Each of these blocks contains the record of the transactions that have been verified and validated cryptographically, making it difficult to tamper with the information stored in these blocks, thus guaranteeing the reliability and immutability of the data.

In this sense, the authors in [5] propose the adoption of this technology as a driver of the digital transformation of companies from 3 perspectives:

1. Building a new digital transaction model: in which Blockchain is considered an electronic database, where each node in the network stores the network's information, breaking with the traditional centralized structure and forming a distributed storage system. This technology makes it difficult to tamper with data, guaranteeing authenticity and reliability. Thus, this new model can help companies cooperate reliably with unknown partners and provide better services to their customers [5].
2. Consolidation of data management capabilities: each node in the blockchain network stores data in the system, helping to solve the problem of "data silos", making the process open, transparent, and improving the efficiency of collaboration. In addition, Blockchain's asymmetric cryptographic technology ensures secure information sharing without revealing private data, helping companies manage their data more securely during the digital transformation process [5].
3. Integration of Smart Contracts for greater efficiency in work processes: companies can establish predetermined conditions and terms in Smart Contracts in code form. When these conditions are verified, the Smart Contract is executed automatically, increasing the efficiency of the process. Because the data on the Blockchain is difficult to tamper with, security and trust in the execution of the process are guaranteed [5].

Despite the advantages mentioned above, some authors also point out that there are some disadvantages to adopting this technology. The authors in [8] point out the technical complexity required to apply this tool and the performance overhead involved in running this solution. In addition, these authors warn of the eventual impossibility of protecting against a malicious node belonging to the network, which could reveal transaction logs.

In addition, the authors in [6] warn of the lack of regulation and standardization in this technology, pointing to the need to work on access control and privacy policies that work in favour of all those involved and that are carried out from a legal point of view. The lack of standardization of protocols in the Blockchain structure is also one of the challenges presented by the authors.

As for the main types of Blockchain in existence, the authors in [7] describe that, based on their access mechanism, they can be public or private. While the article [5] suggests adopting the Private Blockchain typology in organisations' internal information management system, [1] indicates a Permissioned Blockchain, based on the underlying assumptions of a private network for the business context, suggesting the use of Hyperledger Fabric as a platform.

3 Blockchain

The Blockchain works like a Ledger which keeps a digital record of all transactions.

Given the etymology of the word, Blockchain: is a chain of blocks - which holds digital information from the record of transactions, stored in a public or private database is kept permanently in a blockchain format, where each block references the previous one through hash algorithms (unique identifier for each set of data). In this way, control information on the overall structure of the chain is held, also in hash format.

Hash calculation algorithms are fundamental in supporting Blockchain technology, especially in guaranteeing immutability and trust in the chain. In this way, the blockchain configuration ensures that blocks cannot be changed in order, thus ensuring their traceability.

On the other hand, if network participants are given permission to visualize transactions, information transparency is achieved.

Communications Peer-to-Peer

By keeping a digital record of all transactions on a peer-to-peer network, the parties involved in the transaction operate directly with each other without needing third-party intervention. All information transferred in this way is encrypted at each occurrence and recorded in the Ledger, making the process secure and reliable.

Decentralized System

As there are no "intermediaries" and no "central authority", decisions are validated by consensus algorithms between Blockchain nodes.

Distributed System

As Blockchain is a decentralized system, the risk of successful attacks is reduced, as there is a copy of the Blockchain on every node. Even if one or more "nodes" are successfully attacked, the network keeps the information replicated, allowing it to be recovered.

Smart Contracts
Traditional contracts are present in our daily lives practically all the time. Since ancient civilizations, different parties have negotiated and stipulated conditions between themselves by establishing clauses to make contracts.

A Smart Contract is nothing more than a contract made by programming that can be processed on a blockchain network, which will be signed if the parties involved in the process agree to the stipulated conditions. As they are written by means of programming, all the contractual conditions and information are recorded in this code.

Since Blockchain is a technology that allows transactions to be concluded from any geographical location, Smart Contracts become faster and more accessible than traditional contracts. Naturally, there are always legal issues, particularly jurisdictional ones, which must be legally assessed.

Blockchain Main Features
Figure 1 below summarizes the main attributes of Blockchain technology. According to the authors in [5], there are three types of Blockchain: Public, Private or Consortium.

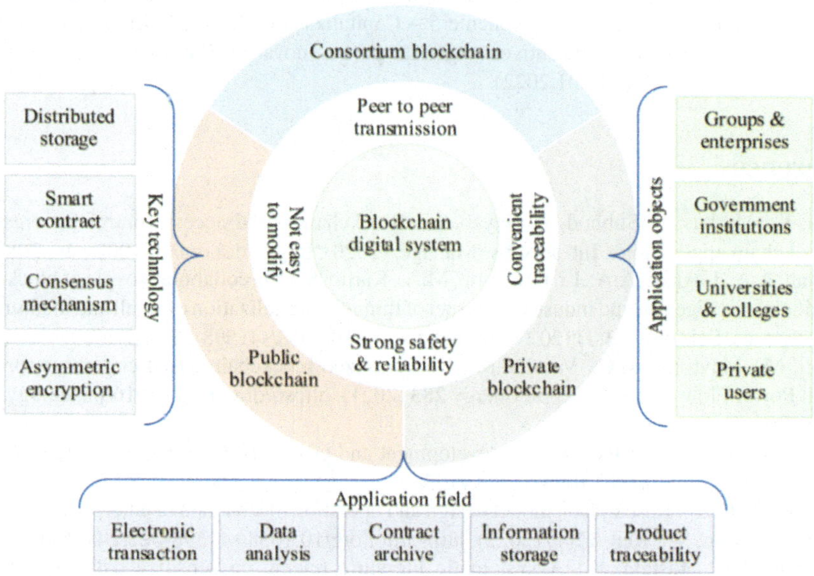

Fig. 1. Main attributes of blockchain technology [5].

4 Conclusion

During the course of this work, we came across some questions that required more in-depth research. In this sense, and through access to various literature sources, we sought answers and made decisions to achieve the objectives proposed in the work.

This use case was revealed to be implementable and improved on the current state, not only because of the increased confidence that customers have in the information processed by the service, the resilience and confidentiality of the data but also because of the ability to automate procedures that are currently manual through the use of Smart Contracts. In the event that data has to be used as evidence, whether for disciplinary proceedings or in court, this technology can increase confidence in the immutability of that evidence.

In addition, we cannot fail to mention the costs and high technical complexity associated with a Blockchain implementation, not only because of the need for highly specialised human resources, but also because of the high investment required in the technological architecture, both in the implementation phase and in the necessary maintenance of the solution.

We, therefore, conclude that Blockchain technology allows us to implement or improve services by increasing confidence in the security of the technology and the information it processes. However, as it is a technology with little maturity, changes are constantly being made to existing technologies and new solutions are being presented, so we need to be attentive and keep a close eye on this evolution.

Acknowledgments. This work was supported by the project BLOCKCHAIN.PT (RE-C05-i01.01 – Agendas/Alianças Mobilizadoras para a Reindustrialização, Plano de Recuperação e Resiliência de Portugal na sua componente 5 – Capitalização e Inovação Empresarial e com o Regulamento do Sistema de Incentivos "Agendas para a Inovação Empresarial", aprovado pela Portaria N.° 43-A/2022 de 19.01.2022).

References

1. Xu, L., Markus, I., Subhod, I., Nayab, N.: Blockchain-based access control for enterprise blockchain applications. Int. J. Netw. Manage. (2020). https://doi.org/10.1002/nem.2089
2. Khan, A.A., Laghari, A.A., Li, P., Dootio, M.A., Karim, S.: The collaborative role of blockchain, artificial intelligence, and industrial internet of things in digitalization of small and medium-size enterprises. Sci. Rep. **13**(1) (2023). https://doi.org/10.1038/s41598-023-28707-9
3. Foti, M., Mavromatis, C., Vavalis, M.: Decentralized blockchain-based consensus for Optimal Power Flow solutions. Appl. Energy **283** (2021). https://doi.org/10.1016/j.apenergy.2020.116100
4. Zhao, W.: Blockchain technology: development and prospects. Natl. Sci. Rev. **6**(2), 369–373 (2019). https://doi.org/10.1093/nsr/nwy133
5. Su, X., Wang, S.: Research on model design and operation mechanism of enterprise blockchain digital system. Sci. Rep. **12**(1) (2022). https://doi.org/10.1038/s41598-022-24796-0
6. Khanum, S., Mustafa, K.: A systematic literature review on sensitive data protection in blockchain applications. Concurr. Comput.: Pract. Exp. **35**(1) (2023). https://doi.org/10.1002/cpe.7422
7. Du, R., Ma, C., Li, M.: Privacy-preserving searchable encryption scheme based on public and private blockchains. Tsinghua Sci. Technol. **28**(1), 13–26 (2023). https://doi.org/10.26599/TST.2021.9010070
8. Joseph, A., Yadav, N., Ganapathy, V., Behl, D., Jayachandran, P.: Data protection in permissioned blockchains using privilege separation. In: 2023 15th International Conference on COMmunication Systems and NETworkS, COMSNETS 2023, pp. 748–756 (2023). https://doi.org/10.1109/COMSNETS56262.2023.10041304

Multiparty Trust Levels in Evidence Management: Ensuring Tamper-Proof Chain of Custody in Blockchain

Nuno Santos[1](✉), Joel Curado[1,3], and Joao C. Ferreira[1,2,3]

[1] ISTAR, ISCTE-Instituto Universitário de Lisboa, 1649-026 Lisbon, Portugal
nuno.mbr.santos@gmail.com

[2] Inov Inesc Inovação—Instituto de Novas Tecnologias, 1000-029 Lisbon, Portugal

[3] Faculty of Logistics, Molde University College, NO-6410 Molde, Norway

Abstract. International Criminal Courts are a guarantor that justice can be achieved for the most egregious crimes against humanity and that surviving victims from those crimes can live with the assurance that the perpetrators are held accountable for their actions. Evidence sits at the core of the judicial process, and the participants rely on it to be authentic, integer and untampered with to ensure the fairness of the proceedings. Without enforcing powers, these International Criminal Courts could rely extensively on evidence provided by other parties to build a strong case from inception to appeals stage. With the increased digitization, size and complexity of evidence, traditional methods of managing the chain of custody are becoming vulnerable to successful challenging of admissibility of evidence. Blockchain could be the answer for strengthening the Chain of Custody in evidence management and mitigate these identified threats, as this technology brings essential characteristics such as time-stamping, authentication, immutably and trust among independent parties. This paper conceives and designs a blockchain-based solution that maintains a tamper proof chain of custody and multilevel trust in evidence management applied to the International Criminal Court. The convergence of International Criminal Justice with Blockchain Technology is notably absent within existing academic discourse, and arguably in one of the most unstable and conflictual periods in human history, this paper offers a timely approach to deliver justice in a more effective and credible way.

Keywords: Blockchain · Chain of Custody · Evidence Management · Court · Judicial · International Criminal Justice · International Criminal Court

1 Introduction

In legal proceedings, the chain of custody of evidence is crucial for establishing its authenticity and reliability. However, challenges like evidence tampering, lack of transparency, and trust issues within the legal ecosystem have led to the exploration of technological solutions. Specific to International Criminal Courts, they are highly scrutinized and are usually involved in high-profile cases, including trying powerful persons for

V. Sakalauskas et al. (Eds.): IBICA 2023, LNNS 1231, pp. 197–209, 2025.
https://doi.org/10.1007/978-3-031-78946-5_19

crimes against humanity. These proceedings are usually public and have a broader audience than the judicial actors themselves, leading to wider cultural and societal narratives about the crimes being tried [1]. The success of International Criminal Courts, in particular, will be perceived based on the fairness of the trials in charging the accused, and this is highly dependent on the ability of the Prosecution, Defence and the Court's Registry to manage evidence properly since collection until disclosure [2]. The cost of submitting, discussing, and deliberating on false or tampered with evidence could be as high as to permanently tarnish the reputation of such Courts at a multitude range of stakeholders, from Member States who fund these institutions to victims of those alleged crimes.

The motivation behind this thesis stems from the increasing need to address the challenges and complexities associated with situations and cases where the evidence produced comes from multiple parties, with varying trust levels. This research takes the International Criminal Court (ICC) as the beneficiary of the solution proposed. The ICC was formed under the guidance of an international treaty known as the Rome Statute [2]. Additional institutional details about the ICC's mission and structure are available through its official public information resources [4, 5]. Its creation was driven by the global aspiration to end impunity for the gravest crimes. The requirements for this solution are based on its public rules and procedures [2, 3], however it could be appropriate in many other related organizations.

In summary, this research has two main objectives: (1) Provide a systematized summary of the existing body of knowledge in the area of blockchain and judicial chain of custody; and (2) Design a Blockchain based system that supports a tamperproof chain of custody during the process of evidence acquisition, usage and disclosure made by independent multiple parties based on the publicly available information of the ICC.

The judicial process works through several iterating phases, where each one could trigger the next one, or end there, if no justification to proceed exists.

Evidence collection and custody assumes a fundamental role in ensuring the Court proceedings are done in a fair and expeditious way. As there is always a time gap between the occurrence of crimes, its assessment and investigation, it is very important to ensure that the evidence collected is captured as early as possible in the process to reduce the chances of tampering with it, and remains authentic and valid until is assessed by the court's Chambers.

2 Related Work

A systematic literature review was made using PRISMA methodology [6]. The selection of the studies to be included in this review, and within the scope of this research, consisted in all studies that focused on Blockchain and the use case of Chain of Custody within the legal or investigative area. The search string chosen, was therefore, the following: Blockchain AND "Chain of Custody" AND (Court OR Judicial OR Evidence). This string was run in 4 databases on the 22nd June 2023, namely Scopus, Web of Science, ACM Digital library and IEEE.

A total of 65 documents were screened, after removing a total of 41 duplicate records from the databases searched. 12 were not able to be retrieved without incurring additional costs. 8 studies with IoT in the title were removed as it deviates from the purpose of this research. Once the first screening of studies was concluded, 45 articles were assessed for eligibility. 15 of these had an abstract which was not relevant, and for other 3, after analyzing the whole article, 2 didn't fit the scope of the research and 1 was in essence a duplication of another relevant article, despite the different title. In summary, 27 studies qualified for the systematic review.

2.1 Literature Review

The literature around this topic is still very nascent, with the first related study being published only in 2018. The studies were assessed on 5 key attributes that fit with the area of this research, being: 1) Blockchain: Does the solution rely on blockchain technology?; 2) Chain of Custody: Is there a mechanism to guarantee a chain of custody for data, in particular evidence?; 3) Evidence Management: Does the solution offer evidence management capabilities, through the evidence lifecycle?; 4) Multi-level trust: Does the solution support data transfer and/or data access by different, independent parties?; and 5) Public Sector: Is there a use case identified for organizations in the public sector?

Once the studies were reviewed, it was possible to group them based on these categories, to more clearly understand the area they focus on and how it relates to this research. As such, and based on the key attributes of comparison with this research, the articles were allocated to one of the five groups identified by a letter:

- A – Research solely blockchain and Chain of Custody attributes;
- B – Research addresses evidence management with Chain of Custody on Blockchain, but not multi-level trust nor the public sector;
- C – Research covering all attributes except 5. Public Sector;
- D – Research covering all attributes except 4. Multi-level trust;
- E – Individual studies which had unique coverage of attributes.

2.2 Literature Review Conclusions and Gaps Identified

The literature available within this topic clearly demonstrates the increased attention to the value of blockchain in maintaining tamperproof chain of custody and supporting the management of evidence in such technology.

It is clear how blockchain technology can improve the resiliency and integrity of a chain of custody for many areas, including forensic analysis or evidence management. The studies analyzed approach this subject in different ways.

No study reviewed would encompass all 5 attributes that this research touches on, and as such this is a good addition to the body of knowledge within this area. Table 1 depicts the articles, grouped by attribute set and compares them with this research.

Other gap that was identified in the literature reviewed is the absence of use cases and solutions tailored for the international legal public sector, or a sector where a network of providers feed the main entity with evidence.

3 MultiTrustBloc (MTB) Framework

In the realm of international jurisprudence, the ICC stands as a beacon for prosecuting heinous crimes that resonate on a global scale. A pivotal component of this mandate is the meticulous management and authentication of evidentiary materials, which are

Table 1. Literature review overview and comparison with this research

Group	Blockchain framework	Chain of Custody	Evidence Management	Multi-level trust	Public Sector	Total
A: [7–10]	Y	Y	N	N	N	4
B: [11–17]	Y	Y	Y	N	N	7
C: [18–25]	Y	Y	Y	Y	N	8
D: [26–30]	Y	Y	Y	N	Y	5
E: [31]	Y	Y	N	Y	Y	1
E: [32]	Y	Y	N	N	Y	1
E: [33]	Y	N	N	N	N	1
This research	**Y**	**Y**	**Y**	**Y**	**Y**	1

frequently procured from a myriad of sources, encompassing national judiciaries, non-governmental organizations, and other international entities. These conventional systems, while comprehensive, occasionally grapple with challenges related to ensuring unbroken chains of custody and the inviolability of evidence. The potential for human oversight, inadvertent errors, or deliberate manipulation cannot be entirely discounted. Given this backdrop, a compelling exigency emerges for an innovative, impregnable, and streamlined system dedicated to evidence management and authentication.

The ICC could grapple with challenges pertaining to evidence management, particularly in ensuring the authenticity, and chain of custody of evidentiary materials. This chapter elucidates the objectives of integrating a blockchain solution, hereinafter called MultiTrustBloc (MTB) within the ICC's operational framework, specifically addressing the identified problems.

- **Ensuring Tamper-Proof Evidence Management:** To establish an immutable and chronological record of evidence, ensuring that it cannot be altered retroactively once an item is entered into the blockchain.
- **Facilitating Transparent Chain of Custody:** To create a transparent and traceable record of every interaction with a piece of evidence, from its inception to its eventual use in court proceedings.
- **Streamlining Collaboration with Third Parties:** To foster a seamless and secure platform for collaboration between the ICC and external entities, ensuring that evidence shared by third parties retains its authenticity and integrity.
- **Enhancing Verification and Authentication Processes:** To expedite the process of evidence verification, ensuring rapid yet robust authentication mechanisms.
- **Bolstering Public Confidence and Accountability:** To enhance the public's trust in the ICC's evidence management processes, ensuring that justice is not only done but is seen to be done.

As the MTB solution aims to support the whole of the Judicial Lifecycle of the ICC, therefore distinct actors are involved at different stages of the process. Table 2 describes the roles each one of them have, together with a description of the privileges they would have at what stage of the process.

Table 2. MultiTrusBloc identified roles

Role	Description	High level view of permissions
Prosecution team	Determines the data deemed as evidence and submitted to the Court Chamber	Until evidence disclosure, the team can view and control any artifact within the MTC
Accredited entity	The entity cooperates with the Prosecution team to identify and gather information, either directly or through trusted partners	it can create and submit evidence records in the MTC, accessing the artifact until the Prosecution team takes ownership
Non-accredited entity	May include civil organizations or citizens seeking to expose committed crimes	Submits potential evidence to an accredited entity or the prosecution team for evaluation
Court Chamber	Assesses submitted evidence and delivers judgement accordingly	Receives and takes control of the evidence once handed over by the prosecution team
Involved third parties	Have different objectives and could represent different people or groups in the specific case	Parties such as Defence or Victims representatives may require evidence disclosure to participate in the proceedings

4 Process Flows Identified

This process starts in the moment that any entity identifies an artifact as potential evidence, and depending on the actor, it is automatically added to the MTB or submitted for review. A visual depiction of the process can be found in Fig. 1, using BPMN [34].

The process can have multiple starting points, and what differentiates them is the stage in the process they start in, and what actor triggers it.

1. *Submits data to accredited entity:* The entity, holding potential valuable information contacts an accredited organization that has direct relationship with the Court.
2. *Validates suitability of data:* When the data is received, the receiver will validate its suitability against ICC's Rome Statute [2] and rules of procedure and evidence [3].
3. *Informs local citizen/entity of its inadmissibility:* If the data provided doesn't meet the minimum admissibility threshold, the evaluator informs the entity that submitted the data, that it cannot be considered as evidence and will not be processed further.
4. *Treats data as evidence and creates snapshot:* If the analysis results are positive to potential evidence admissible to the ICC, the accredited entity will ensure the record is ready to be submitted to the MTB for Prosecution's assessment. The evidence is stored and baselined to ensure no future tampering possibilities. This could be the first activity of the process, if the Accredited Entity is the one acquiring the evidence.

5. *Creates records in MTB:* as soon as the data is ready for submission, the accredited entity will create a record in the MTB, meaning the first block for this artifact is created. A notification will be sent later to the Prosecution team for review.
6. *Reviews Evidence:* As the Prosecution team is notified, it will review the submission by the accredited entity, based on the current status of the existing or potential case. During the review process, the MTB is accessed and any changes to the record will be reflected through the creation of new blocks. The outcomes of this review will decide on the suitability for the case.
   a. Updates Record in MTB setting status as inadmissible: If the review outcomes are unsatisfactory, a new block in the MTB is created to reflect this update.
   b. Informs accredited entity of its inadmissibility: The prosecution team informs the accredited entity with the conclusions of its review.
7. *Creates record in MTB:* Derives from the Prosecution's Team direct action, which has acquired evidence at the source and creates the record immediately.
8. *Updates record in MTB and retrieves evidence to its control:* The record in the MTB is enriched with additional relevant metadata that is required to fulfil the acceptable criteria for judicial evidence. The status of the artifact is updated and committed with the creation of a new block in MTB, and the record is finally accepted as evidence.

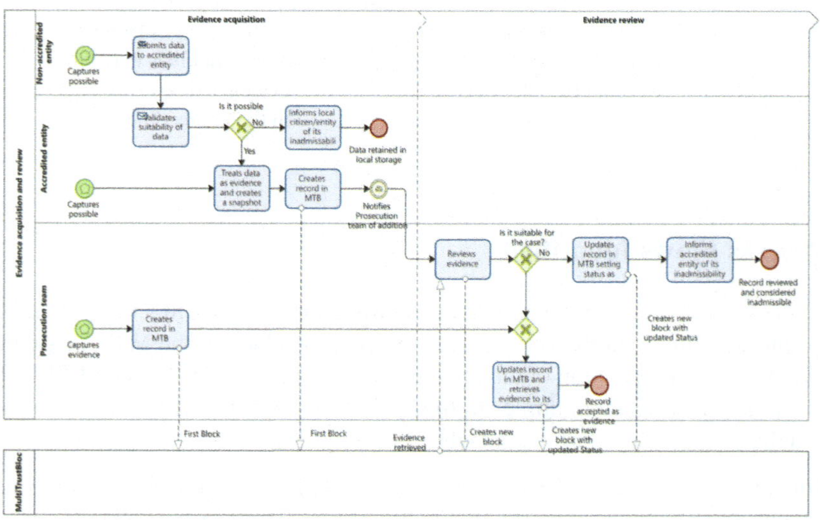

Fig. 1. Evidence acquisition process

Once the submission is analyzed and considered as evidence, it resides under the Prosecution team's custody for review and for building a case against a suspect. This process starts with the need to evaluate specific evidence, visually depicted in Fig. 2.

1. *Reviews evidence:* The Prosecution team fetches the evidence record from the MTB and does the analysis it needs within the context of its mandate.
   a. Reads the data: In case the review doesn't require an update to the record, the action is only to read the data within the record.

   b. Updates the record: If there is the need to update the record with the outcomes of the review, the changes are written and committed into the MTB by creating a new block to represent such action.
2. *Closes record:* Once the actions are done and the judicial process remains in the same stage, the record is closed, and the process finishes then.
3. *Prepares disclosure:* If the evidence becomes relevant to advance to another state and be shared with other parties, the Prosecution team will prepare its disclosure.
4. *Selects privileged participants:* as the disclosure process requires the sharing of evidence to third parties, the Prosecution team identifies which participants will assume specific privileges. For example, which (Pre-)Trial Chamber will assume ownership of the record and evidence and which other teams can access it, such as Defence or Victims representatives' teams.
5. *Submits record disclosure:* The Prosecution team commits the updates and a new block is created to reflect the ownership and access rights changes.
   a. Hands over ownership of record and data: after submitting the record disclosure, the prosecution team transfers the evidence to the Chambers data repository and relinquishes its ownership.
   b. Receives ownership of record and data: the Judges Chamber is notified of evidence transfer and custody in their control.
   c. Receives notification of access to the record: Involved parties are notified that new evidence is available for their review.
6. *Accesses the record*: All three actors, throughout the different judicial stages, access the record as required and if they have the rights, they will add new blocks to reflect changes in the record. This could be actions like presentation in court, link to another case, etc.

Once the different Judicial Stages are concluded, the process ends.

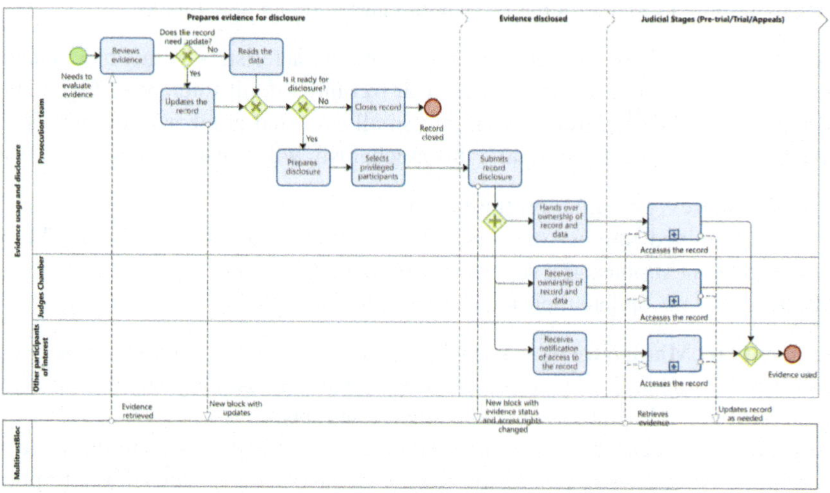

Fig. 2. Evidence usage and disclosure process

5 High Level System Overview

As designed, the MTB is a registry for the status of an artifact with potential evidentiary value for a crime within ICC's jurisdiction which is decentralized, thus tamper-proof. As such, it allows holders of potential evidence to submit their evidence to an entity accredited by the ICC, meaning that it is allowed to add a record in the MTB to register it as potential evidence, before evaluation. The ICC's Prosecution teams could have the role of Accredited Organization, for allowing a direct contact with sources.

When the accredited entity receives the submission and deems it suitable to share with the Prosecution team, a new record is created with the possible evidence baselined.

The accredited entity further validates the submission, and it is moved to the respective repository, which the prosecution team has access to.

Should the Prosecution team identify the artefact as valid evidence for their case, they will reflect that in the MTB and proceed to transfer the evidence into their custody. As the evidence is used, the MTB will be creating new blocks in the chain to reflect the change in status of that same evidence. Participants in the case will have different types of access to the records, depending on the state of that evidence.At the solution's foundation, there is a consortium between Prosecution, Judiciary and other interested parties to make it robust, credible and tamper-proof. Figure 3 provides a high-level view of how the system will interact with evidence and multiple parties.

Roles and Responsibilities - The MTB is supported by two groups of entities that bring different roles to the solution. The first and critical group is represented by the Court's Chambers and the Prosecution (Teams). These are the main decision makers on the future of the solution and permanent members of it. They propose and validate transactions, manage and change smart contracts, and have a final say in any disputes or changes to the blockchain's protocol. The other is the secondary entities group, consisting of any accredited entity by the primary entities that wishes to participate in the advancement of justice, helping to maintain the blockchain and validate new blocks.

Consensus Mechanism - Given the consortium characteristics of this blockchain, the consensus mechanism selected was practical byzantine fault tolerance [35], due to its performance in validating transactions and small expected number of malicious actors involved in the network. In summary, this mechanism will consist of three main stages:

1) Proposal phase: A member proposes a block
2) A pre-determined number of consortium members validate the block. The block moves to the next phase if a supermajority (e.g., 2/3) agrees.
3) All organizations update their local copy of the blockchain.

Smart Contract Management - The whole process is ruled by Smart Contracts, ensuring adherence to the judicial process and proper chain of custody. Only primary entities have the right to deploy new versions of smart contracts.

Any member can execute a new smart contract, which will go to its initial stage of the process, however only the Prosecution team will have sufficient privileges to promote any new artifact to evidence and under its custody.

The primary entities can only terminate a smart contract, which will be permanently stored in the blockchain for transparency and accountability reasons.

Access Control - All members in the consortium have read access to the blockchain, however some parts of the blockchain, including block specific data, will not be available to them, depending on the privileges each one has. They can also propose and add new blocks.

Dispute Resolution - In the event of disputes that impact the functioning of the MTB, the primary entities will have a key role in its resolution, particularly if those arise from secondary entities. Should the dispute arise between Prosecution and the Court Chambers, elected representatives from the state parties decide on how to settle the dispute.

Upgrades and Forks - Protocol Upgrades to the MTB are proposed jointly by the primary organizations and require a majority consensus from all consortium members.

In the unlikely event of a fork, the chain supported by the primary organizations is considered legitimate.

Audits and compliance - Ensuring the compliance of the MTB with legal standards is paramount, therefore regular audits should be conducted, mandated by the ASP to guarantee the integrity of the blockchain and smart contracts.

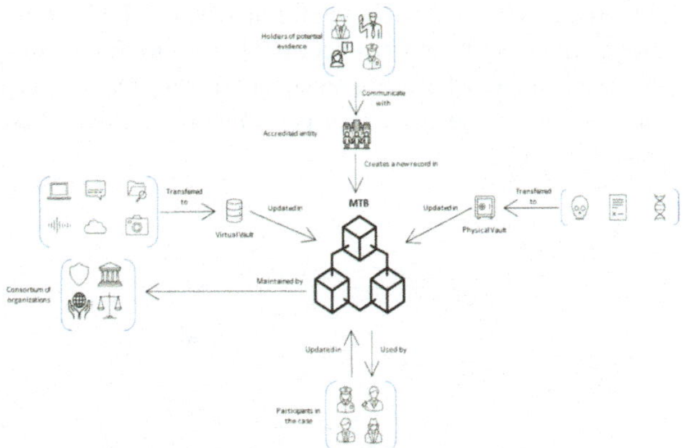

Fig. 3. High level view of MultiTrustBloc

Evidence Records State lifecycle - As depicted in Fig. 4, with the progress in the judicial process, from non-existent to appeals and reparations, the evidence evolves through different states, according to its validity as evidence within a specific case. The major lifecycle states from evidence are:

1. *Pre-Artifact Creation:* Before an artifact is formally recognized and documented within the ICC's system, the potential evidence or information resides externally.
2. *Artifact Created:* This is the inception point where a piece of evidence or an artifact is first identified and documented as such.
3. *Submitted for Prosecution Review:* The artifact, is presented for preliminary assessment by the prosecution team, determining its relevance to the case.

4. *Pending Review:* The evidence is under active scrutiny by the prosecution. It's being evaluated for authenticity, relevance, and overall value to the proceedings.
5. *Approved - Pending Transfer:* The evidence has been deemed valuable and relevant by the prosecution and is awaiting transfer to the next process phase.
6. *Rejected:* Evidence that does not meet the necessary criteria or is deemed irrelevant to the case is rejected at this stage.
7. *In Prosecution Custody:* The approved evidence is now formally in the custody of the prosecution, ready for further analysis, presentation, or other relevant activities.
8. *In Utilization:* This stage signifies the active use of the evidence in the legal proceedings, be it investigations, pre-trial hearings, or main trial sessions.
9. *Archived (from Prosecution):* Post utilization, if the evidence is not immediately required, it is archived for potential future use or reference.
10. *Submitted to the Judiciary:* Evidence that needs to be presented before the judicial chambers is submitted for their review and consideration.
11. *Evidence in Judiciary Custody:* The evidence is now under the formal custody of the judicial chambers, indicating its active consideration in judicial deliberations.
12. *Disclosed to Parties:* Relevant evidence is disclosed to all involved parties, ensuring transparency and adherence to the principles of a fair trial.
13. *Retained:* Post-disclosure, evidence is retained for a specified duration.
14. *Archived (from Retained):* After the retention period, the evidence is moved to long-term storage, ensuring it remains accessible for any future suitable purposes.
15. *Disposed:* The final stage where evidence, after serving its purpose and being archived, is either returned, destroyed, or permanently archived based on its nature and the policies of the ICC.

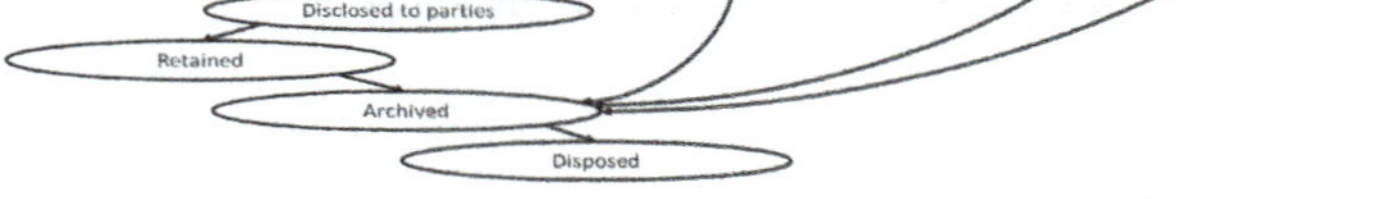

Fig. 4. Evidence Lifecycle stages

6 Conclusion

This paper underscores the profound impact that blockchain could have in reshaping how international legal entities handle and safeguard evidence with a particular look into the ICC. Ensuring the integrity of the chain of custody is an imperative in the realm of law, especially in criminal trials. The journey of evidence, from the moment it's gathered to its presentation in a courtroom, must remain unblemished.

The ICC's mandate often leads it to tread on complex terrains, handling cases of immense sensitivity that draw global scrutiny. In these intricate settings, the Court interacts with a plethora of stakeholders ranging from nation-states to non-governmental bodies. Building and maintaining trust among these diverse entities is paramount.

The solution proposed meets all identified requirements and leverages on Blockchain technology to provide an immutable, decentralized ledger, where the records are impervious to unilateral alterations. The use of smart contracts, can automate certain procedural aspects related to evidence, such as defining who can access certain pieces of evidence and when. This not only makes the process more efficient but also minimizes human errors, which can sometimes be costly in a legal context.

This area of research can be further expanded by identifying the most fit-for-purpose Blockchain platforms, as well as by creating a Proof-of-Value based on the solution conceived in this paper. Furthermore, identifying the specific attributes that would describe the evidence and the different stages they would be participating in would enable enriching the solution proposed.

References

1. Cogan, J.K.: The problem of obtaining evidence for international criminal courts. Hum. Rights Q. **22**(2), 404–427 (2000). https://doi.org/10.1353/hrq.2000.0021
2. Rome Statute of the International Criminal Court
3. Rules of Procedure and Evidence of the International Criminal Court
4. About the Court. International Criminal Court. https://www.icc-cpi.int/about/the-court. Accessed 10 Sept 2023
5. Public Information and Outreach Section, understanding-the-icc.pdf. Oude Waalsdorperweg 10, 2597 AK, The Hague, The Netherlands. Available: https://www.icc-cpi.int/sites/default/files/Publications/understanding-the-icc.pdf. Accessed 10 Sept 2023
6. Page, M.J., et al.: The PRISMA 2020 statement: an updated guideline for reporting systematic reviews. BMJ n71 (2021). https://doi.org/10.1136/bmj.n71
7. Ehrenberg, A.J., King, J.L.: Blockchain in context. Inf. Syst. Front. **22**(1), 29–35 (2020). https://doi.org/10.1007/s10796-019-09946-6
8. Olukoya, O.: Distilling blockchain requirements for digital investigation platforms. J. Inf. Secur. Appl. **62** (2021). https://doi.org/10.1016/j.jisa.2021.102969
9. Khan, A.A., Uddin, M., Shaikh, A.A., Laghari, A.A., Rajput, A.E.: MF-ledger: blockchain hyperledger sawtooth-enabled novel and secure multimedia chain of custody forensic investigation architecture. IEEE Access **9**, 103637–103650 (2021). https://doi.org/10.1109/ACCESS.2021.3099037
10. Jaquet-Chiffelle, D., Casey, E., Bourquenoud, J.: Tamperproof timestamped provenance ledger using blockchain technology. FORENSIC Sci. Int.-Digit. Investig. **33** (2020). https://doi.org/10.1016/j.fsidi.2020.300977
11. Liu, G., He, J., Xuan, X.: A data preservation method based on blockchain and multidimensional hash for digital forensics. Complexity **2021** (2021). https://doi.org/10.1155/2021/5536326
12. Ali, M., Ismail, A., Elgohary, H., Darwish, S., Mesbah, S.: A procedure for tracing chain of custody in digital image forensics: a paradigm based on grey hash and blockchain. Symmetry **14**(2) (2022). https://doi.org/10.3390/sym14020334

13. Sathyaprakasan, R., Govindan, P., Alvi, S., Sadath, L., Philip, S., Singh, N.: An implementation of blockchain technology in forensic evidence management. In: Naranje, V., Singh, B., Velan, S. (eds.) Proceedings IEEE International Conference on Computational Intelligence and Knowledge Economy, ICCIKE, pp. 208–212. Institute of Electrical and Electronics Engineers Inc. (2021). https://doi.org/10.1109/ICCIKE51210.2021.9410791
14. Burri, X., Casey, E., Bollé, T., Jaquet-Chiffelle, D.-O.: Chronological independently verifiable electronic chain of custody ledger using blockchain technology. Forensic Sci. Int. Digit. Investig. **33** (2020). https://doi.org/10.1016/j.fsidi.2020.300976
15. Alruwaili, F.E.: CustodyBlock: a distributed chain of custody evidence framework. Information **12**(2), 88 (2021). https://doi.org/10.3390/info12020088
16. Kusuma, R.S.: Digital evidence security system design using blockchain technology. Int. J. Saf. Secur. Eng. **13**(1), 159–165 (2023). https://doi.org/10.18280/ijsse.130118
17. Gopalan, S.H., Suba, S.A., Ashmithashree, C., Gayathri, A., Jebin Andrews, V.: Digital forensics using blockchain. Int. J. Recent Technol. Eng. **8**(2 Special Issue 11), 182–184 (2019). https://doi.org/10.35940/ijrte.B1030.0982S1119
18. Zhang, Y., Wu, S., Jin, B., Du, J.: A blockchain-based process provenance for cloud forensics. In: IEEE International Conference on Computer Communication, ICCC, pp. 2470–2473. Institute of Electrical and Electronics Engineers Inc. (2018). https://doi.org/10.1109/CompComm.2017.8322979
19. Manjre, B.M., Goyal, K.K.: A novel and custom blockchain approach for the integrity assurance of the digital evidences extracted during the extraction and decoding of mobile artifacts from the mobile forensic tools. In: AIP Conference on Proceedings. American Institute of Physics Inc. (2023). https://doi.org/10.1063/5.0127910
20. Bonomi, S., Casini, M., Ciccotelli, C.: B-CoC: a blockchain-based chain of custody for evidences management in digital forensics. In: Danos, V., Herlihy, M., Potop-Butucaru, M., Prat, J., Tucci-Piergiovanni, S. (eds.) OpenAccess Series Informatics. Schloss Dagstuhl- Leibniz-Zentrum fur Informatik GmbH, Dagstuhl Publishing (2020). https://doi.org/10.4230/OASIcs.Tokenomics.2019.12
21. Yan, W., Shen, J., Cao, Z., Dong, X.: blockchain based digital evidence chain of custody. In: Proceedings of the 2020 the 2nd International Conference on Blockchain Technology, ICBCT 2020, pp. 19–23. Association for Computing Machinery, New York (2020). https://doi.org/10.1145/3390566.3391690
22. Ahmad, L., Khanji, S., Iqbal, F., Kamoun, F.: Blockchain-based chain of custody: towards real-time tamper-proof evidence management. In: ACM International Conference on Proceedings Series. Association for Computing Machinery (2020). https://doi.org/10.1145/3407023.3409199
23. Chopade, M., Khan, S., Shaikh, U., Pawar, R.: Digital forensics: maintaining chain of custody using blockchain. In: Proceedings of the International Conference on I-SMAC IoT Social, Mobile, Analytics and Cloud, I-SMAC, pp. 744–747. Institute of Electrical and Electronics Engineers Inc. (2019). https://doi.org/10.1109/I-SMAC47947.2019.9032693
24. Lone, A.H., Mir, R.N.: Forensic-chain: Blockchain based digital forensics chain of custody with PoC in Hyperledger Composer. Digit. Investig. **28**, 44–55 (2019). https://doi.org/10.1016/j.diin.2019.01.002
25. Santamaria, P., Tobarra, L., Pastor-Vargas, R., Robles-Gomez, A.: Smart contracts for managing the chain-of-custody of digital evidence: a practical case of study. Smart Cities **6**(2), 709–727 (2023). https://doi.org/10.3390/smartcities6020034
26. Awuson-David, K., Al-Hadhrami, T., Alazab, M., Shah, N., Shalaginov, A.: BCFL logging: an approach to acquire and preserve admissible digital forensics evidence in cloud ecosystem. Future Gener. Comput. Syst. **122**, 1–13 (2021). https://doi.org/10.1016/j.future.2021.03.001

27. Al-Khateeb, H., Epiphaniou, G., Daly, H.: Blockchain for modern digital forensics: the chain-of-custody as a distributed ledger. In: Advanced Sciences and Technologies for Security Applications, pp. 149–168. Springer (2019). https://doi.org/10.1007/978-3-030-11289-9_7
28. Li, M., Lal, C., Conti, M., Hu, D.: LEChain: a blockchain-based lawful evidence management scheme for digital forensics. Future Gener. Comput. Syst.- Int. J. Esci. **115**, 406–420 (2021). https://doi.org/10.1016/j.future.2020.09.038
29. Shahaab, A., Hewage, C., Khan, I.: Preventing spoliation of evidence with block-chain: a perspective from South Asia. In: 2021 the 3rd International Conference on Blockchain Technology, ICBCT 2021, pp. 45–52. Association for Computing Machinery, New York (2021). https://doi.org/10.1145/3460537.3460550
30. Tsai, F.-C.: The application of blockchain of custody in criminal investigation process. In: Watrobski, J., Salabun, W., Toro, C., Zanni-Merk, C., Howlett, R.J., Jain, L.C. (eds.) Knowledge-Based and Intelligent Information & Engineering Systems (kse 2021), pp. 2779–2788. Elsevier Science Bv, Amsterdam (2021). https://doi.org/10.1016/j.procs.2021.09.048
31. Sanda, P., Pawar, D., Radha, V.: Blockchain-based tamper-proof and transparent investigation model for cloud VMs. J. Supercomput. **78**(16), 17891–17919 (2022). https://doi.org/10.1007/s11227-022-04567-4
32. Akello, P., Vemprala, N., Lang Beebe, N., Raymond Choo, K.-K.: Blockchain use case in ballistics and crime gun tracing and intelligence: toward overcoming gun violence. ACM Trans. Manag. Inf. Syst. **14**(1), 7:1–7:26 (2023). https://doi.org/10.1145/3571290
33. Martin, T., Hammoudeh, M.: Data preservation system using BoCA: blockchain-of-custody application. In: The 5th International Conference on Future Networks & Distributed Systems, ICFNDS 2021, pp. 70–77. Association for Computing Machinery, New York (2022) https://doi.org/10.1145/3508072.3508084
34. BPMN Specification - Business Process Model and Notation. https://www.bpmn.org/. Accessed 02 May 2023
35. Castro, M., Liskov, B.: Practical byzantine fault tolerance (1999)

A Survey on Mutation Testing

Moez Krichen[1,2(✉)]

[1] Al-Baha University, Al-Baha 6552, Saudi Arabia
moez.krichen@redcad.org
[2] ReDCAD Laboratory, National School of Engineers of Sfax,
University of Sfax, Sfax, Tunisia
https://www.redcad.org/members/mkrichen/

Abstract. This survey explores mutation testing, a technique widely used to assess software testing quality. It covers the formal definition, steps for generating and executing mutants, different mutation classes, types, and industrial applications. Mutation testing plays a crucial role in identifying weaknesses in test suites and improving overall fault-detection capability. The survey emphasizes the significance of mutation testing in software development and offers valuable insights to researchers and practitioners. It discusses theoretical foundations, including mutation operators and equivalent mutants, and practical aspects such as generating and executing mutants. Various mutation types are categorized based on their characteristics and scope. Real-world applications in industry highlight its effectiveness in enhancing software quality. This survey serves as a comprehensive resource for understanding mutation testing and provides directions for future research and development.

Keywords: Mutation Testing · Principles · Types

1 Introduction

Software testing, which aims to ensure the quality and dependability of software systems, is a crucial component of software development [4]. Over the years, numerous software testing methods have been created, each with unique advantages and disadvantages. Mutation testing (MT) is one such method that has gained popularity recently as a reliable way to assess the quality of test cases [15]. MT is a fault-based testing technique that involves creating a set of mutants, which are versions of the program that contain one or more faults. The goal of MT is to measure the effectiveness of test cases by calculating the mutation score, which is the percentage of mutants that are killed by the test suite. If a mutant is not killed by any test case, it is considered to be equivalent to a real fault that went undetected, and the mutation score is reduced. Mutants are created by making small changes to the original program's code. The goal is to introduce faults into the program that are representative of common programming errors. This survey paper provides an overview of MT, covering its principles, types, innovations, industrial applications, and future directions.

V. Sakalauskas et al. (Eds.): IBICA 2023, LNNS 1231, pp. 210–219, 2025.
https://doi.org/10.1007/978-3-031-78946-5_20

2 Formal Definition

Let P be a program, and let T be a set of test cases that are designed to test P. Let M be a set of mutants that are created by applying one or more mutation operators to the code of P. Each mutant m in M is a variation of P that contains one or more faults.

The goal of MT is to evaluate the effectiveness of T by measuring its ability to detect faults in the mutants in M. For each mutant m in M, let $f(m)$ be a function that represents the behavior of m. The function $f(m)$ takes the inputs of P and produces the corresponding outputs of m.

The mutation score, $MS(T)$, is defined as:

$$MS(T) = \frac{(|M| - |K(T, M)|)}{|M|} * 100\%$$

where $|M|$ is the total number of mutants in M, and $|K(T, M)|$ is the number of mutants in M that are killed by T. A mutant m is considered to be killed by T if there exists at least one test case in T that produces a different output for m than for P.

Intuitively, the mutation score measures the ability of T to detect faults in the program. If T is effective, it should be able to detect most of the faults in the mutants, resulting in a high mutation score. If T is ineffective, it will fail to detect many of the faults, resulting in a low mutation score.

3 Steps for Generating and Executing Mutants

There are several steps involved in generating mutants for MT, including identifying mutation operators, applying the operators to the original program, compiling and executing the mutants, and evaluating the effectiveness of the test suite in detecting the faults. These steps provide a general framework for generating mutants in MT and can be adapted to suit the specifics of the testing environment and the program being tested. More details about these steps are provided below (Fig. 1:

1. **Identify Mutation Operators:** Identify the mutation operators that will be used to generate the mutants. Different mutation operators can be used to generate different types of mutants, such as weak or strong mutants.
2. **Apply Mutation Operators:** Apply the mutation operators to the original program to generate a set of mutants. For traditional MT, a single mutation operator is applied to the original program at a time to generate mutants. For other types of MT, such as weak or strong MT, multiple mutation operators may be applied to the original program to generate a larger set of mutants.
3. **Compile and Execute Mutants:** Compile and execute the mutants to determine their behavior. The mutants should be compiled and executed in the same environment as the original program, using the same input data and runtime conditions. The behavior of each mutant should be compared to

the behavior of the original program to determine if the mutant introduces a fault.

4. **Evaluate Test Suite Effectiveness:** Evaluate the effectiveness of the test suite in detecting the faults introduced by the mutants. The test suite should be run on each mutant, and the number of mutants that are detected by the test suite should be recorded. The effectiveness of the test suite can be measured using metrics such as mutation score, which is the percentage of mutants that are detected by the test suite.

These steps provide a general framework for generating mutants in MT. Depending on the specifics of the testing environment and the program being tested, additional steps or modifications to these steps may be necessary.

Fig. 1. Steps for Generating and Executing Mutants.

4 Different Classes of MT

There are various classes of MT, such as strong and weak MT as well as conventional MT. In conventional MT, only one mutation operator is used at a time to create mutants. The process of creating mutants for weak MT requires combining different mutation operators. Making mutations that are functionally identical to the original program but with a different implementation is a part of strong MT. There are more, less popular types of MT, including comparable MT, testing based on mutant schemas, and testing based on behavior. We give further information about these various classifications below:

- **Traditional MT:** Traditional MT is the most widely used class of MT, where a single mutation operator is applied to the original program at a time to generate a set of mutants. A mutation operator is a rule or algorithm that modifies the original program by introducing a fault (e.g., changing a relational operator, or deleting a conditional statement). The goal of traditional MT is to evaluate the effectiveness of a given test suite in detecting the faults introduced by the mutation operators. Traditional MT is computationally expensive, as it requires generating and executing a large number of mutants.
- **Weak MT:** Weak MT is a variation of traditional MT, where multiple mutation operators are applied to the original program to generate mutants. The idea behind weak MT is that it can increase the number of mutants generated and reduce the number of tests required to evaluate the test suite, while still being computationally feasible. Weak MT can also help to detect faults that are missed by traditional MT.

- **Strong MT:** Strong MT is another variation of traditional MT, where the mutants generated are functionally equivalent to the original program, but with a different implementation. The goal of strong MT is to evaluate the ability of a test suite to detect faults in the program's implementation, rather than just its syntax. Strong MT is more difficult to implement than traditional or weak MT, as it requires generating mutants that are functionally equivalent to the original program.
- **Equivalent MT:** Equivalent MT is a less common class of MT that involves generating mutants that are semantically equivalent to the original program, but with different syntax. The goal of equivalent MT is to evaluate the ability of a test suite to detect faults that are related to the program's structure and syntax.
- **Mutant Schema-Based Testing:** Mutant schema-based testing is another less common class of MT that involves generating mutants based on a schema that defines the structure and syntax of the mutants. The goal of mutant schema-based testing is to generate mutants that are more relevant to the program's structure and syntax, and to reduce the number of irrelevant mutants that need to be generated.
- **Behavior-Based MT:** Behavior-based MT is a more recent class of MT that involves generating mutants by modifying the behavior of the program, rather than its syntax. The goal of behavior-based MT is to generate mutants that are more representative of the program's actual behavior, and to reduce the number of irrelevant mutants that need to be generated.

5 Comparison with Other Types of Testing

In addition to MT, there are several other types of testing that developers can use to ensure the quality and reliability of their software:

- **Unit Testing** [7]: It involves testing individual components of the code to ensure that they are working correctly. The focus of unit testing is on identifying bugs in small, isolated pieces of code. While unit testing is an important part of any testing strategy, it may not be sufficient to catch all faults in the code. Unlike MT, which evaluates the effectiveness of the test cases themselves by introducing faults into the code, unit testing only verifies that the individual components are working correctly.
- **Integration Testing** [5]: It consists in testing the interactions between different components to ensure that they are working together correctly. The focus of integration testing is on identifying bugs that arise when components are combined. While integration testing is important, it may not catch all faults that arise from individual components. MT can identify faults in both individual components and their interactions with other components.
- **System Testing** [18]: It involves testing the entire system to ensure that it meets the requirements and performs as expected. The focus of system testing is on identifying bugs that arise from the system as a whole. While

system testing is important, it may not catch all faults that arise from individual components or their interactions. MT can complement system testing by identifying weaknesses in the test cases themselves.

- **Acceptance Testing** [14]: It consists in testing the software to ensure that it meets the requirements and satisfies the needs of the users. The focus of acceptance testing is on evaluating the software from the perspective of the end-users. While acceptance testing is important, it may not catch all faults that arise from individual components or their interactions. MT can complement acceptance testing by identifying weaknesses in the test cases themselves.
- **Fault Injection Testing** [12]: It involves intentionally introducing faults into the code or the environment to observe the behavior of the system under those conditions. The focus of fault injection testing is on evaluating the resilience of the system to unexpected faults. While fault injection testing can be useful for identifying weaknesses in the system, it may not be as effective as MT for evaluating the effectiveness of test cases. Fault injection testing is more focused on identifying faults that are already known to be present in the system, whereas MT is designed to identify the weaknesses in the test cases themselves.
- **Model-Based Testing** [8–10]: It consists in creating models of the system and using those models to generate test cases. Model-based testing can be useful for generating a large number of tests and verifying that the system meets the requirements. While model-based testing can be effective, it may not catch all faults that arise in the actual code. MT can complement model-based testing by identifying weaknesses in the test cases themselves. However, developing accurate models can be time-consuming and may require specialized skills and tools.

Each type of testing has its own strengths and weaknesses, and developers need to choose the appropriate type of testing based on the specific requirements of the software. By using a combination of these types of testing, developers can ensure that their software is thoroughly tested and reliable.

6 Different Types of MT

MT can be classified into different types based on the criteria used to generate the mutants:

- **Statement MT:** Statement MT involves modifying individual statements in the program by introducing faults. This can be done by changing the values of constants, variables, or operators, or by deleting or inserting statements. The goal of statement MT is to evaluate the effectiveness of test cases in detecting faults in individual statements.
- **Decision MT:** Decision MT involves modifying the control flow of the program by changing the branching conditions of if statements and loops. This can be done by changing the relational operators, logical operators, or Boolean

expressions used in the conditions. The goal of decision MT is to evaluate the effectiveness of test cases in detecting faults in the control flow of the program.
- **Condition MT:** Condition MT is a more fine-grained version of decision MT that involves modifying the Boolean expressions used in the conditions of if statements and loops. The mutations can be created by changing the Boolean operators, the operands of the operators, or the negation of the operators. The goal of condition MT is to evaluate the effectiveness of test cases in detecting faults in the Boolean expressions used in the conditions.
- **Data MT:** Data MT involves modifying the input data that is used to run the program. This can be done by changing the values of constants, variables, or data structures. The goal of data MT is to evaluate the effectiveness of test cases in detecting faults in the input data of the program.
- **Object-Oriented MT:** Object-oriented MT involves modifying the structure of the program, such as changing the inheritance hierarchy or the method calls between objects. The mutations can be created by changing the method invocations, modifying the access modifiers of the methods and fields, or changing the inheritance relationships. The goal of object-oriented MT is to evaluate the effectiveness of test cases in detecting faults in the object-oriented design and implementation of the program.

Overall, each type of MT provides a different perspective on the effectiveness of test cases and can be useful in identifying different types of faults in the program. The choice of type depends on the goals of the testing and the characteristics of the program. By using a combination of different types of MT, a comprehensive evaluation of the quality of test cases and the reliability of software systems can be achieved.

7 Industrial Applications of MT

MT has been successfully applied in various industrial and real-world applications to improve software quality. Some examples include:

- **Linux kernel:** The main contribution of the work presented in [2] was to demonstrate the usefulness of mutation analysis in large systems, specifically on the Linux kernel's RCU module. The authors adapted existing techniques to address the computational requirements and high numbers of false positives typically associated with mutation analysis. The study showed that mutation analysis was able to identify gaps in the RCU test harness and uncovered two previously masked bugs in the RCU module. The authors argue that MT should be more extensively used in practice as it can be a valuable tool for measuring the completeness and quality of a test suite and identifying gaps in even well-tested modules.
- **Web Applications:** MUTTA, an automated tool for MT web applications in end-to-end (E2E) web testing, was proposed by the authors of [11]. The target web application's server source files are modified by MUTTA, which then runs the E2E test suite against the modified web applications and gathers test

results. The program generates a large number of mutants, uses coverage data to take into account just the mutants ran, and records test suite results in an acceptable manner. The MT process of E2E web test suites may be automated using MUTTA, and it can be used to assess E2E web test suites, according to a case study the authors did to compare two E2E web testing methodologies. The article presents a novel technique for automatically evaluating the efficacy of test suites to enhance web application quality.

- **Spark Programs:** The study presented in [17] proposed TRANSMUT-Spark, a tool for automating MT of Spark programs' big data processing code. The study emphasizes the complexity of parallel large data programming and the need to smartly combine Spark built-in functions within programs to exploit computational resources needed for big data processing and avoid significant output losses. MT is used to test Spark applications, and TRANSMUT-Spark automates mutant generation, test execution, and adequacy analysis. The publication also covers the tool's breadth and limitations based on validation studies. The paper introduces a new method to assess test sets and find code errors in Spark programs.
- **Distributed Applications:** MuTomVo, an MT framework for distributed applications in simulated environments was presented in [3]. The study underlines the constraints of testing distributed applications in highly distributed systems during development and proposes using simulation platforms to model a wide range of distributed system configurations and execute these applications in the modeled system. A test suite can be run against the set of altered models to assess its error detection ability using the proposed framework. The paper explains the framework's implementation in MuTomVo and gives a case study of three applications running in diverse distributed systems to demonstrate its practicality. Overall, the research proposes a unique framework that can improve distributed application quality by offering a reliable and automated mechanism to evaluate test suites' error detection in simulated situations.
- **Large-Scale Software Projects:** The study [16] proposed HadoopMutator, a cloud-based MT platform that employs MapReduce to accelerate mutant creation and testing. The paper proposes a solution to improve automated MT and provide a scalable solution for large-scale software projects. In two application situations, the suggested system improves automated MT by a factor of 10. By treating source code as data, newer data-centric frameworks offer additional reuse options, according to the report. The paper introduces an MT framework that helps increase software project quality by evaluating testing methodologies faster and more scalable.
- **Java Programs**: There are several different MT systems available for Java [1, 6], including:
  - **MuJava (μJava)**: MuJava is a tool designed specifically for Java application MT [13]. The tool creates mutants by using a variety of mutation operators on the original source code, including altering method calls, changing arithmetic or logical operators, and deleting statements. MuJava supports a variety of MT techniques, such as strong, weak, and conven-

tional MT. Along with support for automated test case generation and coverage analysis, MuJava also offers a complete set of tools for mutant generation, execution, and analysis.

- **Jester:** Jester is a popular Java, C, and C++ MT tool. To create mutants, it changes relational operators or deletes conditional statements. Jester then performs the test suite on each mutant and calculates a mutation score, the percentage of mutations discovered. Jester is extremely adjustable, letting users set mutation operators and code mutation levels.
- **Simple Jester:** Simple Jester is an MT tool that was developed to be a lighter-weight alternative to the original Jester tool. Like Jester, Simple Jester is designed to work with Java code and applies mutation operators to generate a set of mutants. It then runs the test suite on each mutant and calculates a mutation score. Simple Jester is designed to be easy to use and to provide quick feedback on the effectiveness of a test suite. It has a simple command-line interface and can be integrated with continuous integration tools like Jenkins. While it may not be as feature-rich as other MT tools, Simple Jester can be a good choice for projects that require a lightweight and efficient MT solution.
- **Jumble:** Jumble is an MT tool that is designed to be highly customizable and extensible. It can be used with Java, .NET, and Ruby codebases and provides a wide variety of mutation operators. Jumble also allows users to create their own mutation operators and to configure the tool to apply mutations selectively based on specific criteria, such as code complexity or test coverage. Jumble has a command-line interface and can be integrated with build tools like Ant, Maven, or Gradle. Jumble has been used in a variety of software development projects and has been found to be effective in identifying faults that are missed by other testing methods.
- **Javalanche:** Javalanche is an MT tool that is designed to be highly parallelizable and scalable. It applies mutation operators to Java code and generates a set of mutants, then runs the test suite on each mutant and calculates a mutation score. Javalanche is designed to work well with large codebases and can be run in parallel across multiple machines. It also provides detailed reports on the results of the testing process, including information on which mutants were killed by the test suite and which were not. Javalanche can be integrated with build tools like Ant and Maven, and also provides a web interface for visualizing the results of MT. Javalanche has been used in a variety of software development projects and has been found to be effective in identifying faults that are missed by other testing methods.
- **PIT:** PIT (short for "Pitest") is a popular Java MT tool. It generates a collection of mutations, runs the test suite on each mutant, and calculates a mutation score. PIT lets users choose mutation operators, classes to test, and other features. Users can also use PIT's reports and visualizations to understand testing findings. "Incremental" MT, which helps

speed up MT on big codebases, is one of PIT's features. PIT is usually connected with build technologies like Maven or Gradle and can be executed automatically in a continuous integration workflow. PIT has been used in many software development projects to find defects missed by conventional testing methods.

These examples demonstrate the effectiveness of MT in identifying weaknesses in the test suite and improving the overall quality, reliability, and performance of the software in various industrial applications.

8 Conclusion and Future Works

In conclusion, this survey paper has provided a comprehensive overview of mutation testing (MT) as a fault-based testing technique for evaluating the quality of test cases. We have explored the principles, types, innovations, industrial applications, advantages, and limitations of MT. By generating mutants, which are modified versions of the original program containing faults, MT assesses the effectiveness of test cases through the calculation of the mutation score. The paper highlighted the importance of MT in software testing and compared it with other testing techniques. Through the discussion of various classes and types of MT, such as statement, decision, condition, data, and object-oriented MT, we have presented insights into the different aspects of this technique. The survey also covered recent advancements in MT techniques and tools, showcasing their potential impact on improving the effectiveness and efficiency of mutation testing. Additionally, we examined the industrial applications of MT, demonstrating its practical relevance and effectiveness in enhancing software quality and reducing undetected faults in real-world systems.

Looking towards future directions, there are several areas in which MT can continue to evolve and improve. One potential avenue for research is the development of more sophisticated mutation operators that can simulate a broader range of faults and programming errors. This could enhance the effectiveness of MT by generating mutants that are more representative of real-world faults. Additionally, exploring the integration of MT with other testing techniques, such as model-based testing or machine learning-based testing, holds promise for enhancing the overall testing process and increasing fault detection capabilities. Furthermore, the automation and scalability of MT can be improved to handle large-scale software systems and reduce the manual effort required for generating and analyzing mutants. This includes the development of efficient mutation testing tools and frameworks that can support the mutation process in different programming languages and environments.

References

1. Abbas, S.T.E., Hassan, R., Abd Halim, S., Kasim, S., Ramlan, R.: Investigation on java mutation testing tools. JOIV: Int. J. Inform. Vis. **6**(2-2), 455–462 (2022)
2. Ahmed, I., Jensen, C., Groce, A., McKenney, P.E.: Applying mutation analysis on kernel test suites: an experience report. In: IEEE International Conference on Software Testing, Verification and Validation Workshops (ICSTW), pp. 110–115 (2017)
3. Cañizares, P.C., Núñez, A., Merayo, M.G.: Mutomvo: mutation testing framework for simulated cloud and HPC environments. J. Syst. Softw. **143**, 187–207 (2018)
4. Coutinho, J., Andrade, W., Machado, P.: A survey of requirements engineering and software testing practices in agile teams. In: Proceedings of the 7th Brazilian Symposium on Systematic and Automated Software Testing, pp. 9–18 (2022)
5. Grechanik, M., Devanla, G.: Mutation integration testing. In: 2016 IEEE International Conference on Software Quality, Reliability and Security (QRS), pp. 353–364. IEEE (2016)
6. Khari, M.: Comparison and validation of mutation testing tools based on java language. In: Optimization of Automated Software Testing Using Meta-Heuristic Techniques, pp. 13–29. Springer, Cham (2022)
7. Khorikov, V.: Unit Testing Principles, Practices, and Patterns. Simon and Schuster, New York (2020)
8. Krichen, M.: Model-based testing for real-time systems. Ph.D. thesis, Universit Joseph Fourier (2007)
9. Krichen, M.: A formal framework for conformance testing of distributed real-time systems. In: International Conference on Principles of Distributed Systems, pp. 139–142. Springer, Cham (2010)
10. Krichen, M.: Contributions to model-based testing of dynamic and distributed real-time systems. Ph.D. thesis, École Nationale d'Ingénieurs de Sfax (Tunisie) (2018)
11. Leotta, M., Paparella, D., Ricca, F.: MUTTA: a novel tool for E2E web mutation testing. Softw. Qual. J. 1–22 (2023)
12. Lojda, J., Panek, R., Podivinsky, J., Cekan, O., Krcma, M., Kotasek, Z.: Testing embedded software through fault injection: case study on smart lock. In: 2021 IEEE 22nd Latin American Test Symposium (LATS), pp. 1–6. IEEE (2021)
13. Ma, Y.S., Offutt, J., Kwon, Y.R.: MuJava: a mutation system for Java. In: Proceedings of the 28th International Conference on Software Engineering, pp. 827–830 (2006)
14. de Oliveira, L.F., Rodrigues, C.L., de Freitas Bulcão-Neto, R.: Characterizing the software acceptance testing and the inclusion of people with disabilities by means of a systematic mapping. IEEE Lat. Am. Trans. **21**(1), 35–46 (2023)
15. Papadakis, M., Kintis, M., Zhang, J., Jia, Y., Le Traon, Y., Harman, M.: Mutation testing advances: an analysis and survey. In: Advances in Computers, vol. 112, pp. 275–378. Elsevier (2019)
16. Saleh, I., Nagi, K.: HadoopMutator: a cloud-based mutation testing framework. In: Software Reuse for Dynamic Systems in the Cloud and Beyond: 14th International Conference on Software Reuse, pp. 172–187. Springer, Cham (2014)
17. de Souza Neto, J.B., Martins Moreira, A., Vargas-Solar, G., Musicante, M.A.: TRANSMUT-spark: transformation mutation for apache spark. Softw. Test. Verif. Reliab. **32**(8), e1809 (2022)
18. Zafar, M.N.: Model-based system testing of safety-critical embedded software. Malardalen University (Sweden) (2022)

FTISI: Folksonomy Based Automatic Tweet Tagging Integrating Community Derived Semantic Intelligence

A. Aravind Krishnan and Gerard Deepak(✉)

Department of Computer Science and Engineering, Manipal Institute of Technology Bengaluru, Manipal Academy of Higher Education, Bengaluru, India
gerard.deepak.christuni@gmail.com

Abstract. There exists a requirement to tag content on the web and specifically tweets as the social web has a lot of activity, an uptick in recent years caused by the introduction of a considerably high number of users on the social web and specifically Twitter which attracts a lot of celebrities. This paper proposes the folksonomy-based Twitter tag recommendation framework which encompasses the prominence of building and generating a folksonomy from the Twitter dataset. The FTISI also focuses on enriching the folksonomy using Linked Open Data Cloud (LOD), eBooks, indexes, and several other news repositories. This model extracts the categories from the dataset and applies the Latent Dirichlet Allocation (LDA). The FTISI model fits in the LDA for enriching the categories in the dataset and it is subjected to TF/IDF to derive the informative term set from the historical documents, and the entities derived from the TF/IDF and the LDA and the Dspace tool are utilised in creating the metadata which is in turn classified through the usage of a strong deep learning BiLSTM model. Subsequently, the latent Twitter Semantic Similarity (TSS) using COSIM rank is achieved by encompassing the top 20% of classified metadata and the generated taxonomies. The entities which come out of the lateral semantic similarity computation using COSIM rank and the rest of the 70% of the metadata are subjected to the computation of the SOCPMI under the stochastic optimization algorithm and TSS is also computed using empirically decided thresholds and values and these units are ranked in the rising order of TSS and is sent for review which is conceived as tags. The highest primary evaluation metrics and lowest FDR has been achieved by the FTISI model.

Keywords: Folksonomy-based · Twitter Semantic Similarity · Deep learning BiLSTM model · Latent Dirichlet Allocation (LDA) · Linked Open Data Cloud (LOD)

1 Introduction

Web 3.0 represents a remarkable evolution of the virtual society, building upon the user empowerment and collaborative nature of Web 2.0. This next phase of the web embraces decentralization, semantic web technologies, and the integration of artificial

V. Sakalauskas et al. (Eds.): IBICA 2023, LNNS 1231, pp. 220–230, 2025.
https://doi.org/10.1007/978-3-031-78946-5_21

intelligence (AI), resulting in a more intelligent, context-aware, and personalized online experience. Web 3.0 enables seamless data exchange between platforms and devices, enhances privacy and security through encryption and distributed ledger technologies, and fosters the development of decentralized applications (dApps) and platforms. As this dynamic transition continues, the virtual society continues to expand rapidly, driving the growth and transformation of the web into a highly interconnected and ever-evolving landscape.

The social web refers to the online platforms and technologies that facilitate social interactions and networking among users. It encompasses a wide range of platforms, including social media sites, online forums, and messaging apps, which enable individuals to connect, share information, and engage in conversations. Twitter is a popular social media platform known for its real-time microblogging format, which enables global users to exchange concise messages known as "tweets". It permits users to follow accounts of interest, engage in discussions, and share updates, opinions, and news on an extensive array of subjects. With its concise and public nature, Twitter has become a significant platform for user data and social engagement. As of 2022, Twitter has approximately 450 million monthly active users which includes verified celebrities, news sources, political figures, etc. along with the average user. Most of the existing models/frameworks are not semantically driven and are not operable or relevant to the transforming structure of Web 3.0 since Web 3.0 is highly hyperlinked. This paper reviews folksonomy-based and semantically driven methods of training models to tag tweets on Twitter in order to make tweets more available to search engines as they grow to be more relevant in the present day.

Motivation: There is a need for tagging entities in the social web as the social web is a major constituent of the current structure of Web 2.0 which is transforming into Web 3.0 which has seen a large number of users in the past decade and the number of users has been increasing exponentially. Entities on the social web also have to be tagged in order to ensure that the social web is also visible and indexed and retrieved by popular search engines. In order to achieve this a tag recommendation framework that is semantically inclined is essential because learning from the present organization of the World Wide Web and having only learning models would make it extremely complex. In order to overcome this, a knowledge-centric, semantically inclined model is essential and is the need of the hour for tagging instances and entities on the social web. Since the web is extremely dynamic and houses large amounts of information, there is an utmost need that every instance on the web and the social web to be tagged. Twitter is one of the most visited social networking sites, Twitter entities have to be tagged as they could be identifiable and retrievable by search engines and henceforth the tag recommendation using semantically inclined models which are much lighter compared to learning-only models are of greater significance.

Contributions: The encompassment of Latent Dirichlet Allocation and the TFIDF on the categories from the dataset and the historical web documents relevant to the domain respectively yield the informative term set novel. Generation of metadata from the informative term set and classification by utilizing a strong deep learning classifier namely the BiLSTM and building the folksonomy by generating it using the dataset entities and enriching it using the open data cloud and enriching it using eBooks, indexes, and several other news repositories and addition of auxiliary knowledge thereby, increasing the information density and provides betterment of results. The subjectivity of COSIM rank to compute the lateral semantic similarity between the entities which comprises about 20% of classified metadata and the entities coming out of the enriched folksonomies is definitely a novel contribution to this work. The usage of stochastic optimization by consideration of SOCPMI as the object of the function and computation of TSS with differential thresholds and step deviance measure with that of 70% of the metadata and the enriched folksonomy instances ensures that the relevance of the notation mechanism in a model is quite strong which further yields in the recommendation of reliable entities for the Twitter document tagging. The overall lowest average recall, precision, accuracy, F-measure, and FDR are achieved by the baseline model of FTISI.

Organisation: Section 2 contains the Related Works, Sect. 3 contains the Proposed System Architecture, Sect. 4 contains the Implementation, Performance Evaluation, and Results, Sect. 5 depicts a Conclusion.

2 Related Works

Ma et al., [1] have proposed two probabilistic PLSA-style models to improve Twitter hashtag annotation incorporating publicly available tag metadata. Weerasooriya et al., [2] have put forth a method for analysis of tweets by enhancing the Stanford CoreNLP POS tagger with a Twitter model, rule-based parsers, and a domain-specific corpus. Gattani et al., [3] have presented an end-to-end industrial system tailored for social data where entities are extracted effectively, linked to a knowledge base, tweets are classified, and descriptive tags are assigned. Albogamy et al., [4] evaluates the performance of POS taggers for Arabic tweets, showcasing a large decline in accuracy compared to standard Arabic text and were able to enhance the taggers performance to 79% accuracy.

Trant et al., [5] has reviewed social tagging and folksonomy research and presents a framework for studying these concepts and highlights key research in the field by focusing on the folksonomy itself, tagging behaviour of users, and the socio-technical aspects of social tagging models. Yu et al., [6] have introduced a method to enhance tagging system by considering the user's tagging status, proposing algorithms to determine the status and strategies to compute tag probability distribution. O'Leary et al., [7] have examined the use of crowdsourcing for tags in accounting and finance while highlighting challenges such as prevalence of redundant tags and variation in tags for the same concept. Xu et al., [8] have discussed the evolution of content organization on the internet, from structured directories to unstructured search engines and have highlighted the issues associated with identifying appropriate tags while reducing noise and spam and proposes a collaborative tag suggestion algorithm. Madhu et al., [9] have presented a survey on the significance of search engines in the Web 3.0 and role of semantic based search technologies in retrieving significant information from the large amount of data residing on the World Wide Web. Ramos et al., [10] have explored the efficacy of the Term Frequency-Inverse Document Frequency (TF-IDF) algorithm in identifying words with high relevance to a document, which improves query retrieval by suggesting documents that may be of interest to users based on the presence of these words. Ngoc et al., [11] have explored paraphrase identification and semantic similarity tasks in NLP, focusing on unstructured social media texts from Twitter and the leveraging of simple features and word alignment techniques from machine translation evaluation metrics to achieve competitive performance on both tasks. Graves et al., [12] have introduced bidirectional LSTM networks and a modified learning algorithm and shows that it outperforms unidirectional LSTM networks. Yang et al., [13] have introduces a hierarchical attention network designed specifically for classification of documents. In [14–17, 18], several semantics-oriented frameworks in support of the literature of the proposed model have been put forth. This gives confidence to realise the proposed model.

3 Proposed System Architecture

Figure 1 depicts the proposed system architecture of the knowledge-centric, folksonomy-driven, semantically inclined, automatic tweet tag generation framework. As the entire framework relies on the Twitter/tweet dataset, it undergoes pre-processing to extract categories as a categorical dataset. However, keywords from tweets are also fed as categories. Subsequently, the tweet dataset, keywords, and categories are inputted into a historical document archive from the World Wide Web (WWW), which is archived based on the domain of the tweet dataset. The historical documents are crawled using a customized agent-driven crawler, and relevant historical documents are loaded into a repository and archived. Based on the categorical terms and their keywords from the tweet dataset, these historical documents are searched, and the TF/IDF (Term Frequency/Document Frequency model) is applied to discover the most informative terms from the historical documents by reading the most frequent and rare terms across the document corpus. This informative term set is merged with the categories extracted from the Twitter dataset, to which the Latent Dirichlet Allocation (LDA) topic modeling is enforced. LDA is a topic modeling framework that discovers relevant yet hidden entities by keeping the World Wide Web as a standard reference corpus. The entities discovered by LDA and TF/IDF are formulated into an informative term set.

The informative term set is used to generate metadata using a tool called Dspace is a metadata generation tool that generates metadata from the structural constituents of the present organization of the World Wide Web, which is an intermediary between Web 2.0 and Web 3.0. The metadata generated from the World Wide Web using Dspace is exponentially large and needs to be handled, for which the BiLSTM classifier is used to classify the metadata. The BiLSTM classifier is chosen since it is a deep learning classifier that is not feature-controlled, meaning features are selected automatically, and the metadata is instantly classified. Owing to the large scale and exponential size of the metadata, the classification is well achieved using the BiLSTM classifier.

Furthermore, the classified metadata is split into the top 30% and the rest 70%. The top 30% is fed into the folksonomy, which is formalized by computing the lateral semantic similarities between the top 30% of entities in the metadata. The correlation is achieved between the course in rank model with a threshold of 0.5 because a large number of associations have to be made, and the folksonomy has to be built. Along with the folksonomy, the information is also merged at the most relevant node, and the

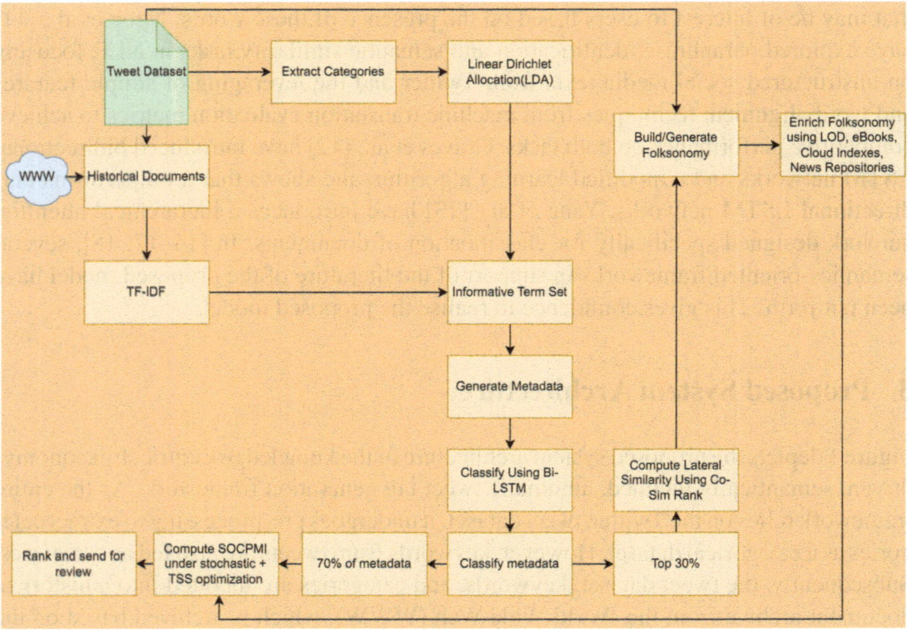

Fig. 1. Architecture of the Proposed Folksonomy based Twitter Tag Generation

folksonomy is generated. However, the generated folksonomy still needs to be enriched since it is still sparse. To enrich the folksonomy, APIs like the Linked Open Data Cloud (LOD Cloud) are used. Apart from that, for the categories in the tweet dataset, eBooks are crawled, and the indices and glossaries from them are extracted and formalized as the folksonomy. News repositories are also given as search terms comprising strategies in the Twitter dataset, and these entities in the news repositories are all aggregated into the folksonomy. However, the folksonomy is generated by matching the terminal node by computing its COSIM rank with the terminal nodes, and a rich folksonomy is formalized.

The spliced 70% of the metadata that came out of the BiLSTM classifier and the top 30% of metadata are subjected to computation of semantic similarity using semantic similarity models, namely, the SOCPMI and (TSS). SOCPMI and TSS are given a threshold of 0.6, which is neither a stringent nor a relaxed threshold. Both SOCPMI and TSS are stronger as individual semantic similarity measures. Initially SOCPMI and TSS are used independently to generate an initial solution set. However, the initial solution set is converted into a most feasible solution set using a stochastic optimisation algorithm which is a metaheuristic optimisation algorithm but at this stage SOCPMI and TSS is set to a threshold of 0.5 and is used as criteria or objective functions and this is done to yield the most relevant entities among relevant entities. The finally yielded principal set is ranked in ascending rank of TSS measure. It is then sent for review as tags to the dataset elements and the mapping is done based on the categories.

The BiLSTM classifier offers a significant benefit by effectively accommodating input sequences of varying lengths. This attribute proves particularly valuable in tasks where the length of the input sequence fluctuates, such as text classification, wherein documents exhibit diverse extents. The BiLSTM classifier can learn to capture pertinent features from the input sequence regardless of its length. The BiLSTM classifier achieves this by incorporating the outputs of the two LSTM layers at each time step, thus accounting for forward and backwards contexts. Recent research has demonstrated that incorporating attention mechanisms into the BiLSTM classifier can yield notable enhancements. The integration of attention mechanisms has been proven to bolster the performance of the BiLSTM classifier across diverse NLP tasks, including machine translation and document classification. In document classification, for example, the BiLSTM classifier with attention has been shown to outperform other models on datasets such as the Reuters-21578 and 20 Newsgroups.

Stochastic optimization and deterministic optimization are two commonly used methods for optimization. Deterministic optimization aims to find the optimal solution for a given problem using a fixed set of rules and parameters. Stochastic optimization, however, involves randomizing and observing probability distributions to find the optimal solution. Different simulations are run and this makes it possible for stochastic optimization to be more diverse in capabilities and it is able to handle problems with higher degrees of uncertainty. Therefore, we use stochastic optimization here.

Traditional optimization methods tend to make inaccurate or large assumptions and pose many limitations in a complex practical setting. When posed with real world scenarios, we seek both high quality and feasible solution. Scenario optimization involves solving deterministic subproblems for different scenarios and merging the optimal solutions of each scenario into a feasible decision. The scenario optimization approach can be outlined through the following steps. Calculate the best possible solution for each deterministic subproblem in every scenario, define the decision variables, objective function, and constraints posed by the scenarios, assign scenarios probabilities and optimize the decision variables, and employ a tracking model to determine a single, viable decision that accommodates all scenarios. For example, a social media company like Twitter may use scenario optimization to decide which hashtags/tags to include in a tweet based on different possible user engagement scenarios. Mathematically, the scenario optimization involves formulating an optimization problem with multiple scenarios. Let X be a vector of tags and F(X, S) be the objective function mapping tags and scenarios to expected user engagement. It can be formulated as in Eq. (1).

$$\max E[F(X, S)] = \sum P_i F(X, S_i), \text{ where } G(X, S) \leq 0 \text{ and } X \in X \tag{1}$$

where S is set of scenarios, G(X, S) represents the constraints that depend on the tags and scenarios, X is the set of feasible tags. The scenario probabilities is represented by the probability vector P, where each element represents the probability of a particular scenario. $F(X, S_i)$ is the objective function that represents the engagement of a tweet based on a hashtag/tag. This problem can now be solved using linear/stochastic programming.

TF-IDF a metric that measures the significance of words in a text corpus. It is able to do so my considering the product of two popular metrics that are the tf and idf. Which are term frequency measures. SOCPMI (Second Order Co-occurrence Pointwise Mutual Information) is a statistical measure used to identify important collocates of a target word. The SOCPMI is an extension of PMI metric and accounts for the co-occurrence of pairs of words with the target word. The PMI function is defined as described by Eq. 2:

$$\text{PMI}(t_i, \text{w}) = log_2 \frac{f^b(t_i, w).m}{f^b(t_i).f^t(w)} \quad (2)$$

The term $f^t(t_i)$ denotes the frequency of the type t_i occurring across the entire corpus, while $f^b(t_i,w)$ indicates the frequency of word t_i appearing close to w in a context window. The variable m represents the number of tokens present. The SOCPMI is given in Eq. 3 and gives the association strength between two words:

$$\text{SOCPMI}(w_1, w_2) = log_2 \frac{P(w_1, w_2)^2}{P(w_1).P(w_2).P(w_1 w_2)} \quad (3)$$

4 Implementation, Performance Evaluation and Results

Implementation of the proposed FTISI framework was carried out using Python, Google Colaboratory was the preferred IDE. The Dspace metadata harvester was used to generate the folksonomy. However, only the top 10% of the tags generated using Dspace were made use of with at least a single link as a lightweight folksonomy in the proposed FTISI. TensorFlow was made use of in order to implement the BiLSTM model. The framework was applied to a hybrid Twitter dataset which is an integrated result of the broad Twitter corpus datasets by Derczynski-Bontcheva-Roberts, the Twitter Dataset by Aditya Goyal (2023), and the Twitter Dataset of 100+ million tweets related to COVID-19 by Ramya (2022). Category enhancement was done using a focused web crawler. Based on the common categories, The other records were prioritized and these three datasets were integrated based on common categories in order to yield a single large hybrid Twitter dataset corpus.

The performance of the FTISI(proposed) is measured using various metrics including precision, accuracy, recall, F-measure as the primary metrics and the FDR as an auxiliary metric.

Table 1. Comparison of Performance of the proposed FTISI with other approaches

Model	Precision %	Recall %	Accuracy %	F-Measure %	FDR
EETSM [1]	85.32	87.08	86.2	86.19	0.15
EEKT [2]	86.45	89.45	87.95	87.92	0.14
TYT [3]	88.45	90.74	89.59	89.58	0.12
PTAT [4]	91.74	92.78	92.26	92.26	0.09
Proposed FTISI	97.83	99.43	98.63	98.62	0.03

According to the results presented in Table 1, the FTISI model demonstrates superior performance compared to four benchmark models, namely EETSM, EEKT, TYT, and

PTAT. The FTISI model achieves the highest primary performance metrics and the lowest FDR.

The outstanding performance of the FTISI model can be attributed to its knowledge-centric approach, which incorporates folksonomies as part of its knowledge model. It aggregates categorical knowledge from the eBook indices and the glossaries of the eBooks, along with related news repositories. Moreover, the categories from the tweet dataset are subjected to Latent Dirichlet Allocation and data interpretation, which harness topics from the external Worldwide Web and reduce the gap between the external Worldwide Web and the knowledge that passes into the localised framework. Most importantly, the TFIDF is applied to the historical tweet documents to yield informative terms. Apart from all this, large amounts of metadata are generated, which is handled efficiently and intelligently using the BiLSTM model, which is a strong deep learning model that helps tune the metadata into capsulated, fragmented, seasoned knowledge. Strong relevance computation mechanisms are yielded, like latent similarity using CoSIM rank and SOCPMI and TSS, and a metaheuristic stochastic optimisation framework is encompassed to optimise the initial solution set into a much more feasible solution set. Owing to all of these novel compositions, which have strong knowledge models, strong topic models that encompass folksonomies and knowledge from other external, secondary sources, and also a strong optimisation record, a series of different relevance computations can be generated in terms of semantic similarity, which facilitates reasoning.

The FTISI model surpasses all the baseline models, achieving the highest average precision, recall, accuracy, and F-measure, while also yielding the lowest false discovery rate. The FTISI model's success can be attributed to its unique design, which incorporates folksonomies as part of its knowledge model, providing valuable insights. In contrast, the EETSM model, which relies on a Wikipedia-based approach to link classification and tagging in social media, faces challenges due to its extensive reliance on Wikipedia knowledge. This large amount of knowledge may not be optimized for the specific task, leading to reduced performance compared to the more specialized FTISI model. Further, the regulation of knowledge is very poor in the model. When a high density of knowledge is not learned properly, it leads to the learning of noise; therefore, high deviations exist in the EETSM framework and thus yield a high FDR.

The EEKT underperforms with respect to the proposed model because it uses keyword extraction for tweets using standard NLP tools, whereas the Stanford POS NLP tagger is used. The POS tagger is one of the only sources used, which yields a lot of ambiguous results. Depending only on a single source, like the POS tagger, doesn't result in a wise model. A wise model aggregates knowledge from the external World Wide Web and reduces the semantic gap, or cognitive gap, between the entities in the external world and the entities in this pipeline. Additionally, there is an absence of auxiliary knowledge and strong relevance computations and sematic similarity models. Therefore, it performs poorly in comparison to the proposed FTISI model.

The TYT framework also lags in performance in comparison to the FTISI model because it is a probabilistic model of hashtags. So, a content-devoted model and a hashtag-devoted model, which is a probabilistic model, are used. Probabilistic models work well, but they cannot handle high densities of data and cannot be deployed as the only standalone model for the proposed Web 3.0, which is highly dense. Learning models are absent in the TYT framework, which increases the overall complexity of the framework. This statistical model can only work for small amounts of data. This is why the TYT model does not perform well.

POS taggers, named entity recognition, and Twitter terms and slang handling have been used along with agreement-based bootstrapping. These models also don't work as expected. While bootstrapping helps in adding legal data, data is added but not knowledge. So, knowledge derivation failed in this case. So, labelling can increase the probability of knowledge conception in the data but cannot be identified as factual knowledge. Therefore, epistemic reasoning cannot be convergent. For this reason, the PTAT model also does not perform as well as the FTISI model. Due to these several reasons, the proposed model performs better than all the baseline models as it is a knowledge-centric framework that not only aggregates knowledge from several sources but also derives it, regulates it, classifies it, encapsulates it, fragments it, and ensures that the cognitive gap between the external web and the knowledge that goes into the proposed local framework is reduced (Fig. 2).

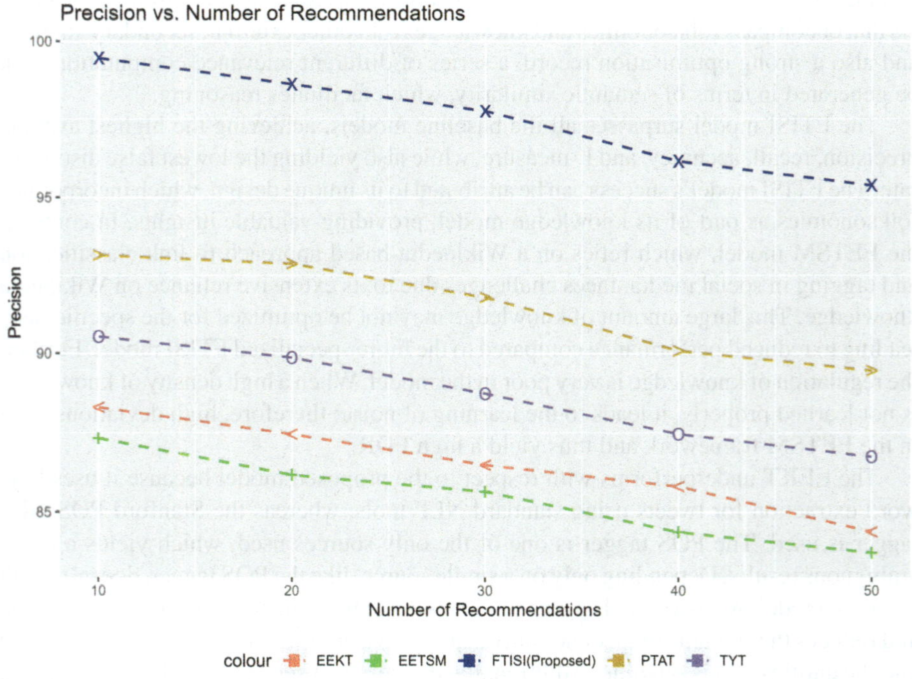

Fig. 2. Precision Vs No. of Recommendations

The line graph shows how the proposed FTISI model occupies the highest percentage of accuracy across any number of recommendations, followed by the PTAT model, while the TYT model occupies the median value of accuracy. The EEKT and EETSM models take the bottom two positions, respectively. The EETSM model uses Wikipedia knowledge but has poor knowledge regulation, leading to high deviations and a high false discovery rate. In contrast, the EEKT model depends on a single source, the Stanford POS NLP tagger, resulting in ambiguous results. While the TYT model performs well, it is not as effective as FTISI, which takes a more knowledge-centric approach and utilises topic modelling. Additionally, the PTAT model performs well, but it is not as comprehensive as FTISI, which employs deep learning for metadata handling and has a more comprehensive knowledge model.

5 Conclusions

This paper proposes the FTISI framework which is a semantically inclined folksonomy-based Twitter tag recommendation model which encompasses knowledge enrichment and strong relevance computation strategies. FTISI generates folksonomy using the terms in the dataset and enriches the folksonomy using linked open data cloud, eBooks, indices, and several other news repositories and discovers informative terms by subjecting the historical documents noted from the World Wide Web to TFIDF and LDA to the categories extracted from the data set. The informative term is further used to generate metadata and is then classified using the BiLSTM which is a strong deep learning model. The classified metadata Is divided into the top 20% which is used to calculate the lateral semantic similarity using the COSIM rank with that of the folksonomy discovered entities. The entities which come out of this pipeline are further subjected to the computation of SOCPMI under stochastic optimization. This TSS with differential thresholds and step deviance methods applied to 70% of the metadata in order to yield the most relevant tags for the Twitter documents which were discovered from the dataset. The FTISI achieves an overall precision of 97.83% with the lowest value of FDR of 0.03.

References

1. Ma, Z., Sun, A., Yuan, Q., Cong, G.: Tagging your tweets: A probabilistic modeling of hashtag annotation in twitter. In: Proceedings of the 23rd ACM International Conference on Conference on Information and Knowledge Management, pp. 999–1008 (2014). (TYT)
2. Weerasooriya, T., Perera, N., Liyanage, S.R.: A method to extract essential keywords from a tweet using NLP tools. In: 2016 Sixteenth International Conference on Advances in ICT for Emerging Regions (ICTer), pp. 29–34. IEEE (2016). (EEKT)
3. Gattani, A., et al.: Entity extraction, linking, classification, and tagging for social media: a Wikipedia-based approach. Proc. VLDB Endow. **6**(11), 1126–1137 (2013)
4. Albogamy, F., Ramsay, A.: POS tagging for Arabic tweets. In: Proceedings of the International Conference Recent Advances in Natural Language Processing, pp. 1–8 (2015). (PTAT)
5. Trant, J.: Studying social tagging and folksonomy: a review and framework (2009)
6. Yu, H., Zhou, B., Deng, M., Hu, F.: Tag recommendation method in folksonomy based on user tagging status. J. Intell. Inf. Syst. **50**, 479–500 (2018)
7. O'Leary, D.E.: Crowdsourcing tags in accounting and finance: review, analysis, and emerging issues. J. Emerg. Technol. Account. **12**(1), 93–115 (2015)
8. Xu, Z., Fu, Y., Mao, J., Su, D.: Towards the semantic web: collaborative tag suggestions. In: Collaborative Web Tagging Workshop at WWW 2006, Edinburgh, Scotland (2006)
9. Madhu, G., Govardhan, D.A., Rajinikanth, D.T.: Intelligent semantic web search engines: a brief survey. arXiv preprint arXiv:1102.0831 (2011)

10. Ramos, J.: Using TF-IDF to determine word relevance in document queries. In: Proceedings of the First Instructional Conference on Machine Learning, vol. 242, no. 1, pp. 29–48 (2003)
11. Ngoc Phuoc An, V., Magnolini, S., Popescu, O.: Paraphrase identification and semantic similarity in twitter with simple features. In: The 3rd International Workshop on Natural Language Processing for Social Media, pp. 10–19 (2015)
12. Graves, A., Schmidhuber, J.: Framewise phoneme classification with bidirectional LSTM and other neural network architectures. Neural Netw. **18**(5–6), 602–610 (2005)
13. Yang, Z., Yang, D., Dyer, C., He, X., Smola, A., Hovy, E.: Hierarchical attention networks for document classification. In: Proceedings of the 2016 Conference of the North American Chapter of the Association for Computational Linguistics: Human Language Technologies, pp. 1480–1489 (2016)
14. Deepak, G., Priyadarshini, J.S.: Personalized and enhanced hybridized semantic algorithm for web image retrieval incorporating ontology classification, strategic query expansion, and content-based analysis. Comput. Electr. Eng. **72**, 14–25 (2018)
15. Deepak, G., Santhanavijayan, A.: OntoBestFit: a best-fit occurrence estimation strategy for RDF driven faceted semantic search. Comput. Commun. **160**, 284–298 (2020)
16. Deepak, G., Ahmed, A., Skanda, B.: An intelligent inventive system for personalised webpage recommendation based on ontology semantics. Int. J. Intell. Syst. Technol. Appl. **18**(1–2), 115–132 (2019)
17. Deepak, G., Rooban, S., Santhanavijayan, A.: A knowledge centric hybridized approach for crime classification incorporating deep bi-LSTM neural network. Multimed. Tools Appl. **80**(18), 28061–28085 (2021)

Sustainable and Green Supplier's Evaluation Using TOPSIS and Fuzzy TOPSIS Methods

S. M. Vadivel[1](✉), Deeksha Sanjay Shetty[2], A. Eswaran[3], K. Loganathan[4], and A. Abhinav[5]

[1] VIT Business School, Vandalur-Kelambakkam Road, Chennai 600127, India
vadivel.sm@vit.ac.in
[2] School of Humanities, Social Sciences and Management, National Institute of Technology Karnataka, Surathkal 575025, India
[3] Department of Robotics and Automation, Rajalakshmi Engineering College, Thandalam, Chennai 602105, India
eswaran.a@rajalakshmi.edu.in
[4] Department of Mechanical Engineeing, Rajalakshmi Engineering College, Thandalam, Chennai 602105, India
loganathan.k@rajalakshmi.edu.in
[5] VIT Law School, Vandalur-Kelambakkam Road, Chennai 600127, India
abhinav.a2023@vitstudent.ac.in

Abstract. This study evaluates the efficacy of substitute selection at ABC Automobile Company in Chennai, Tamil Nadu, using a realistic fuzzy multi-criteria method. While management provided quantitative data, employees provided qualitative data. Fuzzy TOPSIS and the TOPSIS technique were used to examine subjective assessments of both qualitative and quantitative data in order to calculate the cost-effectiveness of sustainable supplier selection. Performance evaluation systems' subjectivity and imprecision are represented by fuzzy numbers that are derived from linguistic notions. Fuzzy MA models are widely used to ascertain the degree of likelihood of each characteristic and sub-attribute, avoiding complex and inaccurate comparisons of fuzzy values.

Keywords: Fuzzy TOPSIS · MCDM · Supplier selection · TOPSIS

1 Introduction

Evaluating the selection of environmentally friendly and sustainable providers requires the use of an efficient process. Here, we need to talk about two crucial subjects. Usually, the grading standards are complex and organised into hierarchies with several levels. Second, because most assessment techniques relied on subjective judgements, they generated qualitative and unreliable data. The fuzzy set theory was developed by Zadeh [16] in 1965 and provided a framework for making decisions in a fuzzy environment using ambiguous and subjective data.

V. Sakalauskas et al. (Eds.): IBICA 2023, LNNS 1231, pp. 231–239, 2025.
https://doi.org/10.1007/978-3-031-78946-5_22

ABC Automobile Company effectively selected the best supplier by utilising fuzzy MA models. Vadivel et al. (2020 & 2023) [17] evaluated Tamil Nadu private bus companies from the perspectives of customers and the site of flat waste disposal using TOPSIS and FTOPSIS [14]. It has been demonstrated that fuzzy numbers (FN), first proposed by Yeh et al. (2000) [15], are useful in transportation and traffic systems. FN can more properly convey qualitative data in spoken formulations when it is utilised to provide subjective and imperfect assessments of the conditions around decision-making. In this study, we investigate one of the top suppliers in terms of sustainable and eco-friendly operations.

Furthermore, the framework of this study was as follows: Sect. 2 covers the literary foundation for TOPSIS and FTOPSIS as they are applied in organisations. In Sect. 3, the recommended applicable methodology is presented. Section 4 discussed the investigation of ABC Company's suppliers. Section 5 discusses the study's limitations and potential future directions.

2 Literature Background

The purpose of this work is to present a comprehensive approach for evaluating technologies and assigning a business-appropriate ranking to them. A hybrid model is proposed based on FANP FTOPSIS. A case study from real life is used to validate the proposed paradigm [1]. Fuzzy DEMATEL, fuzzy ANP, and fuzzy TOPSIS were employed in a novel way in the study to increase the efficacy of concept selection. Finding the best idea design for a vehicle component is the aim of the study. The case organisation put the chosen design into practise [2]. The goal of this paper was to provide two innovative concepts: fuzzy logic-based MCDM and fuzzy TOPSIS. The proposed concepts aim to mitigate the elevated levels of uncertainty arising from divergent expert judgements and imprecise evaluations [3]. Presented a thorough comparison of these crucial stages in this work, highlighting differences in a systematic, step-by-step fashion. We also point out and talk about the drawbacks, problems, and difficulties with the current FTOPSIS technique [4]. The development of a dispersed yet cohesive network within a hospital to deliver its services to the target population was the topic of this study. We propose a novel MCDM process that integrates hesitant fuzzy sets (HFSs) and TOPSIS [5]. This study suggests using the TOPSIS and the Fuzzy AHP to identify gas turbine component defects. The management of the electronic control, hydraulic-pneumatic, and bearing equipment present in gas turbine components is the aim of this study [6]. The goal of that work is to offer a cutting-edge DM to effectively address these MCDM problems. First, we develop the fuzzy rough set model. Second, by combining the principles of the TOPSIS method with the suggested CPFRS model, a novel strategy is proposed to address the MCDM challenges in PF settings [7]. Using the fuzzy TOPSIS method and a few key concepts, we examined fuzzy sets in this article. We introduced the TOPSIS approach in a fuzzy environment, where each criterion's weight and each alternative's rating are represented by a triangular fuzzy integer. Ultimately, we selected a supplier by applying the recommended procedure for garment industry decision-making [8]. The decision-making committee in this study consisted of twenty active decision makers. The fuzzy TOPSIS MCDM model was used to analyses the data, and four possible locations were assessed in total based on the technical, economic, social, and environmental aspects [9]. This article demonstrated how fuzzy logic can be used to evaluate risks to safety, health, and the environment. It also suggests a strategy for choosing materials that is appropriate for the manufacturing industry that is based on the fuzzy-TOPSIS MCDM model [10]. The aim of this work is to make three group MCDM techniques and then use financial data and expert subjective evaluations to apply them to the identification of undervalued companies. Three different forms of fuzzy TOPSIS are proposed in this study: modified TOPSIS (M-TOPSIS), adjusted TOPSIS (A-TOPSIS), and conventional TOPSIS (C-TOPSIS). The suggested techniques now include fuzzy performance evaluations and a new fuzzy distance metric that is based on the confidence level of the experts [11]. This study included sensitivity analysis based on = 0.1, 0.5, and 0.9 values. Three of the cutting levels had the same sequence of possibilities. Turkey's preference rankings for renewable energy sources are primarily based on energy production, with subsequent considerations including land use, installed capacity, efficiency, payback periods, investment costs, job creation, and the value of CO2 emissions [12].

3 Research Study

(See Fig. 1).

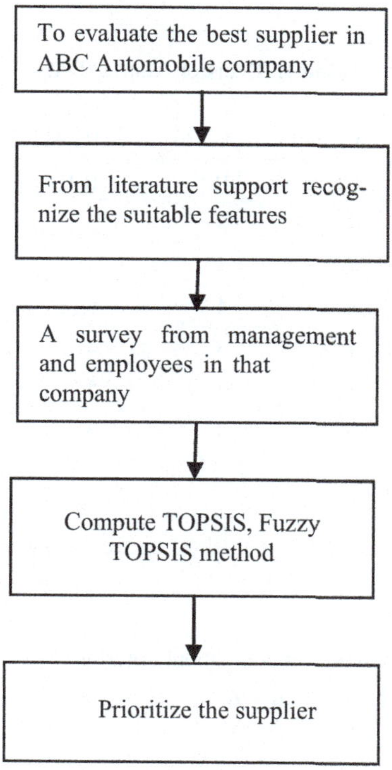

Fig. 1. Proposed Study

4 Case Study

This study intent to evaluate the best supplier in ABC Automobile Company in Chennai, Tamil Nadu. The company has many suppliers in terms of supplying the automobile ancillary parts such as brakes, gears, turbo engine, shock absorbers and fasteners etc. The company wants to recognize and giving importance to prompt and sustainable also green imitative suppliers in their company. For that, they have perform the following attributes such as Financial Perspective (FP), Customer Perspective (CP), Process Perspective (PP), Innovation and Learning Perspective (ILP), Environmental Perspective (EP) and Social Perspective (SP). For more information, read the articles from **Vadivel et al. 2013** [13].

4.1 TOPSIS and Fuzzy TOPSIS – Application Methodology

According to **Hwang and Yoon (1981)** [14, 18], the TOPSIS technique requires that the chosen alternative be the closest to the ideal positive solution and the farthest from the ideal negative solution. The advantage of using a fuzzy method is that FNs are precise values utilized to assign relative value to attributes. To reflect decision-maker preferences, the fuzzy set theory incorporates language terms (Fig. 2 and Tables 1, 2, 3, 4, 5, 6, 7, 8 and 9).

Table 1. Decision Matrix (DM)

Weightiness	0.1389	0.1549	0.1701	0.2411	0.1207	0.1743
Attributes	FP	CP	PP	ILP	EP	SP
X1	5	4	3	3	5	4
X2	3	4	5	4	4	4
X3	4	3	4	5	3	3
X4	4	4	5	4	4	3
X5	2	3	4	5	4	3
X6	3	5	4	3	5	4

Alternative	6	m
Attribute	6	n
Xij	The score of option i about criterion j	
X	xij	6 × 6

Table 2. Normalization of the DM

Weightiness	0.1389	0.1549	0.1701	0.2411	0.1207
X1	0.56	0.42	0.29	0.30	0.48
X2	0.34	0.42	0.48	0.40	0.39
X3	0.45	0.31	0.39	0.50	0.29
X4	0.45	0.42	0.48	0.40	0.39
X5	0.23	0.31	0.39	0.50	0.39
X6	0.34	0.52	0.39	0.30	0.48

Table 3. Results based on TOPSIS method

Suppliers	Si'	Si*	Ci*	TOPSIS Priorities
X1	0.0584	0.0606	0.4908	3
X2	0.0521	1.9344	0.0262	6
X3	0.0598	0.0502	0.5436	2
X4	0.0551	0.0404	0.5772	1
X5	0.0523	0.0638	0.4505	5
X6	0.0502	0.0598	0.4564	4

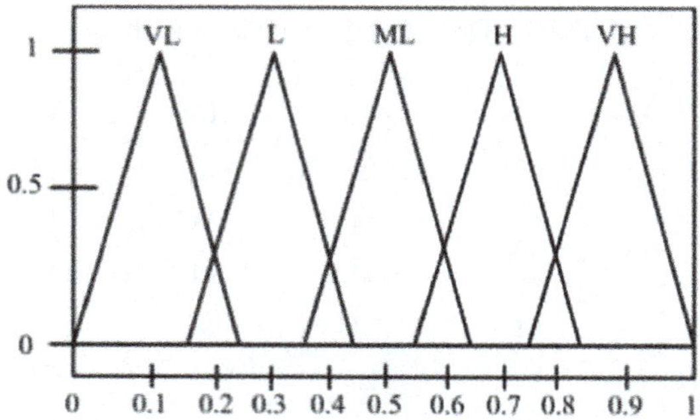

Fig. 2. Fuzzy triangular membership (FTM) functions

Table 4. FTM function

Rank	Grade	Membership function
Very Low (VL)	1	(0.00, 0.10, 0.25)
Low (L)	2	(0.15, 0.30, 0.45)
Medium (M)	3	(0.35, 0.50, 0.65)
High (H)	4	(0.55, 0.70, 0.85)
Very High (VH)	5	(0.75, 0.90, 1.00)

Table 5. DM in the form of Normalization

Suppliers	C1	C2	C3	C4	C5	C6
X1	0.56	0.42	0.29	0.30	0.48	0.46
X2	0.34	0.42	0.48	0.40	0.39	0.46
X3	0.45	0.31	0.39	0.50	0.29	0.35
X4	0.45	0.42	0.48	0.40	0.39	0.35
X5	0.23	0.31	0.39	0.50	0.39	0.35
X6	0.34	0.52	0.39	0.30	0.48	0.46
Weight	0.14	0.15	0.17	0.24	0.12	0.17

Table 6. DM - Fuzzy linguistic variables

Suppliers	C1	C2	C3	C4	C5	C6
X1	H	M	L	L	M	M
X2	L	M	M	M	L	M
X3	L	L	M	M	L	L
X4	L	M	M	M	L	L
X5	VL	L	L	M	M	L
X6	L	M	L	L	M	M
Weight	VL	VL	L	L	VL	L

A*	1	1	1	1	1
A-	0	0	0	0	0
W	(0.00, 0.10, 0.25)	(0.00, 0.10, 0.25)	(0.15, 0.30, 0.45)	(0.15, 0.30, 0.45)	(0.00, 0.10, 0.25)

Table 7. Weightage of Fuzzy DM

Suppliers	C1	C2	C3	C4	C5
X1	(0.55, 0.70, 0.85)	(0.35, 0.50, 0.65)	(0.15, 0.30, 0.45)	(0.15, 0.30, 0.45)	(0.35, 0.50, 0.65)
X2	(0.15, 0.30, 0.45)	(0.35, 0.50, 0.65)	(0.35, 0.50, 0.65)	(0.35, 0.50, 0.65)	(0.15, 0.30, 0.45)
X3	(0.15, 0.30, 0.45)	(0.15, 0.30, 0.45)	(0.35, 0.50, 0.65)	(0.35, 0.50, 0.65)	(0.15, 0.30, 0.45)
X4	(0.15, 0.30, 0.45)	(0.35, 0.50, 0.65)	(0.35, 0.50, 0.65)	(0.35, 0.50, 0.65)	(0.15, 0.30, 0.45)
X5	(0.00, 0.10, 0.25)	(0.15, 0.30, 0.45)	(0.15, 0.30, 0.45)	(0.35, 0.50, 0.65)	(0.35, 0.50, 0.65)
X6	(0.15, 0.30, 0.45)	(0.35, 0.50, 0.65)	(0.15, 0.30, 0.45)	(0.15, 0.30, 0.45)	(0.35, 0.50, 0.65)
Weight	(0.00, 0.10, 0.25)	(0.00, 0.10, 0.25)	(0.15, 0.30, 0.45)	(0.15, 0.30, 0.45)	(0.00, 0.10, 0.25)

Table 8. Fuzzy TOPSIS Analysis

Suppliers	C1	C2	C3	C4	C5
X1	(0.00, 0.07, 0.21)	(0.00, 0.05, 0.16)	(0.02, 0.09, 0.20)	(0.02, 0.09, 0.20)	(0.00, 0.05, 0.16)
X2	(0.00, 0.03, 0.11)	(0.00, 0.05, 0.16)	(0.05, 0.15, 0.29)	(0.05, 0.15, 0.29)	(0.00, 0.03, 0.11)
X3	(0.00, 0.03, 0.11)	(0.00, 0.03, 0.11)	(0.05, 0.15, 0.29)	(0.05, 0.15, 0.29)	(0.00, 0.03, 0.11)
X4	(0.00, 0.03, 0.11)	(0.00, 0.05, 0.16)	(0.05, 0.15, 0.29)	(0.05, 0.15, 0.29)	(0.00, 0.03, 0.11)
X5	(0.00, 0.01, 0.06)	(0.00, 0.03, 0.11)	(0.02, 0.09, 0.20)	(0.05, 0.15, 0.29)	(0.00, 0.05, 0.16)
X6	(0.00, 0.03, 0.11)	(0.00, 0.05, 0.16)	(0.02, 0.09, 0.20)	(0.02, 0.09, 0.20)	(0.00, 0.05, 0.16)

5 TOPSIS, FTOPSIS - Results

Table 10 summarizes the conclusions of the comparative analysis. According to the TOPSIS analysis, the best provider is supplier 4 (X4), followed by supplier 3 (X3). In contrast, fuzzy TOPSIS prioritizes supplier 2 (X2), followed by supplier 1 (X1) and supplier 4 (X4). TOPSIS and FTOPSIS have the highest global weight (0.5772, 0.5436: 0.1298, 0.1239). Overall, we feel that Supplier 4 will be the best based on a comparison of

Table 9. Results of Fuzzy TOPSIS

Suppliers	Di*	Di'	Cci	Rank
X1	5.417538381	0.76636189	0.1239	2
X2	5.368635565	0.800539048	0.1298	1
X3	5.447985662	0.70603331	0.1147	4
X4	5.425920803	0.736986793	0.1196	3
X5	5.505762118	0.642725324	0.1045	6
X6	5.461141182	0.704388019	0.1142	5

Table 10. Relative study in Fuzzy TOPSIS, TOPSIS Prioritize Ranking

Suppliers	TOPSIS	Fuzzy TOPSIS
X1	3	2
X2	6	1
X3	2	4
X4	1	3
X5	5	6
X6	4	5

both TOPSIS and FTOPSIS methodologies. This work's absence of sensitivity analysis is seen as a shortcoming.

6 Conclusion

It is difficult and deceptive to evaluate the efficacy of supplier selection in the context of environmental and sustainability activities. The most obvious and effective method for the evaluation process is the DM's use of vague assessments that are stated in language. ABC Automobile Company, Chennai, Tamil Nadu, thoroughly examined six supplier profiles in order to present the procedure. When evaluating suppliers, the TOPSIS & Fuzzy TOPSIS approach is useful. In general, Supplier 4 emerged as the most optimal provider for both Fuzzy TOPSIS and TOPSIS methodologies. There is only one area in Chennai, Tamil Nadu, where the ban is in effect. Moreover, future results improvements and more economical MCDM methodology adoption may be made possible by the addition of additional crucial aspects.

References

1. Aliakbari Nouri, F., Khalili Esbouei, S., Antucheviciene, J.: A hybrid MCDM approach based on fuzzy ANP and fuzzy TOPSIS for technology selection. Informatica **26**(3), 369–388 (2015)

2. Vinodh, S., Sai Balagi, T.S., Patil, A.: A hybrid MCDM approach for agile concept selection using fuzzy DEMATEL, fuzzy ANP and fuzzy TOPSIS. Int. J. Adv. Manuf. Technol. **83**, 1979–1987 (2016)
3. Jumarni, R.F., Zamri, N.: An integration of fuzzy TOPSIS and fuzzy logic for multi-criteria decision making problems. Int. J. Eng. Technol. **7**(2), 102–106 (2018)
4. Madi, E.N., Garibaldi, J.M., Wagner, C.: An exploration of issues and limitations in current methods of TOPSIS and fuzzy TOPSIS. In: 2016 IEEE International Conference on Fuzzy Systems (FUZZ-IEEE), pp. 2098–2105. IEEE (2016)
5. Senvar, O., Otay, I., Bolturk, E.: Hospital site selection via hesitant fuzzy TOPSIS. IFAC-PapersOnLine **49**(12), 1140–1145 (2016)
6. Balin, A., Demirel, H., Alarcin, F.: A novel hybrid MCDM model based on fuzzy AHP and fuzzy TOPSIS for the most affected gas turbine component selection by the failures. J. Marine Eng. Technol. **15**(2), 69–78 (2016)
7. Jin, J., Zhao, P., You, T.: Picture fuzzy TOPSIS method based on CPFRS model: an application to risk management problems. Sci. Program. **2021**, 1–15 (2021)
8. Zulqarnain, R.M., et al.: Generalized fuzzy TOPSIS to solve multi-criteria decision-making problems. J. New Theory **32**, 40–50 (2020)
9. Noori, A., Bonakdari, H., Morovati, K., Gharabaghi, B.: The optimal dam site selection using a group decision-making method through fuzzy TOPSIS model. Environ. Syst. Decis. **38**, 471–488 (2018)
10. Rahim, A.A., Musa, S.N., Ramesh, S., Lim, M.K.: Development of a fuzzy-TOPSIS multi-criteria decision-making model for material selection with the integration of safety, health and environment risk assessment. Proc. Inst. Mech. Eng. Part L: J. Mater.: Design Appl. **235**(7), 1532–1550 (2021)
11. Hatami-Marbini, A., Kangi, F.: An extension of fuzzy TOPSIS for a group decision making with an application to Tehran stock exchange. Appl. Soft Comput. **52**, 1084–1097 (2017)
12. Şengül, Ü., Eren, M., Shiraz, S.E., Gezder, V., Şengül, A.B.: Fuzzy TOPSIS method for ranking renewable energy supply systems in Turkey. Renew. Energy **75**, 617–625 (2015)
13. Vadivel, S.M., Palanivelu, S., Sequeira, A.H., Chandana, V.: apartment waste disposal sustainable facility location using ENTROPY method. In: Intelligent Systems Design and Applications, ISDA 2022. Lecture Notes in Networks and Systems, vol. 646. Springer (2023)
14. Vadivel, S.M., Sakthivel, V., Praveena, L., Chandana, V.: Apartments waste disposal location evaluation using TOPSIS and fuzzy TOPSIS methods. In: Innovations in Bio-Inspired Computing and Applications. IBICA 2022. Lecture Notes in Networks and Systems, vol. 649. Springer (2023)
15. Yeh, C.H., Deng, H., Chang, Y.H.: Fuzzy multi-criteria analysis for performance evaluation of bus companies. Eur. J. Oper. Res. **126**, 459–473 (2000)
16. Zadeh, L.A.: Electrical engineering at the crossroads. IEEE Trans. Educ. **8**(2), 30–33 (1965)
17. Vadivel, S.M., Sequeira, A.H., Jauhar, S.K., Chandana, V.: Tamilnadu omnibus travels evaluation using TOPSIS and fuzzy TOPSIS methods. In: Innovations in Bio-Inspired Computing and Applications, IBICA 2021. Lecture Notes in Networks and Systems, vol. 419. Springer (2022)
18. Hwang, C.L., Yoon, K.: Methods for multiple attribute decision making. In: Multiple Attribute Decision Making, pp. 58–191. Springer, Heidelberg (1981)

A Transformer Models Customization Pipeline for Radiology Text Reports Prediction Based on Frontal and Lateral Chest-X-Ray Images

Abdelfettah Elaanba(✉), Mohammed Ridouani, and Larbi Hassouni

RITM Laboratory, Hassan II University, CED Engineering Sciences, Casablanca, Morocco
Abdelfettah.elaanba@gmail.com

Abstract. A chest X-ray image report includes a text description of the abnormalities and normalities of a patient lung, heart, and chest. The report is an image-to-text translation based on the radiologist's diagnosis of the X-ray view. It helps to communicate the diagnosis to no radiologist or medical experts and track the patient's health condition in the future. The manual writing of reports is a tedious task that consumes time and is vulnerable to possible human errors. In this work, we propose a pipeline for building based-transformer models to process the task of automatic report writing. We included the experiment that justified the choice of the models and configurations of the feature extractor encoder, transformer attention layers, and training. We show that: 1) First, the pre-trained ViT feature extractor performs better than the CNN-based encoder (DensNet121). 2) Second, the dual-view input (Frontal and Lateral images) give better result than one input (Frontal or Lateral).

Keywords: Transformers · Radiology reports · CNN encoder · Chest X-Ray

1 Introduction

On average an expert radiologist spends more than 10 min [1] to write a radiology report. This issue consumes the expensive time of experts. Moreover, expert radiologist is considered a scarce resource. The stress caused by the workload and the task complexity can open a space for human errors. Besides those reasons an automatic radiology report generation system will save time and help in areas with fewer experts to automate the task, identify all possible normalities or abnormalities (normalities or abnormalities present in training dataset), and avoid the radiologist bias to look and search for a specific problem. In this work, we propose a pipeline to train transformer models as a step toward the development of such a system. The development of a general and efficient system for automatic radiology reports generation is retarded by several humpstones. First, the datasets imbalances. The datasets contain reports and radiology images full of normalities versus limited numbers of diseases (abnormalities) in boat images and text reports. Likewise, the available datasets suffer from the scarcity of diseases diversity, which prevent the generalization of report generation for all kind of health

V. Sakalauskas et al. (Eds.): IBICA 2023, LNNS 1231, pp. 240–249, 2025.
https://doi.org/10.1007/978-3-031-78946-5_23

issues. Datasets also contains poor chest-X-Ray image qualities, in addition X-Ray images are very similar [2]. This similarity adds complexity to the models' training. The second hump-stone is the bias introduced by the sample studied and the methodology of radiologists and researchers in writing and processing a radiology report. The sample can concern a limited number of patients from one country, one city, or just one hospital sharing some comment living conditions. So, the results can't generalize to the wider population. In addition, there is no standard methodology for the task of writing a radiology report. Besides this, a model trained in a specific study cannot directly applied to another context. The third problem is the nature of radiology reports with long and repetitive sentences, adds complexity to the problem compared to the normal captioning task requiring only the model to predict one sentence for an input image. For those reasons, we contribute with the trains of research to fix those issues, and we propose a flexible pipeline to train customized transformers for the text report prediction based on chest x-ray images. The pipeline will help to transcend the methodology and train models unto other private or public datasets for the given task of radiology report generation. As an application of the proposed method, we apply the pipeline using the IU-X-Ray dataset. The IU dataset is a known benchmark used for research evaluation and comparisons. In this paper, we adopt the BLUE metric to evaluate our results. We present below the detailed descriptions for steps to use the proposed pipeline to customize and train a transformer model for similar tasks. This paper is structured as follows: Sect. 2 presents the related works to the automatic reports' generation task. Section 3, introduce the task of reports generation. In Sect. 4, we introduce the transformer models. Section 5 introduces the experiments. Section 6 presents the results and the analysis. Finally, in Sect. 7, we introduce the conclusion.

2 Related Works

Image captioning is the base-related task to radiology reports generation. Gaung Li et al. [3] used a transformer for the image captioning task. The authors customized the attention and proposed an EnTangled Attention (ETA) to permit the model to capture semantic and visual information. The captioning consists of attributing one sentence to an image. It is a particular case for radiology reports with long paragraphs and multi-sentences. Each sentence describes a specific medical normality or abnormality for a precise zone in the image. Concerning chest X-Ray images report generation Jianbo Yuan et al. [4] Propose an encoder-decoder model test on the IU dataset for multi-view chest X-Ray images (Frontal and Lateral). The proposed encoder is a CNN model for feature extraction. The author tested with a pre-trained CNN on ImageNet and CheXpert datasets. The author reported that pre-training on domain-specific data performs better. For the decoder side, the author uses a hierarchical LSTM. The author noticed that the proposed model performs like the state-of-the-art related baseline approaches. Using a different approach Omar et al. [5] proposed a conditioned transformer for radiology report generation. The authors fine-tuned a pre-trained CheXnet [6] for disease tag generation on the IU dataset. The tags were used later to fine-tune a pre-trained transformer GPT-2 for the task of report generation based on tags associated with the extracted decoder visual features from the Chest images. The authors report that the proposed solution performs more than transformers-based models in quantitative metrics. Moreover, the solution is faster to train. In this work, the use of a large NLP pre-trained model like GPT-2, eliminates the need to specify a vocabulary for the model. In a different strategy instead of predicting the radiology report directly from the patient's chest image, Koji et al. [7] proposed a progressive framework starting with extracting global context text from the images in the first stage. Then, reforms the text with a transformer model (Text-to-Text) to generate the full radiology report. The authors report that the progressive approach surpasses standard and single Visual models in BLUE-1, 2, 3, and ROUGE Metrics. The metrics measures were performed on two benchmarks: IU-X Ray and MIMIC-CXR [8] datasets. Meanwhile, the model is under the baseline stat-of-the-art models in precision.

3 Radiology Reports Generation Task

Given the use of attention mechanism by transformer models and their comparable or higher results on large image processing tasks compared to the CNN [9–15] models, we choose to tackle the task of medical radiology images reports generation task with such models.

3.1 Radiology Report Structure

A medical report is a text document written by a radiologist to diagnose a patient's health condition based on his results analyses. The report's audience is other radiologists, doctors, patients, and their families. A chest X-ray report is a semantic description of one or several chest X-ray images using text. The report is composed of a secession of sentences. Each sentence highlights an abnormality or a normality. The length of a report can be long (more than 1000 words) or short, depending on the problems present in the X-ray image of a given patient. The report is subdivided into three parts. 1) impression which serves as a report title. 2) the finding contains the full report, and 3) the manual tags which consist of a keywords list used as a report summary. Depending on the report format a recommendation or conclusion paragraph can included and serve as a guide to how the report is understood [16].In general, report writing follows a known pattern but there is no standardized unique format to write the report [17]. The image below shows an example of a chest X-ray report from the IU-Dataset [18] (Fig. 1).

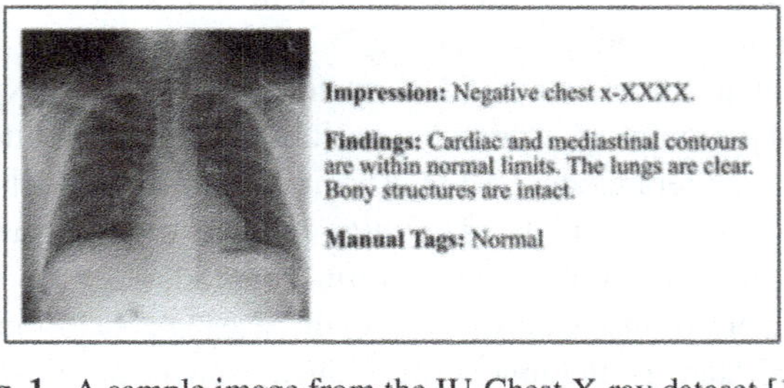

Fig. 1. A sample image from the IU-Chest X-ray dataset [18]

3.2 Radiology Report Generation Methods

In typical work conditions, a radiology report is written by a radiologist within an average of 10 min. The radiologist looks at the patient's x-ray images to deduce abnormality or normality to report. In general, a radiologist doesn't study at school a standard method for writing a typical report. So, the practice and experiment are the way to produce the report text. The consequence of this, is that reports are not standardized. From a deep learning view, it is difficult to train models when data is not homogenous. In practice, the dataset for the task of automatic report generation is annotated by a group of radiologists with a defined and unified methodology. The second method to resolve the problem is with the application of deep-leaning Image to text Models, in particular transformer models which is the object of this paper. The principle is to use supervised learning to train a model on X-Ray images with reports as a label and use the Gradient to minimize the loss function.

4 Sequence to Sequence Models

In deep learning, a sequence-to-sequence model takes a bench of tokens input and outputs a sequence of tokens. The input and output can be homogeneous (Text to Text, Image to image, speech to speech.) or heterogeneous (text to image, image to speech…). If we consider an input X (x1, x2, x3…, xn) and Y(y1, y2…, yn), the output sequence of the model can be represented as a function F(X) = Y. The function F represents the task performed by the S-T-S model. This task can be translation, image captioning, or text summarization to name a few. F is predicted by the minimization of the loss function during training using the gradient descent [19]. Such models are used first for natural language processing such as translation. The task consists of taking a words sequence from a source language and seeking to get the corresponding output using a target language. One of the used models is RNNs [20].

4.1 Transformers

Because of the recurrent property of RNN models, the long path can cause the loss of relevant information along the way (gradient vanishing). Moreover, the calculation cannot be paralleled. To fix this problem, an attention mechanism is introduced to give the decoder the possibility to look at all encoder hidden states. This fixes the issue of information loss. For the parallel calculation, the recurrent connection is dropped giving the encoder the ability of independent processing. Besides the absence of the recurrence in transformers, the sequence order is not tracked in the transformer models. This problem is mitigated by adding a position encoding layer for the transformer's encoder input. Inspired by RNNs and intention mechanism, transformer models were first proposed by Google in the famous paper 'Attention is all you need' [21].The model in the paper is designed to fix the comment issues in NLP models. The authors also mentioned that the model can be good at other tasks like image processing. Which was experimented with and proved a few years after the big success of NLP. The Transformer models become the gold standard in computer vision by 2020. The transformers are maintaining the

Encoder-decoder architecture Fig. 2 enhanced by the attention mechanism. The name of the transformer came from the action made by the attention layer, transforming a given sequence to an output sequence. We will briefly describe the transformer architecture and blocs in the paragraph below.

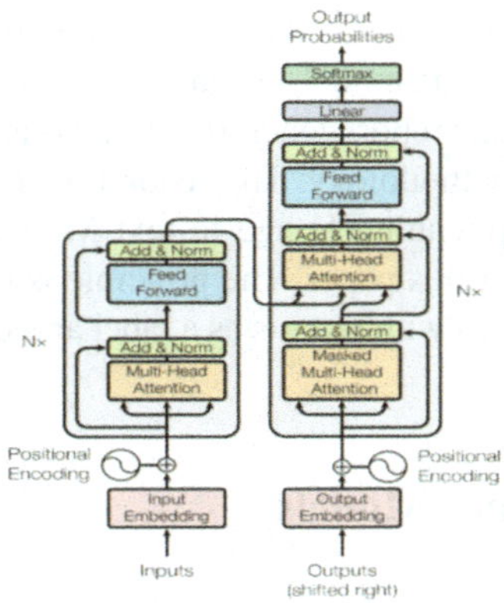

Fig. 2. Transformer model architecture [21]

4.1.1 Attention Mechanism

Inspired by the human ability to process reel time data stream by using attention to focus on relevant information and ignore the unimportant parts. The attention mechanism was first introduced by [22] as "Banhdanau Attention". Which is applied at the beginning for the language-translation task. The attention is used to find the dependencies between the elements of an input sequence (Sentences in NLP). Two attention types persist: First, the self-attention where this dependency is computed between the encoder input sequence elements (Intra-sequence dependency). Second, the attention where the dependence computed between the encoder output sequence and the decoder input sequence (Inter-sequence dependency).

The attention is calculated as a vectorial projection of a transformed input vector X (weighted by the encoder weights) into tree vectors Q, V, and K using the equation below (Eq. 1). The Q is the query vector Q = Wq * X.K is the Key Vector K = Wk*X, and V is the Value vector V = Wv * X.

$$Attention(Q, K, V) = Softmax(\frac{Qk^T}{\sqrt{d_k}})V \tag{1}$$

The difference when computing the self-attention and the attention is; for the first, the V, K, and Q derived from the same encoder input X. Otherwise, for the second, the V and K derived from the encoder source input X, while the Q derived from, the target, the decoder input sequence.

5 Experiments

5.1 Dataset Presentation

IU-dataset is considered one of the few benchmarks publicly available to evaluate the automatic reports generation task. We used the IU dataset to test our transformer models and compare our results. The Indian Dataset is available for research and downloadable from the National Library of Medicine[1]. The dataset contains 7470 Chest X-ray images (Frontal and Lateral images) collected from Indiana Network for Patient Care. With 3955 corresponding radiology reports. Each report in the dataset is associated with two chest X-ray images, one frontal and a lateral view. To preserve data privacy, de-identification is applied to all reports and DICOM images. This operation removes any information that can identify a patient's identity.

5.2 Metrics for Evaluating Text Generated Quality

Evaluating how good a model in the chest x-ray medical reports generation is to find a metric to measure predicted and original report resemblance. The idea is to find a metric that compares on a semantic level the two reports. This kind of metric can be calculated only by humans looking manually and comparing the reports. In practice, there are several metrics that look to exact or synonym words into source and target text. An example is the BLEU [23] metric which calculates the number of exact words and gives a score between 0 and 1. This is the BLUE-1 metric. There are other variants of BLUE that track the number of identical sequences with N successive words in the compared reports named BLUE-N. N can be an integer number > 0. Two metrics designed first for language translation are ROUGE-L [24] and METEOR [25]. Practically ROUGE-L metric is used for the estimation of text summary quality.

5.3 Methodology

For this work, we propose to use a transformer model for the task of medical images reports generation. The proposed architecture is formed with a feature extractor model followed by an encoded-decoder transformer. For the Input, we train our model using a single (Frontal image) or dual-view input (Frontal and lateral chest x-ray images). The model architecture is reported in the diagram below Fig. 3.

For the training and the test procedures, we train our model variants on the training dataset by minimizing the loss function. Then, we compute the BLUE-1 metric on the test dataset. The objective is to compare the efficiency of different feature extractor encoders and different transformer model variants. The variant differs by the number of attention layers and the number of heads par attention layers. For the training hyperparameters, we test different configurations for the learning rate, batch size, input images, and optimizers. The best parameters are reported in the table below Table. 1.

[1] (http://openi.nlm.nih.gov/).

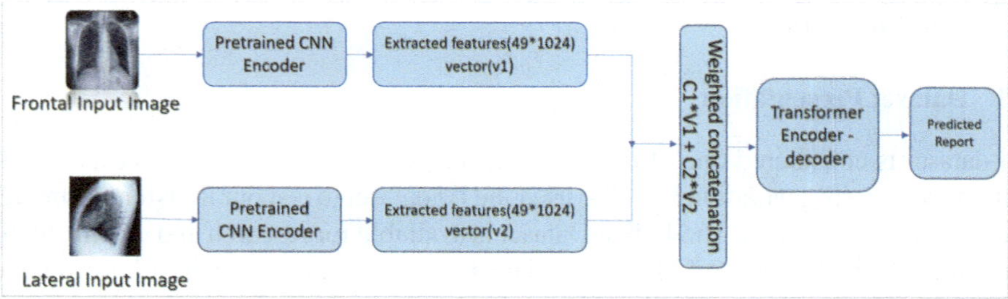

Fig. 3. Multi-inputs transformer architecture

Table 1. Training Hyperparameters

Training HyperParameters	
Initial Learning rate	1e−4
Image Input	(224, 224, 3)
Augmentations	RandomContrast, RandomRotation
Batch size	16
Loss function	SparseCategoricalCrossentropy

6 Results and Analyses

The table below shows the results of different tested feature extraction encoders with our proposed transformer model. The test aims to compare the pre-trained cnn-encoders with pre-trained ViT Models. After the training using the same transformer model configurations, we noticed that the ViT encoder trained on imageNet performs better than all tested CNN-based encoders. Furthermore, the CNN-encoders pre-trained on the imageNet dataset outperform the one trained on the chest X-ray 8 dataset. Besides the chest X-ray8 contains chest images similar to those of the processed problem in this paper, we aim to get a higher result than the training on imageNet containing images for general context. However, the experimental result shows that the Densnet121 encoder pre-trained on ImageNet (BLUE = 0.15) outperforms the pre-trained Resnet121 on the Chest X-ray 8 database (BLUE = 0.13). We explain these results by the fact, that even if the Chest X-ray 8 countians X-rays images, the diseases processed with this dataset are diffident from those processed in the IU dataset. So, the pertained model on ImageNet is more general and can used for the specific task of X-rays report generation. We know that transfer learning [26] is efficient from a general to specific tasks not between two specifics tasks (Table 2).

After the selection of the feature extractor encoder model, we fix the chosen model in our architecture, and we start testing different transformer configurations for the encoder-decoder model. The table below shows the results of our tests (see Table 3). For each test, we change the number of blocks and the number of heads per block. The 3 blocks with 16-heads transformer model perform better than all tested versions.

Table 2. Features extraction Models

Features extraction Models	BLUE-1
ViT-Encoder (Image Net)	0.26
InceptionV3 (ImageNet)	0.19
DensNet121 (ImageNet)	0.15
DensNet121 (Chest X-ray 8)	0.13

Table 3. Transformer models configurations tests

Transformer model	Number of attention layers	Number of heads per attention layer	BLUE-1
Transformer1–2	1 layer	2 heads	0.13
Transformer1–4	1 layer	4 heads	0.15
Transformer1–8	1 layer	8 heads	0.22
Transformer1–16	1 layer	16 heads	0.23
Transformer2–16	2 layers	16 heads	0.23
Transformer3–16	3 layers	16 heads	0.26

The result shows that the precision augments broadly for the one attention bloc transformer model when we augment the number of heads (2, 4 and 8 blocs). The improvement became small with big number of heads and more blocs (transformer1–16, transformer2–16, and transformer3–16). The explanation for this behavior is that with the larger models, the training tends to overfit because of the small size of the training dataset (IU-dataset) in this work.

After the selection of the best model from the tests above. We proceed to test the model using our proposed architecture. First, with two images as input (frontal and lateral images). The input is the sum of the decoder extracted features for the frontal (F1) and the lateral image (F2). To simplify the test, we suppose that the frontal and the lateral images contribute equally to the final result (Eq. 2). Then we compare the results obtained with a single image input (Frontal or Lateral).

$$Input = C1 * F1 + C2 * F2(C1 = C2 = 1) \quad (2)$$

The table below shows the result for different possible inputs (Frontal only, Lateral only, weighted sum of lateral and frontal images) (Table 4).

The table shows that dual input performs better than a single input image (frontal or lateral X-ray images). This portends that the combination of two inputs gives the model the possibility to find additional information that is present in one image and not in the other. Or to reinforce the redundant information found in both inputs.

Table 4. Weighted sum input tests

Model input	BLUE-1
Frontal images	0.26
Lateral Images	0.23
Weighted feature sum F1 + F2	0.27

7 Conclusion

The proposed pipeline is a step toward the development of an efficient automatic system for report generation tasks based on chest X-ray inputs. Such a system will considerably reduce the load work charge of radiologists and help them inspect the outputs of the system to reduce possible diagnosis errors. We proposed the use of attention-based models such as transformers, which are state-of-the-art captioning models. The development of the report generation task needs more annotated data and a global standardization for how to write reports. We consider for our future works to experiment with the coefficients of the dual view input to find the exact contribution of the frontal and the lateral inputs to in the result.

References

1. Cowan, I.A., MacDonald, S.L., Floyd, R.A.: Measuring and managing radiologist workload: measuring radiologist reporting times using data from a Radiology Information System. J. Med. Imaging Radiat. Oncol. **57**(5), 558–566 (2013). https://doi.org/10.1111/1754-9485.12092. Epub 2013 Jul 12 PMID: 24119269
2. Wang, Z., Han, H., Wang, L., Li, X., Zhou, L.: Automated radiographic report generation purely on transformer: a multicriteria supervised approach. IEEE Trans. Med. Imaging **41**(10), 2803–2813 (2022). https://doi.org/10.1109/TMI.2022.3171661. Epub 2022 Sep 30 PMID: 35507620
3. Li, G., Zhu, L., Liu, P., Yang, Y.: Entangled Transformer for Image Captioning. In: 2019 IEEE/CVF International Conference on Computer Vision (ICCV), Seoul, Korea (South), pp. 8927–8936 (2019). https://doi.org/10.1109/ICCV.2019.00902
4. Yuan, J., Liao, H., Luo, R., Luo, J.: Automatic radiology report generation based on multi-view image fusion and medical concept enrichment. In: Medical Image Computing and Computer Assisted Intervention–MICCAI 2019: 22nd International Conference, Shenzhen, China, 13–17 October 2019, Proceedings, Part VI 22, pp. 721–729. Springer (2019)
5. Alfarghaly, O., Khaled, R., Elkorany, A., Helal, M., Fahmy, A.: Automated radiology report generation using conditioned transformers. Inform. Med. Unlock. **24**, 100557 (2021)
6. Rajpurkar, P., et al.: ChexNet: Radiologist-level pneumonia detection on chest x-rays with deep learning. arXiv preprint arXiv:1711.05225 (2017)
7. Nooralahzadeh, F., Gonzalez, N.P., Frauenfelder, T., Fujimoto, K., Krauthammer, M.: Progressive transformer-based generation of radiology reports. arXiv preprint arXiv:2102.09777 (2021)
8. Johnson, A.E., et al.: MIMIC-CXR, a de-identified publicly available database of chest radiographs with free-text reports. Sci. Data **6**(1), 317 (2019)

9. Rabbah, J., Ridouani, M., Hassouni, L.: A new classification model based on transfer learning of DCNN and stacknet for fast classification of pneumonia through X-ray images. Int. J. Reliable Qual. E-Healthc. 12(1) (2023)
10. Elaissaoui, K., Ridouani, M.: Application of deep learning in healthcare: a survey on brain tumor detection. In: International Conference on Connected Object and Artificial Intelligence (COCIA 2023) ITM Web Conference, vol. 52 (2023)
11. Elaanba, A., Ridouani, M., Hassouni, L.: Automatic diagnosis chest-X-ray-based-framework for semantic segmentation and placement errors detection of catheters and tubes. Int. J. Comput. Inf. Syst. Industr. Manage. Appl. **2023**(15), 257–265 (2023)
12. Techa, C., Ridouani, M., Hassouni, L., Anoun, H.: Automated Alzheimer's disease classification from brain MRI scans using ConvNeXt and ensemble of machine learning classifiers. In: Abraham, A., Hanne, T., Gandhi, N., Manghirmalani Mishra, P., Bajaj, A., Siarry, P. (eds) SoCPaR 2022. LNNS, vol. 648, pp. 382–391. Springer, Cham (2023). https://doi.org/10.1007/978-3-031-27524-1_36
13. Elaanba, A., Ridouani, M., Hassouni, L.: Automatic diagnosis framework for catheters and tubes semantic segmentation and placement errors detection. In: Abraham, A., Bajaj, A., Gandhi, N., Madureira, A.M., Kahraman, C. (eds.) IBICA 2022. LNNS, vol. 649, pp. 176–188. Springer, Cham (2023). https://doi.org/10.1007/978-3-031-27499-2_17
14. Techa, C., Ridouani, M., Hassouni, L., Anoun, H.: Alzheimer's disease multi-class classification model based on CNN and StackNet using brain MRI data. In: Hassanien, A.E., Snášel, V., Tang, M., Sung, TW., Chang, KC. (eds.) AISI 2022. LNDECT, vol. 152, 248–259. Springer, Cham (2023). https://doi.org/10.1007/978-3-031-20601-6_23
15. Elaanba, A., Ridouani, M., Hassouni, L.: A Stacked Generalization Chest-X-Ray-Based Framework for Mispositioned Medical Tubes and Catheters Detection. Biomed. Signal process. Control **79**, 104111 (2023)
16. Hartung, M.P., Bickle, I.C., Gaillard, F., Kanne, J.P.: Radiological Society of North America, Inc. (2020). https://doi.org/10.1148/rg.2020200020
17. Brady, A.P.: Radiology reporting—from hemingway to HAL? Insights Imaging **9**, 237–246 (2018). https://doi.org/10.1007/s13244-018-0596-3
18. Demner-Fushman, D., et al.: Preparing a collection of radiology examinations for distribution and retrieval. JAMIA **23**(2), 304–310 (2016)
19. Ruder, S.: An overview of gradient descent optimization algorithms. arXiv preprint arXiv:1609.04747 (2016)
20. Rumelhart, D.E; Hinton, G.E., Williams, R.J.: Learning internal representations by error propagation. Technical report ICS 8504. San Diego, California: Institute for Cognitive Science, University of California (1985)
21. Vaswani, A., et al.: Attention is all you need. In: Advances in Neural Information Processing Systems, vol. 30 (2017)
22. Bahdanau, D., Cho, K., Bengio, Y.: Neural machine translation by jointly learning to align and translate (2014). https://doi.org/10.48550/ARXIV.1409.0473
23. Papineni, K., et al.: BLEU: a method for automatic evaluation of machine translation. In: Proceedings of the 40th Annual Meeting of the Association for Computational Linguistics (2002)
24. Lin, C.-Y.: ROUGE: a package for automatic evaluation of summaries. Text Summarization Branches Out (2004)
25. Banerjee, S., Lavie, A.: METEOR: an automatic metric for MT evaluation with improved correlation with human judgments. In: Proceedings of the ACL Workshop on Intrinsic and Extrinsic Evaluation Measures for Machine Translation and/or Summarization (2005)
26. Weiss, K., Khoshgoftaar, T.M., Wang, D.: A survey of transfer learning. J. Big Data **3**(1), 1–40 (2016)

Subjective Question Bank Generation Using Large Language Models with Custom Knowledge Base

Amaan Sayed(✉), Mahendra Kanojia, and Subhashish Nabajja

Sheth L.U.J and Sir M.V. College, Mumbai, Maharashtra, India
amaansayed1510231@gmail.com

Abstract. Systems of education and assessment are mostly based on questions and answers all over the world. They are used to assess and confirm a student's comprehension of academic subject. In this study we have developed a system that generates Subjective questions based on user query and the document uploaded by the user. The proposed models use a custom dataset, the user must upload their file to obtain the desired result. We started this procedure with a dataset that contained a variety of file types, including PDFs, text files (TXT), CSVs, and few others. Three of the specific gpt4all models—Falcon, Llama-2-7b, and Mini Orca (Large)—was used in this investigation. We get a high-level summary of these models' capabilities from the comparison of the models. The comparison of the models was done with the help of a field expert who hand annotated each question generated by the proposed models. When comparing the models, we discovered that Llama-2-7b produces questions that are the most relevant to the topic the user has asked about.

Keywords: Large language models · generative pretrained transformers · question generation · natural language processing

1 Introduction

All around the world, systems of education and assessment rely primarily on questions. They are employed to evaluate and validate a pupil's understanding of academic material. Questions are crucial for assessing information retention and critical thinking abilities [1], whether in traditional classroom settings, online learning environments, or standardized testing. The demand for high-quality questions covering a range of topics and levels of difficulty has significantly increased as educational practices change [2]. Due to the time and resource requirements involved in manually creating question banks, this growing demand presents a problem for educators and organizations. Some computer-assisted language learning programs have exam features that offer questions about a story's content. However, because most of them prepare questions in advance, they have limited ability to generate questions that are adaptive depending on learners' comprehension states [3]. Manually generating question banks for educational purposes

V. Sakalauskas et al. (Eds.): IBICA 2023, LNNS 1231, pp. 250–259, 2025.
https://doi.org/10.1007/978-3-031-78946-5_24

requires a lot of work and takes a lot of time. Creating a diverse and complete set of questions can be extremely difficult for educational institutions, teachers, and material authors. These difficulties consist of the requirement for subject-matter expertise, the sizeable time commitment necessary, and the potential for biases or inconsistent 2 question design. The urgent need to discover more effective and scalable ways for question bank production has also been highlighted by the rise of online learning platforms and the desire for customized assessment materials. Natural language processing (NLP) has made huge advances in recent years because of the development of large language models like the Generative Pre-trained Transformer (GPT-3) [3–6] Modern neural-network-based LMs train on immense datasets, such as approximately a terabyte of text in English, and massive model architectures (e.g., 175 billion parameters) [7]. In this regard, utilizing Large Language Models' (LLMs') capabilities in combination with unique knowledge bases appears to be a potential way to revolutionize this procedure. Because huge language models offer such a diverse range of applications, it has been determined that this is a potential area of study. Individuals at all stages of education, including primary, secondary, tertiary, and professional development, may be able to improve their learning and teaching experiences using these models [2, 5]. The primary goal of this research is to develop a method for automated question bank generation and assist in the learning [5].

Large language models can help teachers create (inclusive) lesson plans and activities for their classes. The document collection on which teachers want to base their course can be entered into the models. An example of the output would be a course curriculum with brief summaries of each topic. Aside from generating questions and prompts that encourage engagement from people with varying levels of knowledge and skill, language models can also foster critical thought and problem-solving [8]. Additionally, they can be used to create customized and focused practice questions and tests, which can ensure that students are learning the topic.

Large language models can help school students learn a language and writing styles for a variety of concepts and areas, such as History, Chemistry, literature, and language learning, among others [5]. The creation of practice questions and tests using these models can aid students in better comprehending, contextualizing, and remembering the subject they are learning. The utilization of these models should be done with precaution because they also have drawbacks that need to be addressed, such as lack of precision, bias possibility, and unintended brittleness in very simple tasks [9].

2 Literature Review

In recent years, several studies have explored the use of technology and artificial intelligence (AI) in the field of education. Liu and Liu (2021) [10] presented an English Grammar Intelligent Question Bank System; it highlighted the integration of technology and custom question banks. The authors collected English grammar data from authoritative grammar books and Baidu Encyclopedia. A five-layer system was used. Kurdi et al. (2020) [11] conducted a systematic review on automatic question generation for educational purposes, emphasizing its relevance in modern education. Zhong, Xiong, and Socher (2017) [12] introduced the Seq2SQL framework, which utilizes reinforcement learning for generating structured queries from natural language. Hussain et al.

(2018) [13] The study showed the development of a dashboard to assist instructors in monitoring student engagement and providing timely interventions. 3 These models can be integrated into virtual learning environment (VLE) systems to evaluate engagement in different activities and materials, potentially improving student performance. Andersson and Palm (2017) [14] In this research paper, the authors study investigates how modifications to instructors' formative classroom instruction that followed professional development feedback had an impact on students' academic progress. Gilson et al. (2023) [15] assessed ChatGPT's performance in the United States medical licensing examination, highlighting the implications of large language models in medical education and knowledge assessment. Thirunavukarasu et al. (2023) [16] reviewed Large Language Models (LLMs) such as ChatGPT in the fields of medicine. Sallam (2023) [17] This study examined ChatGPT as an outstanding example of LLMs in academic/scientific writing, healthcare practice, and education based on the existing research. The study conducted by Wei et al. (2022) [18] explored the recent discovery of emergent skills in language models resulting from scaling up and the potential future research areas to understand their origin and scalability. Using the VNHSGE English dataset, Dao (2023) [19] evaluated the performance of three well-known LLMs on tasks involving natural language processing, assisting researchers and programmers in better understanding the usefulness of each LLM in real-world settings. The assessment of big language models was thoroughly analyzed by Chang et al. (2023) [20], who also emphasized the need for ethical frameworks and rigorous evaluation.

3 Dataset Description

The dataset is generated from a file that the user uploads to the model, the model then uses this dataset to generate the desired output. For this specific study the file is the book "Linux Server Administration" by Wale Soyinka [21] in pdf format. The book consists of a total of 665 pages, and is divided into 30 chapters. The entire embedding process for a given chapter takes around 2 min. The topics on which the questions were generated was from a university level computer science course.

Text data is extracted from the uploaded files. Once the data is extracted from the source document, we need to split the text into smaller tokens called chunks, so they can fit in the model's context. We used a recursive character text splitter to split the text by different characters – starting with "\n\n", then "\n", then "". Important parameters that are related to chunks are "chunkSize" and "chunkOverlap". "chunkSize" controls the max size (in terms of number of characters) of the final documents. "chunkOverlap" specifies how much overlap there should be between chunks. The chunks are then passed onto to and embedding model which converts the chunks into their respective numerical representation, which are called embeddings. These embeddings are then store in a database, which is known as the vector database. This vector base is the dataset that LLM uses to extend its context. The dataset is only generated once when the user uploads a file into the model [22].

4 Research Methodology

With the rise of open-source and commercially free LLMs, GPT4ALL offers various trained language models based on the GPT-3.5-Turbo dataset [23]. We selected three GPT4All models for this study.

4.1 Falcon

Although there is a smaller version of Falcon LLM available with only 7 billion parameters, it has been trained on 1 trillion tokens and contains 40 billion parameters [24]. The encoder consists of six identical layers, each with two sub-layers: multi-head self-attention and position wise fully connected feed-forward networks. A detailed block diagram of Falcon is show in Fig. 1.

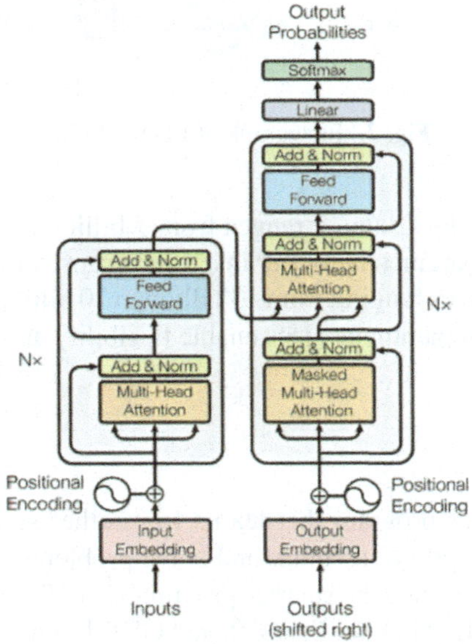

Fig. 1. Architecture of Falcon [24]

Residual connections and layer normalization are applied to each sub-layer. All sub-layers and embedding layers produce 512-dimensional outputs for efficient information flow. The decoder in this architecture consists of six layers. The decoder features a third sub-layer for multi-head monitoring of the output from the encoder, opposite of the encoder. Each sub-layer receives residual connections as well as layer normalization. Additionally, the decoder's self-attention sub-layer is modified to stop positions from acquiring over positions in the future. This, along with an offset in output embeddings, ensures that predictions for a position only depend on known outputs at earlier positions. The weight of each value is established by the query's compatibility function with its corresponding key, and the result is produced is the sum of the values, weighted.

4.2 Llama-2-7B

An improved transformer architecture is used by the auto-regressive language model. Llama 2. The modified versions adapt to users' preferences for safety and helpfulness. using supervised fine-tuning (SFT) and reinforcement learning with human feedback. (RLHF), the flowgraph for Llama-2 is shown in Fig. 2.

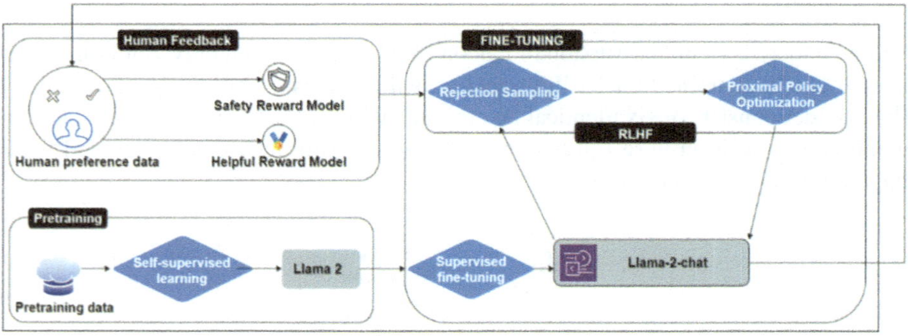

Fig. 2. Flowgraph of Llama-2 [25]

The parameter range for Llama 2 ranges from 7 billion to 70 billion. However, we decided to use 7 billion parameters for our instance because of its affordability and ease of use. With parameter sizes ranging from 13 billion to 70 billion, Llama 2 offers a wider selection. These different parameter sizes enable flexibility and customization based on needs and goals [26].

4.3 Mini Orca (Large)

Orca is a smart computer program that learns from other smart computer programs. Orca watches how other programs think and solve problems, and then it learns to do the same things. In our instance the smarter program is GPT-4. Orca has been evaluated and tested against other LLMs like ChatGPT and GPT-4 in terms of producing toxic or hate speech, making it the least toxic [27]. The architecture of Orca is not yet known. Orca is a version of LLaMA that Microsoft developed with adjustments based on GPT-4 explanation traces.

5 Proposed Model

The process starts with the user uploading the required document to the LLM interface which is a web client specifically built for this study. The text data is then extracted from the document and the split into smaller tokens. These tokens are then passed onto the embedding model which converts them into their numerical equivalent vector. These vectors are then stored into the Vector Database which forms the Knowledge Base. The overall process is shown Fig. 3(a). This process is a one-time process and the knowledge base formed, is used to extend the LLM's context window.

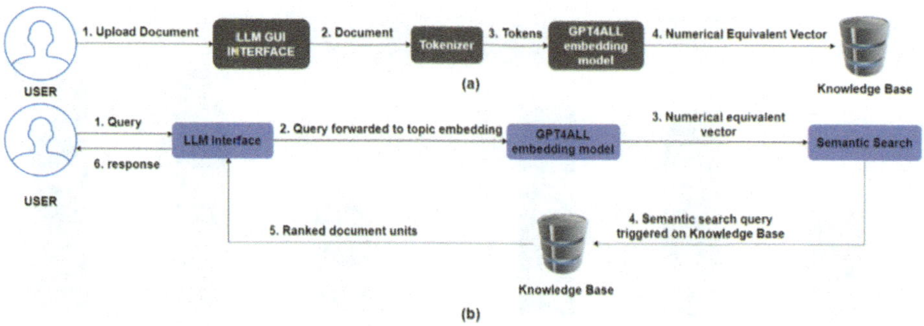

Fig. 3. (a) Knowledge Base uploading process. (b) Flowgraph of Proposed Model

When the user wants to generate Subjective Questions, the user sends a query to the LLM interface. The LLM interface then passes the query to the embedding model which converts it to its equivalent numerical vector. This vector is then used to perform semantic search on the Knowledge Base which returns relevant ranked document units. The retrieved results are then inserted into the prompt along with the user query and is sent to the LLM. The LLM then generates relevant questions with the aid of extended context. This process is depicted in Fig. 3(b).

6 Results

Automatic evaluation metrics are inadequate to assess the models, therefore a manual human review of each question was performed to determine the results. The results were calculated using the following expressions for rating and relevancy.

$$Y_{rating} = \frac{\sum_{i=1}^{n} A_i}{n}$$

$$Y_{relevancy} = \frac{\sum_{i=1}^{n} B_i}{n}$$

where Yrating denotes the average rating, Yrelevancy denotes the average relevancy factor, *Ai* denotes the rating and *Bi* denotes the average relevancy of question *i* respectively, and *n* is the total number of questions generated. The specified evaluation metrics to assess the proposed model due to the unsuitability of automatic evaluation metrics. All of the models in the proposed systems were assessed and executed on a MSI laptop with an Intel i5-11400H. The models were executed on CPU which gave an average response time of 150.95 s. Comparing all the 3 models reveals that Llama-2 generates questions that are more relevant and usable. The average rating is shown below in Fig. 4(a), and the average relevancy is shown in Fig. 4(b). By the help of graphs, it is clearly visible that Falcon generated that are least relevant and also lowest in ratings.

The average rating is shown below in Fig. 4(a), and the average relevancy is shown in Fig. 4(b). The experts hand annotated the questions based on two factors viz. Ratings which lies between 0 to 5 and indicates the quality of the question generated. Relevancy which indicates whether the question is relevant to the topic asked and it is a Boolean value. Table 1 shows the numerical value of all the models.

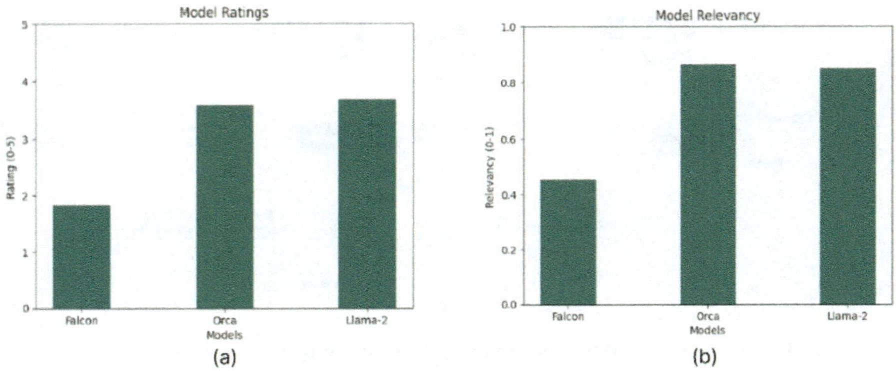

Fig. 4. Ratings and Relevancy Graphs.

Table 1. Quantitative comparison of the models.

Models	Average Ratings	Average Relevancy
Falcon	1.82	0.45
Orca	3.58	0.87
Llama-2	3.68	0.85

7 Conclusion

In this study we proposed a system based on LLM that generates Subjective Questions that will help students and teachers to improve their academic performance. The proposed model is built using GPT4ALL models and a comparison of all these models is shown in Table 1.

Table 2 provides a comparison of research articles and their respective contributions in the field of LLM. Each model is associated with some dataset which is used to finetune or extend the context of the LLM. It can be noted that these systems are evaluated by performing human review and automatic evaluation of these models are not suitable [1, 28].

Table 2. Comparison of the proposed systems with recent models

LLM	Data	Proposed System	Results
[1] GPT-3 model	EQG-RACE	Fine-tuned GPT-3 on macaw I I-b and EEQG	Manual human evaluations were done which suggests that the model generates high quality questions
[6] gpt-3.5	MultispanQA, QUOREF, DROP	Developed a prompting system named FBPrompt	Improved the performance of the LLMs
[28] ChatGPT	None	Personalized adaptive practicing system using ChatGPT	Correct feedback by ChatGPT 99%, rest 1% human intervention needed
[29] GPT-3.5, GPT-4, and Google Bard	149-question Self-Assessment Neurosurgery Exam (SANS) Indications Exam was used	Performance analysis of LLMs	correct feedback percentage GPT-3.5 = 62.4%, GPT4 = 82.6%, Google Bard = 44.2%
[Proposed model] Falcon, LLAMA2, Orca	University Textbook	LLM with extended context using custom knowledge based	Manual Human evaluation was performed on the generated questions and received average rating of 3.03

8 Future Discussion

Future research can use the findings of this study as a benchmark. Models can be developed to assess the answers to the generated questions. The current system is hardware-intensive hence, there is a need of LLMs that could function with efficient hardware requirements. This system can be implemented at numerous educational institutions, allowing all stakeholders to gain from it.

References

1. Dijkstra, R., Genç, Z., Kayal, S., Kamps, J.: Reading comprehension quiz generation using generative pre-trained transformers (2022)
2. Ayanwale, M.A., Sanusi, I.T., Adelana, O.P., Aruleba, K.D., Oyelere, S.S.: Teachers' readiness and intention to teach artificial intelligence in schools. Comput. Educ.: Artif. Intell. **3**, 100099 (2022)
3. Kunichika, H., Katayama, T., Hirashima, T., Takeuchi, A.: Automated question generation methods for intelligent English learning systems and its evaluation. In: Proceedings of ICCE, vol. 670 (2004)

4. Tinn, R., et al.: Fine-tuning large neural language models for biomedical natural language processing. Patterns **4**(4) (2023)
5. Kasneci, E., et al.: ChatGPT for good? On opportunities and challenges of large language models for education. Learn. Individ. Differ. **103**, 102274 (2023)
6. Huang, Z., Zhou, J., Xiao, G., Cheng, G.: Enhancing in-context learning with answer feedback for multi-span question answering. arXiv preprint arXiv:2306.04508 (2023)
7. Carlini, N., et al.: Extracting training data from large language models. In: 30th USENIX Security Symposium (USENIX Security 21), pp. 2633–2650 (2021)
8. Touvron, H., et al.: LLaMA: open and efficient foundation language models (2023). https://arxiv.org/abs/2302.13971
9. Moore, S., Nguyen, H.A., Bier, N., Domadia, T., Stamper, J.: Assessing the quality of student-generated short answer questions using GPT-3. In: European Conference on Technology Enhanced Learning, pp. 243–257. Springer, Cham (2022)
10. Liu, X., Liu, H.: Design and application of english grammar intelligent question bank system. Sci. Program. **2021**, 1–10 (2021)
11. Kurdi, G., Leo, J., Parsia, B., Sattler, U., Al-Emari, S.: A systematic review of automatic question generation for educational purposes. Int. J. Artif. Intell. Educ. **30**, 121–204 (2020)
12. Zhong, V., Xiong, C., Socher, R.: Seq2SQL: generating structured queries from natural language using reinforcement learning. arXiv preprint arXiv:1709.00103 (2017)
13. Hussain, M., Zhu, W., Zhang, W., Abidi, S.M.R.: Student engagement predictions in an e-learning system and their impact on student course assessment scores. Comput. Intell. Neurosci. **2018** (2018)
14. Andersson, C., Palm, T.: The impact of formative assessment on student achievement: a study of the effects of changes to classroom practice after a comprehensive professional development programme. Learn. Instr. **49**, 92–102 (2017)
15. Gilson, A., et al.: How does ChatGPT perform on the United States medical licensing examination? The implications of large language models for medical education and knowledge assessment. JMIR Med. Educ. **9**(1), e45312 (2023)
16. Thirunavukarasu, A.J., Ting, D.S.J., Elangovan, K., Gutierrez, L., Tan, T.F., Ting, D.S.W.: Large language models in medicine. Nat. Med. 1–11 (2023)
17. Sallam, M.: The utility of ChatGPT as an example of large language models in healthcare education, research and practice: systematic review on the future perspectives and potential limitations. medRxiv, pp. 2023–02 (2023)
18. Wei, J., et al.: Emergent abilities of large language models. arXiv preprint arXiv:2206.07682 (2022)
19. Dao, X.Q.: Performance comparison of large language models on vnhsge english dataset: openAI chatGPT, microsoft bing chat, and google bard. arXiv preprint arXiv:2307.02288 (2023)
20. Chang, Y., et al.: A survey on evaluation of large language models. arXiv preprint arXiv:2307.03109 (2023)
21. Soyinka, W.: Linux Administration: A Beginner's Guide. McGraw-Hill Education, New York (2020)
22. Recursively split by character 🦜🔗Langchain. https://python.langchain.com/docs/modules/data_connection/document_transformers/text_splitters/recursive_text_splitter. Accessed 11 Oct 2023
23. GPT4All: An ecosystem of open-source assistants that run on local hardware. https://gpt4all.io/reports/GPT4All_Technical_Report_3.pdf. Accessed: 26 Sept 2023
24. Vaswani, A., et al.: Attention is all you need. In: Advances in Neural Information Processing Systems, 30 (2017)
25. Touvron, H., et al.: LLaMA 2: open foundation and fine-tuned chat models. arXiv preprint arXiv:2307.09288 (2023)

26. Anonymous4chan/LLAMA-2–70B hugging face anonymous4chan/llama-2-70b Hugging Face. https://huggingface.co/anonymous4chan/llama-2-70b
27. Mukherjee, S., Mitra, A., Jawahar, G., Agarwal, S., Palangi, H., Awadallah, A.: Orca: Progressive learning from complex explanation traces of GPT-4. arXiv preprint arXiv:2306.02707 (2023)
28. Kabir, M.R., Lin, F.: An LLM-powered adaptive practicing system (2023)
29. Ali, R., et al.: Performance of ChatGPT, GPT-4, and Google bard on a neurosurgery oral boards preparation question bank. Neurosurgery 10–1227 (2022)

Wine Quality Assessment Using Machine Learning

Tiasha Dasgupta[1(✉)] and Ajith Abraham[2]

[1] Machine Intelligence Research Labs (MIR Labs), Scientific Network for Innovation and Research Excellence Auburn, Washington 98071, USA
tiashadasgupta01@gmail.com

[2] School of Computer Science Engineering and Technology, Bennett University, Greater Noida 201310, Uttar Pradesh, India
ajith.abraham@ieee.org

Abstract. Evaluation of wine quality has a significant impact on both production methods and consumer preferences in the wine business. This article provides a summary of machine learning-based methods for estimating wine quality that leverage the use of data-driven methodologies to improve the precision and effectiveness of quality assessment. We have considered three types of wines: French Bordeaux wine, red and white variants of the Portuguese "Vinho Verde" wine. There are existing models that incorporate one or two types of wine at a time and perhaps with lower accuracy, which makes it difficult to assess wine quality more objectively and efficiently. Expert opinions on quality are highly individualized and may not reflect consumer preferences. Furthermore, it is possible that the experts will not always be available for the wine tasting. We attempted to design a wine quality prediction approach by applying principles of machine learning, which incorporates easy wine quality assessment and in turn saves time and cost. This project not only emphasizes prediction precision but also opens the possibility of easily incorporating other types of wine. Machine learning algorithms were trained and tested on a holdout original sample. The accuracy ranges around 85% and above which has surpassed the previous works.

Keywords: Wine Quality · Random Forest Classifier · Bordeaux wine

1 Introduction

Wine quality is a multidimensional, complicated subject that may be handled from both the consumer and the wine business perspectives. Today, all industries are advancing by means of the adoption and widespread use of new technologies. These technologies aid in improving production and streamlining the entire procedure. But some fields, including ensuring the quality of products, require human skill. Since the demand for the product keeps rising, it is now a costly process. In this project, the product we have considered is wine. Western countries' dining traditions have historically included wine in large quantities. The inherent quality of wine is what distinguishes it, and specific

V. Sakalauskas et al. (Eds.): IBICA 2023, LNNS 1231, pp. 260–268, 2025.
https://doi.org/10.1007/978-3-031-78946-5_25

applications over the years have helped spread information about how to judge wine quality. It is said that the quality of wine has been impacted due to an increase in supplies sold at lower rates worldwide. In contrast to their marketing efforts, the wine industry has traditionally been required to pay close attention to wine quality. The consumption of wine has risen even further in recent decades because of the rising economies of Asian nations. Various wine varieties have different applications and are based on chemical testing. While most chemical elements are the same across all wine kinds, the quantity of each chemical constituent varies by wine variety. These days, wine classification is essential for quality assurance. Understanding the wine's quality and establishing a consistent production are crucial industry goals from the perspective of the maker. Winemakers can adapt and intervene as necessary by estimating wine quality early in the production process. By eliminating the creation of subpar wine that would be difficult to market, this can result in cost savings. However, predicting the quality of different types of wine is complex and diverse. This study examines various machine learning methods for predicting wine quality, such as Random Forest Classifier, CatBoost, and Ensemble machine learning approaches. These methods automate the quality assurance process by using the product's existing characteristics and minimize human interference. This work is all about the first step towards predicting quality of different wines using machine learning algorithms. Rest of the article is organized as follows. Section 2 illustrates related literature, followed by methodologies and experimental results in Sects. 3 and 4. Some Conclusions are also provided towards the end.

2 Related Literature

The analysis of earlier models that were put forth to forecast the wine quality connected to our proposed endeavor is covered in this Section. Dong et al. [1], computed the various classification metrics and obtained accuracy of 87.3% using Naïve Bayes algorithm and 86.9% using Support Vector Machine. Gupta et al. [2] used machine learning methods on White wine and red wine datasets and concluded that more accurate predictions significantly rely on choosing the right set of feature variables. Authors compared the accuracy of different algorithms such as K-Nearest Neighbour, Support Vector Machine, Random Forest and achieved an accuracy of 60%, 63% and 76% respectively for White wine and 61%, 62% and 73% respectively for red wine. Kumar et al. [3] demonstrated the use of machine learning algorithms on red wine data extracted from the UCI machine learning repository. Authors calculated the classification metrics such as precision, recall, f1-score, accuracy as well as error of machine learning for predicting wine quality using Naïve Bayes algorithm, Support Vector Machine algorithm and Random Forest algorithm. Additionally, they compared the accuracy of the algorithms Naïve Bayes (56%), Support Vector Machine (69%) and Random Forest (65%). Chiu et al. [4], used Support Vector Machine, Decision Tree, and Random Forest algorithms, measuring precision, accuracy, recall and f1-score metrics for both red wine and White wine datasets and achieved an accuracy of 57%, 53%, and 58% respectively for red wine and 51%, 58%, and 66% respectively for White wine dataset. Gupta et al. [5] achieved a root mean square error of 54% for red wine and 58% for white wine using Random Forest and K-nearest neighbour algorithm. Er et al. [6] achieved an accuracy of 72% for red wine dataset and

70% for white wine dataset when trained and tested using the Random Forest algorithm. Malfeito et al. [10] states that the classical tasting methods begins with visual appraisal, followed by smell and taste or mouthfeel, and ends with an overall evaluation. The general classical tasting method encourages top-down processes through mental imaging, which may result in a lack of consistency in experts' overall perceptions of good wines. Another problem is the emergence of sporadic fallacies regarding the assessment of superb wines, which likely explains why traditional, excellent reds and whites are overlooked when blind tasting. Koranga et al. [14] demonstrates the better performance of the J48 algorithm using the measures Root Mean Square Error (RMSE) score of 0.0324 and Root Relative Squared Error (RRSE) of 26.3684. Shaw et al. [16] Computed 3 algorithms SVM, RF and MLP to find the most accurate result in the analysis of wine quality. During the comparative study, RF yields an accuracy percentage of 81.96, MLP yields a percentage of 78.78, and SVM yields a percentage of 57.29. Kothawade et al. [17] used the artificial neural network (ANN) which achieved 85.16% accuracy from the red wine and 88.28% accuracy from the white wine, the naïve Bayes (NB) algorithm yielded 46.33% accuracy from the red wine and 46.68% accuracy from the white wine, and the support vector machine (SVM) algorithm yielded 83.52% accuracy from the red wine and 86.86% accuracy from the white wine. Ribeiro et al. [19] Presented a study on the prediction of the organoletic parameters (foam, color, flavor, and savour) from the vinification process's chemical parameters. To build models for the classification and regression goals of the Data Mining concepts application, we employ Decision Trees, Artificial Neural Networks, and Linear Regression as Data Mining approaches. The open-source WEKA tool and the recently released Microsoft Business Intelligence Studio 2008 were used in the studies. Accuracies between 86% and 99% were obtained. Radosavljević et al. [21] used six classification algorithms: k *, J48, Random Forest, Multilayer perceptron, Support Vector Machine, and Naïve Bayes. The Random Forest algorithm produced the best results, according to an analysis of the data using statistical indicators such as mean absolute error (MAE), root mean square error (RMSE), recall, and ROC area. Aich et al. [23] Computed the red and white wine datasets using SVM classifiers and obtained accuracies ranging from 95.23% to 98.81% with different feature sets. Clarin et al. [24] demonstrated that there is a moderately positive correlation between the physicochemical characteristics of the wine. Random Forest produced the best results ($r = 0.7459$), followed by k-Nearest Neighbor ($r = 0.6225$), Decision Tree ($r = 0.5500$), Linear regression ($r = 0.5257$), multi-layer perceptron ($r = 0.5236$), and Support Vector Machine ($r = 5225$). Tingwei et al. [25] estimated the quality of red wines by an active learning method combined with K-Nearest Neighbor algorithm and ranked batch-mode sampling. The active learner attained an accuracy of 88.2% in predictions after training on the first 60 training data and querying for 15 iterations. The total labelled data obtained surpasses 30% of the total number of samples as the number of iterations rises. Liu et al. [27] predicted wine quality through sensory characteristics. The wine quality prediction and assessment model developed in this study passed the PLS-SEM test and demonstrated its out-of-sample predictive ability, which is supported by prior research on sensory aspects related to wine quality assessment and prediction. Zaza et al. [31] computed the wine quality prediction using five machine learning algorithms namely Support Vector Machine, Decision Tree, K-Nearest Neighbour, Gradient

Boosting, and Random Forest. With 78% accuracy, Random Forest produced the best results.

3 Methodology

3.1 Machine Learning Algorithms

We several machine learning-based classification algorithms and ensemble techniques to predict the results. Some introduction to the best methods we implemented are given below:

- **Classification Algorithms:**
  - **Random Forest Classifier:** It is a supervised algorithm that is used for both regression and classification problems. Random Forest classifier uses the average of many decision trees on different subsets of a given dataset to improve the projected accuracy of that dataset. Rather than depending on a single decision tree, the random forest collects the results of each tree and predicts the eventual output based on most prediction votes.
  - **CatBoost Classifier:** "CatBoost" name comes from two words "Category" and "Boosting". It handles categorical features by allowing us to incorporate non-numerical factors.
  - **LightGBM Classifier:** The Light GBM framework is a gradient boosting framework that employs a tree-based learning method. Light GBM grows trees vertically, but other algorithms grow trees horizontally, implying that Light GBM grows tree leaf-by-leaf, whereas other algorithms grow level-by-level. It will grow the leaf with the greatest delta loss. Leaf-wise algorithms can minimise more loss than level-wise algorithms when expanding the same leaf.

- **Ensemble Classifier techniques:**
  - **Stacking Ensemble Classifier:** Stacking involves combining predictions from numerous machine learning models on the same dataset, similar to bagging and boosting.

    A stacking model's architecture consists of two or more base models, also known as level-0 models, and a meta-model that aggregates the predictions of the base models, known as a level-1 model.

    Level-0 Models (Base-Models): These are models that are fitted to the training data and whose predictions are compiled.
    Level-1 Model (Meta-Model): Learns the optimum way to combine the predictions of the base models.

  - **Max Voting Ensemble Classifier:** One of the simplest methods of integrating predictions from different machine learning algorithms is max voting, which is commonly used for classification tasks. Each base model predicts and votes for each sample in max-voting. The final prediction class includes only the sample class with the most votes.

- **Bagging meta-estimator ensemble classifier:** Bagging entails averaging the predictions from many decision trees that have been fitted to various samples of the same dataset. Here, various base models are independently and concurrently trained using various subsets of the training data. Using bootstrap sampling, each subset is created by randomly selecting and replacing data points.

3.2 Cross Validation

Cross validation is a method for checking overfitting. It is done by dividing the dataset into K folds. During the first iteration, first fold is used for testing purpose and rest K – 1 folds are employed for training purpose. Similarly, for second iteration, the second fold is used for testing purpose and the remaining K – 1 folds are used for training purpose. In this manner, the process is repeated for K iterations.

4 Experimental Results

4.1 Data Preparation

We used three datasets: The 21st-century Bordeaux wine dataset from 2000 to 2016 is obtained from IEEE DataPort. The two datasets which are related to red and white variants of the Portuguese "Vinho Verde" wine are obtained from the UC Irvine Machine Learning Repository. The Bordeaux wine dataset comprises 14349 instances and 989 features. Both red and white wine consists of 12 features, and 1599 and 4898 instances, respectively.

4.2 Data Preprocessing

Data pre-processing is carried out to clean the data to use it in an efficient format. Most of the data contains redundant, duplicate, or missing values. So, if such data is used to construct a model, it provides incorrect results. Hence, data must be cleaned before being fed to a model to get accurate results.

- **Missing/NaN values:** In the Bordeaux wine Dataset, the "Price" attribute consists of some NA values. But for these null values to not interfere with the accuracy of the result, we have dropped those corresponding rows.
- **Correlation Matrix:** In order to comprehend the correlation between more than just one pair of variables, a correlation matrix is mapped for each of the four datasets. Values can vary between -1 to 1 where, a value of 1 denotes a completely negative linear correlation between two variables, 0., is the linear correlation between two variables and a value of 1 denotes a perfect positive linear correlation between two variables.
- **Outlier check:** A data item or object that considerably differs from the other objects is called an outlier. We use the boxplot representation to check for outliers in the DataFrame of all three datasets. We implement the InterQuartile Range (IQR) Method to remove the detected outliers in the datasets.

- **Imbalanced dataset:** In the case of Red and White wine quality prediction, the target feature in both the datasets can be seen to be imbalanced since there is huge parity between the number of 0s and 1s. As a result, the model will be biased towards the majority class only. In order to handle the imbalanced dataset, we have made use of SMOTE (Synthetic Minority Oversampling Technique).

For the Bordeau wine dataset, we have incorporated the use of inbuilt conventional preprocessing tool that is using for data preprocessing of very large datasets.

4.3 Results Obtained

The algorithms used in this research are Random Forest Classifier, LightGBM, CatBoost, Ensemble Classifier techniques and the accuracies are calculated using the cross-validation with factor cross validation as 10. The datasets of diseases are divided into training and test datasets for classification.

Table 1. Accuracy for Bordeaux Wine Quality

Technique Used	Accuracy	Precision	Recall	F1-score
Naïve Bayes [1]	87.32%	81.94%	73.52%	77.49%
SVM [1]	86.92%	80.13%	74.45%	77.18%

Table 2. Red wine (Portuguese) Quality Prediction

Technique Used	Accuracy
Random Forest Classifier [2]	73.25%
Support Vector Machine [3]	68.64%
Naïve Bayes Classifier [3]	55.91%
Random Forest Classifier [3]	65.46%
Support Vector Machine [4]	57%
Random Forest Classifier [4]	58%
Decision Tree [4]	53%
Random Forest Classifier [6]	71.87%

Tables 1, 2 and 3 report some of the previously best reported results. The highest accuracies achieved by our work for Bordeaux Wine, Red Wine, and White Wine are as follows 87.9%, 92.6%, and 90.5% respectively using CatBoost for Bordeaux Wine, Max Voting Ensemble Classifier for Red wine, and Random Forest Classifier for White Wine. Tables 4 and 5 illustrate the obtained results. As evident, the suggested ensemble method seems to work better than individual approaches.

Table 3. White Wine (Portuguese) Quality Prediction

Technique Used	Accuracy
Random Forest Classifier [2]	76.39%
Support Vector Machine [4]	51%
Random Forest Classifier [4]	66%
Decision Tree [4]	58%
Random Forest Classifier [6]	70.37%

Table 4. Performance of Different Algorithms (For Bordeaux Wines)

Performance Parameter	Technique Used	Best Performance Obtained
Accuracy	Catboost	87.9%
Recall	Stacking ensemble classifier	90.6%
Precision	Random Forest using Bagging meta-estimator ensemble classifier	87.7%
F1 score	CatBoost	82.1

Table 5. Red and White Wines (Portuguese)

	Technique Used	Best Performance Obtained			
		Accuracy	Precision	Recall	F1-score
Red Wine	Max Voting Ensemble	92.6%	87.9%	97.7%	92.5%
White Wine	Random Forest	90.5%	91.4%	89.5% (using Bagging meta-estimator ensemble classifier)	90.5%

5 Conclusions

The wine sector has seen an increase in interest recently, which is driving desire for market expansion. To improve the quality and sales of wine, the businesses are also investing in new technology. The main objective of this project is to create a system that would predict wine quality and do so with high accuracy. The proposed approach is more accurate than other existing approaches, according to experiments on the datasets. For this purpose, we have used various machine learning algorithms like Random Forest, CatBoost, and Ensemble classifier techniques to achieve maximum accuracy.

Some of the future aspects for this project work include:

- We can incorporate other different kinds of wine.
- Develop a user-friendly GUI which will demonstrate the wine quality percentage.

- Other varieties of evolutionary algorithms may be used with the suggested framework to produce a classifier that is more dependable.
- To assess whether the model's prediction capability may be improved, broaden the current study to incorporate feature development methods.

References

1. Dong, Z., Atkison, T., Chen, B.: Wineinformatics: using the full power of the computational wine wheel to understand 21st century Bordeaux wines from the reviews. Beverages **2021**(7), 3 (2021)
2. Gupta, M., Vanmathi, C.: A study and analysis of machine learning techniques in predicting wine quality. Int. J. Recent Technol. Eng. **10** (2021)
3. Kumar, S., Agrawal, K., Mandan, N.: Red wine quality prediction using machine learning techniques. In: 2020 International Conference on Computer Communication and Informatics (ICCCI), pp. 1–6. IEEE (2020)
4. Chiu, T.H.Y., Wu, C., Chen, C.H.: A generalized wine quality prediction framework by evolutionary algorithms. Int. J. Interact. Multimed. Artif. Intell. **6**(7) (2021)
5. Mahima, Gupta, U., Patidar, Y., Agarwal, A., Singh, K.P.: Wine quality analysis using machine learning algorithms. In: Micro-Electronics and Telecommunication Engineering: Proceedings of 3rd ICMETE 2019, pp. 11–18. Springer, Singapore (2020)
6. Er, Y., Atasoy, A.: The classification of white wine and red wine according to their physicochemical qualities. Int. J. Intell. Syst. Appl. Eng. **4**(Special Issue-1), 23–26 (2016)
7. Gupta, Y.: Selection of important features and predicting wine quality using machine learning techniques. Procedia Comput. Sci. **125**, 305–312 (2018)
8. Sun, L.X., Danzer, K., Thiel, G.: Classification of wine samples by means of artificial neural networks and discrimination analytical methods. Fresenius' J. Anal. Chem. **359**, 143–149 (1997)
9. Cortez, P., Teixeira, J., Cerdeira, A., Almeida, F., Matos, T., Reis, J.: Using data mining for wine quality assessment. In: Discovery Science: 12th International Conference, DS 2009, Porto, Portugal, 3–5 October 2009, pp. 66–79. Springer, Heidelberg (2009)
10. Malfeito-Ferreira, M.: Fine wine flavour perception and appreciation: blending neuronal processes, tasting methods and expertise. Trends Food Sci. Technol. **115**, 332–346 (2021)
11. Cardebat, J.M., Livat, F.: Wine experts' rating: a matter of taste? Int. J. Wine Bus. Res. **28**(1), 43–58 (2016)
12. Hu, G., Xi, T., Mohammed, F., Miao, H.: Classification of wine quality with imbalanced data. In: 2016 IEEE International Conference on Industrial Technology (ICIT), pp. 1712–1217. IEEE (2016)
13. UC Irvine Machine Learning Repository. Wine Quality Datasets-Red and White. https://archive.ics.uci.edu/dataset/186/wine+quality
14. Koranga, M., Pandey, R., Joshi, M., Kumar, M.: Analysis of white wine using machine learning algorithms. Mater. Today Proc. **46**, 11087–11093 (2021)
15. Trivedi, A., Sehrawat, R.: Wine quality detection through machine learning algorithms. In: 2018 International Conference on Recent Innovations in Electrical, Electronics & Communication Engineering (ICRIEECE), pp. 1756–1760. IEEE (2018)
16. Shaw, B., Suman, A.K., Chakraborty, B.: Wine quality analysis using machine learning. In: Emerging Technology in Modelling and Graphics: Proceedings of IEM Graph 2018, pp. 239–247. Springer, Singapore (2020)
17. Kothawade, R.D.: Wine quality prediction model using machine learning techniques (2021)

18. Smith, B.C.: Getting more out of wine: wine experts, wine apps and sensory science. Curr. Opin. Food Sci.. Opin. Food Sci. **27**, 123–129 (2019)
19. Ribeiro, J., Neves, J., Sanchez, J., Delgado, M., Machado, J., Novais, P.: Wine vinification prediction using data mining tools. In: ECC 2009 Proceedings of the 3rd International Conference on European Computing Conference. Computing and Computational Intelligence. WSEAS, pp. 78–85 (2009)
20. Mohana, R., Sharma, P., Sharma, A.: Ensemble framework for red wine quality prediction. Food Anal. Methods **16**(1), 30–44 (2023)
21. Radosavljević, D., Ilić, S., Pitulić, S.: A data mining approach to wine quality prediction (2019)
22. Zhang, H., Wang, Z., He, J., Tong, J.: Construction of wine quality prediction model based on machine learning algorithm. In: 2021 5th International Conference on Artificial Intelligence and Virtual Reality (AIVR), pp. 53–58 (2021)
23. Aich, S., Al-Absi, A.A., Hui, K.L., Sain, M.: Prediction of quality for different type of wine based on different feature sets using supervised machine learning techniques. In: 2019 21st International Conference on Advanced Communication Technology (ICACT), pp. 1122–1127. IEEE (2019)
24. Clarin, J.A.: Comparison of the performance of several regression algorithms in predicting the quality of white wine in WEKA. Int. J. Emerg. Technol. Adv. Eng. **12**(7), 20–26 (2022)
25. Tingwei, Z.: Red wine quality prediction through active learning. In: Journal of Physics: Conference Series, vol. 1966, no. 1, p. 012021. IOP Publishing (2021)
26. Shruthi, P.: Wine quality prediction using data mining. In: 2019 1st International Conference on Advanced Technologies in Intelligent Control, Environment, Computing & Communication Engineering (ICATIECE), pp. 23–26. IEEE (2019)
27. Liu, S., Vega, A.R., Dizy, M.: Assessing ultrapremium red wine quality using PLS-SEM. LWT **177**, 114560 (2023)
28. Ji, H., Pu, D., Yan, W., Zhang, Q., Zuo, M., Yuyu, Z.: Recent advances and application of machine learning in food flavor prediction and regulation. Trends Food Sci. Technol. (2023)
29. Bhardwaj, P., Tiwari, P., Olejar, K., Jr., Parr, W., Kulasiri, D.: A machine learning application in wine quality prediction. Mach. Learn. Appl. **8**, 100261 (2022)
30. Sagi, O., Rokach, L.: Ensemble learning: a survey. Wiley Interdisc. Rev. Data Min. Knowl. Discov. **8**(4), e1249 (2018)
31. Zaza, S., Atemkeng, M., Hamlomo, S.: Wine feature importance and quality prediction: a comparative study of machine learning algorithms with unbalanced data. arXiv preprint arXiv:2310.01584 (2023)

Sentiment Analysis of Online Product Reviews Using CNN-LSTM Cascaded Deep Learning Architecture

Ancy Rominus, Chinju John(✉), and Jayakrushna Sahoo

Department of Computer Science and Engineering, Indian Institute of Information Technology Kottayam, Kottayam, Kerala, India
Chinjuj.phd201001@iiitkottayam.ac.in

Abstract. Identifying the underlying sentiments, attitudes, and emotions in product reviews, and comments is a challenging but vital aspect of market research, brand monitoring, social media monitoring, and customer service. Here we have performed a comparative study of long short-term memory (LSTM), convolutional neural network (CNN), and combinations of CNN with LSTM (CNN-LSTM) and Gated Recurrent Unit (GRU), to recommend the most suitable deep learning architecture for the sentiment analysis of product reviews thereby revealing the emotional undertone of a group of words. Our study employed the TF-IDF (Term Frequency-Inverse Document Frequency) technique to compute a weight for each word, indicating its importance in the document and corpus. The results indicate that the CNN-LSTM cascaded approach significantly boosts the accuracy of sentiment analysis on the latest dataset with a promising accuracy of 98.3. This study can assist researchers in selecting an appropriate sentiment analysis model for their use case and in applying effective pre-processing techniques to enhance the accuracy of their model.

Keywords: Deep neural network · Deep Learning · Sentiment analysis · Convolutional neural networks (CNN) · Long short-term memory (LSTM) · CNN-LSTM

1 Introduction

E-commerce is making leaps and bounds in today's modern era in bringing things within reach of everyone. It is highly convenient for consumers because they are not even forced to leave the comfort of their homes to make purchases. Additionally, there are many different brands from which to choose. Ratings are becoming more valuable as consumers increasingly rely on Internet shopping sites. People just rely on the product reviews that are offered to purchase these goods. It is necessary to perform sentiment analysis on these reviews, which can be helpful for both consumers and manufacturers [1].

As the internet changed from read-only (mainly) to read-write (largely), users have been keener to express their feelings and opinions in public [2]. The exponential growth of

V. Sakalauskas et al. (Eds.): IBICA 2023, LNNS 1231, pp. 269–278, 2025.
https://doi.org/10.1007/978-3-031-78946-5_26

social messaging applications, such as Facebook, Twitter, and YouTube, has generated a huge volume of user-generated data that organizations leverage to gain valuable insights, optimize business decisions, and enhance profitability [3]. These sites offer evaluations and comments about their products and services, enabling the audience to rate [4]. Sentimental analysis (SA) starts with cleaning the data and generating the variance using a variety of pre-processing techniques like Normalization, Tokenization, and Duplication [5].

Machine Learning models are widely used for sentiment analysis tasks. However, these models' accuracy is largely dependent on how well the text data was cleaned up and converted into an analysis-ready format during data pre-processing. Effective pre-processing techniques can significantly improve the sentiment analysis model accuracy. Due to sentiment analysis's drawbacks, it has previously been suggested as a method to tell humans and machines apart [6]. Ashima Yadav and Dinesh Kumar's comprehensive survey emphasized the role of deep learning-based sentiment analysis, which is crucial for businesses to extract insights from the vast amount of unstructured information shared on social media.[7]. Apart from the selection of appropriate models, effective data pre-processing techniques are essential for sentiment analysis. Pang and Lee (2008) proposed a simple approach to prepare the text data by excluding the stop words, punctuation, and HTML tags [8].

The CNN model proficiently extracts advanced features by leveraging convolutional layers and max-pooling layers, while the LSTM model exhibits the ability to capture enduring relationships among word sequences [9]. Furthermore, the study also proposes a novel approach to improve sentiment analysis performance by implementing effective data pre-processing techniques, including the processing of emojis/emoticons, misspelt words, contractions, lemmatization, and the stemming of reviews. To mitigate the pitfalls in conventional workflows, we discuss the importance of feature extraction, encoding of the target variable (sentiment), and the application of the TF-IDF (Term Frequency Inverse-Document Frequency) technique to improve the accuracy of sentiment analysis models.

2 Literature Review

The growth of online review platforms and social media induces a proportional growth in sentiment analysis research. It is ubiquitous in various domains such as product reviews, political analysis, and healthcare. Artificial Intelligence models have been the primary work-horse here, ranging from traditional classifiers such as Naive Bayes and Support Vector Machines (SVM) to more advanced computing concepts such as Recurrent Neural Networks (RNN) and Convolutional Neural Networks (CNN). Machine Learning (ML) and Deep Learning models (DL) on sentiment analysis tasks is an actively explored research platform. Deep Learning models are outperforming the traditional ML-based classifiers in the domain with automated extraction of emotional keywords in an efficient manner. Wang et al. observed the Attention-Based RNN model's performance over the DL models in terms of accuracy [10].

Big data is frequently utilized to extract attitudes, emotions, and summaries, and researchers can use this data to generate certain predictions [6]. The most significant component of sentiment analysis is examining the text to identify and comprehend the viewpoint stated considering numerous elements like mood and modality. Sentiment can be expressed explicitly (as in the subjective statement) and implicitly (by implying an opinion). Zhao et. al introduced a novel method for analyzing client feelings from reviews on e-commerce websites [5]. He introduced a new optimized ML algorithm, LSIBA-ENN for this. The experimental results demonstrate that LSIBA-ENN outperforms existing algorithms in sentiment classification, with higher recall rates. Naveen et.al Gathered data from Amazon Reviews (2018) online cell phone and accessory category and used it to create an LSTM model for classification [1]. A long short-term memory (LSTM), variant of recurrent neural network (RNN) employs an internal circuitry to control the information flow.

Deceiving reviews of products can manipulate the user's thoughts and impact adversely. Muhammad Eid et. al proposed a study that utilizes LSTM and CNN-LSTM architectures for analysis of e-commerce reviews at the sentiment level, achieving high accuracies of 94% and 91% respectively [9]. A study was carried out by Sitaula Et al. on people's sentiments during the COVID-19 pandemic by classifying tweets collected from Twitter in Nepal using an ensemble of CNN models. The experimental results on the newly created Nepali Twitter sentiment dataset demonstrate the effectiveness of the suggested key term identification methods for CNN architectures, making them suitable for sentiment classification. Additionally, the dataset can serve as a benchmark for studying sentiment analysis of COVID-19-related data in the Nepali language [11].

People employ various "emoticons" on social media and e-commerce websites to convey their sentiments. Therefore, it will be more successful to use both the sentiment lexicon along emoticons in a single algorithm to mine customer evaluations on the Web. In the same spirit, Wolska demonstrates that adding emoticons to sentiment analysis produces a more thorough and accurate result [12]. Abdulganiyu et. al proposed an emoticon-included sentiment analysis model. The study's findings demonstrate that the model performs better when emoticons are used and implicitly mentioned factors are considered, with overall accuracy, precision, and recall values of 88.5%, 88.1%, and 84.6% [13]. Shelke N. et.al used SentiWordnet 3.0 to assign polarity to customer evaluations on several amazon.com domains, including digital camera, restaurant, and automobile [14]. The Lingo correction module is used to repair spelling errors and replace abbreviated phrases like gr8, gm, etc. after tokenizing reviews.

Using a sizable dataset of Amazon reviews, Haque et. al employed supervised learning models for sentiment analysis [15]. In this work, a unique technique incorporating data gathering, preprocessing, feature encoding, and classification using three variations of LSTM models is proposed. The results demonstrate the importance of sentiment analysis in deriving valuable insights from consumer reviews and social media content, with the proposed technique showing promising results in terms of accuracy and computational complexity.

Overall, this literature review highlights the importance of selecting appropriate models and applying effective data pre-processing techniques in sentiment analysis tasks. The subsequent sections of this paper present a novel approach that integrates several data pre-processing techniques to improve the accuracy of sentiment analysis models.

3 Material and Methods

3.1 Deep Learning Models for Sentiment Analysis

The deeply trained neural networks in polymorphic behaviour are commonly used artefacts in Natural Language Processing. From the very Multilayer Perceptron to the deeply stacked Transformer models, a bunch of Deep Learning-inspired trained models are available these days. The following section deals with the most employed DL algorithms in the Sentiment Analysis dimension.

Convolutional Neural Networks (CNN)
A Convolutional Neural Network (CNN) or "covnets", is a type of feedforward neural network that comprises four layers. The input layer represents sentences that are represented as matrices with a dimension of n*k. The subsequent layers include a convolutional layer, a global max pooling layer, and a fully connected layer, which is responsible for output generation [16]. The convolution-incorporated layer is a fundamental component of a CNN, it serves as a feature extraction layer, where features are extracted locally using filters. The convolution kernel function applies convolution to compute a feature map, which is then passed on to the pooling layer to reduce volume and prevent overloading. This layer protects the most important information while downsampling the result of the rendering process. A fully connected layer that calculates class scores and creates an equally sized 1-dimensional array between classes is then used for the output of the pooling operation.

Long Short-Term Memory (LSTM)
LSTM has some similarities with Recursive Neural Networks. RNNs use the outcome of one state as input to the next state. Hochreiter and Schmidhuber reported LSTMs [17]. It solves the long-term RNN dependency problem, RNNs cannot work efficiently when the distance difference increases [18]. Unlike traditional RNNs, which can only consider input at any given time, LSTMs can store data longer. Four neural networks and special memory centres form the chain model of LSTM. The number of gates and parameters is higher for LSTM compared to Gated Recurrent Units (GRUs). Data processing is the responsibility of the cell and memory management is the responsibility of the gateway. Input gates provide important information to cells, memory gates help extract irrelevant information, and exit gates provide relevant information from cells.

CNN–LSTM
CNNs are designed to learn spatial properties such as edges, shapes, and textures from input data. Our hybrid proposal includes feeding these features to the LSTM component, which can learn about the environment in the data. The cascading of CNN and LSTM in a single architecture allows spatial and temporal learning from input data. This makes the model more powerful than CNN or LSTM alone. The CNN-LSTM model has proven effective in many applications such as image and video recognition, and text and audio recognition.

3.2 Proposed Approach

In this study, we evaluated commonly employed DL algorithms to identify the architecture that could come up with more accurate and sensible results. Figure 2 depicts the holistic methodology of the proposed project. The process consists of several stages, including data collection, data analysis, data prioritization, weighting time, feature selection, sentiment scoring, identifying polarity, applying the CNN-LSTM model, and statistical and probability analysis. Each step is given in detail in the following sections.

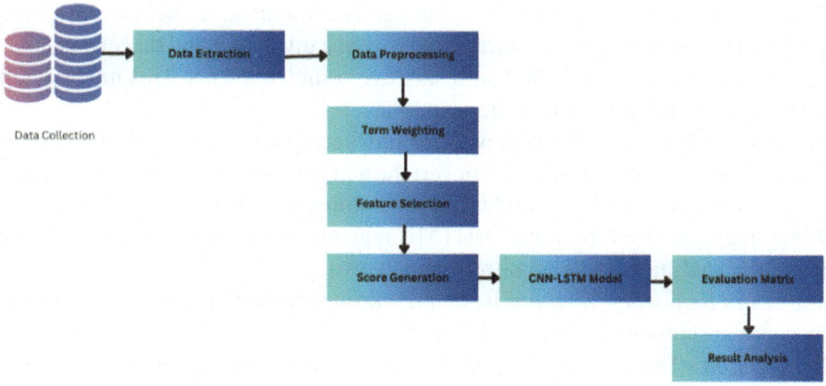

Fig. 1. Workflow of the proposed Sentiment Analysis Model

Data Collection and Data Extraction

Amazon (Amazon, 2018) provides the datasheet used in this study. Reviews are for the products found on Amazon.com. The review includes reviews from customers from 1995 to 2018. The cell phone and accessories Product reviews are selected for this sentimental analysis model. Each line of the json file specifies each review. The file contains the following 9 attributes: reviewerID, asin, reviewerName, helper, review text, general, summary, unixReviewTime, and Review time.

The cell phone and accessories Products reviews dataset comprises a total of 24340 records, out of which 19986 were labelled as positive and 4354 with negative labels. For a rating value less than three the sentiment rating was set as negative; otherwise, it belongs positive category.

Data Preprocessing

We have employed multiple data preprocessing stages to elevate the data to a high standard of cleanliness and ease of processing, ensuring accuracy is paramount.

Null values removal - Data points that are either missing or empty are systematically removed as part of the dataset-cleaning process.

Duplicates removal - recognizing and eradicating instances of data that either match precisely or closely resemble each other within the dataset.

Emoticon processing - An emoticon-centric methodology has been integrated, involving the creation of an emoticon dictionary encompassing various smileys along with their respective polarities. Within this framework, smileys within the text are cross-referenced with the emoticon dictionary, resulting in their classification as either positive or negative according to the established polarity. The emoticon dictionary was meticulously constructed through a synthesis of diverse sources, including Emojipedia.org, kaggle.com, etc., and additional contributions were incorporated from various other outlets.

Misspelt words correction - The correction of misspelt words entails a process wherein the misspelt words are juxtaposed with a dictionary or a corpus of words, and subsequently replaced with their correct counterparts.

Contraction Removal - Contraction removal is a linguistic refinement process characterized by the discernment of contractions within a given text and the subsequent substitution of said contractions with their respective complete forms. This method helps to enhance the semantic precision of the text.

Punctuation Removal - We removed the unwanted punctuations or numbers which does not add any value to our model. Convert the text to lowercase, remove text in -square brackets, links, and punctuation and words containing numbers.

HTML tags and URL Removal - HTML tags and URL removal is the process of eliminating HTML tags and URLs from review text.

Stopwords Removal - Stopwords, as they do not impart substantive meaning to a sentence, are systematically excluded from the text.

Tokenization - Her long texts are broken into tokens restricted by the white spaces. It also analyses the word sequence to interpret the context.

Stemming Review - The words which compose the review are extracted and traced back to their root words. PorterStemmer was used here which stems the words to their more accurate stem using a probabilistic set of adjustments to preserve the meaning of a word's final sequence.

Term Weighting - Next was the term weighting ω_i of the input, was performed. We have used TF-IDF ("Term Frequency - Inverse Document Frequency") for extracting pivotal features from the input dataset. The extracted keywords were converted to corresponding vectors and embedded using the embedding layers of our proposed architecture.

$$I(\omega_i) = \log \frac{|D|}{\left|\left\{j|\omega_i \in D_j\right\}\right|} \tag{1}$$

The D in Eq. (1) denotes the input text sample, D_j represents the jth text and ω_i indicates the ith word in the vocabulary. TF-IDF sorts different words according to their significance in various texts. The word weight can change from one text to another, hence the words with a higher degree of distinguishability are considered more important.

Score Generation - The score was generated after evaluating the review text based on sentiment. For calculating the sentiment score, the overall rating was considered. The review text was labelled as positive if the overall rating was > 3; otherwise, it was labelled as negative.

CNN-LSTM Model

An overview of the proposed CNN-LSTM model is given in Fig. 1. The first layer of the CNN-LSTM model, called the embedding layer, converts each word in the training dataset into a number or real-value vector. The set of emotional expressions produced becomes the intake for the next layer. The dropout layer is designed to prevent the model from overfitting, main function of this method is not to disturb the set of neurons in the embedding process to make the difference between thought points in the examination. An output value of 0.2 is set for this layer.

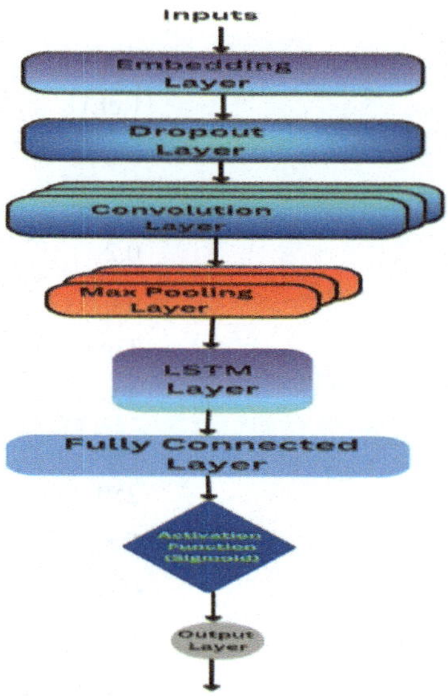

Fig. 2. CNN-LSTM Model

The system is configured to use a convolution layer with 256 filters having a filter core size of 3×3. We have employed 2 consecutive LSTM layers for further feature extraction. A dense layer, often called the fully integrated layer, consists of 250 interconnected components that communicate all the neurons of the network used to flatten the output. The activation function named Sigmoid was used in the last layer of the CNN-LSTM model. It detects the output and categorizes it as good or bad. We have detailed the hyperparameter which we used in Table 1.

Evaluation Matrices

The metrics chosen to measure the model predictions are Accuracy, Precision, Recall, and F1 Score, as mentioned in Eqs. (2–5). Where TP stands for True Positive, FP for False Positive, TN for True Negative and FN for False Negative.

$$Accuracy = \frac{TP + TN}{TP + FP + TN + FN} \tag{2}$$

$$Precision = \frac{TP}{TP + FP} \tag{3}$$

$$Callback = \frac{TP}{TP + FN} \tag{4}$$

Table 1. Hyperparameters and values

Hyperparameter	Value
Embedding input_dim	20000
Embedding output_dim	32
Embedding input_length	100
Conv1D filters	256
Conv1D kernel size	3
Conv1D activation	ReLU
Conv1D padding	Same
MaxPool1D_size	2
Dropout value	0.2
LSTM Layer1 units#	64
LSTM Layer 1 return_sequences	True
LSTM Layer 2 units#	32
Flatten	N/A
Dense 1 unit #	250
Dense 1 activation	ReLU
Dropout rate	0.2
Dense 2 unit #	1
Dense 2 activation	Sigmoid
Optimizer	Adam
Loss function	Binary Crossentropy
Maximum Length	100
Epochs #	20
Hidden activation	ReLU

$$F_1 \text{ score} = 2\frac{Precision * Recall}{Precision/Recall} \tag{5}$$

4 Results and Discussion

We aimed to contribute to the Sentiment Analysis sector NLP tasks, for which we have compared the outcomes of the most exploited and pioneer deep learning algorithms in the area - CNN, LSTM, CNN-LSTM, and CNN-GRU. To ensure that LSTM can outperform GRU, we have included CNN-GRU hybrid in our experiment. The measured outcomes reflecting the efficiency of these models for classifying the reviews according to their sentiment context are listed in Table 2. On analyzing this performance, CNN-GRU has the highest accuracy in both the training and testing phases, followed closely by CNN-LSTM. LSTM achieves the highest values for precision, recall, and F1-score in this dataset but it's comparatively less when analysed with other Amazon product datasets. However, CNN-LSTM also performs well in these metrics, especially in recall value, where it outperforms other models. CNN and CNN-GRU have the lowest training losses, while LSTM has a relatively higher training loss. CNN-LSTM strikes a balance with lower training loss than LSTM and competitive accuracy with CNN-GRU.

Table 2. Comparison of accuracies of Deep Learning models

Model	Training		Validation		Testing		Metrics		
	loss	Acc	loss	Acc	loss	Acc	Precision	Recall	F1-Score
CNN	0.005	0.995	0.124	0.982	0.124	0.982	0.82	0.89	0.85
LSTM	0.031	0.991	0.152	0.960	0.134	0.961	0.97	0.98	0.98
CNN-LSTM	0.008	0.997	0.114	0.983	0.114	0.983	0.82	0.96	0.88
CNN-GRU	0.005	0.998	0.159	0.985	0.159	0.985	0.77	0.91	0.83

After applying this model to multiple diverse datasets, the observed consensus points towards the CNN-LSTM model emerging as a robust contender. The key strength lies in its ability to strike a balance between accuracy and precision-recall metrics, a crucial consideration in various applications. This consistency across multiple datasets suggests the model's adaptability and effectiveness across different domains, reinforcing its viability as a versatile solution for tasks where a harmonious blend of overall accuracy and nuanced evaluation of precision and recall is imperative.

5 Conclusion

Sentiment Analysis and Opinion Mining examines people's thoughts, feelings, and opinions about various things. Information is analyzed to determine the relationship between thought and what is said. Here we have conducted the sentiment analysis of Amazon product reviews using four different learning models: CNN, LSTM, CNN-LSTM, and CNN-GRU. Our tests show that all four are compatible with the data set with the highest accuracy and precision-recall metrics produced by the CNN-LSTM combination. The enumerated results indicate the suitability of DL models for online product identification perception. Future work may explore using this model for emotional analysis of reviews in other fields and languages.

References

1. Gondhi, N.K., Chaahat, Sharma, E., Alharbi, A.H., Verma, R., Shah, M.A.: Efficient long short-term memory-based sentiment analysis of e-commerce reviews. Comput. Intell. Neurosci. **2022**, 1–9 (2022)

2. Kiran, R., Kumar, P., Bhasker, B.: OSLCFIT (organic simultaneous LSTM and CNN fit): a novel deep learning based solution for sentiment polarity classification of reviews. Expert Syst. Appl. **157**, 113488 (2020)
3. Paliwal, S., Kumar Khatri, S., Sharma, M.: Sentiment analysis and prediction using Neural Networks. In: 2018 International Conference on Inventive Research in Computing Applications (ICIRCA) (2018)
4. Dong, J.: Financial investor sentiment analysis based on FPGA and convolutional neural network. Microprocess. Microsyst. 103418 (2020)
5. Zhao, H., Liu, Z., Yao, X., Yang, Q.: A machine learning-based sentiment analysis of online product reviews with a novel term weighting and feature selection approach. Inf. Process. Manag. **58**, 102656 (2021)
6. Kumar, R.S., Saviour Devaraj, A.F., Rajeswari, M., Julie, E.G., Robinson, Y.H., Shanmuganathan, V.: Exploration of sentiment analysis and legitimate artistry for opinion mining. Multimed. Tools Appl. **81**, 11989–12004 (2021)
7. Yadav, A., Vishwakarma, D.K.: Sentiment analysis using deep learning architectures: a review. Artif. Intell. Rev. **53**, 4335–4385 (2019)
8. Pang, B., Lee, L.: Opinion mining and sentiment analysis. Found. Trends® Inf. Retrieval **2**, 1–135 (2008)
9. Alzahrani, M.E., Aldhyani, T.H., Alsubari, S.N., Althobaiti, M.M., Fahad, A.: Developing an intelligent system with deep learning algorithms for sentiment analysis of e-commerce product reviews. Comput. Intell. Neurosci. **2022**, 1–10 (2022)
10. Wang, B., Fan, B.: Attention-based hierarchical LSTM model for document sentiment classification. In: IOP Conference Series: Materials Science and Engineering, vol. 435, p. 012051 (2018)
11. Sitaula, C., Basnet, A., Mainali, A., Shahi, T.B.: Deep learning-based methods for sentiment analysis on Nepali covid-19-related tweets. Comput. Intell. Neurosci. **2021**, 1–11 (2021)
12. Wegrzyn-Wolska, K., Bougueroua, L., Yu, H., Zhong, J.: Explore the effects of emoticons on Twitter sentiment analysis. Comput. Sci. Inf. Technol. (CS & IT) (2016)
13. Harazeem, A.O., Kabir, U., Kakudi, H.A.: Emoticon Aware aspect based sentiment analysis of online product review. Dutse J. Pure Appl. Sci. DUJOPAS **7**, 166–179 (2021)
14. Shelke, N., Deshpande, S., Thakare, V.: Domain independent approach for aspect oriented sentiment analysis for product reviews. In: Proceedings of the 5th International Conference on Frontiers in Intelligent Computing: Theory and Applications. LNCS, vol. 515, pp. 651–659. Springer, Singapore (2017)
15. Haque, T.U., Saber, N.N., Shah, F.M.: Sentiment analysis on large scale Amazon product reviews. In: 2018 IEEE International Conference on Innovative Research and Development (ICIRD) (2018)
16. Rani, S., Kumar, P.: Deep learning based sentiment analysis using convolution neural network. Arab. J. Sci. Eng. **44**, 3305–3314 (2018)
17. Hochreiter, S., Schmidhuber, J.: Long short-term memory. In: Neural Computation, vol. 9, pp. 1735–1780. MIT Press, Cambridge (1997)
18. Jha, S., et al.: Deep learning approach for software maintainability metrics prediction. IEEE Access **7**, 61840–61855 (2019)

An ANN Approach for Measuring Operational Performance Efficiency in India Post Service Industry

S. M. Vadivel[1(✉)], Sunil Khosla[2], C. Devanathan[3], A. H. Sequeira[4], and A. Abhinav[5]

[1] VIT Business School, Chennai, Vandalur-Kelambakkam Road, Chennai 600127, India
vadivel.sm@vit.ac.in
[2] VIT-AP School of Social Sciences and Humanities, VIT-AP University, Amaravati 522237, Andhra Pradesh, India
sunil.khosla@vitap.ac.in
[3] Department of Mechanical Engineering, Rajalakshmi Engineering College, Thandalam, Chennai 602105, India
devanathan.c@rajalakshmi.edu.in
[4] National Institute of Technology Karnataka, Surathkal 575025, India
[5] VIT Law School, Vandalur-Kelambakkam Road, Chennai 600127, India
abhinav.a2023@vitstudent.ac.in

Abstract. In this study, the authors proposed Artificial Neural Network (ANN) as a decision making tool for India post service operational excellence, Karnataka, southern part of India. Though the concept of lean system bundles has been excellent results in the service sector, implementing in the postal service has been taken up as a new challenge. The performance has been validated based on Lean Service Practices (LSPs), Lean Workplace Environmental Practices (LWEPs), and Lean Social Practices (LSoPs) on Operational Performance (OP). Results can be deriving specific conclusions. As per our expectation, it is observed that ANN gives promising of ranking the decision making units. Here, we have used radial based function model. The expense of implementing lean was around INR 2 lakhs.

Keywords: Lean services · Radial based function neural network · ANN · operational performance

1 Introduction

Lean thinking is another managerial philosophy focusing on enhancing the efficiency and reducing wastes (**Womack et al. 1990**) [13]. One might believe that lean implementation is simple in service businesses after reading the extensive literature on lean implementation and successes in manufacturing or service industries. The difficulty lies in the fact that the service industry's production measurements require rigorous analysis and are intangible, making it difficult to quantify efficiency. Developed countries like Japan, the US, the UK, Germany, and the Italy has significantly contributed to the

V. Sakalauskas et al. (Eds.): IBICA 2023, LNNS 1231, pp. 279–287, 2025.
https://doi.org/10.1007/978-3-031-78946-5_27

success companies of LM implementation (Abu et al. 2019) [14]. Slowly, developing countries have taken effort to implement LM in Manufacturing and service firms. Interestingly, LM implementation in Indian postal service is a new area none of them have been tried to address it. Evidence indicating that, Iberahim et al. 2016) [15] improved the production and decrease the inventory of mails in leading Malaysian postal services. Three variables leadership, process design and standard operating procedure (SOP) are correlated with productivity but no relationship on labor productivity. Yet, LM is a new service paradigm especially for the public postal service industries, India.

India Post is one of the largest public sectors and covered widest network of all regions of India and utilizing maximum human resources. The key objective of this research includes analyzing the operational performance after implementing lean service in NSH Mangalore, Karnataka, and Southern part of India.

Additionally, this paper is structured as follows: The literature support of ANN applied in industries described in Sect. 2. The recommended methodology is in Sect. 3. The evaluation of ANN method in India Post service covered in detail in Sect. 4, and the Sect. 5 finishes with a discussion of the study's shortcomings and potential future advantages.

2 Literature Support

In an effort to fill this gap, the current study suggests a novel, objective, and cognitive strategy for locating appropriate sites where the most wave energy can be generated. With the use of MCDM cascaded to ANN approaches, the suggested method was able to forecast an index that exemplified a location's suitability for wave energy production. Two alternative locations, each with a different potential for wave energy, were subjected to the index. The outcomes inspire the writers to use the strategy more widely [1]. To make the model predictions accurate from the perspective of materials engineering, this paper aims to add previous information into the artificial neural network (ANN) model for the prediction of the continuous cooling transformation (CCT) diagram of steel [2]. In order to identify locations that are susceptible to landslides, this study analyses the use of Artificial Neural Net-work (ANN) and Rank Reciprocal (MCDM) analysis methodologies. The results showed that the accuracy provided by the ANN approach is 18.18% higher than that of the MCDM methodology. [3]. This research presents a methodology that uses Artificial Neural Networks (ANNs) and Multi Criteria Decision Making (MCDM) to prioritise creative and entrepreneurial indicators while concurrently accounting for three dimensions: organisational, cultural, and individual. This method's two primary advantages are its high precision and absence of calculating errors, together with its ease of use and quickness in the accounting process. Consequently, a range of results were considered and categorised into groups based on the Meta synthesis framework's instructions. [4]. Several pre-injection methods are suggested in this work to lower nitrogen oxides in the Wärtsilä 6L 46 marine engine. To describe the engine's emissions and consumption, a numerical model was used. A number of pre-injection amounts, lengths, and starting instants were examined. It was discovered that nitrogen oxide emissions may be significantly decreased, albeit at the expense of higher consumption and additional pollutants including carbon monoxide and hydrocarbons. This hybrid MCDM ANN methodology is an effective tool for creating new marine engines, as it was proposed in the current work [5]. In the current work, a more accurate estimation of the wave energy potential has been obtained. The suggested selection criteria are based on a number of factors, including the physical site qualities, environmental circumstances, and socioeconomic regional state. As a result, we looked at a hybrid ANP method to determine the best spot for a wave energy potential site [6]. In this paper, a system for diagnosing heart illness is developed using multiple-criteria decision-making (MCDM), feature selection (FS), and artificial neural networks (ANN). Five feature selection strategies are used to choose the appropriate features. The performance of three artificial neural networks for CVD prediction was then evaluated using eight reliable criteria [7]. The current study provided a novel approach of liability analysis in which each key element was assigned a distinct weight based on literature, expert opinion, and local surveys. A financial liability index was additionally suggested to represent the project's financial burden. Hydropower plants with varying capacities and efficiencies used the index. The findings are discovered to be consistent with the real-world circumstances [8].

The current study's goal was to use ANN, TOPSIS, and fuzzy methods to deter mine the most effective way to clean up a soil oil spill. In 2017, the investigation was conducted in an oil pumping complex in Iran's southwest. An ANN of the radial basis function (RBF) type was employed. The most effective extraction technique was bio logical. The outputs of RBF and the TOPSIS-prioritized results were perfectly consistent [9]. This study examined the mechanical characteristics and component weight of an ABS part made using FFF. They combined use of response surface methodology (RSM), ANN, GA, and MCDM techniques to mathematical modelling and multi objective optimization [10]. This study offered a decision support framework for inventory management that uses artificial neural network (ANN) and genetic algorithms (GA-ANN) to estimate demand [11].

In this proposed model investigation for requirements prioritization, a case study from Ozcan et al. was used. An ANN and a logarithmic fuzzy analytic hierarchy pro cess (LOG FAHP) based model were used to select the best supplier firm. Three sup plier selection criteria based on LOGANFIS and FAHP as well as a fuzzy comparison matrix were used to evaluate the test findings, which were conducted using the MATLAB software. This proved that the test's requirements prioritisation results were better than those of earlier approaches that gave more priority to requirements [12].

3 Research Methodology

(See Figs. 1 and 2).

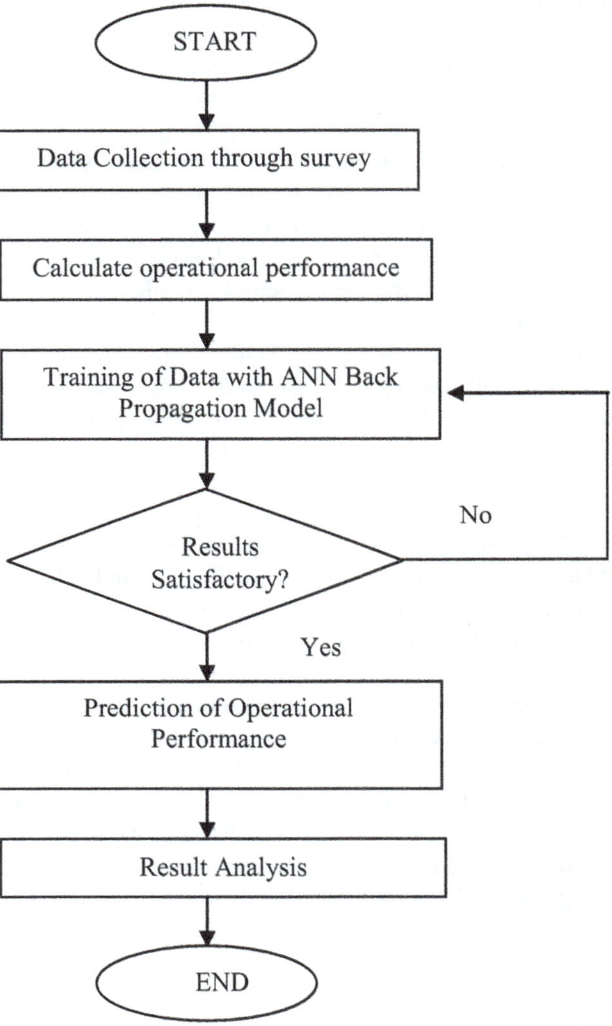

Fig. 1. Proposed Research Methodology

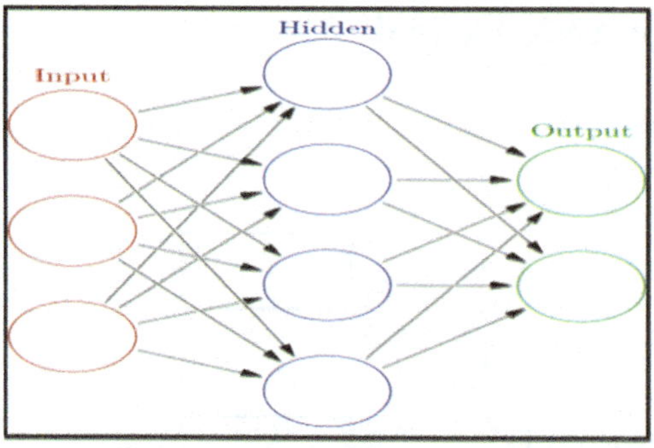

Fig. 2. ANN Model – RBF Neural Network

4 Case Study – Indian Postal Service, Mangalore, Karnataka

This case study conducted in NSH Mangalore in the year 2019. The following reasons have given why we have done this research study.

- Growing competition from the private postal service, rapid changes in communication changes, globalization, customer expectations, and other issues that the postal service must deal with in order to be viable in the long run.
- Internal activities are time-consuming due to the long procedures/rules in place. As, personnel errors are on the rise, standardization of operating procedures requires special attention.
- The post office's working environment need to be improved significantly from ergonomics point of view, insufficient lighting, outdated tools, decaying old building, dull appearance, and other issues.
- Almost all international postal authorities of developed countries already started renovating postal facilities and enlarged customer satisfaction. Their administrative practices have undergone a significant transformation to provide practical, fast, and sensitive counter services.
- Customers are concerned about receiving high-quality service from the Indian Postal Service in the shortest period possible. This issue is addressed by lean system, which boosts back-end efficiency by sending mail as quick as possible to the addressee.
- Improving physical facilities such as providing ventilation, lighting facilities, 5S, and quality work culture, enhances public perception. It also improves an operational performance.

The survey covered by the questionnaire in this study are: 1. Lean Service Practises (LSPs), 2. Lean Workplace Environment Practises (LWEPs), 3. Lean Social Practises (LSoPs), and 4. Operational Performance (OP). As previously stated, the measures for these constructs were adapted from literature review investigations. These constructions' items were scored on a 5-point Likert scale, with weights ranging from strongly disagree to strongly agree to neutral. Following the adoption of lean, data were collected from 163 postal employees via a survey approach. The valid 150 replies were then chosen for analysis.

5 Application of ANN Method

(See Fig. 3 and Tables 1 and 2).

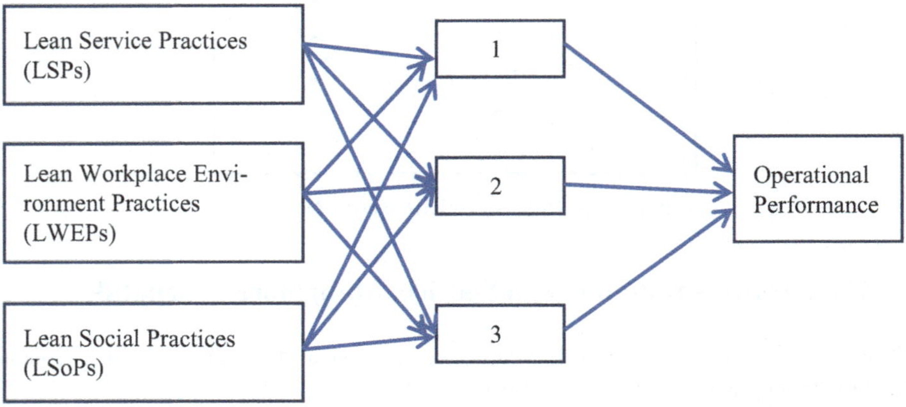

Fig. 3. Recommended Network Topology

Table 1. Summary of Sample data

		N	Percent
Sample	Training	100	66.7%
	Testing	50	33.3%
Valid		150	100.0%
Excluded		0	
Total		150	

6 Comparative Study

The South Karnataka Post Master General (PMG) has given permission and motivated to implement lean service concept with the assurance fund allotment INR 2 Lakh in order to improve the operational performance.

The comparative analysis results are presented in Table 3. According to our research, LSPs having most priority value than LWEPs and LSoPs. In ANN method, both in testing and training data, LSoPs is better results than SEM analysis result. The reason may be cleared the noisy data in that survey of 150 respondents.

Table 2. Results of ANN method

Training	Sum of Squares Error		43.980
	Average Overall Relative Error		.222
	Relative Error for Scale Dependents	LSP	.442
		LWEP	.124
		LSoP	.168
		OP	.155
	Training Time		0:00:00.61
Testing	Sum of Squares Error		23.123[a]
	Average Overall Relative Error		.279
	Relative Error for Scale Dependents	LSP	.530
		LWEP	.212
		LSoP	.232
		OP	.212

a. The number of hidden units is determined by the testing data criterion: The "best" number of hidden units is the one that yields the smallest error in the testing data.

Table 3. Comparative study results

Lean System	SEM Results [16]	ANN Results Training	ANN Results Testing
LSPs	0.330	0.442	0.530
LWEPs	0.281	0.124	0.212
LSoPs	0.266	0.168	0.232

7 Conclusion

The empirical findings are encouraging, and they can help reduce inventory, cycle time, space use, and workforce, among other things. This study precisely studies the lean system implementation in the postal service industry, setting a new path-breaking method to the best of the author's knowledge. The decision made by an Indian postal service industry to adopt lean bundles is a bold step towards supporting the postal systems to sustain in the long run. In the present study, an effort has been made to evaluate in ANN – RBF model. The proposed ANN is more flexible and able to handle noisy data better than traditional methods. The present study can be analyzed with sensitivity analysis and more sample size with addition of customers consideration as a future research direction.

References

1. Ghosh, S., Chakraborty, T., Saha, S., Majumder, M., Pal, M.: Development of the location suitability index for wave energy production by ANN and MCDM techniques. Renew. Sustain. Energy Rev. **59**, 1017–1028 (2016)
2. Chakraborty, S., Das, P., Kaveti, N.K., Chattopadhyay, P.P., Datta, S.: MCDM towards knowledge incorporation in ANN models for phase transformation in continuous cooling of steel. Multidiscip. Model. Mater. Struct. **15**(1), 170–186 (2018)
3. Salleh, S.A., Abd Rahman, A.A., Othman, A.N., Mohd, W.N.W.: Comparative study of landslides susceptibility mapping methods: multi-criteria decision making (MCDM) and artificial neural network (ANN). In: IOP Conference Series: Earth and Environmental Science, vol. 117, no. 1, p. 012035. IOP Publishing (2018)
4. Karimi, M., Namamian, F., Vafaei, F., Moradi, A.: Entrepreneurship policy and innovative indicators of industrial companies: evaluation by MCDM and ANN Methods. J. Ind. Strateg. Manag. **4**(1), 66–93 (2019)
5. Rodriguez, C.G., Lamas, M.I., Rodriguez, J.D., Abbas, A.: Analysis of the pre injection system of a marine diesel engine through multiple-criteria decision-making and artificial neural networks. Pol. Mar. Res. **4**, 88–96 (2021)
6. Ghosh, S.: Wave energy potential site selection based on MCDM and neural network analysis. In: Application of Geographical Information Systems and Soft Computation Techniques in Water and Water Based Renewable Energy Problems, pp. 107–120 (2018)
7. Najafi, A., Nemati, A., Ashrafzadeh, M., Zolfani, S.H.: Multiple-criteria decision making, feature selection, and deep learning: a golden triangle for heart disease identification. Eng. Appl. Artif. Intell. **125**, 106662 (2023)
8. Majumder, P., Saha, A.K.: Development of financial liability index for hydro power plant with MCDM and neuro-genetic models. Application of geographical information systems and soft computation techniques in water and water based renewable energy problems, pp. 71–105 (2018)
9. Shafie, A., Fard, N.J.H., Monavari, M., Sabzalipour, S., Fathian, H.: Artificial neural network and multi-criteria decision-making methods for the remediation of soil oil pollution in the southwest of Iran. Model. Earth Syst. Environ. 1–8 (2023)
10. Dev, S., Srivastava, R.: Influence of process variables on mechanical properties and material weight of acrylic butadiene styrene parts produced by fused filament fabrication. Progress Additive Manuf. **8**(2), 143–158 (2023)
11. De Paula Vidal, G.H., Caiado, R.G.G., Scavarda, L.F., Ivson, P., Garza-Reyes, J.A.: Decision support framework for inventory management combining fuzzy multicriteria methods, genetic algorithm, and artificial neural network. Comput. Ind. Eng. **174**, 108777 (2022)
12. Singh, Y.V., Kumar, B., Chand, S.: A hybrid approach for requirements prioritization using LFPP and ANN. Int. J. Intell. Syst. App. **11**(1), 13 (2019)
13. Womack, J.P., Jones, D.T., Roos, D.: The Machine that Changed the World: The Story of Lean Production. Rawson Association, USA (1990)
14. Abu, F., Gholami, H., Saman, M.Z.M., Zakuan, N., Streimikiene, D.: The implementation of lean manufacturing in the furniture industry: a review and analysis on the motives, barriers, challenges, and the applications. J. Clean. Prod. **234**, 660–680 (2019)

15. Iberahim, H., Mazlinda, H., Marhainie, M.D., Hidayah, A.N.: Determinants of sustainable continuous improvement practices in mail processing service operations. Procedia - Soc. Behav. Sci. **219**, 330–337 (2016). Int. J. Serv. Technol. Manag. **5**(5-6), 545–564
16. Vadivel, S.M., Sequeira, A.H., Sakkariyas, R.R., Kirubaharan, B.: Impact of lean service, workplace environment, and social practices on the operational performance of India post service industry. Ann. Oper. Res. **315**, 2219–2244 (2022). https://doi.org/10.1007/s10479-021-04087-z

Enhancing Epilepsy Care in Resource-Constrained Settings Through Streamlined EEG Data Analysis

Romel Thommeykutty, Anwesh Roy, and K. Asha(✉)

Department of Advanced Computing, School of Information Technology, St. Joseph's University, Bangalore, Karnataka, India
ashak@sju.edu.in

Abstract. Epilepsy, a debilitating neurological disorder characterized by recurring seizures, affects millions of people around the world. This condition poses significant challenges for patients and healthcare systems. Shockingly, there are approximately 65 million cases of epilepsy globally with varying rates of occurrence in high-income and low-middle-income nations, ranging from 45 to 82 cases per 100,000 people annually. Alarmingly, more than 90% of epilepsy cases in these areas are undiagnosed and untreated, contributing to what is known as the epileptic diagnostic gap. An electroencephalogram, or EEG, is a medical test that captures electrical activity in the brain. EEG enables medical professionals to track brain activity and identify conditions like epilepsy. In countries with low incomes, epilepsy often goes unnoticed because there is a shortage of healthcare professionals and the expenses involved in acquiring EEG recording equipment are quite high. In this paper, an analytical method has proposed for the detection of epilepsy using dataset from Nigeria. The signal data trained the model with various machine learning algorithms, applied dimensionality reduction and implemented hyperparameter tuning. The paper can bridge the gap in epilepsy diagnosis and potentially revolutionize healthcare provision by offering trustworthy and accessible diagnostic procedures.

Keywords: Machine learning · signal processing · epilepsy detection · Principal component Analysis · bigdata analytics

1 Introduction

Epilepsy is a widespread condition that affects people all around the world. Its prevalence is not limited to Nigeria. Being one of the most populated nations in the world, India also faces substantial difficulties in the diagnosis and treatment of epilepsy. When it comes to cutting-edge medical facilities and diagnostics; India faces a serious challenge of resource limitation similar to those experienced by many other nations. Particularly in rural and underdeveloped locations, access to sophisticated diagnostic equipment for epilepsy, such as high-end EEG machines may be restricted. Furthermore, the challenges associated with EEG recordings are compounded by the inherent complexities,

V. Sakalauskas et al. (Eds.): IBICA 2023, LNNS 1231, pp. 288–297, 2025.
https://doi.org/10.1007/978-3-031-78946-5_28

such as the unpredictable nature and vulnerability to discrepancies among observers. To effectively identify individuals at high risk of epilepsy, maximize the use of limited resources, and support community-based healthcare services, it is essential to establish dependable procedures that do not require expert involvement. The objective of the paper is to establish a robust framework that can be effectively utilized in challenging healthcare circumstances. This accomplished through conducting practical research on different data segmentation methods, feature extraction approaches, and implementation of machine learning algorithms. The performances of the classifiers have been analyzed, and an optimum classifier has been found. Principal compound analysis and hyperparameter tuning was used to enhance the performance of SVM and random forest classifiers. EEG relative power band was used as a feature extraction technique.

1.1 Pervasiveness of Seizure Epilepsy in Realtime

In India, the cost of epilepsy diagnosis and treatment is a significant consideration. Access to epilepsy care can be significantly improved by using low-cost or affordable diagnostic techniques and instruments, especially for groups who are economically deprived. India has a thriving telemedicine sector that could benefit from low-cost and portable epilepsy diagnosis technologies. Remote monitoring and diagnosis can help people in remote or underserved areas gain access to healthcare. Local researchers and inventors in India can be inspired by studies using datasets such as the Nigerian dataset to produce low-cost and novel solutions for epilepsy diagnosis and management. This could include the creation of low-cost EEG sensors as well as machine learning methods for early detection. Not much research has been done on the EEG dataset from Nigeria. Hence, the paper addresses not just Nigeria but areas with socio-economic factors analogous to Nigeria. The paper has implemented fine tuning techniques and dimensionality reduction techniques, along with techniques to treat imbalances in the data. The proposed study is significant because they provide a viable solution for improving epilepsy care in resource-constrained settings via faster EEG data analysis, leveraging machine learning models and big data techniques.

2 Notion of Bigdata in Classifying Epilepsy

Bigdata can provide significant benefits for classifying epilepsy by offering access to a vast amount of patient data and enabling advanced analysis techniques. Bigdata allows for the collection of extensive patient records, including EEG data, medical histories, medication responses, and more. With the rich dataset, machine learning algorithms can be trained to identify subtle patterns and markers of epilepsy, leading to a more accurate and timely diagnosis. By analyzing historical data, big data analytics can help identify early signs and predictive factors associated with epilepsy. This can lead to earlier interventions and improved outcomes for patients. Bigdata can facilitate the development of personalized treatment plans by considering individual patient characteristics, such as genetics, lifestyle, and responses to various medications.

IoT devices and wearables can collect continuous data from epilepsy patients, such as heart rate, sleep patterns, and movement. Through predictive analytics, big data can help healthcare systems identify high-risk individuals and provide targeted interventions, reducing hospital admissions and healthcare costs associated with epilepsy management.

3 Literature Review

Numerous studies have evolved in epileptic detection using EEG signals. Research work depicts the usage of a dataset from Nigeria and applies various statistical and feature extraction methods [1]. Study used TUH EEG seizure corpus to accurately classify multiclass seizure classifications [2]. Another work uses 54-DWT mother wavelets for the analysis of EEG signals along with four machine learning classifiers [3]. P-1D-CNN is basically a pyramidal, one-dimensional CNN that uses consumes less memory and has a significantly shorter detection time [4]. In another paper, a machine learning-based model using ANN was proposed [5]. Deep learning models were used to detect epilepsy and a hardware implementation of STFT (Short-Time Fourier Transform) [6]. Genetic algorithms and particle swarm optimization (PSO) were used to improve the parameters of SVM [7]. Detection of meaningful patterns in EEG signals were used with machine learning algorithms for timely seizure detection [8]. Another study integrates XLtek EEG data and local epilepsy center signals into the CHB-MIT dataset, emphasizing dominant channel selection [9]. To understand the non-linearity of the data and down-sample the convolutional layer's output a deep-learning model with a number of layers including activation and max-pooling 1D layers, is utilized [10]. Preprocessing the EEG signals to extract the required properties for categorization was the suggested method.

To automatically predict epileptic seizures, research applies deep learning to the spectral analysis of EEG signals [11]. The one-dimensional Convolutional Neural Network (CNN) was used in a study to automatically detect epilepsy seizures in EEG readings [12]. The CNN model comprises an embedding layer, dropout layer and a single-unit output layer and achieved 82.818% accuracy [13]. To identify epileptic seizures, decision tree (DT), k-nearest neighbor (KNN), naive Bayes (NB), support vector machine (SVM), and artificial neural network (ANN) classifiers were utilized [14]. A specialized CNN was proposed and evaluated for EEG feature extraction and classification, using fivefold cross-validation [15]. CHB-MIT Scalp EEG (24 recordings, 23 patients) analyzed for seizure detection. Deep learning effective with raw EEG or coefficients; shallow networks need features [16]. A 13-layer CNN achieves 88.67% accuracy, surpassing manual methods with 90.00% specificity and 95.00% sensitivity, showcasing the potential of CNNs for precise and efficient EEG-based epilepsy diagnosis [17]. An efficient epilepsy detection, utilizing Delta sub-band features via Discrete Wavelet Transform on subsets A (healthy) and E (seizures) from a public EEG dataset [18]. The paper improves epilepsy detection using SPPCA and SUBXPCA for dimensionality reduction achieving 94% and 96% accuracy respectively on EEG signals from the University of Bonn's dataset (groups A to E) [19]. FT-VGG16 achieves 99.21% accuracy with signal-to-image conversion through Continuous Wavelet Transform, highlighting key frequency components (30–60Hz) via SHAP analysis for enhanced classification.

4 Methodology

4.1 Data Acquisition

The dataset used in the research paper is based on the people of Nigeria, where the data was collected using the EMOTIV headsets which record brain activity in a very detailed way with a sampling rate of 128 times per second and a high accuracy of 16-bit resolution. Several participants were asked to sit for five minutes while wearing a wireless headset. The two most important sensors were common modsense (CMS)S) and another known as driven right leg (DRL). The electrodes were placed at anterofrontal (AF3, AF4, F3, F4, F7, F8), frontocentral (FC5, F6), occipital (O1, O2), parietal (P7, P8) and temporal sites (T7, T8).

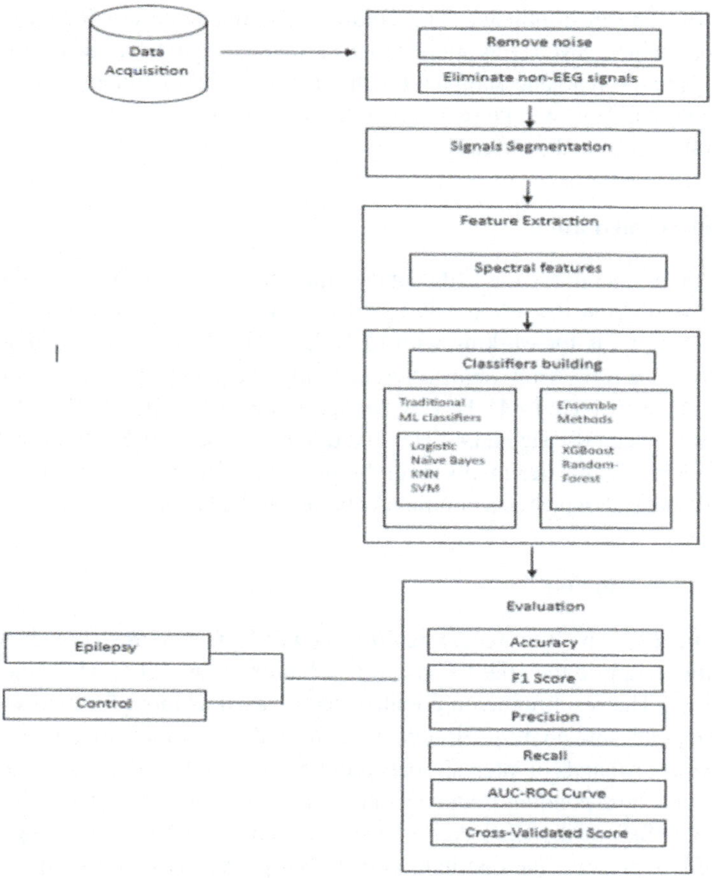

Fig. 1. Proposed framework of Epilepsy Detection

The data was collected for five minutes when the participants were sitting with their eyes closed. Anything unusual was also recorded and stored in a specific format.

But the data recorded was not that accurate. Simple methods to smooth out the errors were performed which included the removal of unwanted data points. The quality of the recorded signals was tested, and the bad signals was removed with immediate effect. The head movements are measured through a gyroscope, and if the head movements were improper, then the recording was deleted. Using the process of wavelet analysis, three different wavelets were used to analyze brain activity. Figure 1 detailing the proposed framework used in the epilepsy detection.

4.2 Pre-processing

EEG data is loaded from multiple files into Python. Specifically, EEG recordings from subjects with epilepsy ("Epilepsy") and control subjects ("Control") are extracted from separate files. These EEG recordings are saved as data frames. The non-EEG channels from the EEG data are eliminated. The stratified sampling to ensure that the class distribution is maintained. SMOTE is applied to the training data to address class imbalances. SMOTE generates synthetic samples for the minority class (Epilepsy) to balance the class distribution. This step helps improve the model's performance, especially when dealing with imbalanced datasets.

4.3 Signal Segmentation

The data frames containing the EEG data are then converted into MNE-Python Epochs objects. MNE-Python is a library designed for efficient analysis of EEG/MEG data, offering various tools for working with EEG data. The data is transposed and used to generate an MNE Raw Array object. Subsequently, the data is filtered within a specific frequency range (0.1 Hz to 45 Hz). A bandpass filter is applied to the data to retain frequencies between 0.1 Hz and 45 Hz. Fixed-length epochs of 5 s duration with a 1-s overlap are created. This step involves combining data from various subjects and sessions into a single dataset, which is a common approach to EEG analysis.

4.4 Feature Extraction

Feature extraction involves extracting EEG power band features from EEG epochs. These features capture the relative power distribution across different frequency bands at each epoch. The resulting feature matrix can be used as input for machine learning models to classify and analyze EEG data based on these extracted features. We define specific frequency bands of interest, such as delta, theta, alpha, sigma, beta, and gamma, along with the corresponding frequency ranges (in Hz). The computing of Power Spectral Density (PSD) of the EEG data in each frequency band is fixed using the Welch method. PSD represents the distribution of signal power across different frequencies. The computed PSD values are normalized by dividing them by the sum of PSD values across all frequencies for each epoch. The concatenation of the computed features is framed as a feature vector for each epoch.

4.5 Classification

Classification is done using traditional machine learning classifiers and ensemble methods. In order to enhance the model's performance, fine-tuning its hyperparameters by performing Grid Search with Cross-Validation was done, and its performance was evaluated on the testing data. The classifiers classified EEG data into two classes: Epilepsy and Control, based on the extracted EEG power band features. The performance metrics provide insights into how well the model can discriminate between these two classes.

5 Results and Implementation

The metadata for all the patients that included subject.id, recorded period, startTime, session.id, first_condition, remarks, group, and csv. file. In this research, we have divided the participants into two distinct groups, namely 'epilepsy' and 'control' subjects, based on the 'Group' activity present in the metadata file. EEG data has stored for each group separately. The task of feature extraction where the important features are kept, and removed the rest making it more clear and easier to work.

EEG data is collected in raw form, so we use the MNE library in Python to streamline it in a more meaningful way. Filtering is applied to focus on the required frequency ranges. Then the EEG data is divided into short, fixed length segments known as epochs, each lasting 5 s with a 1-s overlap. Later, we remove the low-quality epochs to ensure data integrity. Finally, the ones that are well organized and cleaned has received. The list of channel names from the epoch objects is 'AF3', 'AF4', 'F3', 'F4', 'F7', 'F8', 'FC5', 'FC6', 'O1', 'O2', 'P7', 'P8', 'T7', 'T8'. We also use 'convertDF2MNE' to reformat the EEG data of both groups (epilepsy and control). The format is a must because MNE requires a special format for detailed analysis of EEG data. The content has move on to label '0' for epilepsy and '1' for control. Six frequency bands are defined, i.e., alpha, beta, theta, delta, sigma, and gamma. To calculate the Power Spectral Density (PSD) of EEG data within a frequency range moved from 0.5 Hz to 45 Hz. PSD is used to measure the power present at different levels of frequency in EEG signal data. A machine learning model has performed PCA to reduce the dimension while retaining 98% of the explained variance. 25 components are retained after PCA is performed.

Table 1. F1-Scores of Controls and Epilepsy

Classifier	Control	Epilepsy
SVM	0.72	0.84

Control, in this context of Table 1 refers to a healthy person who does not have epilepsy, whereas epilepsy is a chronic neurological disorder that is characterized by recurrent seizures. So here we have taken the f1 score as a measure to distinguish between epilepsy and control classes using the SVM classifier. F1-Score for "Epilepsy" is 0.84, indicating strong performance in correctly identifying epilepsy cases, while the F1-Score for "Control" is 0.72, suggesting good accuracy in classifying non-epilepsy cases. Both scores balance precision and recall in their respective classes.

6 Performance Analysis

The overall process uses 80% of the data for training and 20% for testing. At first, Random Forest classifier is used in the model, and for hyperparameter tuning, Grid Search with cross validation are performed. The best hyperparameter that best fits the model. The model achieved a cross validated accuracy of 0.827 (approx.). The applied SVM, and the hyperparameters found for this algorithm are 'C'(100), 'gamma'('scale') and 'kernel'('rbf') and the model gave a cross validated accuracy of 82.08% approximately. The SMOTE technique to balance the data is used since the dataset was imbalanced. Table 2 provides the details on the comparison between various ML classifiers performed.

Table 2. Comparison between different ML classifiers

		0	1	Accuracy
	precision	0.74	0.83	
SVM	recall	0.7	0.86	0.8
	f1 score	0.72	0.84	
	precision	0.69	0.82	
Logistic Regression	recall	0.7	0.81	0.77
	f1 score	0.7	0.81	
	precision	0.57	0.73	
Naïve Bayes	recall	0.54	0.76	0.68
	f1 score	0.56	0.75	
	precision	0.72	0.84	
XGBoost	recall	0.74	0.83	0.79
	f1 score	0.73	0.83	
	precision	0.73	0.83	
Random Forest	recall	0.71	0.84	0.79
	f1 score	0.72	0.83	

Figure 2 indicates the ROC curve of SVM and achieved the AUC of 0.85. As compared to RF, SVM is able to more distinctly classify between two classes with a high degree of accuracy.

Table 3 detailing the confusion matrix, that demonstrates the SVM classifier properly categorizes 277 healthy individuals and 95 patients with illnesses. Additionally, it designated 118 individuals with disease as having no disease and 561 patients without disease.

Sensitivity (TPR) is the proportion of people with epilepsy who are correctly classified as having epilepsy. Specificity (TNR) is the proportion of people without epilepsy

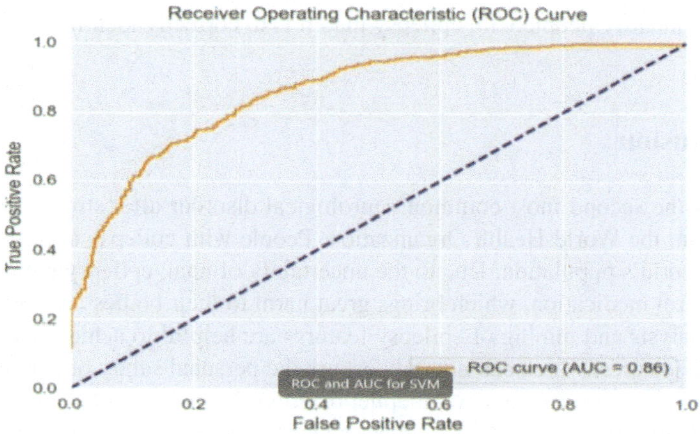

Fig. 2. ROC curve

Table 3. Confusion Matrix for SVM

Confusion Matrix		Predicted	
		Class 0	Class1
Actual	Class0	277	118
	Class1	95	561

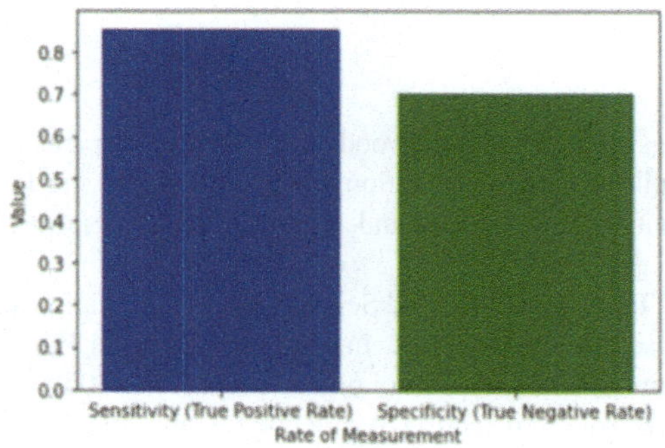

Fig. 3. True positive and True Negative Rate of Classification

who are correctly classified as not having epilepsy. Sensitivity and specificity are important metrics for evaluating the performance of a diagnostic test. A test with high sensitivity is good at identifying people with the condition, while a test with high specificity is good at identifying people without the condition. Figure 3 pointing to the measure of true positive rate achieved 86% and true negative achieved 70%. Based on the evaluation, it's been identified that, SVM classifier have high sensitivity and specificity. This means that the classifier is very good at detecting epilepsy and avoiding false positives. The classifier is more sensitive than specific. This means that the classifier is more likely to correctly identify people with epilepsy than it is to correctly identify people without epilepsy.

7 Conclusion

Epilepsy is the second most common neurological disorder after stroke, according to a report from the World Health Organization. People with epilepsy account for about 1% of the world's population. Due to the uncertainty of ictal, epilepsy patients need to take long-term medication, which brings great harm to their bodies and minds. Therefore, the analysis and mining of epilepsy features are helpful to achieve early warning of epileptic seizures, which can not only ensure the personal safety of patients but also remind them to choose emergency antiepileptic drugs. It was found that various machine learning methods have been used to detect epilepsy, but not much work has been done using consumer-grade devices. With affordable devices and seamless detection methods, this could become a reality for such sections of society. This paper primarily focuses on extracting spectral features from EEG data, specifically relative power in specific frequency bands. These spectral features are related to the frequency domain of EEG signals. Future improvements can include the use of temporal features, which are related to the time domain of EEG signals, such as mean, variance, skewness and kurtosis. These statistics need to be computed directly from the EEG signals over time. This perhaps could provide a more comprehensive analysis of EEG signals. Also, deep learning methods such as LSTM and CNN with appropriate fine-tuning may effectively classify subjects. Deep neural networks like FT-VGG16 can provide enhanced classification.

References

1. Alyas, N., Hastings, D.P., Mehrabidavoodabadi, A.: A comparative analysis of classifier performance for epileptic seizure detection using EEG signals. In: International Conference on Pattern Recognition Applications and Methods. https://doi.org/10.5220/0011631600003411
2. Roy, S., Asif, U., Tang, J., Harrer, S.: Seizure type classification using EEG signals and machine learning: setting a benchmark. In: 2020 IEEE Signal Processing in Medicine and Biology Symposium (SPMB). https://doi.org/10.1109/SPMB50085.2020.9353642
3. Mardini, W., et al.: Enhanced detection of epileptic seizure using EEG signals in combination with machine learning classifiers. IEEE Access **8**, 24046–24055 (2020)
4. Ullah, I., et al.: An automated system for epilepsy detection using EEG brain signals based on deep learning approach. Expert Syst. Appl. https://doi.org/10.1016/j.eswa.2018.04.021
5. Chakraborti, S., Choudhary, A., Singh, A., Kumar, R., Swetapadma, A.: A machine learning based method to detect epilepsy. Int. J. Inf. Technol. **10**(3), 257–263 (2018). https://doi.org/10.1007/s41870-018-0088-1
6. Beeraka, S.M., Kumar, A., Sameer, M., Ghosh, S., Gupta, B.: Accuracy enhancement of epileptic seizure detection: a deep learning approach with hardware realization of STFT. Circuits Syst. Signal Process. (2022). https://doi.org/10.1007/s00034-021-01789-4
7. Subasi, A., Kevric, J., Abdullah Canbaz, M.: Epileptic seizure detection using hybrid machine learning methods. Neural Comput. Appl. (2019). https://doi.org/10.1007/s00521-017-3003-y

8. Savadkoohi, M., Oladunni, T., Thompson, L.: A machine learning approach to epileptic seizure prediction using Electroencephalogram (EEG) Signal. https://doi.org/10.1016/j.bbe.2020.07.004
9. Alharthi, M.K., Moria, K.M., Alghazzawi, D.M., Tayeb, H.O.: Epileptic Disorder Detection of Seizures Using EEG Signals. https://doi.org/10.3390/s22176592
10. Sagga, D., Echtioui, A., Khemakhem, R., Ghorbel, M.: Epileptic Seizure Detection using EEG Signals based on 1D-CNN Approach. https://doi.org/10.1109/STA50679.2020.9329321
11. Singh, K., Malhotra, J.: Deep learning based smart health monitoring for automated prediction of epileptic seizures using spectral analysis of scalp EEG. https://doi.org/10.1007/S13246-021-01052-9
12. Kaushik, M., Singh, D., Kishore-Dutta, M., Travieso, C.M.: A deep learning approach for epilepsy seizure detection using EEG signals. https://doi.org/10.18845/tm.v35i8.6461
13. Shiva Shankar, R., Raminaidu, C.H., Sivarama Raju, V.V., Rajanikanth, J.: Detection of Epilepsy based on EEG Signals using PCA with ANN Model. https://doi.org/10.1088/1742-6596/2070/1/012145
14. Joy, A.D., Sarkar, S., Azad, A.K.: Detection of epileptic seizures from EEG signals using machine learning classifiers. Bangladesh J. Med. Phys. (2022). https://doi.org/10.3329/bjmp.v15i1.63560
15. Boonyakitanont, P., Lek-Uthai, A., Chomtho, K., Songsiri, J.: A review of feature extraction and performance evaluation in epileptic seizure detection using EEG
16. Rajendra Acharya, U.., Oh, S.L., Hagiwara, Y., Tan, J.H., Adeli, H.: Deep convolutional neural network for the automated detection and diagnosis of seizure using EEG signals. https://doi.org/10.1016/j.compbiomed.2017.09.017
17. Lasefr, Z., Shiva, S., Ayyalasomayajula, V.N.R., Elleithy, K.: Epilepsy Seizure Detection Using EEG signals
18. Zubair, M., et al.: Detection of Epileptic Seizures from EEG Signals by Combining Dimensionality Reduction Algorithms with Machine Learning Models. https://doi.org/10.1109/JSEN.2021.3077578
19. Rashed-Al-Mahfuz, Md., Moni, M.A., Uddin, S., Alyamia, S.A.: Deep Convolutional Neural Network Method to Detect Seizures and Characteristic Frequencies Using Epileptic Electroencephalogram (EEG) Data. https://doi.org/10.1109/JTEHM.2021.3050925

ASDS: Ontology Synthesis for Space Science as a Strategic Domain Using Semantics Oriented AI

Haiya Shah[1], Gerard Deepak[1(✉)], and A. Santhanavijayan[2]

[1] Department of Computer Science and Engineering, Manipal Institute of Technology Bengaluru, Manipal Academy of Higher Education, Manipal, India
gerard.deepak.christuni@gmail.com

[2] Department of Computer Science and Engineering, National Institute of Technology, Tiruchirappalli, Tiruchirappalli, India

Abstract. In response to the burgeoning data sets dedicated to space science within the web 3.0 framework, this paper addresses the need for a strategically devised semi-automatic ontology generation model. This intricate framework focuses on extracting data from datasets dedicated to space science, meticulously classifying it using a deep learning transformer-based classifier and utilizing the top 10% of classified instances as a federated feature for another auxiliary classifier, XGBoost. The model seamlessly integrates knowledge resources beyond metadata, drawing from semantic wikis like Google and Wikidata. Employing the encoding similarity index with empirically determined thresholds and the Flower Pollination Algorithm as the selected metaheuristics technique, the model generates optimal solution terms. Achieving an impressive overall performance characterized by a notable recall and low False Discovery Rate (FDR), this work establishes the model as a best-in-class solution for ontology generation, particularly within the space science domain in the web 3.0 context.

Keywords: Ontology Synthesis · Space Science · Web 3.0 · Semantic Artificial Intelligence · Knowledge Resources · Deep Learning

1 Introduction

In an age where the digital universe expands at an unprecedented pace, the domain of space science stands as a treasure trove of knowledge within the ever-evolving web 3.0 framework. The sheer volume of data available within this domain underscores the critical need for a refined, semi-automatic approach to ontology synthesis. This paper encapsulates a diligent exploration into the nuanced world of ontology generation, carefully calibrated to harness the potential of web 3.0's knowledge abundance, driven by a semantic knowledge-oriented paradigm.

At the heart of this endeavor lies a meticulous data extraction process, curated from a rich tapestry of documents dedicated to space science. This extensive dataset is subjected to the discerning scrutiny of a deep learning transformer-based classifier, a cornerstone of

V. Sakalauskas et al. (Eds.): IBICA 2023, LNNS 1231, pp. 298–306, 2025.
https://doi.org/10.1007/978-3-031-78946-5_29

our model. Notably, the model orchestrates a seamless flow of knowledge, as the top 10% of classified instances are judiciously integrated into an auxiliary classifier, XGBoost, through the prism of cross-entropy, bestowing a unique dynamic upon the process. The potency of this model is further amplified by the dexterity of the encoding similarity index, finely calibrated to empirically derived thresholds, and the artful deployment of the Flower Pollination Algorithm as the chosen metaheuristics technique.

As the capstone of this intricate process, our model bequeaths optimal solution terms, which undergo an exhaustive review and meticulous refinement, culminating in their realization as comprehensive ontologies. This tireless dedication to precision has yielded remarkable outcomes, showcased by a remarkable recall and a notably low False Discovery Rate (FDR), cementing our model's position as a pioneering solution for ontology synthesis. It particularly shines in the context of the space science domain within the expansive tapestry of web 3.0, affirming the essence of our research endeavor.

Motivation: The core motivation behind our research is to advance the field of space science ontology synthesis through the strategic integration of semantic artificial intelligence. Our contributions encompass the development of a comprehensive model designed to streamline the generation of ontologies. This model strategically leverages structured metadata derived from carefully curated datasets, effectively extracting and categorizing information. The heart of our approach lies in the use of the Transformer classifier, a powerful deep learning tool, to meticulously classify the dataset. The top 10% of classified instances, once refined through cross-entropy-based selection, serve as a federated feature for the XGBoost classifier, renowned for its feature control capabilities, enabling the ultimate classification of the dataset. Our model seamlessly integrates a wealth of auxiliary knowledge from Google and Wikidata, further enriching the data with community contributions and verifications. The model excels in semantic relatedness computation through an empirically determined threshold using a similarity measure, and the integration of the Flower Pollination Algorithm, a metaheuristics driven optimization mechanism, positions it as a best-in-class solution, yielding highly optimal solution terms from initial inputs. The entire process culminates in a meticulous and reasoned final ontology. Our research establishes the model as a premier solution for ontology generation, addressing the essential need for speed and precision within the domain of space science ontology synthesis.

Contributions: Our work introduces a substantial contribution to the field of ontology synthesis within the domain of space science. At its core, we present a comprehensive framework, meticulously designed for the semi-automatic modeling and generation of ontologies. This framework notably incorporates a fundamental feature, material generation, drawn from an extensive dataset of documents exclusive to space science. The pivotal classification process of the dataset is executed by the XGBoost classifier, which effectively selects features from metadata, classified using the Transformers classifier and refined through cross-entropy. Moreover, our work underscores the integration of Wikidata and Google's Knowledge Graph API as crucial resources for knowledge synthesis and derivation. These entities harmoniously converge to enrich the structural metadata of the web 3.0, thereby endowing our framework with a semantics-oriented, cognitive foundation. Notably, our contribution excels in semantic reasoning, made possible through the computation of encoding similarity, further enriched by the innovative

Flower Pollination Algorithm for metaheuristics-driven optimization. This transformative process elevates initial solutions to highly optimal solution sets, culminating in the creation of ontologies that represent real entities and undergo rigorous activities and reasoning. Ultimately, our work culminates in the development of best-in-class ontologies tailored for the domain of space science, symbolizing a significant stride in the field of ontology synthesis.

Organization: This paper is organized as follows. Section 2 provides an overview of the Literature Survey. Section 3 presents the Proposed System Architecture. Implementation is provided in Section 4. Results and Performance Evaluation and discussed in section 5. This paper is concluded in section 6.

2 Literature Survey

Tijerino et al. [1] introduced a novel approach "Towards ontology generation from tables," focusing on extracting ontological information from tabular data. They aimed to create ontologies from structured data, providing valuable insights into information representation. Wang et al. [2] presented a formal ontology generation method employing deep machine learning, discussed in their paper "Formal ontology generation by deep machine learning." This paper emphasized the use of advanced machine learning techniques for ontology creation. Pivk [3] explored "Automatic ontology generation from web tabular structures" in AI Communications. This work highlighted automated methods to construct ontologies from web-based tabular data, aiming to facilitate information retrieval and knowledge representation. Lonsdale et al. [4] investigated the topic of "Toward Ontology Generation from Tables," offering insights into the generation of ontologies from tabular data. The paper discussed methodologies and challenges in this domain.

Kineman [5] introduced "Relational science: a synthesis" in Axiomathes, focusing on the development of ontologies to represent relationships and connections in various domains, with potential implications for ontology creation. Portelli et al. [6] delved into "Ontology development for knowledge-driven distributed space mission systems engineering" in the context of space missions. This work highlighted ontology development as a key component in space engineering and decision support systems.

Kureichik and Safronenkova [7] discussed the creation of an "Ontology-Based Decision Support System for the Choice of Problem-Solving Procedure of Commutation Circuit Partitioning." Their paper demonstrated how ontologies can be used to support decision-making processes. Villa et al. [8] presented "Modelling with knowledge: A review of emerging semantic approaches to environmental modelling." The authors discussed semantic approaches in environmental modeling and the role of ontologies in representing domain knowledge.

Deepak et al. [9] introduced "An intelligent system for modeling and evaluation of domain ontologies for Crystallography as a prospective domain with a focus on their retrieval." The paper emphasized the importance of domain-specific ontologies in knowledge retrieval. Sim and Duffy [10] explored "Towards an ontology of generic engineering design activities" in Research in Engineering Design. The authors discussed the development of an ontology for engineering design activities, emphasizing its potential applications in design processes. Gruber and Olsen [11] presented "An ontology for engineering mathematics" in the context of knowledge representation. This work discussed the creation of an ontology for engineering mathematics, focusing on mathematical knowledge. Smiraglia [12] addressed "Domain analysis for knowledge organization: tools for ontology extraction." This paper explored tools and methodologies for extracting ontologies from domain-specific knowledge structures, contributing to the field of knowledge organization. Deepak et al., [13–15] have demonstrated several ontology based frameworks that are semantic in nature.

3 Proposed System Architecture

In the pursuit of ontology synthesis tailored to the realm of space science, we present the system architecture outlined in Figure 1. Our approach involves the analysis of a collection of documents, transforming them into a categorized dataset, with the aim of extracting meaningful information from this specialized domain. To this end, we employ the Term Frequency-Inverse Document Frequency (TF-IDF) method to select and prioritize informative terms. TF-IDF measures the importance of a term in a document relative to its occurrence across the entire document corpus. The initial term pool, generated through TF-IDF, is subjected to metadata generation. This metadata not only enhances the auxiliary knowledge but also serves to bridge the cognitive gap between existing knowledge structures on the World Wide Web and the domain specific knowledge of space science. The generation of metadata is facilitated by a tool known as "MarkEdit."

Following metadata generation, the metadata, often voluminous and directly derived from the web's structure, necessitates classification before it can be effectively utilized. For this purpose, we employ transformer models to classify the metadata into distinct categories. The top 10% of classified metadata, along with their corresponding classes, are subjected to the computation of cross-entropy, a measure for feature selection. The selected features are then passed to the Exabit classifier, a robust yet lightweight machine learning classifier.

The rationale for using a powerful machine learning classifier is rooted in its feature control capabilities, coupled with its efficiency when compared to deep learning models. Furthermore, the utilization of a strong classifier helps conserve computational resources during the initial classification of the dataset and mitigates domain divergence. Upon the classification by the Exabit classifier, the classified instances undergo appraisal for domain divergence. Subsequently, these instances are integrated into the Google Knowledge Graph using the Knowledge Graph API. This Knowledge Graph serves as a repository of community-verified knowledge, aggregating information from diverse sources into a single comprehensive graph.

Entities that emerge from the Knowledge Graph and subgraphs are further processed within the Wikidata pipeline. This stage enables the harvesting of community contributed, community-verified, and open-linked metadata, enriching the ontology with hierarchical knowledge. Given the magnitude of data from various sources, we evaluate semantic similarity using the Cosine Similarity measure. However, the encoding similarity index, even when a high threshold is set (e.g., 0.80), may not be sufficient. To

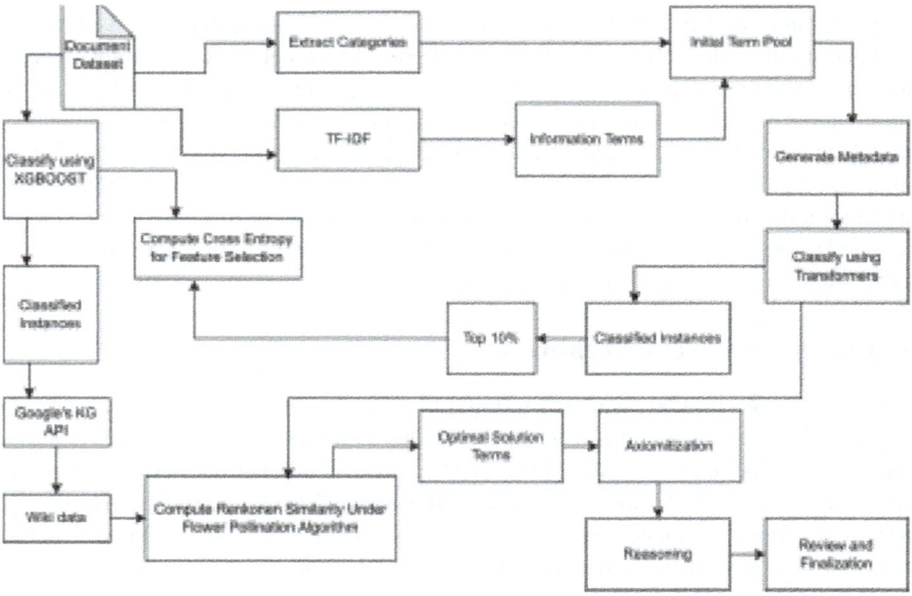

Fig. 1. Proposed System Architecture for Ontology Synthesis in Space Science

address this, we introduce a metaheuristic optimization model, specifically, the Flower Pollination Algorithm. This innovative algorithm leverages the concept of the flower pollination process to optimize the encoding similarity index with the objective function set at 0.80 as the threshold.

The classified instances resulting from this optimization process contribute to the optimal solution, which is further organized in ascending order of the encoding similarity index. Basic rules are then injected for the purpose of harmonization. In summary, our proposed system architecture encompasses a comprehensive approach to ontology synthesis in the context of space science, utilizing a combination of powerful classifiers, metadata generation, and semantic similarity measures to enrich the knowledge base with community-verified and open-linked data.

4 Implementation

The implementation was conducted within the Python Collaboratory as the chosen Integrated Development Environment (IDE). The implementation agents were not only geared for reasoning but also for data accumulation. The agent was meticulously designed utilizing the JAD TFIDF model, further enhancing the model's capabilities through a multi-agent setup. The Google Knowledge Graph API played a pivotal role in the implementation and was accessed through standard APIs. Transformer configurations were established using the Keras platform, while XGBoost was effectively implemented using its Python libraries. The National NLP task was effectively executed using Python's Natural Language Toolkit, further enhancing the robustness of the implementation process.

The implementation phase of this research encompassed four distinct datasets, specifically the National Space Science Data Center Master Catalog Data Set 2018, the 2022 Impact Factors of Astrophysics and Space Science Dataset, the Work with Data 2000 Space Science Dataset, and the 2022 Work with Data Space Science Exploration Policy Dataset. These datasets were systematically integrated based on their categories and annotations, facilitated by a customized annotator. Additional annotations were thoughtfully incorporated into the datasets, and the merging process was meticulously executed, considering categories, annotations, and data flow, resulting in the formation of a comprehensive, unified dataset through the utilization of a customized crawler. Additional entities were judiciously added to expand the dataset, enabling a more comprehensive and expansive experimentation process.

5 Results and Performance Evaluation

In our endeavor to assess the efficacy of our proposed Information Space Domains (ISDs) framework in the domain of space science ontology synthesis, we conducted a rigorous analysis, the outcomes of which are expounded upon below. The distribution curve of recommendations, as presented in Figure 2, offers a visual representation of our proposed ISDs framework's performance, juxtaposed against the baseline models. We've opted for a graphical depiction to complement the data provided in Table 2. This representation emphasizes the framework's superiority over the baseline models, further substantiating its effectiveness.

One compelling observation stems from the hierarchy inherent within the ontology. This hierarchical structure assigns the utmost significance to the highest level, with descending tiers progressively diminishing the importance of elements. Essentially, this hierarchy stratification ranks concepts based on their level of significance, with the most pivotal concepts situated at the uppermost echelons. Furthermore, it's worth noting that the lower hierarchy levels tend to encompass concepts of comparatively lesser importance. The hierarchical arrangement is driven by specific criteria, with typically two or three negative aspects leading to an element's relegation to the lower hierarchy levels.

This hierarchical organization collectively defines the ISDs framework as a best-in class model for ontology synthesis in space science, effectively amalgamating the positive attributes of elements across various hierarchy levels.

In order to quantitatively assess the performance of our ISDs framework, we employed a range of evaluation metrics, encompassing precision, recall, accuracy, and the false discovery rate (FDR). The comprehensive metrics evaluation, detailed in Table 1, sheds light on the framework's effectiveness in ontology synthesis. Furthermore, we conducted a comparative analysis between our ISDs framework and three distinct baseline models, as detailed in Table 1. This comparison offers a foundational understanding of the framework's performance when contrasted with other models. Table 2, representing 'Precision Percentage vs No. of Recommendations,' supplements our quantitative evaluation, shedding light on the framework's precision under varying recommendation scenarios.

Table 1. Comparison of Performance of the proposed ASDS with other approaches

Model	Average Precision %	Average Recall %	Average Accuracy %	Average F Measure %	FDR
TOGT [1]	88.15	91.43	89.79	89.76	0.12
FOGD [2]	90.77	92.62	91.70	91.68	0.10
AOGW [3]	92.16	94.09	93.13	93.11	0.08
Proposed ASDS	**95.18**	**97.09**	**96.14**	**96.12**	**0.05**

While acknowledging certain limitations, our ISDs framework excels in the structured organization of existing knowledge from web-based tabular structures into ontological schemas. It is important to note that the focus remains on organizing pre-existing data rather than generating new knowledge. Furthermore, the model demonstrates a strong inclination towards semantics and knowledge integration. Nevertheless, it still faces challenges in effectively integrating diverse, heterogeneous auxiliary knowledge, resulting in a cognitive gap.

Comparatively, the baseline models, as Table 1 highlights, lack the capability to synthesize new knowledge, emphasizing structured representations of readily available data. Their dependence on conceptual modeling techniques may impede their semantic comprehension and adaptability to new knowledge sources.

Table 2. Precision Percentage vs No. of Recommendations

No. of Recommendations	TOGT	FOGD	AOGW	Proposed ASDS
10	90.47	92.42	94.08	**97.09**
20	89.45	91.45	93.47	**96.08**
30	88.15	90.09	92.09	**95.54**
40	87.49	89.15	91.44	**94.47**
50	86.43	88.11	90.48	**93.08**

To summarize, our ISDs framework, tailored for ontology synthesis in the field of space science, unquestionably outperforms all baseline models, providing a comprehensive and effective solution with a central focus on semantics and knowledge integration. The significance of our model is underscored by its ability to leverage diverse knowledge sources, expand data sets, classify metadata through a transformer classifier, and employ federated learning for enhanced classification accuracy. Furthermore, the framework effectively harnesses Google's Knowledge Graph API and Wikidata, enriching the knowledge base.

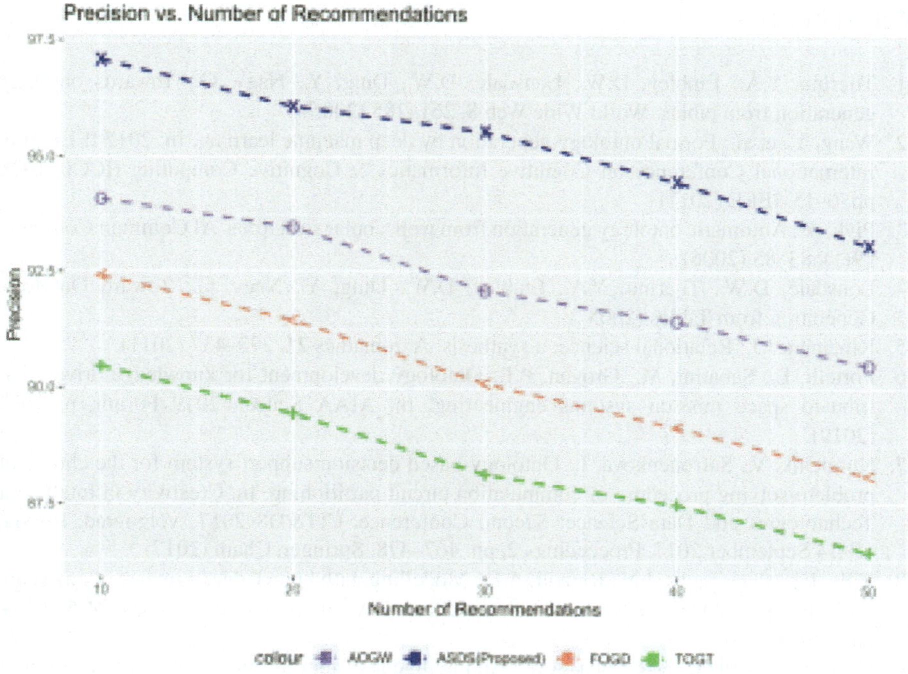

Fig. 2. Precision Percentage Vs Number of Recommendations

Notably, the model excels in semantic reasoning, driven by the computation of semantic similarity. The integration of the Flower Pollination Algorithm for optimization is a cornerstone of its performance. Within the realm of space science, our ISDs framework stands as a pioneering solution, emerging as one of the first semi-automatic models for generating ontologies. It not only performs admirably but also holds the potential to transform the field through advanced information synthesis techniques.

6 Conclusion

An Intelligent Integrative approach for Knowledge Centric web page recommendation has been proposed. The IISWS system initially recommends diversified queries and then the web pages based on the knowledge rendered by the heterogenous SPARQL Endpoints. The IISWS intelligently encompasses an initial classification of the dataset in order to choose the top 25% of each of the classification, which are further converted into RDF which is further interlinked with real-world knowledge stores through the Multisource SPARQL Endpoints based on selective domain integration. The dynamic generation of the query relevant entities, synonymization, and an efficient semantic similarity-based recommendation of the webpages makes IISWS as the best in class web page recommendation system. Moreover, the enrichment of knowledge into the system ensures diversified and yet query relevant recommendations. An overall accuracy of 0.91 with a very low FDR of 0.1 and a nDGD of 0.91 has been achieved by the proposed IISWS Framework for the WebKB Corpus dataset.

References

1. Tijerino, Y.A., Embley, D.W., Lonsdale, D.W., Ding, Y., Nagy, G.: Towards ontology generation from tables. World Wide Web **8**, 261–285 (2005)
2. Wang, Y., et al.: Formal ontology generation by deep machine learning. In: 2017 IEEE 16th International Conference on Cognitive Informatics & Cognitive Computing (ICCI* CC), pp. 6–15. IEEE (2017)
3. Pivk, A.: Automatic ontology generation from web tabular structures. AI Commun.Commun. **19**(1), 83–85 (2006)
4. Lonsdale, D.W., Tijerino, Y.A., Embley, D.W., Ding, Y., Nagy, G.: Toward Ontology Generation from Tables (2005)
5. Kineman, J.J.: Relational science: a synthesis. Axiomathes **21**, 393–437 (2011)
6. Portelli, L., Sabatini, M., Grogan, P.T.: Ontology development for knowledge driven distributed space mission systems engineering. In: AIAA Scitech 2019 Forum, p. 1032 (2019)
7. Kureichik, V., Safronenkova, I.: Ontology-based decision support system for the choice of problem-solving procedure of commutation circuit partitioning. In: Creativity in Intelligent Technologies and Data Science: Second Conference, CIT&DS 2017, Volgograd, Russia, 12–14 September 2017, Proceedings 2, pp. 467–478. Springer, Cham (2017)
8. Villa, F., Athanasiadis, I.N., Rizzoli, A.E.: Modelling with knowledge: a review of emerging semantic approaches to environmental modelling. Environ Model Softw.Softw. **24**(5), 577–587 (2009)
9. Deepak, G., Gulzar, Z., Leema, A.A.: An intelligent system for modeling and evaluation of domain ontologies for Crystallography as a prospective domain with a focus on their retrieval. Comput. Electr. Eng.. Electr. Eng. **96**, 107604 (2021)
10. Sim, S.K., Duffy, A.H.: Towards an ontology of generic engineering design activities. Res. Eng. Des. **14**, 200–223 (2003)
11. Gruber, T.R., Olsen, G.R.: An ontology for engineering mathematics. In: Principles of Knowledge Representation and Reasoning, pp. 258–269. Morgan Kaufmann (1994)
12. Smiraglia, R.: Domain Analysis for Knowledge Organization: Tools for Ontology Extraction. Chandos Publishing (2015)
13. Gerard, D., et al.: Bi-LDR: a bi-classification model for legal document recommendation using knowledge synthesis approach. NeuroQuantology **20**(12), 321 (2022)
14. Adithya, V., Deepak, G., Santhanavijayan, A.: OntoIntAIC: an approach for ontology integration using artificially intelligent cloud. In: Advances in Data Computing, Communication and Security: Proceedings of I3CS2021, pp. 3–13. Springer, Singapore (2022)
15. Manoj, N., Deepak, G.: ODFWR: an ontology driven framework for web service recommendation. In: Data Science and Security: Proceedings of IDSCS 2021, pp. 150–158. Springer, Singapore (2021)

A Novel Signature Based Reconnaissance Attack Detection and Threat Identification System

T. Raj Kumar[1(✉)], M. C. Aswathy[1], and N. V. Sobhana[2]

[1] College of Engineering Kallooppara, Thiruvalla, Kerala, India
{rajkumar,aswathy}@cek.ac.in
[2] Rajiv Gandhi Institute of Technology, Kottayam, Kerala, India
sobhana@rit.ac.in

Abstract. In the dynamic landscape of cybersecurity, detecting reconnaissance attacks at an early stage is critical in preventing potential security breaches. This paper presents a comprehensive approach to identify reconnaissance activities through a signature-based detection system. The system employs a combination of network and Windows OS system information gathering techniques to discern patterns indicative of reconnaissance attacks. In order to obtain relevant data, methods including port scanning, packet sniffing, DNS analysis, and WHOIS lookups are laid out in the section on network information collection. In a similar vein, the Windows system information collection section describes how Registry analysis and Windows Management Instrumentation (WMI) are used to collect system-related data. The development and use of detection mechanisms based on signatures creation and comparison forms the basis of the proposed system. Signatures are carefully constructed patterns that represent characteristics of reconnaissance activities. By using signature-based techniques, this approach provides a comprehensive strategy for detecting reconnaissance attacks and identifying threats. The study also showed a noteworthy advancement in improving cybersecurity procedures.

Keywords: Cyber security · Reconnaissance attacks · Signatures · Information gathering · threat detection

1 Introduction

In the rapidly evolving realm of cybersecurity, reconnaissance attacks stand as the quiet but crucial precursor to more sinister breaches. These attacks, often unnoticed, involve the systematic gathering of information about a target system, network, or organization. As the initial step in the cyber-attack lifecycle, reconnaissance [9] provides assailants with critical insights that facilitate subsequent stages, such as exploitation and infiltration. The ability to detect reconnaissance activities is paramount in safeguarding digital assets, pre-emptively thwarting potential breaches, and fortifying overall cybersecurity defenses. Detecting reconnaissance attacks holds immense significance for several reasons. The most important one is Early Threat Mitigation. Reconnaissance attacks are

V. Sakalauskas et al. (Eds.): IBICA 2023, LNNS 1231, pp. 307–318, 2025.
https://doi.org/10.1007/978-3-031-78946-5_30

typically the initial steps in a cyber-attack. Reconnaissance activities often involve identifying system vulnerabilities and weaknesses. Detecting such activities allows organizations to proactively patch vulnerabilities and bolster their security posture. Recognizing reconnaissance attempts aids in tracing the origins of potential threats. This knowledge is essential for law enforcement and targeted response efforts. By identifying reconnaissance, organizations can prevent attackers from gathering the necessary intelligence to mount a successful attack. This reduces the potential impact of a breach. Detecting reconnaissance provides invaluable information for incident response teams, allowing them to take swift and informed actions to contain and mitigate potential threats.

Existing Network Intrusion Detection (NID) Techniques look for anomalies in traffic that suggest that malicious activity has occurred, or they match patterns of known assaults (also known as abuse detection) [2]. There are basically two forms of network reconnaissance, passive and active [1, 7, 13]. Gathering network data indirectly or directly, without explicitly probing the target network, is known as passive reconnaissance. Using publicly accessible datasets and internet domain registration services are examples of indirect passive strategies.

The landscape of reconnaissance attack detection has witnessed significant advancements driven by the escalating threat landscape. Researchers and practitioners have explored diverse methodologies and technologies to identify these subtle but critical activities. Existing work encompasses signature-based detection, anomaly detection, and machine learning-based techniques. Signature-based methods involve identifying patterns [11] associated with reconnaissance activities, much like fingerprints [14]. Anomaly detection seeks to identify deviations from normal behavior, flagging suspicious activities. Machine learning leverages data patterns to predict and detect reconnaissance attempts. There is no accepted standard for classifying risks and attacks on computers, despite the efforts of numerous academics to divide them into relevant categories [8, 15, 17]. The K. Kendall [6] taxonomy of attacks is the most commonly utilized in the research communities. This classification categorizes the computer attacks into DOS Attacks, User to Root, Remote to User and probes attacks. Furthermore tools like network scanners, packet analyzers and specialized frameworks also aid in gathering and analyzing information related to reconnaissance.

This paper dives into the heart of reconnaissance attack detection using signature-based method by analyzing the network traffic. Also the proposed work acts as a tool for gathering information from both the system and the network. Additionally, the proposed system provides a user interface through which the user can monitor the system and network at the same time.

Section 2 discusses the back ground and related work. The Methodology used in the proposed system is described in Sect. 3. Section 4 describes the native language inheritance. The proposed work is discussed in Sect. 5. The workflow of the proposed system is described in Sect. 6. The Sect. 7 includes the Results and Discussions. Performance evaluation is exhibited in Sect. 8. The Sect. 9 discusses the conclusion and future scope of this system.

2 Background and Related Works

Researchers and practitioners proposed different approaches for detecting reconnaissance attacks. William H. Allen, et al. [1] proposed an approach to detect the malicious reconnaissance automatically by using a tool Xprobe2. Xprobe2 [3, 4] is an ICMP reconnaissance tool which detects the ICMP packets. This approach involves the following sequence.

1. Record the network traffic produced by a particular reconnaissance tool.
2. Examine the intercepted traffic to identify important identifiers.
3. Create the NIDS rules to detect the malicious reconnaissance traffic.
4. Verify those guidelines by testing in an active network setting.

This approach mainly focuses on the ICMP attacks and how the Snort rules can be used to detect ICMP attacks.

Ahana Roy et al. proposed a network based passive information gathering tool using Java platform [9]. The tool is used for identifying the organizational vulnerabilities. This tool is used for gathering passive information only with the help of Google Cashe system. Muhammad Qasim Ali [11] proposed a Firewall Policy reconnaissance techniques which reconstructed the firewall policy by intelligently choosing the probing packets based on the responses of previous probes. This proposed a hybrid approach including scanning and probing.

Omar Al-Jerrah et al. [12] proposed a Network Intrusion Detection System to recognize network attacks by embedding the temporal behavior of the attack into TDNN Neural Network structure. Siti Nur Hidayah et al. [13] conducted a social engineering case study on passive and active reconnaissance which describes the social media and social engineering techniques used for reconnaissance.

3 Methodology

This paper uses an approach that employs tools and modules for network intelligence collection and reconnaissance. The work comprises of multiple crucial stages.

1. **Requirements Analysis**: Determining the precise network information required and the objectives of the reconnaissance was the first step. This step involves figuring out what data needs to be collected, such as open ports, DNS records, WHOIS information, and network traffic.
2. **Selection of Tools:** The Python programming language's tools were utilized. Based on the needs, the right libraries and tools are chosen for each data gathering task. Nmap, dnspython and scapy, are selected for port scanning, DNS analysis, and packet sniffing respectively.
3. **Code Development:** Scripts are utilized to implement every method of acquiring information. This entails writing code for port, DNS, and packet sniffing operations. Classes and functions are used in the code to provide modularity and maintainability. This integration makes it possible to handle data collection centrally and offers an easy-to-use interface for carrying out reconnaissance operations.

4. **User Interface:** To make the tool user-friendly, appropriate GUI package (such as Tkinter) is used to construct a graphical user interface (GUI). The GUI offers choices for entering target IP addresses or domains, choosing particular information collection methods, and presenting the outcomes.
5. **Testing and Validation:** Extensive testing is carried out to guarantee the precision and dependability of the data collection methods. Testing against different network configurations and situations is a part of this process to find and fix any problems or constraints.

Updates may include adding new features, supporting additional protocols, or addressing security vulnerabilities.

The advantages of the proposed system over the existing system is summarized as

1. A user interface for collecting information from both the system and network.
2. Helps the user to analyze himself if there is any vulnerabilities in the system by examining the ports as well as the registry
3. The system can be used as a packet sniffing tool to analyses the network traffic to find any malicious activity occurrences.
4. DNS analysis.
5. Signature based reconnaissance attack detection and alert generation.

4 Native Language Feature Inheritance

The work makes use of Python programming's rich features. This work on network information collection and reconnaissance heavily rely on Python and its libraries.

Utilized feature sets include:

User-Friendly Interfaces: The native languages GUI libraries, such as Tkinter and PyQt, enable the creation of user-friendly graphical interfaces for network reconnaissance tools. This makes it accessible to a wider audience, including non-technical users.
Scapy: This powerful library enables packet manipulation, capture, and network analysis. It also allows to craft custom network packets, capture network traffic, and perform various network-related tasks.
Dnspython: This library provided DNS (Domain Name System) functionality. It allows to perform DNS queries, resolve domain names, and retrieve DNS-related information.
whois: whois library helped in retrieving WHOIS information for domain names. It enabled to query WHOIS servers to obtain domain registration details, including domain ownership and contact information.
Nmap: This library allows to integrate Nmap functionality into the used scripts. This versatile network scanning tool helped in host discovery, port scanning, and service enumeration.
PCAP (pycap): This library provides access to PCAP (Packet Capture) files, which store network packet captures. It enables to read and process PCAP files generated during network packet captures.

5 Proposed Work

Two sections are created for the accomplishment of this work. INFORMATION GATHERING and ATTACK DETECTION (Fig. 1).

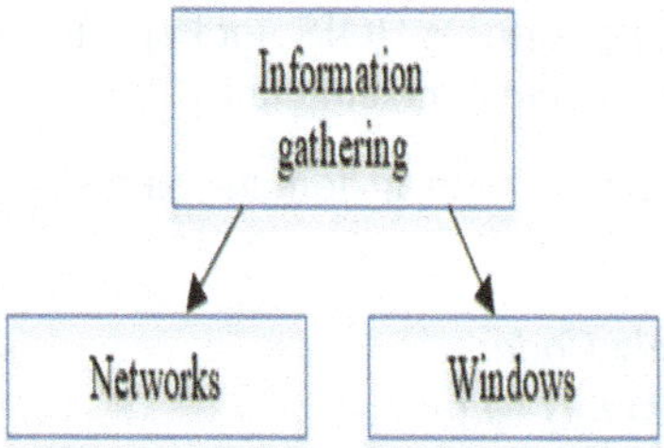

Fig. 1. Collection of Information

5.1 Information Gathering Module

5.1.1 Network Information Gathering

5.1.1.1 Port Scanning

Description: Port scanning is the process of probing a network host for open ports to identify services running on those ports.

Library: The nmap library is used to perform comprehensive port scans.

Usage: Specify target hosts or IP ranges, and nmap will scan them for open ports, determine the service running on each open port, and provide additional information like service version.

Sample Code:

```
import nmap
nm = nmap.PortScanner() nm.scan(hosts="192.168.1.1", arguments="-p
1- 1024") print(nm.all_hosts())
```

5.1.1.2 DNS Analysis

Description: DNS (Domain Name System) analysis involves querying DNS servers to retrieve information about domains, such as IP addresses (A records), mail servers (MX records), and text records (TXT records). *Library:*The dnspython library used to perform DNS queries programmatically.

Usage: Resolve domain names and retrieve different types of DNS records.

Sample Code:

```
import dns.resolver
resolver = dns.resolver.Resolver()
answers = resolver.query("example.com", "A") for rdata in aswers:
print(f"IP Address: {rdata.address}")
```

5.1.1.3 WHOIS Lookups

Description: WHOIS is a protocol for querying databases that store registration information about domains and IP addresses.

Library: Who is used to perform WHOIS lookups to retrieve domain ownership information, registration dates, and contact details.

Usage: Query a WHOIS server with a domain name or IP address to get detailed registration information.

Sample Code:

```
import whois
domain=whois.whois("example.com")
print(domain)
```

5.1.1.4 Packet Sniffing

Description: Packet sniffing is the process of capturing and analyzing network traffic, allowing you to inspect packets for information such as source/destination IP, ports, and even payload data.

Library: The scapy library is a powerful tool for packet manipulation and analysis.

Usage: Sniff packets on a network interface, inspect packet headers, and even craft custom packets for network reconnaissance.

Sample Code:

```
import scapy.all as scapy
def packet_callback(packet):
if packet.haslayer(scapy.IP):
print(f"Source IP: {packet[scapy.IP].src},
Destination IP: {packet[scapy.IP].dst}")
scapy.sniff(iface="eth0", prn=packet_callback, count=10)
```

5.1.2 Windows Information Gathering

5.1.2.1 Windows Management Instrumentation (WMI)

Description: WMI is a powerful management technology for Windows systems. It provides access to a wide range of system data, such as hardware, software, processes, and more.

Library: The pywin32 library enables scripts to interact with WMI and retrieve system information.

Sample Code:

```
import win32com.client
wmi=win32com.client.GetObject("winmgmts:\\\\.\\root\\cimv2")
processes  = wmi.ExecQuery("SELECT * FROM  Win32_Process")
for process in processes:
print(f" ProcessName:{process.Name},PID:{process.ProcessId}")
```

5.1.2.2 Registry Analysis

Description: The Windows Registry contains critical system configuration and application settings. Analyzing registry keys and values can provide insights into system behavior and installed software.

Library: The reg module or third-party libraries like winreg can be used for registry interaction.

Sample Code:

```
import winreg
key_path = r"SOFTWARE\Microsoft\Windows\CurrentVersion\Uninstall" with
winreg.OpenKey(winreg.HKEY_LOCAL_MACHINE, key_path) as key:
for i in range(winreg.QueryInfoKey(key)[0]): subkey_name
  winreg.EnumKey(key, i) print(subkey_name)
```

5.1.2.3 File System Analysis

Inspecting the file system can uncover information about installed applications, user profiles, and potentially malicious files.

Python's built-in file and directory manipulation functions can be used for file system analysis.

Sample Code:

```
import os import pathlib
directory_path = "C:\\Program Files" for item I os.listdir(directory_path):
  item_path = os.path.join(directory_path, item) if os.path.isfile(item_path):
    print(f"File: {item}")
elif os.path.isdir(item_path): print(f"Directory:{item}")
```

By employing these methods, we can gather a wide array of information from both network and Windows systems, enabling you to detect potential reconnaissance activities and safeguard your digital assets effectively.

5.2 Attack Detection

5.2.1 Signature Based Attack Detection

This work encompasses a proactive cybersecurity strategy called signature-based detection that compares gathered data to predefined patterns or signatures linked to known reconnaissance attacks. It works on the premise that different activities leave unique traces that may be recognized by particular patterns. Security systems are able to recognize possible risks early in their lifespan by identifying these recognized scan patterns.

6 Workflow

The entire system follows a sequential workflow as detailed below

Signature Creation: Historical reconnaissance attacks are recognized by security experts and common characteristics, behaviors, and patterns exhibited by such attackers are identified. These patterns may include specific port scans, unusual DNS queries, or queries for sensitive system information.
Pattern Definition: The identified patterns are transformed into signatures, which are essentially digital fingerprints of reconnaissance activities. These signatures may include combinations of network behaviors, system interactions, or specific data queries.
Data Collection: The system collects data from network traffic, Windows systems, or other relevant sources. This data encompasses network communication, system configurations, and application interactions.
Comparison and Analysis: The collected data is compared against the predefined signatures. If a match is found, it indicates that the data exhibits characteristics consistent with known reconnaissance activities.
Alert Generation: When a match is detected, the system generates alerts or notifications. These alerts can be sent to security personnel, administrators, or automated response systems for further investigation and mitigation.

Sample Code:

```
# Predefined signatures for reconnaissance activities recon_signa-
tures =
{
"Port_Scan": ["SYN Scan", "NULL Scan", "XMAS Scan"], "Unu-
sual_DNS_Queries":["example.com","malicious.com"],
"WHOIS_Query":["malicious_domain.com", "suspicious_domain.net"],
"WMI_Process_Query":    ["SELECT    *    FROM Win32_Process
WHERE Name = 'cmd.exe'"],}
def compare_signatures(activity_list): alerts = []
  for activity in activity_list:
  for signature, patterns in recon_signatures.items(): for pattern in patterns:
        if pattern in activity:
          alerts.append(f"Suspicious activity    detected: {signature}")
          break return alerts
def raise_alerts(alerts): for alert in alerts:
    print(alert)
```

7 Results and Discussions

The network and Windows information in this work were gathered using a variety of tools and modules. The signature-based detection technique is utilized to distinguish between various attacks. An easy-to-use framework is put in place for collecting data so that anyone can examine and evaluate the outcome and quickly spot any harmful activities (Fig. 2).

The packet sniffing requires an in-depth knowledge about the various protocols. In packet sniffing the user can also limit the number of packets for analysis. The port

Fig. 2. Packet Sniffing

number along with the target IP address is inputted to check it's status. Port scanning gave the status of ports i.e. if any ports are open or closed (Figs. 3 and 4).

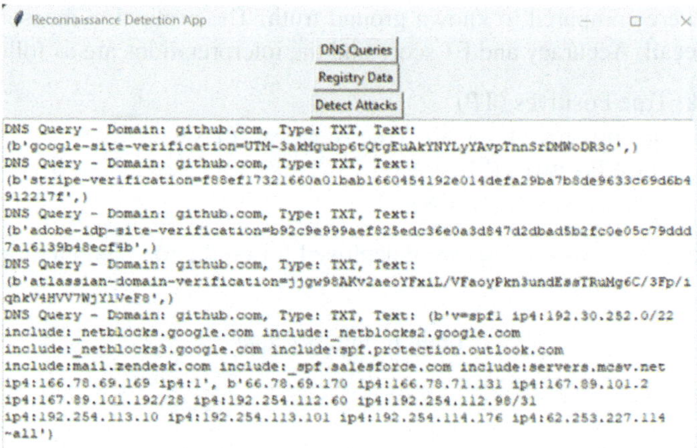

Fig. 3. DNS Query Result

In the attack detection module real time packets are captured and the signatures are compared with the known attack signatures. If any malicious behavior is found, an alarm is generated.

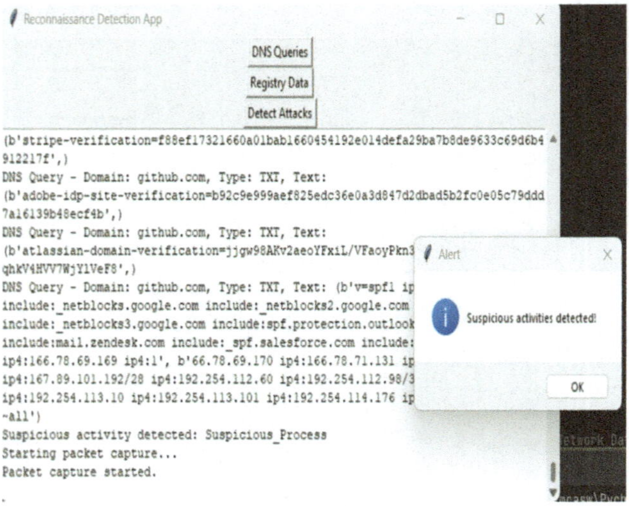

Fig. 4. Threat Detection

8 Performance Evaluation

The results were analyzed from the real world data samples (from network) and the predictions were compared to known ground truth. The evaluation measures used are Precision, Recall, Accuracy and F1 score and the interpretations are as follows.

Actual attack: True Positives (TP)
Legitimate traffic correctly classified: True Negatives (TN)
False Alarms: False Positives (FP)
Missed Attacks: False Negatives (FN)

The following infrastructures were deployed for prediction computation (Table 1).

Table 1. Infrastructure

Server	Client
Windows 11 Cloud Server	Windows 10
Victim (Norton Antivirus updated)	Attacker

Method: The server and client are connected to the network. Network data is analyzed. The ping attacks (ICMP Echo) generated by the client are identified and it is notified to the client IP address. Different attacks were hosted into the network under surveillance. This is done for validating the attack detection model. The experiment was conducted throughout a variety of time periods, and the following are the findings.

Table 2. Evaluation Findings and Performance Metrics Results

Total Samples taken	56350	**Performance Metrics Results**	
Attack Correctly Detected (TP)	3820	Precision	88.02%
Traffic Correctly Classified (TN)	51990	Recall	99.48%
False Alarms(FP)	520	F1 Score	93.40%
Missed Attacks (FN)	20	Accuracy	99.04%

The performance of the system is calculated based on the above data and the findings are given in Table 2.

Precision is computed as **True Positives/(True Positives + False Positives)**

A High value of precision indicates that the model makes positive predictions with a low rate of false alarms, meaning that when it predicts an attack, it is likely correct. In this work a high value is obtained (88.02%) indicates guaranteed prediction of an attack.

Recall is computed as **True Positives/(True Positives + False Negatives)**

High recall indicates that this model can effectively detect most of the actual positive instances, meaning it has a low rate of missed detections. Here a high value is obtained which further indicates the low rate of missed detections.

Accuracy: Accuracy is a measure of the overall correctness of a model's predictions. It calculates the ratio of correct predictions (both true positives and true negatives) to the total number of instances in the dataset.

Accuracy is computed as **(True Positives + True Negatives)**/Total Instances

F1 Score: The harmonic mean of recall and precision is the F1 score. It offers a fair evaluation of a model's effectiveness by taking into account both false positives.

F1 Score is computed as 2*(Precision*Recall)/(Precision + Recall)

High F1 score indicates that this model is achieving both high precision and high recall simultaneously.

9 Conclusion and Future Scope

Signature-based detection allows organizations to identify reconnaissance threat activities before they escalate into more significant attacks. It helps in preventing breaches by identifying potential threats at an early stage. Since this method relies on specific patterns, it can be efficient and accurate in detecting well-defined attacks. This work shows that signature-based detection is particularly effective against known threats and attack patterns. It provides a reliable defense against previously identified reconnaissance activities. Inspite of all the above benefits, it also imposes some challenges. This method is only effective against reconnaissance activities with known patterns. It may miss new or evolving attack patterns that deviate from existing signatures.

There may be chances of occurrences of False positives (legitimate activities identified as attacks) and false negatives (attacks not identified). These challenges require constant tuning and refinement of signature databases. Also Signature-based detection is less effective against zero-day attacks and sophisticated attackers who modify their tactics to avoid detection. In the current scenario, signature based detection serves as a valuable tool in a layered cybersecurity defense strategy. While it may not be the sole solution, it contributes to identifying and mitigating well-known reconnaissance attacks efficiently, allowing organizations to respond swiftly and protect their digital assets.

References

1. Allen, W.H., Marin, G.A., Rivera, L.A.: Automated detection of malicious reconnaissance to enhance network security. IEEE 0-7803-8865-8/05
2. Axelsson, S.: Research in intrusion detection systems: a survey. Technical report 98-17, Chalmers University of Technology (1998)
3. Arkin, O., Yarochkin, F., Kydyraliev, M.: The Present and Future of XProbe2-The Next Generation of Active Operating System Fingerprinting (2003). Sys-Security. www.sys-security.com
4. Arkin, O., Yarochkin, F.: XProbe2-A 'Fuzzy' Approach to Remote Active Operating System Fingerprinting (2003). Sys-Security. www.sys-security.com
5. Jones, A.K., Sielken, R.S.: Computer System Intrusion Detection: a Survey. University of Virginia, Computer Science Department Technical report, September 2000
6. Kendall, K.: A Database of Computer Attacks for the Evaluation of Intrusion Detection Systems. Masterthesis, Massachusets Institute of Technology (1999)
7. Zalewski, M.: Silence on the Wire: A Field Guide to Passive Reconnaissance and Indirect Attacks
8. Kazienko, P.D.: Intrusion Detection Systems (IDS) Part I - network intrusions; attack symptoms; IDS tasks; and IDS architecture. Accessed 2006
9. Roy, A., Majia, L., Helling, P.: Automation of Cyber-Reconnaissance-a Java based Open Source Tool for Information Gathering. IEEE 978-1-908320/93/3 (2017)
10. Lyon, G.F.: Nmap network scanning: The Official Nmap project guide to Network Discovery and Security Scanning
11. Ali, M.Q.: Firewall Policy Reconnaissance Techniques: Techniques and Analysis. IEEE 1556-6013 (2014)
12. Al-Jerrah, O., Arafat, A.: Network Intrusion Detection System using Attack Behavior Classification. IEEE 978–1–4799–3023–4/14
13. Zulkiffli, S.N.H., Zawawi, Md.N.A., Rahim, F.A.: Passive and active reconnaissance: a social engineering case study. In: 8th ICIMU, pp. 1–2. IEEE
14. Beecroft, A.J., Michael, J.B.: Passive Fingerprinting of Network Reconnaissance
15. Hollingworth, D.: Towards threat, attack and vulnerability taxonomies. In: 44th IFIP WG 10.4 Meeting. USA, June 2003

Blockchain Technology in Healthcare: Unifying Patient Medical Records - A Survey

Sudip Phuyal[1], Luís B. Elvas[1,2], João C. Ferreira[1,2,4](✉), and Rabindra Bista[3]

[1] Instituto Universitário de Lisboa (ISCTE-IUL), ISTAR, 1649-026 Lisbon, Portugal
sudip.phuyal@iscte-iul.pt, joam@himolde.no

[2] Inov Inesc Inovação—Instituto de Novas Tecnologias, 1000-029 Lisbon, Portugal

[3] Department of Computer Science and Engineering, Kathmandu University, Dhulikhel-4, Kavre, Nepal

[4] Faculty of Logistics, Molde University College, NO-6410 Molde, Norway

Abstract. In the era of digital transformation, healthcare faces a critical challenge in managing and centralizing patient medical records efficiently and securely. Traditional data management systems, including paper-based records and fragmented electronic health records (EHR) systems, have demonstrated limitations in terms of data security, interoperability, and patient privacy. Blockchain technology has become known as a possible solution to all of these problems, one that has an opportunity to completely transform the administration of healthcare data management. This paper presents a comprehensive exploration of the application of blockchain technology in healthcare, specifically focusing on centralizing patient medical records. Through a literature review and analysis of real-world initiatives, we examine the potential of blockchain to address the shortcomings of existing systems. Blockchain's key characteristics, such as decentralization, data immutability, and cryptographic security, offer a unique framework for enhancing data security, improving interoperability among healthcare providers, and empowering patients to take control of their health information. The literature review highlights previous research and case studies that illustrate how blockchain has been successfully implemented in healthcare settings. We discuss notable initiatives like MedRec and Medicalchain, showcasing their impact on data security, patient care coordination, and data sharing within the healthcare ecosystem. Consequently, this article offers insightful information about how blockchain technology could revolutionize the handling of healthcare data. It underscores the importance of blockchain as a solution to the challenges associated with centralizing patient medical records, ultimately contributing to improved healthcare outcomes, enhanced patient experiences, and the future evolution of healthcare delivery. This research serves as a resource for healthcare professionals, researchers, and policymakers seeking innovative solutions to enhance data management and patient-centric care in the ever-evolving healthcare landscape. Blockchain technology adoption in the healthcare industry has the ability to completely transform security of data and accessibility, resulting in a more effective and patient-centered healthcare system.

Keywords: Blockchain · Healthcare · Medical Records · Data Security · Interoperability · Patient-Centric Care · Electronic Health Records (EHR) · Decentralization

V. Sakalauskas et al. (Eds.): IBICA 2023, LNNS 1231, pp. 319–332, 2025.
https://doi.org/10.1007/978-3-031-78946-5_31

1 Introduction

The healthcare industry is on the brink of a profound transformation, driven by advancements in technology and a growing demand for secure and efficient healthcare services. One of the critical challenges facing healthcare systems worldwide is the management and accessibility of patient medical records [1]. The traditional paper-based and fragmented electronic health record (EHR) systems have proven to be inefficient, error-prone, and susceptible to security breaches [2].

In response to these challenges, blockchain technology has emerged as a promising solution that holds the potential to revolutionize healthcare data management. Blockchain offers a decentralized and immutable ledger that can securely centralize patient medical records, ensuring data integrity, privacy, and accessibility [3]. This paper explores the application of blockchain technology in healthcare for the centralization of patient medical records.

In this paper, we will delve into the core concepts of blockchain technology and its inherent characteristics that make it suitable for healthcare data management. We will discuss the current state of healthcare data management, highlighting the limitations of existing systems and the pressing need for a more secure and interoperable solution. Furthermore, we will examine real-world use cases and initiatives that have already leveraged blockchain to centralize patient medical records, showcasing the tangible benefits and challenges encountered. The objective of this paper is to provide a comprehensive overview of how blockchain technology can address the long-standing issues related to patient data management in healthcare. We analyze the potential impact of blockchain on data security, interoperability, and patient-centric care, ultimately contributing to improved healthcare outcomes and patient experiences.

As we navigate the evolving landscape of healthcare, it is crucial to explore innovative solutions that can empower patients, enhance healthcare provider efficiency, and promote data integrity. Blockchain technology stands at the forefront of this transformation, offering a decentralized, secure, and transparent approach to centralizing patient medical records. In the following sections, we will unravel the intricacies of blockchain's role in healthcare data management and its potential to reshape the future of healthcare delivery.

The healthcare industry is witnessing a paradigm shift in data management, driven by advancements in technology. Managing and centralizing patient medical records has been a long-standing challenge due to the limitations of traditional systems. This paper explores the potential of blockchain technology to revolutionize healthcare data management by centralizing patient medical records securely.

2 Methodology

2.1 Search Strategy and Inclusion Criteria

This literature analysis was done in October 2023, and the search strategy and inclusion criteria were created to guarantee that pertinent, current, English-language publications and reviews about blockchain in healthcare were chosen. The search was restricted to journal papers released in 2019–2023, and it used data from three reliable databases: Web of Science Core Collection, IEEEXplore, and Scopus. Duplicate articles were eliminated from the initial search results in order to preserve data integrity.

The search term was carefully selected to find articles that addressed blockchain in the context of electronic health records (EHR) or electronic medical records (EMR) with an emphasis on decentralization and interoperability in healthcare. The "Healthcare" industry was the population of interest, and the query added phrases like "Interoperability" and "Decentralization" as further filters to make sure the articles were relevant. "("Blockchain") AND ("Electronic Medical Record" or "EMR" or "Electronic Health Record" or "EHR") AND ("Healthcare") AND ("Interoperability" or "Decentralization")" was the last search query that was used. The objective of our search approach was to locate an extensive range of current, relevant publications about the relationship between blockchain technology and healthcare, with a particular emphasis on decentralization and interoperability.

2.2 Search Results

The specified query was applied to the aforementioned Core Collection databases, and 167 articles were found. Removing 59 duplicates, total 108 articles were screened. The chosen publications were then subjected to a systematic analysis using the PRISMA (Preferred Reporting Items for Systematic Reviews and Meta-Analyses) technique, as stated in reference [4]. PRISMA is a well-respected methodology that guarantees the transparent and methodical analysis of research papers in systematic reviews and meta-analyses. This technique is intended to assist researchers evaluate the quality and relevance of the papers they are examining by streamlining the review process and maintaining high standards. Figure 1 shows the PRISMA Workflow.

2.3 Data Extraction and Synthesis

Mendeley, Microsoft Excel, Scopus, and WoSCC web interfaces served as essential tools for managing and storing data pertinent to the articles during the systematic literature review conducted using the PRISMA method. This data encompassed various aspects, including article title, author, publication year, subject area, keywords, and abstract. Additionally, to facilitate comprehensive data analysis, a qualitative assessment was performed using these criteria.

3 State of the Art

The use of blockchain technology is evolving and emerging technology in healthcare industry due to its feature of data security and integrity, patient data privacy, interoperability, consent management, data exchange and sharing, reduced data silos, transparency and trustworthiness, decentralization and many more. The purpose of the research carried out in [3] is to make it possible for various blockchain networks to communicate with one another. Additionally, it has shown a proof-of-concept for Hyper Ledger Fabric-based data exchange across two separate blockchain networks. In the years since Ethereum and Hyperledger first appeared, more research has gone into creating an interoperability standard that works with a variety of networks. According to the research done in [5], there is a need for a general inter-blockchain communication protocol that may

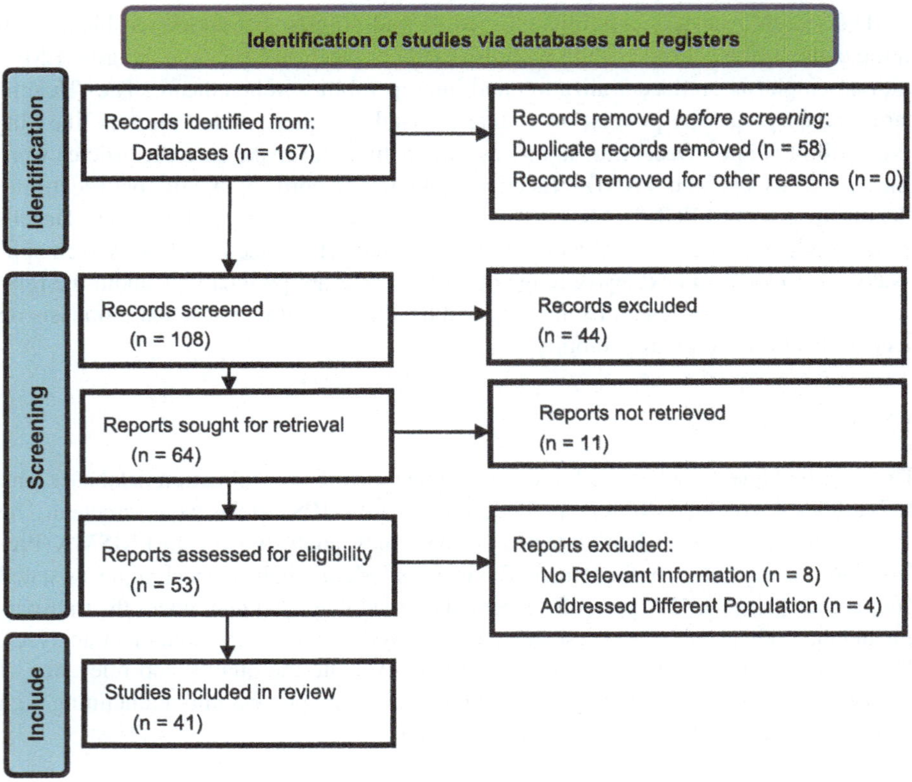

Fig. 1. PRISMA Workflow

securely and decentralized transfer arbitrary data across blockchains, including token exchanges and smart contract interactions. The article in ref [6] presents the appXchain solution for cross-chain interoperability to achieve real-world solutions for healthcare and patient medical records management. Ref [7] suggested managing the synchronization and interoperability of the decentralized e-Health system based on blockchain technology.

Using the public Ethereum blockchain as an interoperability layer across current vendor solutions, ref [8] offers an interoperable OpenPharma blockchain solution for blockchain-based healthcare applications. Doctors who receive the patient's encrypted authentication ID can access encrypted patient information URLs stored on the blockchain. Interoperability between various patient-centered solutions is not appropriate for the suggested solution. Patients and physicians are required to share an endless encrypted authentication ID, which is vulnerable to theft by any other member. Furthermore, for membership authentication and encryption, the architecture depends on centralized services like SAAVHA and Amazon Web Services - Key Management Service (AWS-KMS) [6, 8].

Blockchain technology is revolutionizing EHR administration with MedRec [9], a 2016 MIT invention. Patient records are safeguarded and decentralized, giving patients authority over data access [10, 11]. This patient-centered strategy makes use of blockchain's tamper-proof structure and encryption to improve privacy and guarantee data security [1, 12]. The unchangeable audit trail encourages openness and confidence among healthcare stakeholders.

Medicalchain, a different well-known platform, uses blockchain technology for patient privacy, EHR administration, and telemedicine [13]. It places a strong emphasis on patient autonomy, safe data exchange, and simple provider-patient communication [14]. It expedites medication monitoring and consultations by compiling medical histories into portable Health Passports, which is especially helpful for managing chronic diseases. Innovation and data exchange in healthcare are encouraged via integration with current systems.

Furthermore, DApp for Smart Health, a decentralized application for smart health is a particular kind of decentralized application that also utilizes blockchain technology to enhance and innovate healthcare services and data management. Patients can give data in exchange for compensation to some DApps that enable data monetization [15]. Healthcare procedures including insurance claims and visits are automated by smart contracts. DApps enable access to anonymised data, which promotes medical innovation and research.

4 Benefits of Blockchain in Healthcare

Blockchain's application to health data interoperability is extensive and has the potential to solve a number of significant problems facing the healthcare sector. It has a lot of potential applications in the healthcare industry, and it might completely change how healthcare data is exchanged, handled, and safeguarded [7]. There is a greater need than ever for safe, interoperable, and patient-centered solutions as healthcare becomes more and more dependent on digital data. The special qualities of blockchain, such as its capacity to produce tamper-proof records, provide patients control over their data, and expedite data transfer, are revolutionizing the healthcare industry [16]. We will explore the many benefits of blockchain technology in healthcare in this talk, including safe and compatible electronic health records (EHRs), data ownership and consent, data standardization, integrity, security, audit trails, cross-border data exchange, clinical trials, telemedicine, supply chain management, and regulatory compliance. When taken as a whole, these advantages offer increased privacy, security, and efficiency in healthcare data management, which can improve patient care and produce more dependable research results.

4.1 Secure and Interoperable EHRs

Blockchain transforms Electronic Health Records (EHRs) and secures patient data sharing, revolutionizing healthcare [17]. It addresses security and interoperability concerns with EHRs, guaranteeing confidentiality and integrity via encryption and tamper-proof records. With the use of this technology, platforms and healthcare providers may transfer data securely and seamlessly, which might improve patient care, operational effectiveness, and administrative ease in the industry [18, 19].

4.2 Patient Data Ownership and Consent

Blockchain gives people additional control over their health data. In contrast to traditional systems, it gives patients control over who may access their data and how. These guidelines are upheld via smart contracts, which guarantee that patient agreement is required for data access. This patient-centered approach greatly improves security and privacy while exchanging medical data [18, 20].

4.3 Data Standardization

Blockchain helps guaranteeing that medical records have uniform data formats and communication protocols, which makes it easier for healthcare systems to understand and share information. In the healthcare sector, this improves interoperability and streamlines data exchange and communication across many platforms [21].

4.4 Data Integrity and Security

Blockchain improves accuracy and dependability in healthcare data by establishing an immutable record that guarantees data integrity. By encrypting data and spreading it among several nodes, it improves security by lowering vulnerability to intrusions. This two-pronged approach greatly strengthens patient data security and promotes confidence in medical privacy [22].

4.5 Health Data Exchange

Blockchain transforms the secure and efficient exchange of data amongst stakeholders in the healthcare industry, including patients, providers, insurers, and researchers. Record exchange, billing accuracy, and safe research access are streamlined by its decentralized structure and data security. This improves the efficacy and quality of the healthcare system as a whole, as well as patient care and administrative efficiency.

4.6 Clinical and Audit Trials

Blockchain transforms clinical trials by reliably and transparently collecting and verifying data at every turn. Transparency and fraud prevention are ensured by the blockchain's storage of all trial data, including patient data and conclusions. This accessibility to authorized parties enhances the reproducibility, uniformity, and efficacy of studies. Improved trial integrity promotes healthcare advancement and speeds up medical research. Audit trails, which offer transparent, traceable data access, are immutable recordings of every system transaction on a blockchain. In the healthcare industry, these trails are essential because they provide patients, regulators, and institutions with openness and accountability.

4.7 Telemedicine and Remote Monitoring

Blockchain protects the privacy and security of data in remote patient monitoring and telemedicine. It protects patient privacy during virtual consultations by securely exchanging patient data. Data gathered from various devices is protected, guaranteeing tamper-proof storage. This dependability improves remote healthcare services by ensuring patients receive high-quality and secure data.

4.8 Cross-Border Data Exchange

Different laws and security concerns make it difficult to share healthcare data across borders. Blockchain provides an answer by establishing a safe, international platform. Patients may access their data from anywhere in the globe, with the assurance of security and adherence to local and international regulations even when traveling.

4.9 Supply Chain Management

Blockchain makes it possible to trace the sources and trajectories of medications in great detail, which transforms pharmaceutical supply chains [23]. Its open ledger, which records every step from manufacturing to distribution to consumption, protects against fake medications. By confirming the authenticity of pharmaceuticals at each stage of the supply chain, this technology enhances patient safety and safeguards its integrity [8, 20].

4.10 Regulatory Compliance

Blockchain offers a solid foundation for adhering to healthcare regulations. HIPAA (US) [24] and GDPR (EU) [25] laws, for example, place strict guidelines on the handling of patient data. By protecting patient data and keeping transparent, unchangeable records of data activity, blockchain's encryption, transparency, and data immutability guarantee compliance [26].

4.11 Decentralization

Because blockchains are decentralized, they prevent breaches and single points of failure that are common in centralized systems, improving the security of health data. The distributed structure of blockchain eliminates single ownership, which improves system resilience and data security in contrast to centralized systems. For patient data to be protected and confidence in healthcare data security to be increased, decentralization is essential [27].

5 Challenges of Blockchain Technology in Health Data Integration

Both paper-based and electronic traditional health record systems have security flaws and inefficiencies [26]. Blockchain has obstacles but has the potential to revolutionize healthcare data management. When it is implemented, data processing may be completely redesigned, improving privacy, security, and interoperability in contexts like as clinical trials, EHRs, and drug supply chains. Scalability, compatibility with current systems, and adherence to laws like GDPR and HIPAA are among the problems. The main challenges include standardizing data, protecting privacy and security, and encouraging wider use. Challenges include dangers related to data loss, user education, and budgetary issues. Blockchain's implementation in healthcare is further complicated by legal responsibility, the young age of the technology, and its smooth interaction with other cutting-edge technologies [28].

5.1 Scalability

Scalability of blockchain technology is a big issue, particularly for Ethereum and Bitcoin-based healthcare applications. The difficulty is in managing large amounts of healthcare data and transactions effectively without incurring undue expenses or delays [29]. The constant flow of medical data and patient information necessitates quick and safe processing. The limits of current blockchain technology, particularly on public networks, might result in slower processing times, higher costs, and inefficiencies, which reduce the technology's usefulness for managing healthcare data. For blockchain to successfully satisfy the expectations of the healthcare industry, scalability is a critical issue [30].

5.2 Integration with Legacy Systems

It is difficult to integrate blockchain into antiquated healthcare systems that handle patient data. These systems need a lot of effort, money, and system upgrades to make them seamlessly compatible with blockchain technology since they don't have the necessary interfaces [31]. Fundamental modifications to data processing and storage techniques are required to be in line with blockchain, which may need the creation of new software interfaces. Hardware, network, and security protocol reconfigurations may also be required. Healthcare organizations strive to surmount these obstacles in order to enhance data security, facilitate interoperability, and reap the rewards of sharing [32].

5.3 Regulatory Compliance

Healthcare is obligated by laws such as GDPR and HIPAA to protect patient data. Notwithstanding its advantages, blockchain's decentralization makes it difficult for it to adhere to these requirements [26]. Dispersed data complicates management and auditing and makes privacy compliance difficult. It is imperative to do a privacy legislation analysis of blockchain technology prior to implementing healthcare data systems. It could be necessary to use anonymization procedures to safeguard patient identities. Adherence is essential, necessitating blockchain solutions that give data security and privacy first priority [17, 31].

5.4 Data Standardization, Privacy and Security

Healthcare organizations and blockchain developers need to collaborate to build solutions that make it easier for standardized healthcare data to be translated and compatible with blockchain systems in order to overcome this obstacle [33]. To bridge the gap between current data formats and blockchain needs, this may entail developing middleware or data transformation procedures. To guarantee that blockchain-based healthcare solutions can successfully communicate and exchange data with traditional healthcare systems and providers, data standards and blockchain compatibility must be achieved [34]. It is an essential step in maximizing blockchain's benefits while preserving healthcare data management continuity and enhancing interoperability.

Blockchain improves data security but raises privacy issues for the medical field. It is still difficult to strike a balance between data interchange and privacy, particularly when it comes to patient data. Blockchain's immutability and openness raise the possibility of patient data vulnerability, necessitating strict access control even in the face of cryptographic security [35]. Smart contracts provide an open data interchange, automate permission, and give patients control over data access within prescribed bounds. It's difficult to develop blockchain-based healthcare systems that guarantee data sharing and privacy. For patient privacy, compliance with data protection rules such as GDPR and HIPAA is essential [22, 36].

5.5 Adoption and Interoperability

The two most important conditions for blockchain integration in healthcare are interoperability and broad acceptability. Blockchain adoption for data management is necessary for the healthcare ecosystem to succeed. It's hard to build an innovative and trusting culture while overcoming organizational and technical constraints. For effective patient data transmission, interoperability necessitates smooth cooperation between various blockchain systems and healthcare organizations. Different blockchain designs and data types provide obstacles, requiring protocols in order to establish communication. Establishing interoperability standards and closing these gaps would need cooperation between government agencies, digital companies, and the healthcare industry. This will enhance patient care [1, 37].

5.6 Cost and Resource Constraints

The use of blockchain technology in healthcare is hampered by significant financial limitations, particularly for smaller institutions, because of the high costs of system development, upkeep, and resource-intensive needs, which call for highly qualified staff and dependable infrastructure [38]. In addition, given the novelty and complexity of the technology, user education continues to be a major barrier that has to be overcome. In-depth training programs are needed to provide patients and healthcare professionals with the information they need to fully utilize blockchain's advantages [39, 40].

5.7 Risk of Data Loss

Cryptographic keys are used in blockchain systems to restrict access to data, and their loss can render data permanently inaccessible. This issue highlights how crucial it is to handle and save private keys securely in order to prevent the irrevocable loss of crucial medical data. In a blockchain-based healthcare setting, it emphasizes the necessity of strong key management techniques and user education to reduce the danger of data loss [35, 41].

5.8 Consensus Mechanisms

Consensus processes are used by blockchain networks to verify transactions and the chain's current state [35]. There are other approaches, such as Proof of Work (PoW) and Proof of Stake (PoS), each with pros and cons of its own. Careful consideration of security, energy efficiency, scalability, and decentralization are necessary when choosing the best approach for healthcare. By adapting the consensus process to healthcare requirements, inefficiencies, excessive energy use, and security threats are avoided [3].

5.9 Legal and Liability Issues

While implementing blockchain in healthcare, legal and liability considerations are crucial. The decentralized nature of blockchain presents issues for health data management in terms of accountability [42]. It becomes difficult to assign blame for mistakes or disagreements because of decentralized networks and cryptographic keys. To resolve this, specific legal frameworks and contracts outlining responsibilities and liabilities are required, guaranteeing that all parties understand and abide by the legal protections that have been designed for blockchain-based healthcare systems [38].

5.10 Technology Immaturity

For blockchain systems to operate well, a strong network infrastructure is essential. In rural or remote places, in particular, limited access to reliable networks reduces the efficacy of the device. Investments in network infrastructure and enhanced internet connectivity are necessary to realize the promise of blockchain in these types of environments [43]. Although blockchain technology is still in its infancy, it has the potential to solve concerns about patient privacy, interoperability, and data security in the healthcare industry [39]. Present concerns include, however, less understandable user interfaces, complicated implementations, and scalability issues. As the technology advances, these early difficulties should disappear and make it easier to employ for healthcare purposes. Healthcare blockchain integration with AI and IoT need compatibility and collaboration. IoT collects real-time medical data, AI improves processing, and blockchain guarantees security. Expert cooperation is essential to building a smooth IT environment [44–47].

6 Discussion

Blockchain technology has an exciting prospect in healthcare, one that may fundamentally change how patient data is handled and medical services are provided. But in order to realize this potential to the fullest, a number of important issues must be resolved. Scalability is a major problem, especially when it comes to open blockchains, and creative solutions are needed to handle the increasing amount of healthcare data in an economical and efficient manner. Because of the strict data protection rules like HIPAA and GDPR, which need blockchain solutions to be carefully aligned with legal standards, regulatory compliance is still a major worry. For smooth interoperability, data must be standardized, and common data formats and protocols must be established. The creation of standardized interfaces is necessary to achieve interoperability, guaranteeing that various blockchain systems can successfully communicate and exchange data. Comprehensive user education initiatives are necessary to ensure that patients and healthcare professionals are aware of the advantages and workings of blockchain technology, which will help to promote trust and acceptance. Well-defined frameworks that provide clear guidance for all parties in the case of errors or conflicts are necessary to handle responsibility in relation to legal and liability concerns. It is anticipated that as blockchain technology develops, it will become more trustworthy and user-friendly, which would make deployment easier. In order to build a smooth and connected healthcare environment, coordinated efforts and the implementation of standardized communication protocols are necessary for the efficient integration of blockchain with other developing technologies, such as AI and IoT. Blockchain technology has the potential to significantly improve healthcare services by providing safe, effective, and patient-centered solutions for data management and care delivery. When these issues are appropriately resolved, blockchain technology may also help to improve the healthcare sector as a whole.

7 Conclusion

This study explores the complex problems associated with centralized patient medical records in the healthcare industry, recognizing persistent concerns about patient privacy, interoperability, and data security. The inadequacies of traditional methods, such as paper-based records and fragmented electronic health records (EHR), in fulfilling industry expectations have led to the revolutionary advent of blockchain technology. This paper investigates the potential of blockchain technology to transform hospital data management through an extensive assessment and analysis. Its decentralized structure, data immutability, and cryptographic security present a viable foundation for enhancing provider interoperability, strengthening data security, and giving patients more control over their health information.

This study highlights how blockchain may revolutionize healthcare settings by examining innovative projects such as MedRec and Medicalchain. These initiatives demonstrate how blockchain may transform data security, expedite the coordination of patient care, and encourage data exchange across stakeholders in the healthcare sector. This research highlights the possibilities of blockchain technology and imagines a more safe, efficient, and patient-focused healthcare environment in the future. The introduction of blockchain technology offers a ray of light for healthcare stakeholders searching for creative solutions, as it promises a system that protects patient data while also promoting proactive patient involvement in their health. The continuous revolution of healthcare is heavily dependent on blockchain's critical role in determining its future.

Acknowledgements. This work was supported by the project "BLOCKCHAIN.PT (RE-C05-i01.01 – Agendas/Alianças Mobilizadoras para a Reindustrialização, Plano de Recuperação e Resiliência de Portugal na sua componente 5 – Capitalização e Inovação Empresarial e com o Regulamento do Sistema de Incentivos "Agendas para a Inovação Empresarial", aprovado pela Portaria N.° 43-A/2022 de 19.01.2022).

This work was also under the Eramus+ NEEM project.

References

1. Alzahrani, S., Daim, T., Choo, K.-K.R.: Assessment of the blockchain technology adoption for the management of the electronic health record systems. IEEE Trans. Eng. Manag. **70**, 2846–2863 (2023). https://doi.org/10.1109/TEM.2022.3158185
2. Sonkamble, R.G.G., Phansalkar, S.P.P., Potdar, V.M.M., Bongale, A.M.M.: Survey of interoperability in electronic health records management and proposed blockchain based framework: MyBlockEHR. IEEE Access **9**, 158367–158401 (2021). https://doi.org/10.1109/ACCESS.2021.3129284
3. Hylock, R.H., Zeng, X.: A blockchain framework for patient-centered health records and exchange (healthChain): evaluation and proof-of-concept study. J. Med. Internet Res. **21**, e13592 (2019). https://doi.org/10.2196/13592
4. Page, M., McKenzie, J., Bossuyt, P., et al.: The PRISMA 2020 statement: an updated guideline for reporting systematic reviews. Elsevier
5. Schulte, S., Sigwart, M., Frauenthaler, P., Borkowski, M.: Towards blockchain interoperability. In: Di Ciccio, C., et al. (eds.) BPM 2019. LNBIP, vol. 361, pp. 3–10. Springer, Cham (2019). https://doi.org/10.1007/978-3-030-30429-4_1
6. Madine, M., Salah, K., Jayaraman, R., et al.: AppxChain: application-level interoperability for blockchain networks. IEEE Access **9**, 87777–87791 (2021). https://doi.org/10.1109/ACCESS.2021.3089603
7. Biswas, S., Sharif, K., Li, F., et al.: Interoperability and synchronization management of blockchain-based decentralized e-health systems. ieeexplore.ieee.org
8. Carter, G., Chevellereau, B., Shahriar, H., Sneha, S.: OpenPharma blockchain on FHIR: an interoperable solution for read-only health records exchange through blockchain and biometrics. Blockchain Healthc Today **3** (2020). https://doi.org/10.30953/bhty.v3.120
9. Ekblaw, A., Azaria, A.: MedRec: medical data management on the blockchain. Viral Commun. (2016)
10. Rghioui, A., Bouchkaren, S., Khannous, A.: Blockchain-based electronic healthcare information system optimized for developing countries. IAENG Int. J. Comput. Sci. **49** (2022)
11. Han, Y.J., Zhang, Y.W., Vermund, S.H.H.: Blockchain technology for electronic health records. Int. J. Environ. Res. Public Health **19**, 15577 (2022). https://doi.org/10.3390/ijerph192315577
12. Tith, D., Lee, J.S., Suzuki, H., et al.: Application of blockchain to maintaining patient records in electronic health record for enhanced privacy, scalability, and availability. Healthcare Inform. Res. **26**, 3–12 (2020). https://doi.org/10.4258/hir.2020.26.1.3

13. Singh, Y., Jabbar, M.A., Kumar Shandilya, S., et al.: Exploring applications of blockchain in healthcare: road map and future directions. Front. Public Health **11**, 1229386 (2023). https://doi.org/10.3389/fpubh.2023.1229386
14. Egala, B.S., Pradhan, A.K., Badarla, V., Mohanty, S.P.: Fortified-chain: a blockchain-based framework for security and privacy-assured internet of medical things with effective access control. IEEE Internet Things J. **8**, 11717–11731 (2021). https://doi.org/10.1109/JIOT.2021.3058946. WE - Science Citation Index Expanded (SCI-EXPANDED)
15. Houtan, B., Hafid, A.S., Makrakis, D.: A survey on blockchain-based self-sovereign patient identity in healthcare. IEEE Access **8**, 90478–90494 (2020). https://doi.org/10.1109/ACCESS.2020.2994090
16. Li, C.-T.T., Shih, D.-H.H., Wang, C.-C.C., et al.: A Blockchain based data aggregation and group authentication scheme for electronic medical system. IEEE Access **8**, 173904–173917 (2020). https://doi.org/10.1109/ACCESS.2020.3025898
17. Manoj, T., Makkithaya, K., Narendra, V.G.G.: A Blockchain based decentralized identifiers for entity authentication in electronic health records. **9** (2022). https://doi.org/10.1080/23311916.2022.2035134
18. Shen, B.Q., Guo, J.Z., Yang, Y.L.: MedChain: efficient healthcare data sharing via blockchain. Appl. Sci.-Basel **9**, 1207 (2019). https://doi.org/10.3390/app9061207
19. Daraghmi, E.-Y., Daraghmi, Y.-A., Yuan, S.-M.: MedChain: a design of blockchain-based system for medical records access and permissions management. IEEE Access **7**, 164595–164613 (2019). https://doi.org/10.1109/ACCESS.2019.2952942
20. Zheng, X., Sun, S., Mukkamala, R.R., et al.: Accelerating health data sharing: a solution based on the internet of things and distributed ledger technologies. J. Med. Internet Res. **21**, e13583 (2019). https://doi.org/10.2196/13583
21. George, M., Mary Chacko, A.: MediTrans—patient-centric interoperability through blockchain. Int. J. Netw. Manag. **32**, e2187 (2022). https://doi.org/10.1002/nem.2187
22. Lakshmanan, M., Anandha Mala, G.S.: Merkle tree-blockchain-assisted privacy preservation of electronic medical records on offering medical data protection through hybrid heuristic algorithm. Knowl. Inf. Syst. **66**, 481–509 (2023). https://doi.org/10.1007/s10115-023-01937-z
23. Sharma, V., Ramirez-Asis, E., Ahmad, A.J., et al.: On the Internet of Things, blockchain technology for supply chain management (IoT). Wirel. Commun. Mob. Comput. **2022**, 9185731 (2022). https://doi.org/10.1155/2022/9185731
24. Nosowsky, R., Giordano, T.J.: The health insurance portability and accountability act of 1996 (HIPAA) privacy rule: implications for clinical research. Annu. Rev. Med. **57**, 575–590 (2006). https://doi.org/10.1146/ANNUREV.MED.57.121304.131257
25. Voigt, P., Guide AV dem B-AP, Ed. 1st, et al.: The EU General Data Protection Regulation (GDPR). Springer
26. Nagasubramanian, G., Sakthivel, R.K., Patan, R., et al.: Securing e-health records using keyless signature infrastructure blockchain technology in the cloud. Neural Comput. Appl. **32**, 639–647 (2020). https://doi.org/10.1007/s00521-018-3915-1
27. Tanwar, N., Thakur, J.: Patient-centric soulbound NFT framework for electronic health record (EHR). J. Eng. Appl. Sci. **70**, 33 (2023). https://doi.org/10.1186/s44147-023-00205-9
28. Hussain, S., Rahman, H., Abdulsaheb, G.M., et al.: A blockchain-based approach for healthcare data interoperability. Int. J. Adv. Soft Comput. Appl. **15**, 85–98 (2023). https://doi.org/10.15849/IJASCA.230720.06
29. Shahnaz, A., Qamar, U., Khalid, A.: Using blockchain for electronic health records. IEEE Access **7**, 147782–147795 (2019). https://doi.org/10.1109/ACCESS.2019.2946373
30. O'Donoghue, O., Vazirani, A.A., Brindley, D., Meinert, E.: Design choices and trade-offs in health care blockchain implementations: systematic review. J. Med. Internet Res. **21**, e12426 (2019). https://doi.org/10.2196/12426

31. Panigrahi, A., Nayak, A.K.K., Paul, R.: HealthCare EHR: a blockchain-based decentralized application. Int. J. Inf. Syst. Supply Chain Manag. **15**, 1–15 (2022). https://doi.org/10.4018/IJISSCM.290017
32. Rao, K.R., Naganjaneyulu, S.: Permissioned healthcare blockchain system for securing the EHRs with privacy preservation. Ingenierie des Systemes d'Information **26**, 393–402 (2021). https://doi.org/10.18280/ISI.260407
33. Deepa, V.V., Thamotharan, B., Mahto, D., et al.: Smart embedded health monitoring system and secure electronic health record (EHR) transactions using blockchain technology. Soft. Comput. **27**, 12741–12756 (2023). https://doi.org/10.1007/s00500-023-08893-4
34. Dubovitskaya, A., Baig, F., Xu, Z.G., et al.: ACTION-EHR: patient-centric blockchain-based electronic health record data management for cancer care. J. Med. Internet Res. **22**, e13598 (2020). https://doi.org/10.2196/13598
35. Yang, J., Onik, M.M.H., Lee, N.-Y., et al.: Proof-of-familiarity: a privacy-preserved blockchain scheme for collaborative medical decision-making. Appl. Sci. (Switzerland) **9**, 1370 (2019). https://doi.org/10.3390/app9071370
36. Barman, S., Chattopadhyay, S., Samanta, D., Barman, S.: A blockchain-based approach to secure electronic health records using fuzzy commitment scheme. Secur. Privacy **5**, e231 (2022). https://doi.org/10.1002/spy2.231
37. Puneeth, R.P., Parthasarathy, G.: Survey on security and interoperability of electronic health record sharing using blockchain technology. Acta Informatica Pragensia **12**, 160–178 (2023). https://doi.org/10.18267/j.aip.187
38. Kumar, T., Ramani, V., Ahmad, I., et al.: Blockchain utilization in healthcare: key requirements and challenges. ieeexplore.ieee.org
39. Zakzouk, A., El-Sayed, A., Hemdan, E.E.-D.E.: A blockchain-based electronic medical records management framework in smart healthcare infrastructure. Multimed. Tools Appl. **82**, 35419–35437 (2023). https://doi.org/10.1007/s11042-023-15152-z
40. Pilares, I.C.A.C.A., Azam, S., Akbulut, S., et al.: Addressing the challenges of electronic health records using blockchain and IPFS **22** (2022). https://doi.org/10.3390/s22114032
41. Patel, V.: A framework for secure and decentralized sharing of medical imaging data via blockchain consensus. Health Inform. J. **25**, 1398–1411 (2019). https://doi.org/10.1177/1460458218769699
42. Uddin, M., Memon, M.S., Memon, I., et al.: Hyperledger fabric blockchain: secure and efficient solution for electronic health records. Comput. Mater. Continua **68**, 2377–2397 (2021). https://doi.org/10.32604/cmc.2021.015354
43. Reegu, F.A., Abas, H., Jabbari, A., et al.: Interoperability requirements for blockchain-enabled electronic health records in healthcare: a systematic review and open research challenges. Secur. Commun. Netw. **2022** (2022). https://doi.org/10.1155/2022/9227343
44. Preetha, A.D., Kumar, T.S.P.: Securing IoT-based healthcare systems from counterfeit medicine penetration using Blockchain. Appl. Nanosci. (Switzerland) **13**, 1263–1275 (2023). https://doi.org/10.1007/s13204-021-01984-4
45. Shukla, M., Lin, J., Seneviratne, O.: BlockIoT: blockchain-based health data integration using IoT devices. In: AMIA Annual Symposium Proceedings, pp. 1119–1128 (2021)
46. Yadav, S.G.S., Gunisetti, L., Koduri, S.B., et al.: Securing electronic health records using blockchain technology for IoT in healthcare domain. Soft. Comput. **28**, 601 (2023). https://doi.org/10.1007/s00500-023-08489-y
47. Gohar, A.N., Abdelmawgoud, S.A., Farhan, M.S.: A patient-centric healthcare framework reference architecture for better semantic interoperability based on blockchain, cloud, and IoT. IEEE Access **10**, 92137–92157 (2022). https://doi.org/10.1109/ACCESS.2022.3202902

Predicting Heart Diseases Using Machine Learning Algorithms: A Survey

Isha Gupta, Anu Bajaj(✉), and Vikas Sharma

Thapar Institute of Engineering and Technology, Patiala, India
er.anubajaj@gmail.com

Abstract. In recent years, health-related diseases have become the top cause of death globally, as the rapid development in Machine Learning (ML) has had a significant impact on the healthcare industry. Misclassification of heart disease can have severe repercussions for patients. Requiring enhanced precision in diagnostic tools, researchers have developed numerous approaches dedicated to early prediction, thereby addressing the imperative need to reduce mortality risks associated with heart disease. This paper explores several machine learning and deep learning models, along with various hybrid algorithms proposed for the early detection of Heart Disease. This study utilizes various available datasets and discusses multiple preprocessing and feature extraction methods, along with their evaluation metrics.

Keywords: Heart Disease · Machine learning · Prediction · Deep Learning · Early Diagnosis · Cardiovascular Disease

1 Introduction

In the realm of healthcare data analytics, the identification of disease is one of the foremost challenges. The availability of diverse medical data raises the question of whether there exist effective methods to analyze this data, get insights, and translate them into knowledgeable action. Amid the array of health issues, heart disease (HD) is a broad term that covers a variety of pathological conditions affecting the heart as well as the blood vessels. It has become a significant global concern due to its alarming frequency. Primary risk factors for this illness include being overweight and obese, hypertension, excessive alcohol intake, and many more [1, 7]. These risk factors emphasize how crucial it is to possess an in-depth awareness of the intricate patterns that underpin heart illness and how to apply those patterns in the process of clinical diagnosis and treatment.

Historically, traditional clinical approaches to predict heart disease have relied upon a limited set of risk factors, including age, gender, blood pressure, cholesterol levels, and family medical history. Although these factors provide valuable insights, they often fall short of capturing the full complexity of heart disease risk. Moreover, conventional clinical approaches are static and lack the adaptability required to keep pace with evolving medical knowledge. Patterns in high-dimensional and non-linear data may not always be easily discerned by healthcare professionals.

V. Sakalauskas et al. (Eds.): IBICA 2023, LNNS 1231, pp. 333–345, 2025.
https://doi.org/10.1007/978-3-031-78946-5_32

Therefore, we need to switch to a new paradigm that drives us toward Machine Learning (ML) techniques. As the name indicates, machine learning involves computers learning independently without explicit programming or human intervention. ML-based models exhibit extraordinary capabilities in effectively analyzing vast datasets and unveiling intricate hidden patterns. They can readily adapt to the latest advancements in medical knowledge, which, in turn, leads to more accurate predictions. ML approaches promise to boost the accuracy of heart disease prediction, thereby reducing the mortality rate.

The growth of machine learning is not the end of the road yet. A branch of machine learning, called Deep Learning (DL), is gaining more and more prominence. DL uses neural network with three or more layers and is a form of computer technology that enables computers to comprehend information in a natural way, just like human brain does. DL holds several advantages over traditional ML techniques such as ability to handle non-linear relationships, manage vast and intricate data, and automatic feature learning. It is crucial to identify and predict cardiac failures since early detection can lead to a greater number of lives saved. This review study intends to answer the following research questions in light of this context.

- **RQ1:** Which machine learning models are commonly used for predicting heart disease?
- **RQ2:** Which datasets are commonly utilized for heart disease prediction?
- **RQ3:** What performance metrics are most frequently used to measure the efficiency of ML-based heart disease prediction?
- **RQ4:** What are the potential areas for future research using machine learning approaches to predict heart diseases?

This extensive work covers a broad range of machine learning (ML) models, which includes Decision Tree (DT), K-Nearest Neighbor (KNN), Extreme Gradient Boost (XGBoost), Random Forest (RF), Support Vector Machine (SVM), Naive Bayes (NB), Logistic regression (LR), and more. Similar to this, we explore a wide variety of data learning (DL) models which includes the following: Convolutional neural network (CNN), Radial Wavelet neural network (RWNN), Generative Adversarial network (GAN), Levy Flight CNN, and many more.

The Fig. 1 depicts the basic flowchart of all the steps involved in creating a model.

The organization of the paper as follows: Sect. 2 of this paper discusses the existing literature. Subsequently, Sect. 3 discusses observations and findings followed by the conclusion in Sect. 4.

2 Literature Review

This section covers an existing study that has employed machine learning (ML) and deep learning algorithms (DL) to forecast cardiac illness in recent years.

Machine Learning Approaches

In the field of predicting heart disease, the search for the best machine learning techniques has led researchers down intriguing paths. This section delves into the exploration of various ML models used for the same purpose.

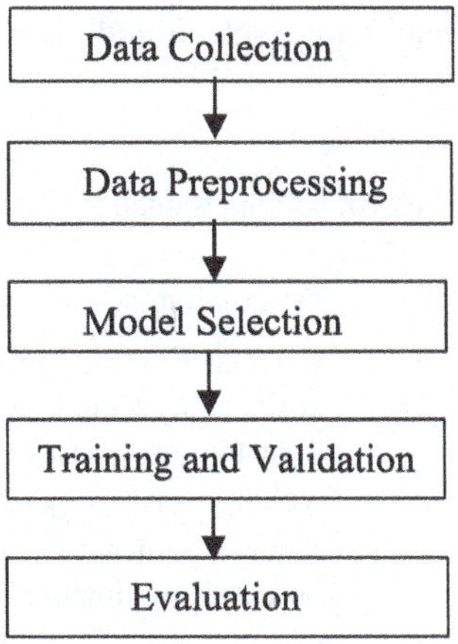

Fig. 1. Basic flowchart of Machine Learning model

Synthetic Minority Over-Sampling Technique (SMOTE)

Yang and Guan [1] introduced Smote XGBoost (Extreme Gradient Boost) based methodology for heart disease prediction. They initiated the process with feature selection based on information gain, followed by SMOTE-ENN, that is, Synthetic Minority Oversampling Technique-Edited Nearest Neighbor algorithm, used to deal with unbalanced data. Then they compared the performance of XGBoost algorithm with five baseline algorithms: K-Nearest Neighbor (KNN), Naive Bayes (NB), Random Forest (RF), Decision Tree (DT), Logistic regression (LR), resulting in an accuracy of 93.44%. This accuracy was further enhanced by Abdellatif et al. [16], who employed SMOTE to address the imbalanced data distribution and then turned to advanced Hyperparameter Optimization (HPO) technique, known as the "Hyperband Algorithm". Their efforts led to an impressive accuracy of 95.73%. In addition, FitriYani et al. [9] introduced a Heart disease Prediction Model, that incorporated several steps, including detection and elimination of outliers by utilizing Density-Based Spatial Clustering of Applications with Noise (DBSCAN). They then subsequently included the XGBoost algorithm after combining a hybrid SMOTE-ENN technique. For Statlog and Cleveland datasets, accuracy values of 95.90% and 98.40% were noted, respectively.

Feature Selection Techniques

Biwas et al. [2] explored three separate feature selection techniques, such as chi-square, ANOVA, and mutual information. ML techniques like RF, SVM, LR, NB, KNN, DT were tested on the selected features and their performances were recorded. SVM and LR yielded better performance. Meanwhile, Qadri et al. [11] introduced an innovative feature engineering technique known as 'Principal Component Heart Failure' (PCHF) to select the most prominent features along with 10-fold cross-validation on various ML models. The Decision Tree outperformed all the other models with an outstanding accuracy of 100%.

Table 1 briefs about different feature selection/dimensionality reduction techniques used by various researchers.

Table 1. Feature Selection/Dimensionality Reduction Techniques

Techniques	Description
Information Gain	Calculates the amount of information provided by an attribute to identify the target value and measures reduction in uncertainty or entropy values
Chi Square	Assesses whether the observed data distribution differs significantly from expected distribution, based on certain assumptions
Analysis of variance (ANOVA)	Calculates variance between features for classification algorithms
Mutual Information (M.I.)	Calculates dependence between two attributes. Higher M.I. means higher dependency
Principal Component Analysis (PCA)	Converts high-dimensional data into a lower-dimensional data

Ensemble Techniques

Itoo and Garg [12] proposed an ensemble model known as the StackingCVClassifier, which comprised three constituent machine learning models: LR, KNN, and NB. They achieved accuracies of 86.66%, 88.33%, and 86.66% respectively. Meanwhile, in a separate approach, Kapila et al. [17] introduced the Quine McCluskey Binary Classifier (QMBC), an ensemble technique that fused Chi-square, ANOVA techniques to remove irrelevant features and PCA for dimensionality reduction. Their achievement was a remarkable accuracy of 98.36%.

Optimization Techniques

A prediction model known as HY_OptGBM was introduced by Yang et al. [5] by using the optimized LightGBM classifier and combining it with the advanced hyperparameter optimization framework, OPTUNA. This approach yielded an Area under curve (AUC) value of 97.8%. In a parallel endeavor, Bertsimas et al. [6] presented a novel methodology for extracting ECG-related features. They used the XGBoost algorithm to train their model and fine-tuned its parameters using the OPTUNA optimization framework, followed by the usage of SHAP (Shapley Additive Explanations) to unveil which features carried the most explanatory power for each label in the training data, thus leading to better results.

Genetic Algorithms

A Clustered Genetic Algorithm approach (C-GA) was suggested by Vijaya et al. [18]. This novel approach significantly improved the prediction rate with an accuracy of 94.56%. Tiwari et al. [19] proposed a new method using spectral, nonlinear techniques, ranking methods, generalized discriminant analysis (GDA), and a 1-NLELM, which is, a 1-Norm linear Extreme Learning Machine (1-NLELM). Their efforts resulted in an outstanding accuracy of 100%. Islam et al. [30] added a new layer by using PCA to reduce data. They then employed Hybrid Genetic Algorithm with K means for final clustering. Their approach achieved an accuracy of 94.06%.

Comparative Analysis

Various researchers carried out comparisons among different ML algorithms on a variety of datasets to find the best model. Ahmad et al. [10] compared several ML algorithms, including KNN, LR, SVM, and GBC (Gradient Boosting Classifier), both with and without using GridSearchCV. Among them, Extreme Gradient Boosting Classifier (XGB) delivered exceptional accuracies of 99.03% and 98.05%, with and without GridSearchCV, respectively. Uddin et al. [4] adopted a different strategy, combining three distinct datasets to generate a powerful arsenal. They trained and tested a multitude of ML techniques such as LR, RF, DT, SVM, XGBoost, KNN, Ada-Boost Classifier (AB), Extra trees Classifier (ET), and Multi-Layer Perception (MLP). Amid this array of choices, DT displayed the highest accuracy of 99.16%. Absar et al. [7] utilized ML techniques, like DT, KNN, RF, and AdaBoost (AB), on Cleveland, Hungary, Switzerland, and Long Beach (CHSLB) datasets and compared their accuracies. Gaikwad et al. [8] embarked on an expedition into the "UCI Machine learning repository for Statlog (Heart) Data Set". Using an array of algorithms, including SVM, GB, DT, RF, and Logistic Regression (LR), they meticulously noted the performance metrics of each model. Jadhav et al. [13] conducted a comprehensive comparison on various machine learning models on the Cleveland dataset from the UCI machine learning repository, analyzing their accuracies. Meanwhile, Sulthana et al. [14] examined the performance of NB (accuracy 85.85%), DT (accuracy 81.84%), and NB with k-means clustering (accuracy 88.06%) on Cleveland's heart disease dataset. Kumar et al. [15], explored fascinating world of the Kaggle dataset, where they compared three approaches to classification: Support Vector Classifier (SVM), K-Nearest Neighbor (KNN), and Multi-Layer Perception (MLP) and found that KNN had the highest accuracy rate of 91.8%. On the other hand, four hybrid classifiers namely, Naive Bayes+ Random Forest, Naive Bayes + J48, Bayes Net + Random Forest, and Naive Bayes+ PART were tested by Prerna et al. [20]. NB+RF combination proved to be a standout performer, achieving an accuracy of 84.44%.

Table 2 provides a summary of ML methods used for heart disease prediction.

Deep Learning Approaches

This section covers several deep learning techniques employed by researchers to enhance the diagnosis of cardiovascular diseases.

Fast Track Gram Matrix-Principal Component Analysis (FTGM-PCA)

Deenathayalan and Narayanan [3] introduced the Fast Track Gram Matrix-Principal Component Analysis (FTGM-PCA) in their hope of increased accuracy. With the support of an Informative Entropy-Based Random Forest (IEB-RF), this approach achieved an amazing 97% accuracy rate. Their methodology of dimensionality reduction, which included the usage of the FTGM strategy to approximate a correlation matrix with a low rank, was the secret to their success. A more precise classification was made possible

Table 2. Machine Learning Approaches for predicting heart diseases

Authors	Datasets	Models used	Models compared	Performance metrics
Yang, J. and Guan, J. (2022)	Pathological data (private)	Smote XGBoost algorithm	XG Boost, DT, LR, NB, RF, KNN	Accuracy, precision, recall, F1-score, and AUC
N. Biwas et al. (2023)	UCI Cleveland dataset	Various feature selection techniques	LR, SVM, KNN, RF, NB, DT	Accuracy, sensitivity, specificity, area under ROC curve, log loss
Yang et al. (2023)	Framingham Heart Institute (FHS)	Light GBM (ensemble learning algorithm based on DT), focal loss function, OPTUNA	ANN, SVM, RF, LR, Epi+ Gen, GPR+KRR	Sensitivity, recall, F-score, accuracy, specificity, precision
Bertsima et al. (2021)	ECG signals from Physio-Net	XG-Boost+ OPTUNA, AP	DT, SVM, LR, KNN	Precision, recall, F1 score, Count
Fitriyani et al. (2020)	2 datasets-Statlog and Cleveland	Heart Disease Prediction Model (HDPM)	NB, LR, SVM, DT, RF, multilayer perceptron	Accuracy, precision, recall, sensitivity, f-score
Ahmad et al. (2022)	Switzerland, Hungary, Cleveland, and VA Long Beach, UCI Kaggle	With and without GridSearchCV	LR, KNN, SVM, GBC	Accuracy
Qadri et al. (2023)	Heart failure dataset from the repository Kaggle	Feature engineering technique- Principle component heart failure (PCHF)	LR, DT, RF, SVM, XGB, NB, KNN, GB, Hyperparameter tuning, Multilayer perceptron (MLP)	Accuracy, F1-score, Precision, Recall
N.N. Itoo and V. Garg (2022)	Cleveland Clinic Foundation (CVF)	Ensemble model: StackingCVClassifier	NB, MLP, KNN+CN, DT, RF, LR	Prediction, accuracy,

(*continued*)

Table 2. (*continued*)

Authors	Datasets	Models used	Models compared	Performance metrics
Kapila et al. (2023)	Cleveland, CVD, HD dataset	Quine McCluskey Binary Classifier (QMBC) with Chi-square, Anova and PCA	DT, RF, SVM, KNN, NB	Accuracy, recall, f1-score, precision, specificity,
J. Vijaya (2023)	UCI heart disease data set	Clustered Genetic Algorithm approach (CGA)	DT, RF, SVM, KNN, NB	Precision, accuracy, error rate, F-score, recall
Tiwari et al. (2023)	Physio-Net ATM	GDA + 1-NLELM binary classifier	KNN, DT, SVM, NB, ANN	Accuracy
Prerna et al. (2023)	Cleveland heart disease dataset	Hybrid Classifier in the Cloud Environment	NB, RF, J48, Bayes-Net	Precision, Recall, Accuracy, F-measure
Islam et al. (2020)	UCI Machine Learning heart disease dataset	Principal Component Analysis, Hybrid Genetic Algorithm with k means		Accuracy

through this feature reduction, and Principal Component Analysis (PCA) added the finishing touch.

Convolutional Neural Network (CNN)

Sharma and Parmar [26] in their research, introduced a deep learning neural network utilizing a novel hyperparameter optimization technique known as 'Talos'. On combining this dynamic duo along with Keras Library, they achieved a commendable accuracy of 90.76%. Meanwhile, Chamundeshwari et al. [23] had their sights set on optimizing Convolutional Neural Networks (CNN) using a distance-based cat swarm optimization approach. Besides this, they also used the Field II program to get rid of annoying speckle noise from echocardiography (ECG) images, and their hybrid pattern extraction technique brought forth substantial gains in performance. The drive for optimization was evolved by Jain et al. [24]. They proposed the Sunflower Optimization Algorithm (SFO) combined with the Levy Flight CNN (LV-CNN) methodology. This combination yielded an accuracy of 95.74% by minimizing the loss function that existed in the CNN model.

Hussain et al. [25] introduced a 1D CNN technique for predicting heart disease. The implementation of a dropout strategy to prevent accuracy resulted in an incredible 97% training accuracy and 96% test accuracy. Meanwhile, Bouny et al. [28] proposed a revolutionary Multi-Level Wavelet Convolutional Neural Network (ML-WCNN). This groundbreaking method extracted discriminative features through the integration of Stationary Wavelet Transform (SWT) and 1D-CNN framework. The application of

10-fold cross-validation technique yielded an astonishing accuracy of 99.57%.

Radial Wavelet Neural Network (RWNN) vs Decision Support System (DSS)
Guillermo et al. [29] proposed a Radial-Wavelet Neural Network (RWNN) classifier, which makes the use of extended Kalman filter (EKF) as its learning algorithm to detect heart murmurs. Bensenane et al. [21] proposed DLCVD (Deep Learning Based Cardiovascular Disease Diagnosis System), a decision support system (DSS). Their two-step deep learning Long short-term memory (LSTM) neural network showed 93% accuracy in detecting irregular heartbeats, or arrhythmias.

Generative Adversarial Network (GAN)
The Generative Adversarial Network (GAN) concept was introduced by Madani and Xiong. Data-efficient deep learning classifiers for prediction have been introduced by Madani et al. [22]. These models had a unique ability to learn both from unlabeled data and labeled data. Two neural networks are utilized by GAN: A generator that aims to produce real pictures and a Discriminator that has the uncanny ability to distinguish between the real and fake [22, 25]. Xiong et al. [32] enhanced this work by proposing a Cascaded multitask Generative Framework, that consists of a 3D multitask generative architecture and 3D nnU-Net for Aortic Dissection (AD) detection.

Hybrid Deep Learning Approaches
K. Verma et al. [35] developed an integrated strategy to predict heart-related diseases that combines an ensemble deep neural network framework with a feature selection method based on genetic algorithms. The usage of Adam optimizer enhanced the model, thereby achieving a remarkable accuracy of 98%. Meanwhile, M. Mohamed Suhail et al. [33] invested in a method that combined the utilization of Nonlinear Vector Decomposed Neural Networks with Discrete Wavelet Transforms to minimize unwanted noise and open up new possibilities for disease detection and prediction.

Weight-Based Approaches
The initial weights used in neural networks play a major role in influencing an artificial neural network's performance. As per the study of Arabasadi et al. [34], a genetic approach was used to identify the neural network's starting weights. Later, the neural network was taught utilizing practice data, and feed-forward topology was used with a single hidden layer to detect coronary artery disease (CAD), without angiography. Xie et al. [27] proposed a strategy built upon the density-based spatial clustering of applications with noise (DBSCAN) algorithm, by utilizing the density information of datasets. This novel approach assigned weights to sample points based on their density distribution. The dataset was then partitioned into three sample categories according to the density weights resulting in creation of a new weight feature and better results.

Table 3 provides a summary of deep learning strategies used to predict heart disease.

Table 3. Deep Learning Approaches for heart disease prediction

Authors	Datasets	Models used	Models compared	Performance Metrics	Data Type
Deenathayalan, and Narayanan (2022)	Cleveland heart disease dataset	FTGM PCA for dimensionality reduction and fusion, and IEB RF for effective classification	NB, SVM, Functional trees, LR, DT, RF	Precision, accuracy, F1-score, sensitivity, AUC (area under curve)	Text
Bensenane et al. (2021)	MIT BIH arrhythmia database	Deep learning approach: decision support system-based system (DSS)	GA+KNN, GA+PNN, ID-CNN, GA_GECS+SVM	Accuracy, specificity and sensitivity	Electrocardiogram (ECG) signals
Madani et al. (2018)	IRB at UCSF (University of California, San Francisco)	Semi supervised generative adversarial network (GAN) model	CNN, CNN with segmentation	Accuracy	Image
Jain et al (2023)		LV-CNN	DT, NB, CNN, RF, SVM, ANN	Accuracy, sensitivity, specificity, MCC	Image (angiography)
Hussain et al. (2021)	Cleveland dataset	1-D CNN	Classical approaches like DT, LR, SVM	Accuracy, recall, f1-score, precision	Text
Sharma and Parmar (2023)	Heart Disease UCI dataset	Deep Neural Network (model "Optimized DNN using Talos")	KNN, SVM, Naïve Bayes, Random Forest	Accuracy	Text
Xie et al. (2021)	Stroke dataset and the University of California, Irvine (UCI) dataset	Weight-learning based on density information	NB, DT, LR, RF	Accuracy, F-measure, precision, sensitivity	Text

(*continued*)

Table 3. *(continued)*

Authors	Datasets	Models used	Models compared	Performance Metrics	Data Type
Bouny et al. (2020)	Standard Database MIT BIH Arrhythmia	Multi-Level Wavelet Convolutional Neural Networks		Accuracy	ECG signal
Guillermo et al. (2015)		(RWNN) Radial Wavelet Neural Network + Extended Kalman Filter (EKF)	Compared with a multilayer perceptron		Cardiac sound

3 Observations and Findings

This section incorporates the responses to the above-listed research questions.

RQ1. Which machine learning models are commonly used for predicting heart disease?

- Researchers have explored a wide range of techniques for heart disease prediction, with a major focus on machine learning (ML) models. Among these models, the most widely used Baseline models are DT, RF, SVM, KNN, and LR.
- Some studies introduced an altered version of these methods, while others have proposed hybrid algorithms such as the Hybrid Genetic Algorithm (HGA) with k means, and Hybrid Classifier in the Cloud Environment to improve the performance. Additionally, researchers have also employed comparative analysis of these models.
- Ensemble models like novel CV Classifier have also been investigated.
- Notably, there seems to be a shift in the field towards deep learning algorithms as many researchers are opting for neural networks like convolutional neural networks (CNN), Levy-flight (LV) CNN, and radial wavelet neural network (RWNN), which have demonstrated remarkable accuracy.

RQ2. Which datasets are commonly utilized for heart prediction disease?

Various datasets have been used in the literature such as Cleveland Heart Disease (CHD), Framingham Heart Institute (FHS), Physio Net, UCI, Cleveland Clinic Foundation (CVF), and HD dataset. Some studies also used the MIT BIH (Massachusetts Institute of Technology Beth Israel Hospital) arrhythmia database. It is observed that the most widely used dataset is the UCI and Cleveland dataset for early prediction.

RQ3. What performance metrics are most frequently used to measure the efficiency of ML-based heart disease prediction?

Performance metrics, in machine learning, are used to measure the efficacy of trained models and to validate the results. Commonly employed evaluation metrics are accuracy,

Specificity, Precision, F1-measure, Area under curve (AUC), Recall, sensitivity, log loss, and area under ROC curve.

RQ4. What are the potential areas for future research using machine learning approaches to predict heart diseases?

- **Multi-Modal Data Integration**: Various data sources can be combined, such as electronic health records (EHRs), genomics, and medical images, to construct a holistic patient profile and enhance prediction accuracy.
- **Unsupervised learning:** A vast majority of studies adopted supervised learning approaches, however sometimes, it is hard to find correctly labeled data, which prompts us to make use of semi-supervised and unsupervised learning strategies [31]. With the limited utilization of unsupervised and semi-supervised methods in existing literature, these domains are yet to be explored.
- **Dimensionality Reduction:** Novel techniques for dimensionality reduction and feature selection can be explored, such as MI-based feature selection, L1 Regularization (Lasso), Recursive Feature Elimination (RFE), and Linear Discriminant Analysis (LDE). These approaches have the potential to reduce overfitting and improve the overall performance of the model.
- **Leveraging extensive and public Datasets:** Not all the datasets used by researchers are publicly available. Many researchers suggested using larger, publicly available datasets. Extensive datasets can provide a diverse range of patient profiles. This can enhance accuracy and decrease the mortality rate.
- **Hybridization:** Since very little work has been done on hybrid models, this area can be explored further. ML models with Genetic Algorithms and gradient Boosting techniques like XGBoost or LightGBM with neural networks can prove to be powerful hybrid models.

4 Conclusion

In this paper, we extensively explored a range of machine learning and deep learning techniques for the early prediction of heart diseases. Our goal was to examine, analyze, and gain insights of all the available existing literature and identify potential future directions. This review study revealed that ML algorithms show remarkable performance in early detection, thereby reducing the mortality rate. Researchers also used deep learning approaches like CNN, LV-CNN, and RWCNN to achieve higher accuracy. Deep learning techniques are now more adept at identifying intricate patterns and features in huge, complex datasets. Thanks to this paradigm shift towards DL, which increases system accuracy and efficiency.

References

1. Yang, J., Guan, J.: A heart disease prediction model based on feature optimization and smote-XG boost algorithm. Information **13**(10), 475 (2022)
2. Biswas, N., et al.: Machine learning-based model to predict heart disease in early stage employing different feature selection techniques. BioMed Res. Int. **2023** (2023)

3. Deenathayalan, D., Narayanan, B.: Predicting heart disease using FTGM-PCA based informative entropy based-random forest. Curr. Appl. Sci. Technol. **23** (2023)
4. Uddin, K.M.M., Ripa, R., Yeasmin, N., Biswas, N., Dey, S.K.: Machine learning-based approach to the diagnosis of cardiovascular vascular disease using a combined dataset. Intell.-Based Med. **7**, 100100 (2023)
5. Yang, H., Chen, Z., Yang, H., Tian, M.: Predicting coronary heart disease using an improved light GBM model: performance analysis and comparison. IEEE Access **11**, 23366–23380 (2023)
6. Bertsimas, D., Mingardi, L., Stellato, B.: Machine learning for real-time heart disease prediction. IEEE J. Biomed. Health Inform. **25**(9), 3627–3637 (2021)
7. Absar, N., et al.: The efficacy of machine-learning-supported smart system for heart disease prediction. Healthcare **10**(6), 1137 (2022)
8. Gaikwad, M.J., Asole, P.S., Bitla, L.S.: Effective study of machine learning algorithms for heart disease prediction. In: 2022 2nd International Conference on Power Electronics & IoT Applications in Renewable Energy and its Control, IEEE Access, pp. 1–6 (2022)
9. Fitriyani, N.L., Syafrudin, M., Alfian, G., Rhee, J.: HDPM: an effective heart disease prediction model for a clinical decision support system. IEEE Access **8**, 133034–133050 (2020)
10. Ahmad, G.N., Fatima, H., Ullah, S., Saidi, A.S.: Efficient medical diagnosis of human heart diseases using machine learning techniques with and without GridSearchCV. IEEE Access **10**, 80151–80173 (2022)
11. Qadri, A.M., Raza, A., Munir, K., Almutairi, M.: Effective feature engineering technique for heart disease prediction with machine learning. IEEE Access **8** (2023)
12. Itoo, N.N., Garg, V.K.: Heart disease prediction using a stacked ensemble of supervised machine learning classifiers. In: 2022 International Mobile and Embedded Technology Conference, pp. 599–604 (2022)
13. Jadhav, S.R., Kulkarni, R., Yendralwar, A., Pujari, P., Patwari, S.: Monitoring and predicting of heart diseases using machine learning techniques. In: 2023 IEEE 8th International Conference for Convergence in Technology, pp. 1–4 (2023)
14. Sulthana, R., Jaithunbi, A.K., Sunraja, P.: Application of machine learning algorithms in predicting the heart disease in patients. In: 2023 Third International Conference on Advances in Electrical, Computing, Communication and Sustainable Technologies, pp. 1–4 (2023)
15. Kumar, K.P., Rohini, V., Yadla, J., VNRaju, J.: A comparison of supervised learning algorithms to prediction heart disease. In: 2023 International Conference on Artificial Intelligence and Knowledge Discovery in Concurrent Engineering, pp. 1–5 (2023)
16. Abdellatif, A., Abdellatef, H., Kanesan, J., Chow, C.O., Chuah, J.H., Gheni, H.M.: An effective heart disease detection and severity level classification model using machine learning and hyperparameter optimization methods. IEEE Access **10**, 79974–79985 (2022)
17. Kapila, R., Ragunathan, T., Saleti, S., Lakshmi, T.J., Ahmad, M.W.: Heart disease prediction using novel quine McCluskey binary classifier (QMBC). IEEE Access **11**, 64324–64347 (2023)
18. Vijaya, J.: Heart disease prediction using clustered genetic optimization algorithm. In: 2023 International Conference on Intelligent and Innovative Technologies in Computing, Electrical and Electronics, pp. 1072–1077 (2023)
19. Tiwari, S.K., Sinha, S., Kaur, K., Singh, R.S.: Prediction of heart diseases using nonlinear techniques and machine learning. In: 2023 10th International Conference on Computing for Sustainable Global Development, pp. 530–535 (2023)
20. Singh, P., Singh, D.P.: A novel framework for prediction of heart disease using hybrid classifier in the cloud environment. In: 2023 10th International Conference on Computing for Sustainable Global Development, pp. 702–707 (2023)

21. Bensenane, H., Aksa, D., Omari, F.W., Rahmoun, A.: A deep learning-based cardio-vascular disease diagnosis system. Indonesian J. Electr. Eng. Comput. Sci. **25**(2), 963–971 (2022)
22. Madani, A., Ong, J.R., Tibrewal, A., Mofrad, M.R.: Deep echocardiography: data-efficient supervised and semi-supervised deep learning towards automated diagnosis of cardiac disease. NPJ. Digit. Med. **1**(1), 59 (2018)
23. Chamundeshwari, Biradar, N., Udaykumar: Adaptive despeckling and heart disease diagnosis by echocardiogram using optimized deep learning model. Comput. Methods Biomech. Biomed. Eng. Imaging Vis. **11**(1), 1–17 (2023)
24. Jain, A., Rao, A.C.S., Jain, P.K., Hu, Y.C.: Optimized levy flight model for heart disease prediction using CNN framework in big data application. Expert Syst. Appl. **223**, 119859 (2023)
25. Hussain, S., Nanda, S.K., Barigidad, S., Akhtar, S., Suaib, M., Ray, N.K.: Novel deep learning architecture for predicting heart disease using CNN. In: 2021 19th OITS International Conference on Information Technology (OCIT), pp. 353–357 (2021)
26. Sharma, S., Parmar, M.: Heart diseases prediction using deep learning neural network model. Int. J. Innov. Technol. Exploring Eng. **9**(3), 2244–2248 (2020)
27. Xie, J., et al.: Prediction of cardiovascular diseases using weight learning based on density information. Neurocomputing **452**, 566–575 (2021)
28. Khalil, M., Adib, A.: An end-to-end multi-level wavelet convolutional neural networks for heart diseases diagnosis. Neurocomputing **417**, 187–201 (2020)
29. Guillermo, J.E., Castellanos, L.J.R., Sanchez, E.N., Alanis, A.Y.: Detection of heart murmurs based on radial wavelet neural network with Kalman learning. Neurocomputing **164**, 307–317 (2015)
30. Islam, M.T., Rafa, S.R., Kibria, M.G.: Early prediction of heart disease using PCA and hybrid genetic algorithm with k-means. In: 2020 23rd International Conference on Computer and Information Technology, pp. 1–6 (2020)
31. Sharma, V., Bajaj, A., Abraham, A.: Machine learning techniques for electronic health records: review of a decade of research. Int. J. Comput. Inf. Syst. Ind. Manag. Appl. **15**, 595–604 (2023)
32. Xiong, X., et al.: A cascaded multi-task generative framework for detecting aortic dissection on 3-D non-contrast-enhanced computed tomography. IEEE J. Biomed. Health Inform. **26**(10), 5177–5188 (2022)
33. Suhail, M.M., Razak, T A.: Cardiac disease detection from ECG signal using discrete wavelet transform with machine learning method. Diabetes Res. Clin. Pract. **187**, 109852 (2022)
34. Arabasadi, Z., Alizadehsani, R., Roshanzamir, M., Moosaei, H., Yarifard, A.A.: Computer aided decision making for heart disease detection using hybrid neural network-genetic algorithm. Comput. Methods Programs Biomed. **141**, 19–26 (2017)
35. Verma, K., Bartwal, A.S., Thapliyal, M.P.: A genetic algorithm based hybrid deep learning approach for heart disease prediction. J Mt Res. **16**(3), 179–187 (2021)

Impact of Contextual Factors on Adaptive Performance: A Study on Indian PSBs Using ANN Analysis

Deeksha Sanjay Shetty(✉) and K. R. Suprabha

School of Humanities, Social Sciences and Management, National Institute of Technology Karnataka, Surathkal 575025, India
{177sm003.deeksha,suprabha}@nitk.edu.in,
deeksha.shetty1@gmail.com

Abstract. The Indian Public Sector Banks (PSBs) are at the cusp of significant structural changes and technological shifts. These radical changes have enormously altered the world of work by endlessly spinning due to the environment's volatility, uncertainty, complexity, and ambiguity (VUCA). The maladaptive conditions created during these circumstances have increased the need for employees with stronger and steadier adaptive performance. Human capital is the asset of any organisation; thus, attracting, retaining, and maintaining these assets becomes the focal function of financial institutions like banks. Training and development activities are ongoing and crucial organizational processes to manage talents. The vitality of these activities is easily traceable in the institute's growth graph. Thus, identifying the contextual competencies and training the employees to enhance these competencies is a challenge to any organization. This study proposes the application of Artificial Neural Network (ANN) as a tool to explore the competency in demand during disruptions in the Indian public sector banks. The use of ANN in accruing answers to research questions about employee competency management to enhance adaptive performance is a rare and inevitable study while dealing with disruptions in the banking sector. This study investigates the influence of Work Autonomy (WA), Emotional Stability (ES), Workplace Spirituality (WS), and Career Adaptability (CA) on Adaptive Performance (AP). The current study results are convincingly satisfactory and are very close to the Structural Equational Modelling (SEM) outcomes when compared with the same.

Keywords: Contextual competencies · Career adaptability · ANN · Adaptive performance · Indian PSBs

1 Introduction

The present-day age of information technology has led to various changes in the occupational hierarchy. The fourth industrial revolution has posed some real-life challenges to the existing nature of the workforce, as stated by a report from the World Economic Forum. Displacement of the workforce is undoubtedly attributed to technological advancements such as Artificial Intelligence (AI), the Internet of Things (IoT), Machine Learning (ML), and digitization in the nation and worldwide [1].

V. Sakalauskas et al. (Eds.): IBICA 2023, LNNS 1231, pp. 346–354, 2025.
https://doi.org/10.1007/978-3-031-78946-5_33

Bankers, specifically, are excavating talents with an adaptive mindset who can deal with uncharted landscapes in the VUCA world [2]. The unprecedented challenges and persistent innovations have increased the appetite for novel and rational approaches to deal with changes in the workplace. Rigidity in work conduct will definitely impose guaranteed struggle in acclimatization to changing work scenarios. While the existing workforce strives hard to retain their positions, the aspiring workforce has to carve a niche to convince the recruiters to make a place for them in a highly demanding and complex job market.

Human capital is the asset of any organisation; thus, attracting, retaining, and maintaining these assets becomes the focal function of institutions. It does not come without the involvement of cost. Training and development activities are ongoing and crucial processes in organizations. The vitality of these activities is easily traceable in the institute's growth graph. Thus, organisations, must let employees evolve, grow and excel every day.

Nonetheless, the transition cannot happen without the involvement of the work place [2]. The need of the hour is a strategically devised training intervention based on individual and organisational requirements relevant to the complex work world. Identifying the skills in demand can be an arduous task for the organisation. Unspotted and unfilled gaps in the training process will not only add to the cost to the company but also give rise to undesirable human challenges.

The Indian banking sector has witnessed a noticeable evolution since it began, especially after India's independence [3]. Banks were founded with varying visions of the founders across the country, which were then taken under the Government's control in the name of nationalisation in episodes in the sixties and eighties. Foreign banks came to India's doorstep due to globalisation, privatisation, and liberalisation, which posed heavy competition to the already existing native bankers. Such structural modifications are always around the corner for the banking industry for the betterment of the sector and the nation at large. The recent Government of India's initiation to merge banks proved to be a bold step in banking history. These developments are a subtle caution for the banks that privatisation, which seems to be a possibility, might be in deed one day. After considering the assorted benefits guaranteed by bank mergers, policymakers will seldom question the outcome before making such decisions when the results are foreseeable. The amalgamation will lead to a synergy benefiting all participating parties. Post-merger analysis revealed that the restructuring has brought immense benefits for the merged banks [3]. These structural changes often come with countless compromises on the part of the employees. To name a few, cultural differences, physical adaptations, challenges posed by leadership changes, abrupt transfer of employees beyond their list of preferences, unthought-off promotions, under-devised training interventions, lack of empathetic take on the situation, and the worst-case scenario is the colossal job cuts due to new workplace demands. Despite all these hardships associated with the change exercise, the policymakers cannot unsee the gigantic advantages that come with structural and other strategic disruptions. Training of bank employees toward change acclimatization becomes an inevitable process altogether. Extant literature upholds the various ways to explore the impact of contextual factors on adaptive performance using qualitative and quantitative inquiries. There is a dearth of ANN usage for exploring the complex problems pertaining to human skill inadequacy identification during organisational disruptions. Machine learning tools like ANN will simplify the process by identifying the patterns, the regression analysis of the identified competencies on the desired performance expectancies of employees, and many more [4]. The focal point of the proposed study is to investigate the impact of four contextual variables on employee adaptive performance in Indian PSBs.

The paper layout is as follows: 2. The review of the literature regarding the variables appearing in the study, ANN application in various extant literature in related fields, and Indian PSB: 3. Research methodology used in this study: 4. ANN application to study the impact of contextual factors on AP: 5. Structural Equational Model ling output and ANN output comparison: 6. Conclusion and future scope of study highlighted by the researcher.

2 Review of Literature

In the banking context, recent studies have generally focused on various factors that affect the changing structure of banking models based on the financial, physical, regulatory, and technological infrastructure. The most important yet least considered infrastructure, i.e., Human resources, draws many researchers' interest in the utmost change-prone and high-pressure sectors like banking. Further, some studies have assessed the impact of employee competencies on the overall performance of an employee. Several studies have scrutinized the forecasters of adaptive performance using various methodologies, but the application of the ANN method is yet to be fully explored in HRM.

Adaptability is one of the most preferred competencies to help employees gain entry into an organization, promote growth, and guarantee sustainable career progression [5]. Until recently, adaptive performance did not have a standalone status; it was usually considered a part of contextual performance or may have camouflaged with some other dimension of performance. The initial study on adaptive performance and its independent existence in incredibly contributing to organizational success was highlighted by Hesketh and Neil in 1999 [6]. Pulakos and her colleagues have contributed hugely to developing the taxonomy of adaptive performance and have successfully paved the way for the researchers for further research to take off from there. Adaptive performance signifies the ability of an individual to modify their behaviour to encounter the demands of changing environmental needs [7].

The uncharted water of positive contributions of spirituality invites scholars to explore the multi-dimensional aspects of workplace spirituality. Ashmos and Duchon (2000) [8] speak about spirituality as the sense of an internal presence, which is rein forced by the exchange of feelings and association among people. The current study is more directed toward exploring the relationship between workplace spirituality and adaptive performance.

Work autonomy denotes the autonomy managers offer staff when executing their work-related errands or making certain crucial decisions for the organisational wellbeing. Work autonomy augments the employee's self-determination, boosting the adaptive capacities and thus exhibiting adaptive performance. Work autonomy has been acknowledged as an imperative predictor of adaptive performance [9].

The career adaptability variable comprises a positive attitude towards concern, curiosity, confidence, and control regarding an employee's career graph [10]. Career adaptability originates from career construction theory and is the willingness to handle the foreseeable tasks of getting ready to participate as per the assigned job role and ready for all the impulsive adjustments provoked by disruptions in work and working environment. Career adaptability is a recent addition to the competence vocabulary where personnel apply resources to effectively tackle new work challenges and disruptive work environments [10].

As per Weisinger (2000), intelligence is the intelligent use of emotions in the workplace. According to the current context, employees with high emotional stability exhibit high adaptive responses to suit the demands of today's labour pool. The change exercise in the organisation is not easy to acclimatize to as it involves a lot of emotions. Humans and emotions are inseparable; thus, dealing with humans without disturbing their emotional state is nearly impossible [12]. This study investigates the impact of emotional stability on Adaptive performance using ANN.

ANNs technique stems from Machin Learning (ML), a Deep Learning (DL) sub field. ANNs have gained popularity recently due to their contribution to the field of enormous data processing and simulating the activities of the human brain [11]. Artificial Neural Network models are the finest when there is an influential form of experimental data but no clear theoretical underpinning exists to develop extrapolative linkages. ANNs impersonate biological neural networks to process information from varied fields across boundaries. Various studies in recent times have highlighted the contribution of ANN and have accepted it as a vigorous method with diverse applications; namely, ANNs were applied in field marketing, supply chain management, performance benchmarking, project management, quality-related studies, time scheduling, and forecasting demand [13]. The current research proposes using ANNs to study contextual factors' impact on adaptive performance during structural changes like mergers and acquisitions in Indian PSBs (Figs. 1, 2 and 3).

3 Research Methodology

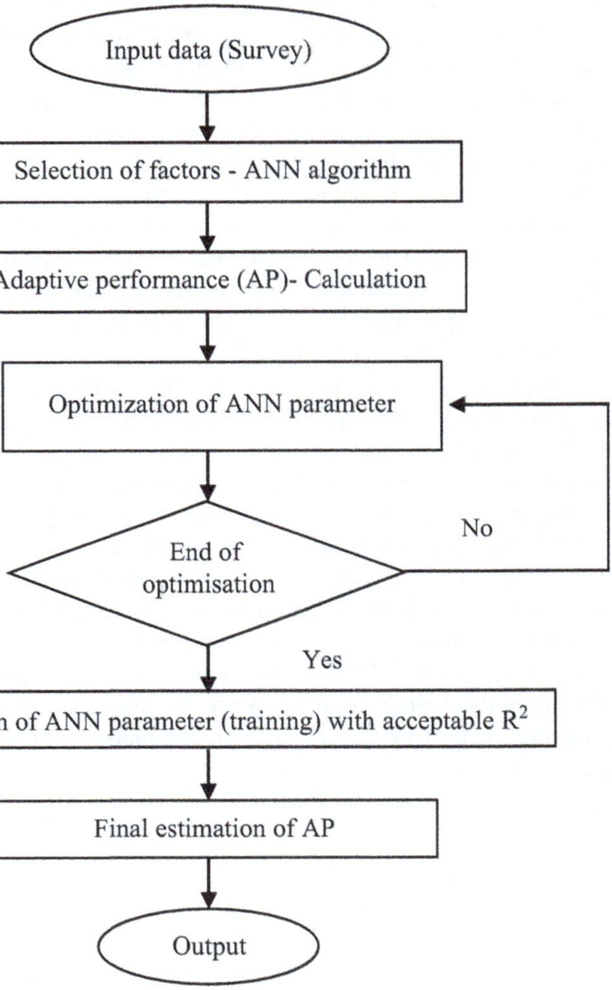

Fig. 1. Artificial Neural Network (ANN) Flowchart

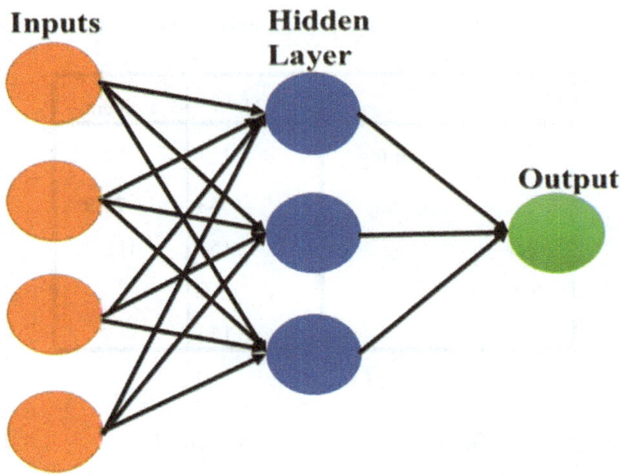

Fig. 2. Basic ANN Model

4 ANN Method for Extracting Study Outcomes

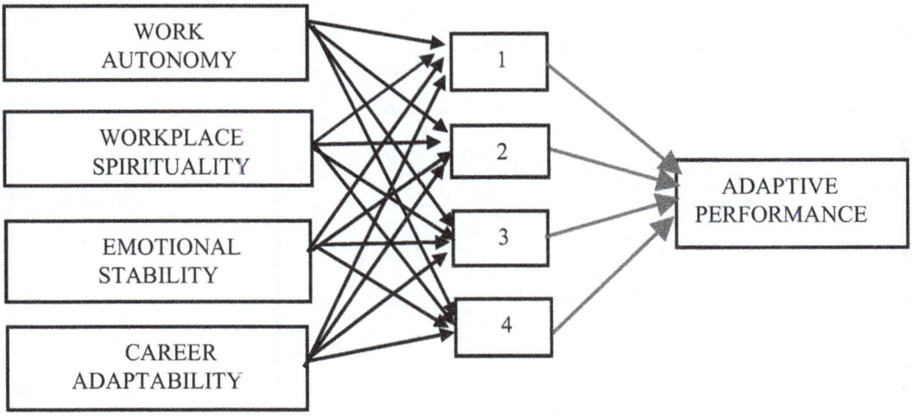

Fig. 3. Proposed ANN Model

Dependent variable: Adaptive Performance (AP)

a. The number of hidden units is determined by the testing data criterion: The "best" number of hidden units is the one that yields the smallest error in the testing data (Tables 1 and 2).

Table 1. Case processing summary

		N	Percent
Sample	Training	111	72.1%
	Testing	43	27.9%
Valid		154	100.0%
Excluded		0	
Total		154	

Table 2. Results of ANN Analysis

Training	Sum of Squares Error		46.820
	Average Overall Relative Error		.182
	Relative Error for Scale Dependents	WS	.422
		WA	.238
		ES	.412
		CA	.222
		AP	.232
	Training Time		0:00:00.00.23
Testing	Sum of Squares Error		21.223[a]
	Average Overall Relative Error		.236
	Relative Error for Scale Dependents	WS	.343
		WA	.118
		ES	.443
		CA	.217
		AP	.282

5 Comparative Study

The impact of contextual factors on adaptive performance was investigated using SEM and ANN analysis. The relationship between the unobserved variables, such as independent and dependent, was analysed with the SEM analysis. The result obtained from the structural equation modeling (SEM) was further compared with the results of ANN [13]. The outputs were relatively in line with each other, excluding some exceptions.

Table 3. Comparison of SEM and ANN outputs

Contextual Factors	SEM Result	ANN Result Training I	ANN Result Testing II
WA	0.379	.238	.118
ES	0.298	.412	.443
WS	0.205	.422	.343
CA	0.323	.222	.217

Source: Result obtained from SEM and ANN analysis (based on author's study)

As per SEM output, workplace autonomy and career adaptability have a more significant influence, whereas emotional stability ranked third in the order, and work place spirituality took fourth place in both SEM and ANN outputs. The data for analysis was collected from Indian PSBs by a self-administered questionnaire. The sample size was 154, which may have also impacted the output.

6 Conclusion

The researcher's initiative to step into the employees' shoes during these restructuring processes and other change dynamics in the bank has taken the work to a different perspective altogether. This study lays the foundation for researchers interested in HR and OB to unearth solutions to all human capital-related complications in multiple ways to explore the factors influencing performance dimension, adaptive performance in particular. The research outcomes, a comparative analysis of both SEM and ANN outputs, would serve as a reference piece to the policymakers, government, banking institutions, academicians, and researchers in the organisational behavior and positive psychology field. The vantage point of this research work is to uphold the impact of contextual competencies on adaptive performance in the banking context and to urge the policymakers, academicians, and researchers to address the importance and identify enablers of adaptive performance and devise the training modules based on the research conclusions. The ANN analysis helps to identify the more impacting factors that influence employee performance, and training interventions can be designed based on these outcomes. Future research in this area can happen by taking into consideration the Feed Forward ANN (FFANN), Deep Neural Network, and Recurrent Neural Network. Rather than comparing the results obtained by SEM and ANN, re searchers should conduct a hybrid SEM-ANN analysis. Emotional Neural Network (ENN) can significantly help researchers in psychology as it successfully captures the emotional side of the research respondents, which can be of excellent value to the researchers.

References

1. George, G., Howard-Grenville, J., Joshi, A., Tihanyi, L.: Understanding and tackling societal grand challenges through management research. Acad. Manag. J.Manag. J. **59**(6), 1880–1895 (2016)

2. Huang, Z., Shahzadi, A., Khan, Y.D.: Unfolding the Impact of Quality 4.0 Practices on Industry 4.0 and Circular Economy Practices: A Hybrid SEM ANN Approach. Sustainability; Multidisciplinary Digital Publishing Institute (2022). https://doi.org/10.3390/su142315495
3. Singh, R., Goel, G., Ghosh, P., Sinha, S.: Mergers in Indian public sector banks: can human resource practices ensure effective implementation of change? Management Decision; Emerald Publishing Limited (2021). https://doi.org/10.1108/md-09-2020-1279
4. Abiodun, O.I., Omolara, A.E., Dada, K.V., Mohamed, N., Arshad, H.: State-of-the-art in artificial neural network applications: A survey. Heliyon; Elsevier BV (2018). https://doi.org/10.1016/j.heliyon.2018.e00938
5. Park, S., Park, S.: Employee Adaptive Performance and Its Antecedents: Review and Synthesis. Human Resource Development Review; SAGE Publishing (2019). https://doi.org/10.1177/1534484319836315
6. Hesketh, B., Neal, A.: Technology and performance. In: Ilgen, D.R., Pulakos, D.P. (eds.), The Changing Nature of Performance: Implications for Staffing, Motivation and Development, pp. 21–55 (1999)
7. Pulakos, E.D., Arad, S., Donovan, M.A., Plamondon, K.E.: Adaptability in the workplace: Development of a taxonomy of adaptive performance. J. Appl. Psychol. **85**(4), 612–624 (2000)
8. Ashmos, D.P., Duchon, D.: Spirituality at work- a conceptualization and measure. In: Journal of Management Inquiry (Vol. 9, Issue 2, pp. 134–145) (2000)
9. Tabiu, A., Pangil, F., Othman, S.Z.: Does Training, Job Autonomy and Career Planning Predict Employees' Adaptive Performance? Glob. Bus. Rev. **21**(3), 713–724 (2020)
10. Neureiter, M., Traut-Mattausch, E.: Two sides of the career resources coin: Career adaptability resources and the impostor phenomenon. Journal of Vocational Behavior; Elsevier BV (2017). https://doi.org/10.1016/j.jvb.2016.10.002
11. Sony, M., Mekoth, N.: The relationship between emotional intelligence, frontline employee adaptability, job satisfaction and job performance. Journal of Retailing and Consumer Services; Elsevier BV (2016). https://doi.org/10.1016/j.jretconser.2015.12.003
12. Hertz, J., Krogh, A., Palmer, R.G.: Introduction to the theory of neural computation. Santa Fe Institute Studies in the Sciences of Complexity (Vol. 1), Redwood City, CA. Addison Wesley (1991)
13. Albahri, A.S., et al.: Hybrid artificial neural network and structural equation modelling techniques: a survey. Complex & Intelligent Systems; Springer Science+Business Media (2021). https://doi.org/10.1007/s40747-021-00503-w

Generalized Skin Cancer Image Classification Performance Using Xception Model

Qurban A. Memon(✉), Ghaya Al Ameri, Namya Musthafa, Aryam AlShamsi, and Aisha AlYaqoubi

UAE University, 15551 Al Ain, United Arab Emirates
qurban.memon@uaeu.ac.ae

Abstract. Skin cancer is well known and regarded as one of the most common types of cancer, killing millions of people worldwide. Detecting and categorizing cancer at an early stage can be advantageous, leading to a faster and higher success rate of therapy. Intelligent technologies are currently being used to classify skin lesions. The fundamental goal of our experimental research is to investigate biomedical skin cancer datasets in order to develop an effective approach for determining whether a cancer is malignant or benign. To train and categorize the dataset images, CNN (sequential), ResNet-50, Inception v3, and Xception models are employed. Two large and balanced datasets are collected for this purpose. One is used to compare the performance of employed model algorithms. Next, the selected model is again retrained on the second dataset for validation and generalization purposes. It turns out that the performance of Xception model is generalized and outperforms other models in accuracy. Experimental results are tabulated and graphed using accuracy and confusion matrix.

Keywords: Skin Cancer Classification · Machine Learning · Xception model

1 Introduction

Skin cancer is a disorder that causes abnormal cells to form in skin tissues. When skin cells age and die, new ones arise to replace them. Cells grow quicker when this mechanism fails, as it occurs after being exposed to ultraviolet radiation from the sun. Symptoms include new bumps or spots on the skin, as well as alterations in the size, form, or color of skin growths. These cells may be noncancerous (benign), meaning they do not spread or harm the body. They could also be malignant. Ultraviolet (UV) radiation from tanning beds and sunlamps can also raise the risk of skin cancer.

Skin cancer is classified into three categories. Types that grow slowly and rarely spread to other regions of the body include Squamous cell carcinoma, which grows in squamous cells in the outer layer of skin, and basal cell carcinoma, which develops in basal cells in the lower epidermis (the outside layer of skin). Melanoma begins in melanocytes cells, a brown pigment, known as melanin, which gives color to skin while also shielding it from part of the sun's harmful ultraviolet radiation. Because it can spread

V. Sakalauskas et al. (Eds.): IBICA 2023, LNNS 1231, pp. 355–365, 2025.
https://doi.org/10.1007/978-3-031-78946-5_34

to other parts of the body, this is the deadliest kind of skin cancer and causes a large number of skin cancer deaths.

Most skin cancers are curable if detected early. In terms of patient well-being and healthcare expenditures, diagnosis and treatment pose a substantial health concern. A full body skin examination is performed on people who are at risk. The vast majority of suspicious lesions found during a physical examination will be subjected to a skin biopsy. The specimen is subsequently delivered to a qualified dermatopathologist for interpretation. If the pathologist's examination of the tissue specimen under the microscope establishes the diagnosis of cutaneous malignancy, further action is usually indicated based on the pathologic assessment and clinical circumstances. The dermatologist will then assess whether the skin lesion is a sign of skin cancer, its type, and the treatment options available.

The current skin cancer detection processes are time-consuming, exhausting, and invasive, and they might result in mental health issues. Timely detection and classification are essential for favorable outcome. Artificial intelligence-assisted or enhanced picture recognition appears to be a viable solution for early detection of different types of cancer [1]. While deep learning algorithms are often used for identification and classification, machine learning and hybrid approaches are also used [2]. These approaches have proven to be excellent classifiers, with promising early detection outcomes.

2 Literature Review

When "skin cancer detection" was typed on google search engine, it showed a total of 129,000,000 results ranging from foundations to articles, support, and resource centers. This shows voluminous information available for general information on research articles, institutes, grants, and support services on preventive approaches, etc. Below, a review of selective articles and papers from conference proceedings is presented.

The authors [3] use raw DNA sequences from matched tumor/normal pairs as input to the proposed technique, which is subsequently processed by sentence transformers to provide DNA representations. Such representations are classified through a series of machine learning methods in order to improve cancer detection. Another paper [4] addresses the challenges and potential of AI and machine learning in cancer imaging, including the use of freely available resources for algorithm development to improve collaboration and openness across centers, as well as the formation of the ecosystem required to encourage artificial intelligence and machine learning adoption in the field of cancer imaging. The authors [5] suggested two approaches for detecting and classifying benign and malignant cancers in dermoscopic imagery. The first method employs k-nearest neighbor (KNN), which utilizes pretrained deep neural networks as feature extractors. Alex Net with grey wolf optimizer is the second. In categorizing skin cancer images, the results are compared to ML and DL techniques. From the ISIC archive dataset with four thousand imagery, the experiments are trained and tested. The results demonstrated that the proposed methods exceeded the other examined techniques.

The study [6] proposes a collaborative learning system based on Convolutional Neural Networks (CNN) and Local Binary Patterns (LBP), followed by the conjunction of all retrieved features using CNN and LBP architecture. To handle multiclass skin-related challenges, the suggested system is trained and tested using a commonly utilized accessible public dataset for skin cancer diagnosis. The architectures and their fusion are compared in terms of results, which demonstrate the fusion architecture's robustness, with an accuracy of 98.60% and a validation accuracy of 97.32%. Another study [7] compares three types of Convolutional Neural Network (CNN) algorithms on a nine-class skin cancer categorization problem, with the model with the highest accuracy being integrated into a web application. VGG-16, VGG-19, and an independently developed network are among the three CNN models that have been compared. The most accurate model, according to test findings, is VGG-19, which obtained 0.9290 in accuracy and 1.2842 in loss, offering a reliable way to aid in skin cancer identification.

Deep learning models have also been explored in skin cancer detection. The research [8] investigates the most recent deep learning-based methods for skin cancer categorization. This research discusses dermatological forms of images with a list of openly accessible datasets related to skin cancer, followed by the successful utilization of convolutional neural networks for skin cancer classification. A final summary of various unresolved issues such as disparity in data, data constraints, domain-specific adaptation, model reliability, and efficiency. The work [9] discusses how a model-based architecture in the cloud employing learning algorithms is utilized to build models that help detect skin cancer. The deep learning models created in this article are tested on common datasets, and the metric area under the curve is measured with 99.77%. Deep learning is also investigated in [10, 11], where two kinds of cancer and one non-cancer form are categorized using a CNN model based on VGG 19 and the transfer Learning utilizing the Human Against Machine (HAM10000) dataset. By assessing the network's overall accuracy and loss, the training approach is presented, tested, and assessed.

The authors [12] present a lightweight and effective deep learning method for classifying skin cancer. The method fine-tunes the pre-trained MobileNet convolution neural network before training it on the HAM1000 skin lesion dataset. Another study [13] employed MobileNetV2 for training to detect skin cancer and U-Net for cancer boundary segmentation, as a skin cancer identification model using the Skin Cancer MNIST dataset. The models were loaded onto a Raspberry Pi to create a useful gadget for pre-screening the skin of the patient. The research [14] tests the HAM10000 dataset on modern models such as AlexNet, VGG-16, and InceptionV3 for skin cancer categorization and compares with a DeepSkinNet model. The experimental experiments revealed that the proposed model outperforms the other models in terms of accuracy.

Based on literature review, it is found that most of the approaches are tested on a single limited dataset size, thus the results may not be generalized. The aim of this study is to propose an automatic customized machine learning based skin cancer detection approach that is applied on two publicly found datasets with images of skin cancer lesion to improve confidence of the physician/dermatologist in detecting cancer in the early stage. The paper is structured as follows. In the next section, the proposed approach is represented, where details of the architecture and publicly available datasets and performance metric are discussed. Section four presents experimental results and comparison done with works found in the literature. In section five, conclusions are discussed.

3 Models and Datasets

The process of training models to recognize patterns in data and make predictions or judgments based on those patterns is central to machine learning. It is based on statistical models and algorithms that allow researchers to investigate the performance of a specific task by learning from available data. As labelled datasets are now becoming easily accessible in public domain, a number of deep learning models have been proposed in the literature with claimed accuracy reaching 90% on selected datasets. With availability of high computational power machines, it seems easy to validate the claimed accuracy on a given dataset. But most of the time, it is not possible to find testing of models on more than one balanced dataset. In this research, the methodology is to investigate deep learning different models (like CNN sequential, Inception v3, ResNet50 and Xception) and calculate their performance during training and testing. Once a model has been selected from this set, its performance will be further investigated on a larger dataset for its generalization performance. The methodology is depicted in shown in Fig. 1.

The loaded dataset 1 requires to be preprocessed and labelled before being supplied to algorithms for model building. It was made sure that the dataset chosen for the model needs to be balanced for accurate prediction. In the first phase, four algorithms mostly found in literature for skin cancer classification are proposed to be trained, cross validated and tested before being declared suitable for further investigation on generalization. The criterion chosen for performance measure was accuracy in classification. In the next phase, the chosen model is selected for generalization, in fact, to be retrained on another balanced and larger dataset. The final performance measure includes computing confusion matrix for ascertaining earlier accuracy measure. Below, the selected models and datasets are discussed briefly.

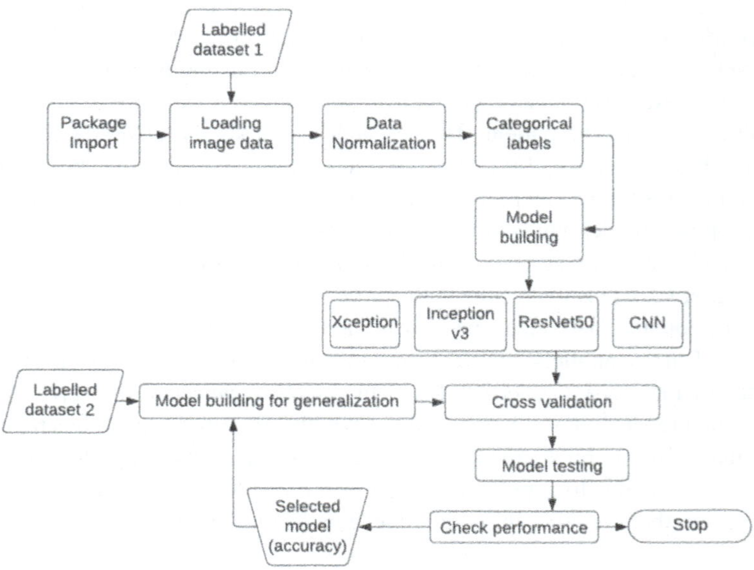

Fig. 1. Skin Cancer Classification Methodology

3.1 Models

A convolutional neural network (CNN) is a form of neural network that uses a grid-like architecture to process data. Simple patterns (lines, curves, and so on) are detected initially, followed by more intricate patterns (faces, objects, and so on). CNN is intended to learn spatial feature hierarchies via backpropagation by utilizing several convolution layers, pooling layers, and a fully connected layer [15].

The study [16] presented several enhancements that improve the accuracy and lower the computational complexity of a deep convolutional network. This work, discussed in the context of the Inception architecture, can result in networks with cheap computing cost while keeping good quality by discussing factorizing convolutions and aggressive dimension reductions inside neural networks. The inception blocks are the primary distinction between Inception models and ordinary CNNs. These involve concatenating the results of many filters on the same input tensor. All enhancements stated for Inception v2 were included in Inception Net v3, as well as the RMSProp Optimizer. Some of the techniques used are factorized 7×7 convolutions, BatchNorm in Auxiliary Classifiers, and Label Smoothing.

Residual Network50, abbreviated as ResNet50 is a CNN architecture of fifty layers. The "Residual" in ResNet refers to the usage of residual blocks in network design, which feature skip connections by enabling the network to learn residual functions. To ease the vanishing gradient problem, these links allow the network to bypass specific levels and directly convey information from earlier layers to later layers [17]. ResNet50 has acquired differentiating features by training on image datasets such as ImageNet, and it can be further tweaked or altered for specific visual identification tasks [18].

Xception, a deep layer neural network developed by Google researchers, is a pre-trained network trained over a million images from the ImageNet repository [19]. It takes the principles of Inception to an extreme. It bases itself on Inception v3, but instead of modules, it utilizes depth wise separable convolutions. As a result, it is more effective and works significantly better than Inception v3 with the same number of variables.

3.2 Datasets

We employ two datasets, one for training of different machine learning models, and another for investigation generalization of the model selected in the first phase. The first freely available dataset [20] contains a balanced total of 2637 training images divided into 1440 benign and 1197 malignant images. The testing set also contained a balanced total of 660 images divided into 360 benign and 300 malignant images. HAM10000, also known as Human against Machine with ten thousand images for training, is another dermatology image dataset [21] that is freely available for building and validating machine learning algorithms for categorizing skin lesions. This dataset, which was released in 2018 at the Medical University of Vienna, consists of over 10,000 dermatoscopic images that can be classified into separate diagnostic categories. In the next section, the results are reported based on training and testing of these models.

4 Experimental Results

Preprocessing is a critical stage in machine learning workflows since it is required for data preparation prior to model training. The primary goals of picture preprocessing are to improve image quality, clarity, and to remove unwanted objects or background noise. Preprocessing in the context of skin cancer classification essentially consists of multiple procedures aimed at preparing and improving the quality of medical pictures and associated data for correct analysis and diagnosis. After loading images from dataset, three steps were performed (i) Standard Scaler processing removed the mean and scales each feature/variable to unit variance (ii) each image was normalized from 0 to 1 (iii) resulting images were labelled.

Four models i.e., Sequential CNN, ResNet50, Inception v3 and Xception were employed on the first public dataset that contained 3297 skin cancer images. These images were divided into two sets: training and testing with ratio of approximately 80:20. The CNN model was trained for fifty epochs with a learning rate of 0.00001. The training and validation were performed threefold, and the model performance turned out to be moderate with an average accuracy of 74.63%. The results are displayed in Table 1, and the training and testing accuracy is graphed in Fig. 2a. The RsNet50 model was also trained for fifty epochs with a learning rate of 0.00001. The training and validation accuracy improved compared to CNN and was recorded as 96.92% and 80.5% respectively. The results are shown in Table 1 and resulting graphs in Fig. 3. The Xception model was applied as well on the same dataset for ten epochs and resulting training and testing accuracy was scored as 97.84% and 86.9% respectively. For comparison purposes, these scores are listed in Table 1, and plot is shown in Fig. 4.

Table 1. Model Accuracy

Model	Accuracy	
	Training	Testing
CNN	78.5%	69.53%
ResNet50	96.92%	82.88%
Inception V3	93.45%	86.7%
Xception	97.84%	86.9%

Based on tabulated results, it appears that Xception has performed better than other models. In order to generalize this performance, another HAM10000 dataset was employed to ascertain the Xception model performance. The model carried the same parameters, and resulting accuracy turned to be 98.18% for training and 89.4% for testing. The resulting plot is graphed is Fig. 3. The results in Figs. 2 and 3 suggest that Xception model performance is superior to CNN (sequential) and ResNet50 models for skin cancer classification. To investigate further, the confusion matrix was used to reveal the model's accuracy in predicting distinct classes, including both accurate and incorrect classifications. Though the testing set is balanced, accuracy measure would be enough. But for the interest of the reader, the confusion matrix on trained Xception model was also computed for one thousand testing images (500 malignant, 500 benign) from dataset 2, and is shown in Fig. 6. Testing results displayed in Table 2 and shown in Fig. 6 reaffirm the better and generalized investigation performance of the Xception model.

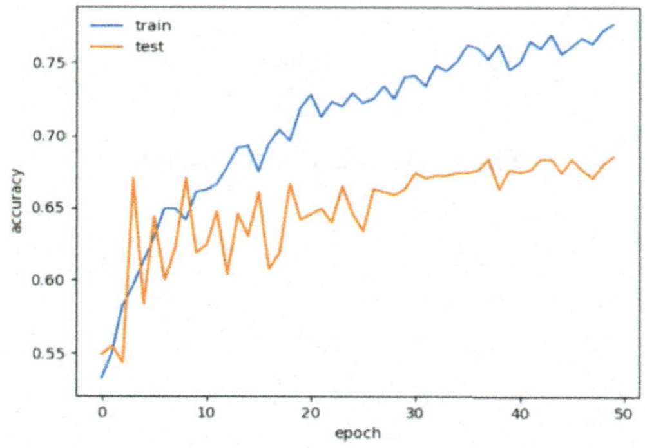

Fig. 2. CNN model accuracy

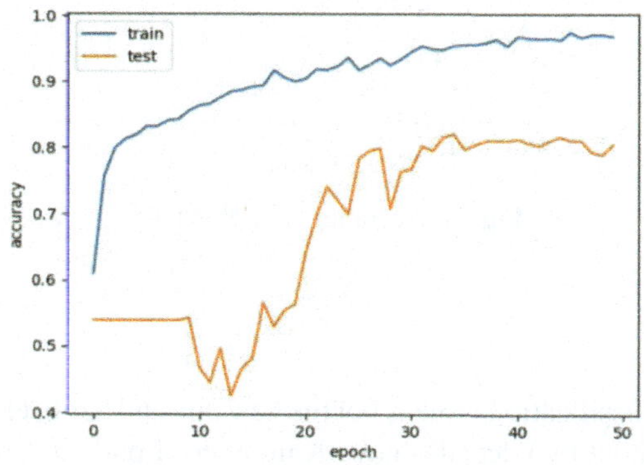

Fig. 3. ResNet50 model accuracy

Table 2. Confusion Matrix Result

Measure	Accuracy	Precision	Recall	F1-Score
Score	89.4%	97.1%	81.2%	88.4%

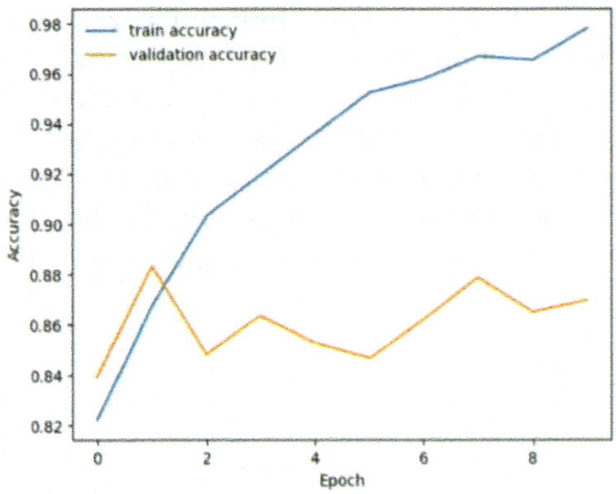

Fig. 4. Accuracy on 1st dataset

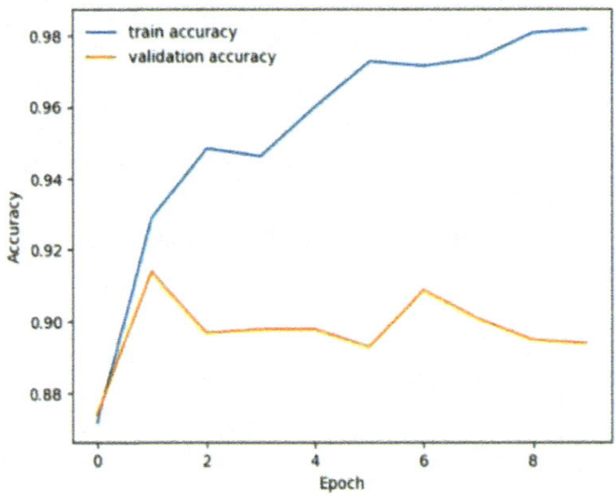

Fig. 5. Accuracy on 2nd dataset

5 Conclusions

Skin cancer image classification issue is not the new one and has been under investigation by researchers for about two decades now. A number of patient care models have been investigated and reported in published literature [22] for classification of human body anomalies. In addition to that, the concerned research groups have facilitated this research by accumulating a number of datasets available in public domain. In this work, the strategy was to investigate a model that does not only provide a good accuracy in detecting benign or malignant cells but also be generalized on different datasets with equivalently

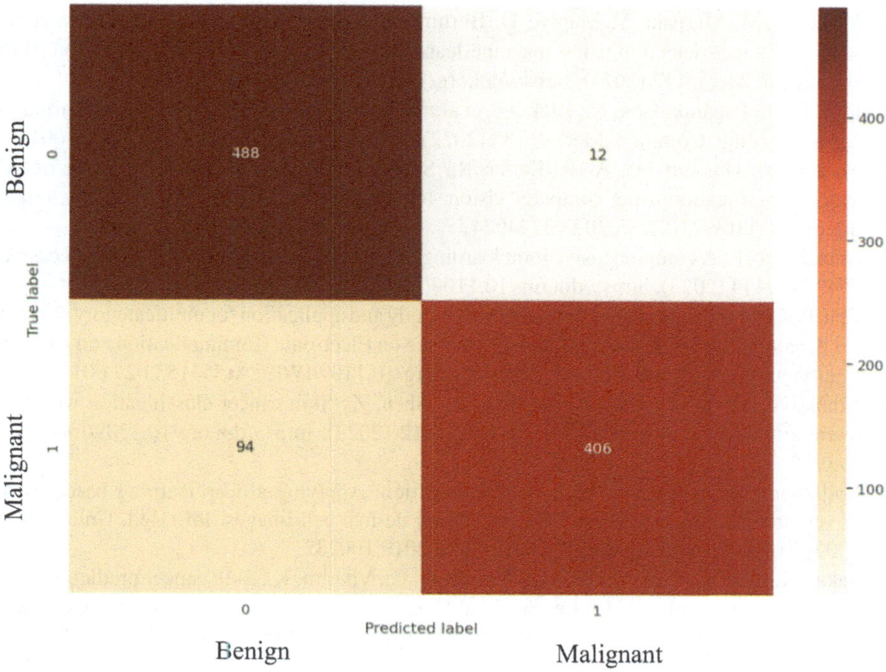

Fig. 6. Confusion matrix for Xception model

comparable results. For this purpose, a number of deep learning models were compared on a large and balanced dataset, and later the better performing model in the first phase was trained again on a larger dataset to ascertain its generalization performance. It turned out that the Xception model performed promising on both datasets with training and testing accuracy reaching 99% and 90% respectively. Though a limited number of epochs were used to train and validate the model, it is believed that the performance of the selected model could be optimized by fine tuning hyperparameters. The confusion matrix developed on one thousand dermatoscopic images displayed similar characteristics.

Acknowledgments. The authors of this work wish to express thanks to United Arab Emirates University for Summer Undergraduate Research Experience Grant No. G00004356, 2023.

References

1. Asadi, B., Memon, Q.: Efficient breast cancer detection via cascade deep learning network. Int. J. Intell. Netw. 4, 46–52 (2023). https://doi.org/10.1016/j.ijin.2023.02.001
2. Memon, Q., Valappil, N.: On multi-class aerial image classification using learning machines. In: Computer Vision and Recognition Systems Using Machine and Deep Learning Approaches: Fundamentals, technologies, and applications, pp. 351–384 (2021). IET Digital Library, https://doi.org/10.1049/pbpc042e_ch15

3. Mokoatle, M., Marivate, V., Mapiye, D., Bornman, R., Hayes, V.M.: A review and comparative study of cancer detection using machine learning: SBERT and SimCSE application. BMC Bioinform. **24**(1), 112 (2023). https://doi.org/10.1186/s12859-023-05235-x
4. Koh, D.M., Papanikolaou, N., Bick, U., et al.: Artificial intelligence and machine learning in cancer imaging. Commun. Med. **2**, 133 (2022). https://doi.org/10.1038/s43856-022-00199-0
5. Magdy, A., Hussein, A., Abdel-Kader, R., Salam, K.: Performance enhancement of skin cancer classification using computer vision. IEEE Access **11**, 72120–72133 (2023). https://doi.org/10.1109/ACCESS.2023.3294974
6. Riaz, L., et al.: A comprehensive joint learning system to detect skin cancer. IEEE Access **11**, 79434–79444 (2023). https://doi.org/10.1109/ACCESS.2023.3297644
7. Zhu, P.: Convolutional neural networks based study and application for multicategory skin cancer detection. In: IEEE International Conference on Electronic Communication and Artificial Intelligence, pp. 558–561 (2022). https://doi.org/10.1109/IWECAI55315.2022.00114
8. Yinhao, C.B., An, Z., Dan, P., Ruixuan, W., Shen, Z.: Skin cancer classification with deep learning: a systematic review. Front. Oncol. **12** (2022). https://doi.org/10.3389/fonc.2022.893972
9. Kadampur, M., Riyaee, S.: Skin cancer detection: Applying a deep learning based model driven architecture in the cloud for classifying dermal cell images. Inf. Med. Unlocked **18**, 100282 (2020). https://doi.org/10.1016/j.imu.2019.100282
10. Sekar, R., Reddy, Nani, Y., Reddy, Chiranjeevi, C., Vikram, K., Skin cancer prediction using deep learning techniques. In: International Conference on Signal Processing and Communication, Coimbatore, India, pp. 28–31 (2023). https://doi.org/10.1109/ICSPC57692.2023.10126035
11. Abuared, N., Panthakkan, A., Al-Saad, M., Amin S., Mansoor, W.: Skin cancer classification model based on VGG 19 and transfer learning. In: International Conference on Signal Processing and Information Security, Dubai, pp. 1–4 (2020). https://doi.org/10.1109/ICSPIS51252.2020.9340143
12. Younis, H., Bhatti, M., Azeem, M.: Classification of skin cancer dermoscopy images using transfer learning. In: International Conference on Emerging Technologies, pp. 1–4 (2019). https://doi.org/10.1109/ICET48972.2019.8994508
13. Jen Ngeh, C., Ma, C., Kuan-Wei Ho, T., Wang, T., Raiti, J.: Deep learning on edge device for early prescreening of skin cancers in rural communities. In: IEEE Global Humanitarian Technology Conference, Seattle, WA, pp. 1–4 (2020). https://doi.org/10.1109/GHTC46280.2020.9342911
14. Abhiram, A., Anzar, S., Panthakkan, A., DeepSkinNet: a deep learning model for skin cancer detection. In: International Conference on Signal Processing and Information Security, Dubai, pp. 97–102 (2022). https://doi.org/10.1109/ICSPIS57063.2022.10002541
15. Yamashita, R., Nishio, M., Do, R.K.G., et al.: Convolutional neural networks: an overview and application in radiology. Insights Imaging **9**, 611–629 (2018). https://doi.org/10.1007/s13244-018-0639-9
16. Szegedy, C., Vanhoucke, V., Ioffe, S., Shlens, J., Wojna, Z.: Rethinking the inception architecture for computer vision. In: IEEE Conference on Computer Vision and Pattern Recognition, Las Vegas, NV, USA, pp. 2818–2826 (2016). https://doi.org/10.1109/CVPR.2016.308
17. Gouda, N., Amudha, J.: Skin cancer classification using ResNet. In: IEEE 5th International Conference on Computing Communication and Automation, Greater Noida, India, pp. 536–541 (2020). https://doi.org/10.1109/ICCCA49541.2020.9250855
18. Zhou, C., Song, J., Zhou, S., Zhang, Z., Xing, J.: COVID-19 Detection Based on Image Regrouping and ResNet-SVM Using Chest X-Ray Images. IEEE Access **9**, 81902–81912 (2021). https://doi.org/10.1109/ACCESS.2021.3086229

19. Chollet, F.: Xception: deep learning with depth wise separable convolutions. In: IEEE Conference on Computer Vision and Pattern Recognition, Honolulu, USA, pp. 1800–1807 (2017). https://doi.org/10.1109/CVPR.2017.195
20. The International Skin Imaging Collaboration. (https://www.isicarchive.com/#!/topWithHeader/wideContentTop/main). Accessed 1 Nov 2023
21. The Kaggle. https://www.kaggle.com/datasets/hasnainjaved/melanoma-skin-cancer-dataset-of-10000-images/data. Accessed 1 Nov 2023
22. Memon, Q.A.: Implementing role based access in healthcare ad hoc networks. J. Netw. **4**(3), 192–199 (2009). https://doi.org/10.4304/jnw.4.3.192-199

Service Migration in Edge Computing: Techniques, Challenges, and Future Directions

Yajnaseni Dash[1](✉), Prateek Yadav[1], and Ajith Abraham[2]

[1] School of Artificial Intelligence, Bennett University, Greater Noida, Uttar Pradesh, India
yajnasenidash@gmail.com, prateek.y089@gmail.com
[2] School of Artificial Intelligence, Sai University, Chennai, India
abraham.ajith@gmail.com

Abstract. Edge computing has emerged as a prominent paradigm to address the challenges posed by the increasing demand for low-latency, high-bandwidth applications and services. It aims to bring computation closer to end-users by deploying resources at the network edge. One crucial aspect of edge computing is service migration, which involves dynamically moving services or their components between different edge nodes based on various factors such as user location, resource availability, and network conditions. Edge computing is an essential strategy for applications that require fast response times and large amounts of data transfer, allowing services to be moved between network nodes flexibly. This study examines different methodologies, challenges, and resolutions for efficient migration, encompassing dynamic resource allocation techniques, live virtual machine migration methods, and containerization strategies. The paper emphasizes the significance of predictive analytics and the validation of post-migration performance. It also addresses critical obstacles such as network connectivity problems and security considerations. Within edge computing contexts, the paper examines various approaches, issues, and possible solutions for effective service migration. Potential future developments encompass the utilization of machine learning methodologies, the integration of hybrid approaches, and the implementation of heightened security measures.

Keywords: Service migration · Fog computing · Edge computing · Internet of Things (IoT) · Cloud computing

1 Introduction

Edge computing is a distributed computing strategy that addresses bandwidth constraints, latency needs, privacy issues, and network congestion by locating computation and data storage close to the network's edge. This lowers latency and boosts performance for Internet of Things devices Reliability improvements through multi-center accounting; optimized bandwidth for constrained networks; decreased latency for applications that are time-sensitive like augmented reality and driverless cars; and enhanced privacy and security by enabling sensitive user data to migrate beyond local boundaries, thereby lowering risks of attack, data breaches, and compliance issues are just a few advantages that edge computing provides[1–15].

V. Sakalauskas et al. (Eds.): IBICA 2023, LNNS 1231, pp. 366–373, 2025.
https://doi.org/10.1007/978-3-031-78946-5_35

1.1 Importance of Service Migration

Organizations in the cloud computing infrastructure, technologies for virtualization, and data centers need to prepare studies concerning service migration in computing at the edge with an emphasis on containerization, and virtualization, which is an adaptive resource allocation approach [16–19].

1) *Dynamic resource allocation techniques:* With the use of dynamic scheduling of tasks and real-time workload models, the authors suggest an adaptive resource allocation technique based on reinforcement learning methods for effective service migrations in edge computing environments [20, 21].
2) *Live VM Migration Approaches:* The authors presented "pre-copy live migration," an optimization method based on machine learning [22].
3) *Service migration using containerization techniques:* Containerization has become an emerging solution for service migration in edge computing systems. Many studies have examined various qualification methods and policies to facilitate effective service migration. The concept of "Fog Computing" has been introduced [23], which uses thin containers to provide service migration at the network edge. Their work established the Fog computing framework and demonstrated the value of placing services in containers close to end users. FORA was proposed by the authors as a computing environment for industrial Internet of Things applications [24]. The authors discussed the efficient distribution of content in the fog-based content delivery network [31].

2 Objectives and Scope of Service Migration in Edge Computing

The purpose and scope of edge computing service migration include smooth transition of apps or services from a centralized cloud environment to dispersed edge nodes ensuring better delivery of services, better resource utilization, reduced latency, scalability scaling up, and increasing productivity are the main objectives of service migration.

A variety of factors are considered when considering service migration in edge computing.

1) *Identify businesses worth migrating:* Based on centralized cloud infrastructure or resource requirements, for example, not all businesses are ideal for migration. A key factor is identifying high-performance projects.
2) *Migration methods:* There are many methods of migrating services, such as continuous migration, pre-migration, and post-migration planning and it is important to consider the best method depending on variables such as service importance, user requirements, network state control, and downtime tolerance
3) *Network configuration and infrastructure design:* Network topology, data routing techniques, load balancing methods, and communication between edge nodes and central cloud all need to be carefully managed when designing a high-performance edge computing architecture
4) *Monitoring tools:* Monitoring tools, such as monitoring performance metrics, application profiling, dynamic provisioning, and fault detection systems, enable you to

effectively manage migrated projects Monitoring helps monitor for optimal performance levels, failure tolerance, and efficient resource allocation in a distributed internal environment.

Service migration is aimed at improving performance, reducing latency, optimizing resources, improving scalability, and driving growth in critical situations such as intermittent connectivity or bandwidth limitations while protecting data confidentiality identifying qualified successful users, migration strategy, network configuration/performance, and deployment of the exempt monitoring mechanisms happen.

3 Challenges in Service Migration

a) Network connection information:

Service migration involves moving services from one location to another, which can cause network connectivity problems. Two major challenges in this area are critical communication delays [25] and limitations in network bandwidth [26].

b) Resource constraints:

1) *Limited resources:* Edge nodes are generally considered low computing devices, and service migrations can require large data transfers. The quality of service during service migrations can be affected by the limited resources at the edge nodes.
2) *Heterogeneous hardware configuration:* Transferring services between various computer environments) including those with varying hardware configurations (requires optimizing service performance to maintain consistency [28].

c) Service migration safety issues:

Ensuring data security and integrity during service migrations is essential to preventing prospective attackers, and putting authentication systems in place to stop unauthorized or altered services [29, 30].

4 Service Migration Techniques and Algorithms

In this section, two aspects of the service migrations, i.e., pre-migration analysis using predictive analytics and post-migration performance metrics validation techniques are discussed.

a) *Pre-migration Analysis Techniques:* Service migration requires the use of pre-migration analytical tools, such as predictive analytics based on user behavior patterns. For cloud workload forecasting and other cloud, edge, fog, and Internet of Things computing paradigms, researchers have developed a framework based on deep learning models [32, 33].
b) *Post-migration Validation Methods:* Measuring performance and resource utilization after migration, post-migration validation techniques employ the Service Quality Index and Resource Utilization Efficiency measures. Such information is crucial for identifying problems such as network delays brought on by more traffic during the phase of transition. Measuring network latency before, during, and after migration enables a thorough analysis of potential disruptions caused by increased network

delays to the end users. The post-migration validation techniques help researchers gather a thorough understanding of performance during service migrations, which, in turn, enables them to redefine algorithms and improve resource allocation. In this, various studies [34–36] have offered valuable perspectives on pre-migration methods and post-migration validation techniques. Using both approaches, service providers can make well-informed decisions about service migrations to achieve higher performance.

5 Implementation Considerations

With the rapid proliferation of edge computing and the ability to support simple, high-bandwidth applications, effective service management strategies are of utmost importance in addressing the challenges of balancing performance, energy consumption reducing workload, and improving quality of service (QoS) in ultra-large-scale edge computing environments Revolving around researchers proposed a new framework that considers criteria when making service migration decisions, with the aim and achieving better resource management and improved user experience. To mitigate latency and alleviate the traffic load associated with data aggregation to centralized cloud computing, multi-access edge computing (MEC) allows end users to access services via physically proximate edge nodes. Figure 1 illustrates an architecture tailored for multi-criteria service migration optimization within ultra-large-scale MEC networks, enabling on-demand service migration/instantiation characterized by low latency and high QoS. More precisely, two new functionalities, decentralized control's global information updating and service migration's initiation, have been integrated into centralized control, overseen by the network operator.

The roles of participants include a) Users who request services from service providers and rely on the network operator to establish their connectivity. b) Service providers: They provide services to users and make decisions about when a service should undergo migration. Optimizing service migration by a combination of centralized and decentralized control mechanisms and thorough network-level data, the network operator oversees migration and service destination selection.

1) *Container Orchestration Frameworks:* With the help of container orchestration frameworks like Kubernetes, content delivery networks (CDNs) may rapidly distribute content among several nodes or edge servers. These frameworks facilitate effective application management and deployment at scale [37].
2) *Virtualization Technologies:* Virtualization technologies, such as Kubernetes and kernel-based virtual machines (KVM), enhance resource efficiency and flexibility by generating virtual environments and automating processes such as container life-cycle management. KVM and similar virtualization technologies enhance resource utilization by enabling multiple CDN components on one server. These technologies offer improved scalability and resilience. However, it is crucial to evaluate the CDN's specific needs and constraints before choosing the right technology mix [38, 39].

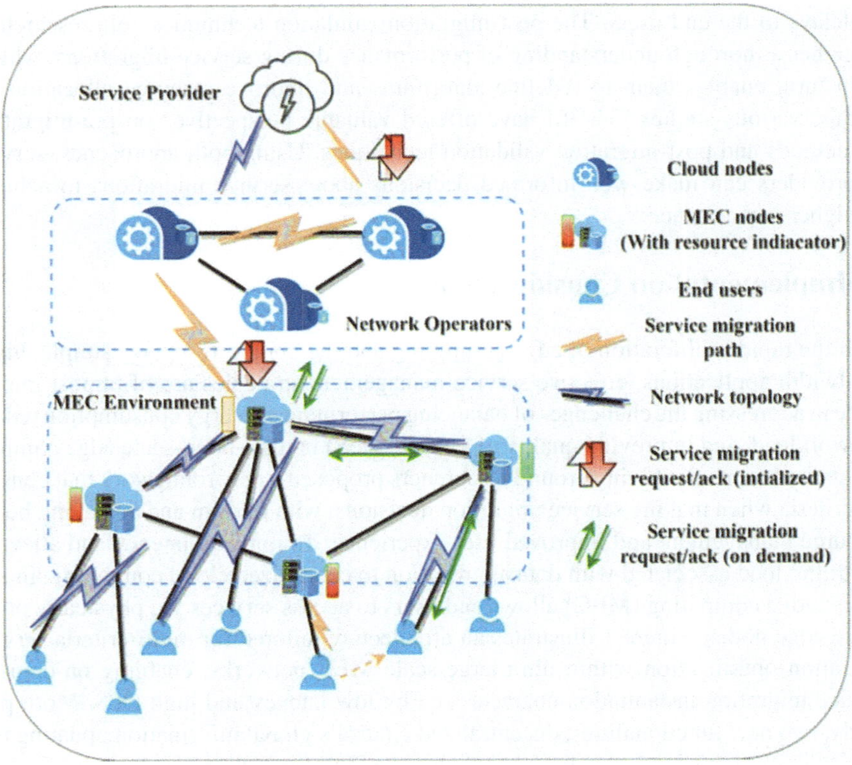

Fig. 1. Optimization architecture of service migration

6 Future Directions in Edge Computing Service Migration

1) Machine learning techniques for better classification

It uses intelligent algorithms to allocate dynamic resources based on business processes, resource availability, and service quality requirements. Reinforcement learning has proven to be effective for efficient resource utilization in edge computing systems [20].

2) Hybrid approaches for combining VMs and containers

It leverages the benefits of both technologies for lightweight deployment and isolation with VMs. Hybrid cloud edge architectures can improve the flexibility and scalability of service provisioning [40].

3) Enhanced safety during migration

Analyze protocol security, cryptography techniques, and access control techniques for secure migration over the network stream. Address challenges such as privacy protection, manipulation detection, and seamless authentication [41, 42]. Dash et al. [43] explored the advancements and potential of edge computing and CDNs in transforming the digital landscape. It highlights their role in improving internet access, scalability, and service delivery, shaping a more networked world. Dash et al. [44] have demonstrated the potential of blockchain technology to enhance Internet of Things (IoT) security, effectively mitigating vulnerabilities and fostering a more secure digital ecosystem.

Future research in edge computing will focus on efficient resource allocation, hybridization approaches, and security enhancements for service migration to the network edge, addressing open challenges and potential future directions.

7 Conclusion

Edge computing is a technological solution that minimizes the delay in data processing and enhances the efficiency of time-critical applications such as augmented reality and autonomous vehicles. Service migration is a process that moves services between nodes in response to factors such as user location, resource availability, and network circumstances. Methods such as dynamic resource allocation, live virtual machine migration, and containerization are effective in resolving network connectivity problems, resource limitations, and security concerns. Subsequent investigations should prioritize machine learning, hybrid methodologies, and sophisticated security measures to enhance quality of service, privacy, and dependability.

References

1. Satyanarayanan, M., Bahl, P., Caceres, R., Davies, N.: The case for VM-based cloudlets in mobile computing. IEEE Pervasive Comput. **8**(4), 14–23 (2009)
2. Shi, W., Cao, J., Zhang, Q., Li, Y., Xu, L.: Edge computing: vision and challenges. IEEE Internet Things J. **3**(5), 637–646 (2016)
3. Khan, W.Z., Ahmed, E., Hakak, S., Yaqoob, I., Ahmed, A.: Edge computing: a survey. Future Gener. Comput. Syst. **1**(97), 219–235 (2019)
4. Abbas, N., Zhang, Y., Taherkordi, A., Skeie, T.: Mobile edge computing: a survey. IEEE Internet Things J. **5**(1), 450–65 (2017)
5. Kumar, S., Sharma, A., Gupta, R.: QoS driven cost-efficient resource allocation in edge computing: a distributed game theoretic approach. J. Comput. Sci. 102106 (2023)
6. Kumar, S., Goswami, A., Gupta, R., Singh, S.P., Lay-Ekuakille, A.: A cost-effective and QoS-aware user allocation approach for edge computing enabled IoT. IEEE Internet Things J. **10**(2), 1696–1710 (2022)
7. Kumar, S., Goswami, A., Gupta, R., Singh, S.P., Lay-Ekuakille, A.: A game-theoretic approach for cost-effective multicast routing in the Internet of things. IEEE Internet Things J. **9**(18), 18041–18053 (2022)
8. Kumar, S., Gupta, R., Lakshmanan, K., Maurya, V.: A game-theoretic approach for increasing resource utilization in edge computing enabled Internet of Things. IEEE Access **10**, 57974–57989 (2022)
9. Kumar, S., Dash, Y.: Caching techniques in edge computing and challenges. In: 3rd International Conference on Advancement in Electronics and Communication Engineering (AECE), IEEE (2023)
10. Dash, Y., Jajoria, P.: Fog computing: applications in smart healthcare. In: 3rd International Conference on Advancement in Electronics and Communication Engineering (AECE), IEEE (2023)

11. Dash, Y., Jajoria, P., Kumar, S., Singh, D., Sarangi, S.C.: Internet of Things: vision, architecture, and applications in healthcare. In: Proceedings of 5th International FTNCT, Institution of Engineer's (IEI) Chandigarh (2023)
12. Dash, Y., Kumar, S., Patle, V.K.: A novel data mining scheme for smartphone activity recognition by accelerometer sensor. In: Proceedings of the 4th International Conference on Frontiers in Intelligent Computing: Theory and Applications (FICTA-2015), vol. 404 of the series Advances in Intelligent Systems and Computing, pp. 131–140 (2015)
13. Sarangi, S.C., Dash, Y.: Sensor-based physical activity monitoring: application of machine learning approaches. In: International Conference on Intelligent Computing, Instrumentation and Control Technologies, Kerala, India, ICICICT, IEEE (2019)
14. Yadav, P., Vidyarthi, D.P.: An efficient fuzzy-based task offloading in edge-fog-cloud architecture, Concurrency, and Computation: Practice and Experience https://doi.org/10.1002/cpe.7843
15. Yadav, P., Kar, S.: Evaluating the impact of region-based content popularity of videos on the cost of CDN deployment. In: 2020 National Conference on Communications (NCC), Kharagpur, India, pp. 1–6 (2020)
16. Katal, A., Dahiya, S., Choudhury, T.: Energy efficiency in cloud computing data centers: a survey on software technologies. Cluster Comput. **26**(3), 1845–1875 (2023)
17. Hines, M.R., Deshpande, U., Gopalan, K.: Post-copy live migration of virtual machines. ACM SIGOPS Oper. Syst. Rev. **43**(3), 14–26 (2009)
18. Jiang, C., et al.: Energy-aware edge computing: a survey. Comput. Commun. **151**, 556–580 (2020)
19. Singh, A., Aujla, G.S., Bali, R.S.: Container-based load balancing for energy efficiency in software-defined edge computing environment. Sustain. Comput. Inf. Syst. **30**, 100463 (2021)
20. Liu, L., Feng, J., Mu, X., Pei, Q., Lan, D., Xiao, M.: Asynchronous deep reinforcement learning for collaborative task computing and on-demand resource allocation in vehicular edge computing. IEEE Trans. Intell. Transp. Syst. (2023)
21. Liu, Z., et al.: Context-aware and adaptive QoS prediction for mobile edge computing services. IEEE Trans. Serv. Comput. **15**(1), 400–413 (2019)
22. Haris, R.M., et al.: A machine learning-based optimization approach for pre-copy live virtual machine migration. Cluster Comput. 1–20 (2023)
23. Ahmed, E., Nashaat, H., Rizk, R.: Machine learning for fog computing: perceiving QoE for IoT applications. In: International Conference on Advanced Intelligent Systems and Informatics (pp. 482–496). Cham: Springer Nature Switzerland (2023)
24. Pop, P., et al.: The FORA fog computing platform for industrial IoT. Inf. Syst. **98**, 101727 (2021)
25. Rahman, S.: Towards responsive user services in edge computing. (Doctoral dissertation, Montana State University-Bozeman, College of Engineering) (2023)
26. Singh, R., Sukapuram, R., Chakraborty, S.: A survey of mobility-aware multi-access edge computing: challenges, use cases and future directions. Ad Hoc Netw. **140**, 103044 (2023)
27. Mansouri, Y., Babar, M.A.: A review of edge computing: features and resource virtualization. J. Parallel Distrib. Comput. **1**(150), 155–183 (2021)
28. Barbalace, A., Karaoui, M.L., Wang, W., Xing, T., Olivier, P., Ravindran, B.: Edge computing: the case for heterogeneous-ISA container migration. In: Proceedings of the 16th ACM SIGPLAN/SIGOPS International Conference on Virtual Execution Environments, pp. 73–87 (2020)
29. Cao, K., Liu, Y., Meng, G., Sun, Q.: An overview on edge computing research. IEEE Access **1**(8), 85714–85728 (2020)
30. Kim, H., Kang, E., Broman, D., Lee, E.A.: Resilient authentication and authorization for the Internet of Things (IoT) using edge computing. ACM Trans. Internet Things **1**(1), 1–27 (2020)

31. Yadav, P., Kar, S.: Efficient content distribution in fog-based CDN: a joint optimization algorithm for fog-node placement and content delivery. In: IEEE Internet of Things Journal. https://doi.org/10.1109/JIOT.2024.3355468
32. S. Ahmad, I. Shakeel, S. Mehfuz, J. Ahmad, "Deep learning models for cloud, edge, fog, and IoT computing paradigms: Survey, recent advances, and future directions," Computer Science Review, 1;49, pp.100568, 2023.
33. Patel, Y.S., Bedi, J.: MAG-D: a multivariate attention network-based approach for cloud workload forecasting. Future Gen. Comput. Syst. **142**, 376–392 (2023)
34. Zhang, F., Liu, G., Fu, X., Yahyapour, R.: A survey on virtual machine migration: Challenges, techniques, and open issues. IEEE Commun. Surv. Tutorials **20**(2), 1206–43 (2018)
35. Alyas et al.: Performance framework for virtual machine migration in cloud computing. Comput. Mater. Continua **74**(3) (2023)
36. He, T.: Migration Management in Software-Defined Networking-enabled Edge and Cloud Computing. School of Computing and Information Systems, THE UNIVERSITY OF MELBOURNE, Environments. degree of Doctor of Philosophy (2021)
37. Chi, H.R., et al.: Multi-criteria dynamic service migration for ultra-large-scale edge computing networks. IEEE Trans. Ind. Inf. (2023)
38. Vyas, J., Love, C.: Core Kubernetes. Simon and Schuster; ISBN13: 9781638350750 (2022)
39. Calheiros, R.N., Ranjan, R., Buyya, R.: Virtual machine provisioning based on analytical performance and QoS in cloud computing environments. In: 2011 International Conference on Parallel Processing, IEEE, pp. 295–304 (2011)
40. Zeng, Y., Chao, M., Stoleru, R.: EmuEdge: a hybrid emulator for reproducible and realistic edge computing experiments. In: 2019 IEEE International Conference on Fog Computing (ICFC), pp. 153–164 (2019)
41. Peng, K., Leung, V., Xu, X., Zheng, L., Wang, J., Huang, Q.: A survey on mobile edge computing: focusing on service adoption and provision. Wirel. Commun. Mob. Comput. (2018)
42. Zhang, J., Chen, B., Zhao, Y., Cheng, X., Hu, F.: Data security and privacy-preserving in edge computing paradigm: Survey and open issues. IEEE Access **6**, 18209–37 (2018)
43. Dash, Y., Kumar, N., Abraham, A.: Edge of innovation: exploring the potential of edge computing and content delivery networks. In: 2024 IEEE International Conference on Computing, Power and Communication Technologies (IC2PCT), Greater Noida, India, pp. 1228–1233 (2024). https://doi.org/10.1109/IC2PCT60090.2024.10486411
44. Dash, Y., Yadav, P.: The synergy of blockchain and IoT: a comprehensive security perspective. In: 2024 IEEE International Conference on Computing, Power and Communication Technologies (IC2PCT), Greater Noida, India, pp. 34–38 (2024). https://doi.org/10.1109/IC2PCT60090.2024.10486687

Analyzing Rainfall Patterns in North Indian States: A Long Short-Term Memory (LSTM) Approach

Jeevika Rajput[1], Yajnaseni Dash[2](✉), and Ajith Abraham[3]

[1] School of Computer Science Engineering and Technology, Bennett University, Greater Noida, UP, India
jeevikarajput0003@gmail.com
[2] School of Artificial Intelligence, Bennett University, Greater Noida, UP, India
yajnasenidash@gmail.com
[3] School of Artificial Intelligence, Sai University, Chennai, India
abraham.ajith@gmail.com

Abstract. This research explores a comprehensive analysis of rainfall patterns specifically within the subdivisions of North Indian states. Focusing on subdivisions such as East Rajasthan, West Rajasthan, East Uttar Pradesh, West Uttar Pradesh, Haryana, Delhi, Chandigarh, Himachal Pradesh, Jammu & Kashmir, Punjab, and Uttarakhand, the study investigates historical meteorological data to discern intricate trends and variations in precipitation. Employing advanced statistical techniques and data visualization tools, this research aims to unveil the characteristics of rainfall, providing insights into the climate dynamics of these aforementioned regions over the studied period. The study examines the precipitation patterns in various states in North India, uncovering unique meteorological attributes. By analyzing historical meteorological data and employing advanced statistical approaches, this study uncovers substantial differences in rainfall patterns across different areas. This facilitates the process of making well-informed decisions for sustainable development and the formulation of plans to adapt to climate change.

Keywords: Machine Learning · Rainfall · North India · Predictive analytics · Regression · LSTM

1 Introduction

This study investigates North Indian rainfall trends from 1901 to 2015 in Jammu & Kashmir, Rajasthan, Haryana, Himachal Pradesh, Uttarakhand, Delhi, Uttar Pradesh, and Punjab. Historical data, statistical models, and data visualization technologies enable it to understand climatic dynamics and manage resources for sustainable development.

V. Sakalauskas et al. (Eds.): IBICA 2023, LNNS 1231, pp. 374–383, 2025.
https://doi.org/10.1007/978-3-031-78946-5_36

a. Purpose of the Research

1) Agricultural Significance

These states depend heavily on agriculture, which contributes significantly to the country's economy, variations in rainfall have a significant impact on soil moisture, crop outcome, and irrigation requirements. The goal of this research is to find past rainfall trends and anomalies that impact agricultural productivity. Results are suitable to help farmers and policymakers in implementing efficient water management plans and climate-dependent agricultural practices.

2) Water Resource Management

The availability of water resources is directly impacted by rainfall trends. To gain insight into the regions' availability of water, possible dry spells, and best practices for managing water resources, a detailed examination of long-term rainfall information has been conducted. The development of sustainable water management plans for home, commercial, and agricultural uses is dependent on water availability.

3) The Impact of Climate Change and Adaptations

Analyzing whether these chosen locations will be affected by changes in climate requires an analysis of the monsoon data of the region. The primary goal of this research is to understand variability in climate and help in the development of policies and methods to adapt in the fields of agriculture, industry, and home use by recognizing long-term trends and changes in rainfall patterns.

Rainfall prediction in India has been extensively studied, investigating the regional climate patterns [1–19]. Northeast India's temperature and rainfall trends, long-term patterns, heavy rains, and dynamism have been investigated [5, 10]. Rainfall prediction has improved using machine learning and artificial intelligence, benefiting academics and decision-makers [11–22]. These discoveries can improve climate forecasting, adaption, and predictive models.

2 Data and Method

1) Data

Kaggle's 'Rainfall in India' dataset [23] provides historical rainfall data for India, supporting climate research, water supply evaluation, agricultural production, and understanding of precipitation's influence on Indian life. We have extracted and processed the raw dataset for the specified northern subdivisions of India, and the resulting processed dataset will be available upon request. The information helps forecasters, climatologists, policymakers, and researchers.

2) Method

LSTM (Long Short-Term Memory) can capture long-term dependencies. It is a form of recurrent neural network, that is excellent at analysing and predicting sequential input. Due to their long-term memory capacity, LSTMs are highly suitable for a variety of applications, including natural language processing and time series prediction. A linear stacking of layers is the basis of the LSTM model, which is referred to as a sequential model. In this paper, we have used two LSTM layers with 50 units each, followed by a dense layer with one unit (for regression tasks). Adam optimizer was used and performance was measured by root mean squared error (RMSE). Training has been carried out on the provided training data (train_X and train_Y) for 100 epochs with a batch size of 32.

3 Result and Analysis

3.1 Calculating Average Rainfall

Analysis of annual rainfall across different subdivisions against the national average of 1411.01 mm.

1) *East Rajasthan:* It experiences a notably lower annual rainfall of 655.22 mm (Fig. 1), indicating a drier climate in this subdivision.
2) *West Rajasthan:* It has a considerably very lower average annual rainfall of 292.67 mm (Fig. 2), signifying a relatively arid climate compared to the national average.
3) *East Uttar Pradesh:* It exhibits an annual average rainfall of 979.21 mm (Fig. 3), which is slightly lower than the national average.
4) *West Uttar Pradesh*: It experiences an annual average rainfall of 827.11 mm (Fig. 4), lower than the national average, suggesting relatively less rainfall in this region.
5) *Haryana, Delhi & Chandigarh:* There is a lower average annual rainfall of 530.50 mm (Fig. 5), showcasing drier conditions compared to the national average.
6) *Himachal Pradesh:* It has an annual rainfall of 1260.35 mm (Fig. 6), which is slightly lower than the national average but significantly higher compared to some other subdivisions.
7) *Jammu & Kashmir:* It has an annual average rainfall of 1139.68 mm (Fig. 7), slightly lower than the national average but still significant which indicates a wetter climate.
8) *Punjab:* It has a lower annual rainfall of 593.54 mm (Fig. 8), indicating drier conditions compared to the national average.
9) *Uttarakhand:* It has a higher annual rainfall of 1465.70 mm (Fig. 9), suggesting a wetter climate compared to the national average.

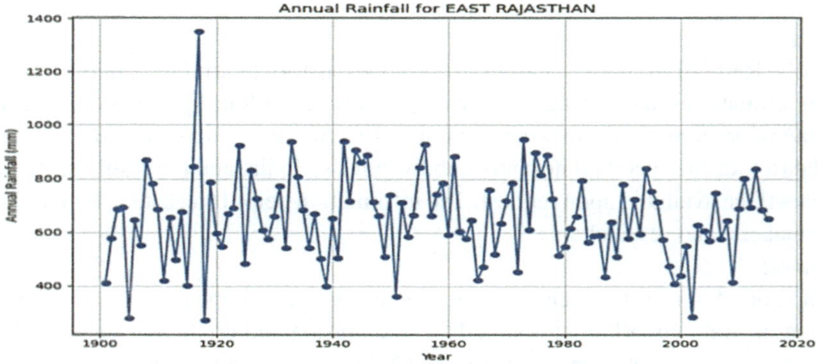

Fig. 1. Annual rainfall for East Rajasthan

These comparisons highlight the diverse rainfall patterns across different subdivisions in India, with some areas experiencing notably lower rainfall than the national average while others exhibit significantly higher rainfall.

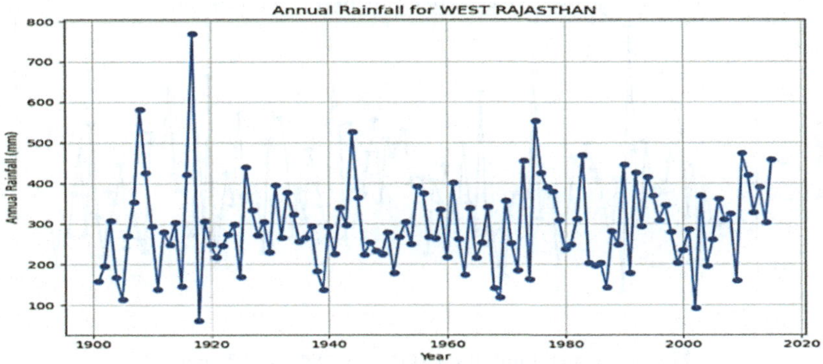

Fig. 2. Annual rainfall for West Rajasthan

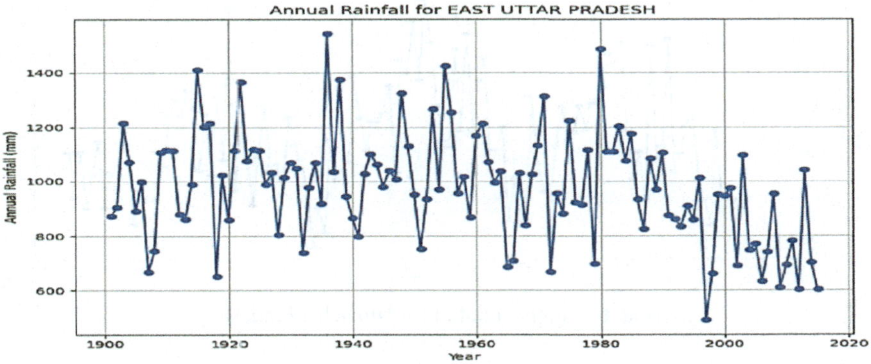

Fig. 3. Annual rainfall for East Uttar Pradesh

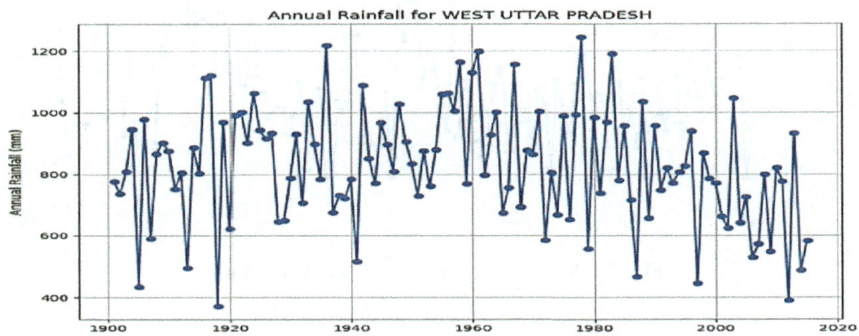

Fig. 4. Annual rainfall for West Uttar Pradesh

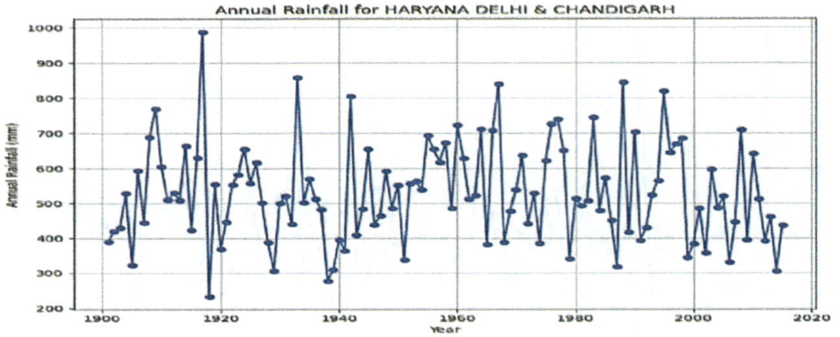

Fig. 5. Annual rainfall for Haryana, Delhi & Chandigarh

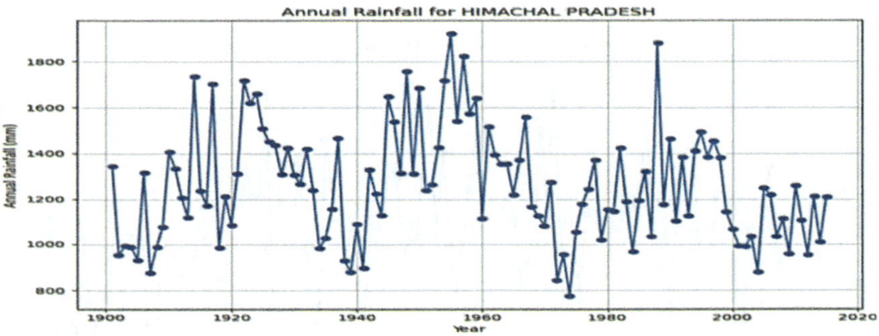

Fig. 6. Annual rainfall for Himachal Pradesh

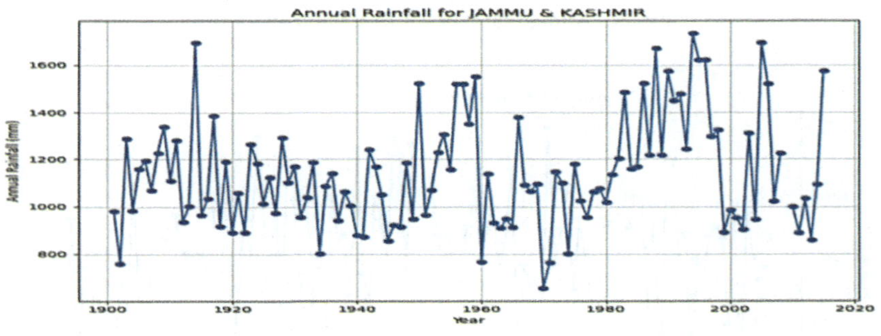

Fig. 7. Annual rainfall for Jammu & Kashmir

3.2 Rainfall Statistics by Subdivision

A comparative analysis and interpretation of the rainfall statistics for the specified subdivisions is presented in Table 1.

The rainfall statistics for North India subdivisions reveal significant variability across different regions. East Rajasthan receives a mean rainfall of 655.2 mm, with a maximum of 1350.4 mm and a minimum of 273.6 mm, indicating considerable fluctuations. West Rajasthan, characterized by its arid climate, shows a much lower mean rainfall

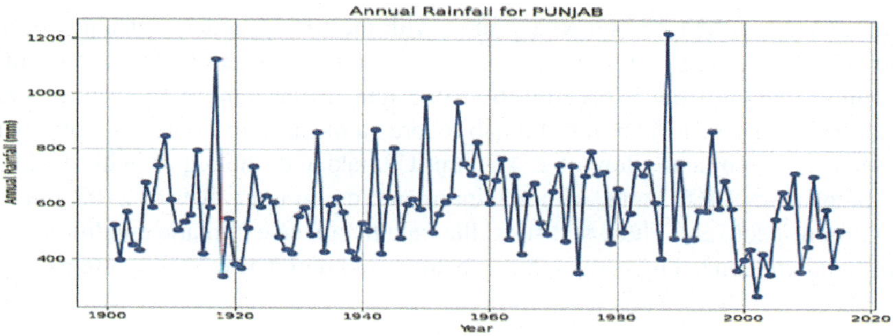

Fig. 8. Annual rainfall for Punjab

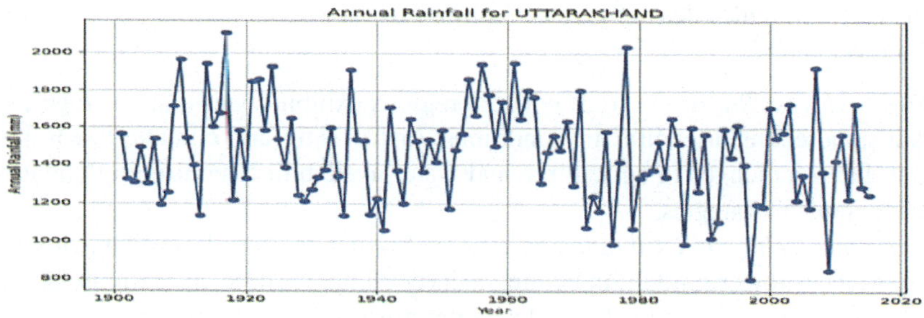

Fig. 9. Annual rainfall for Uttarakhand

Table 1. Rainfall Statistics by North India Subdivisions

Subdivision	Mean	Median	Max	Min
East Rajasthan	655.2	655.4	1350.4	273.6
West Rajasthan	292.7	279.9	768.8	62.3
East Uttar Pradesh	979.2	988.8	1545.5	493.3
West Uttar Pradesh	827.1	809.2	1244.2	371.9
Haryana, Delhi & Chandigarh	530.5	512.0	986.7	234.7
Himachal Pradesh	1260.3	1236.8	1919.2	776.1
Jammu & Kashmir	1139.7	1098.1	1732.5	657.0
Punjab	593.5	586.6	1222.6	274.7
Uttarakhand	1465.7	1478.6	2102.9	803.4

of 292.7 mm, with values ranging from 62.3 mm to 768.8 mm. East Uttar Pradesh experiences higher rainfall, averaging 979.2 mm, with a peak of 1545.5 mm and a minimum of 493.3 mm. West Uttar Pradesh follows with a mean of 827.1 mm, ranging from 371.9 mm to 1244.2 mm. The combined region of Haryana, Delhi, and Chandigarh has a mean rainfall of 530.5 mm, with a maximum of 986.7 mm and a minimum of 234.7 mm.

Himachal Pradesh and Jammu & Kashmir, with their mountainous terrains, receive a higher average rainfall of 1260.3 mm and 1139.7 mm respectively. Himachal Pradesh sees values ranging from 776.1 mm to 1919.2 mm, while Jammu & Kashmir's range is from 657.0 mm to 1732.5 mm. Punjab records a mean rainfall of 593.5 mm, with a peak of 1222.6 mm and a low of 274.7 mm. Uttarakhand, the region with the highest rainfall, averages 1465.7 mm, with maximum and minimum values of 2102.9 mm and 803.4 mm respectively. These statistics illustrate the diverse climatic conditions across North India, influenced by geographical features and regional weather patterns.

3.3 RMSE Obtained by Using LSTM

The Root Mean Squared Error (RMSE) values between the actual and predicted data for various North Indian states are given as follows in Table 2.

1) Lower RMSE

i) *East and west Rajasthan (0.21)*: These regions exhibit lower-than-average rainfall ratios, indicating drier climatic conditions. Arid or semi-arid climates like Rajasthan may face challenges in agriculture and water availability, requiring efficient water management strategies.
ii) *Haryana, Delhi, and Chandigarh (0.22):* Similar to East Rajasthan, this region experiences lower rainfall, indicating a dryer climate. These areas might face water scarcity issues and need enhanced water conservation methods.
iii) *East Uttar Pradesh (0.23) and West Uttar Pradesh (0.23) along with Punjab (0.24):* There is relatively wetter climate in these regions as compared to East Rajasthan, Haryana, Delhi & Chandigarh. Here rainfall can support more extensive agricultural activities, but effective water management remains vital.

Table 2. RMSE scores of used subdivisions

Subdivisions	RMSE
East Rajasthan	0.21
West Rajasthan	0.21
East Uttar Pradesh	0.23
West Uttar Pradesh	0.23
Haryana Delhi & Chandigarh	0.22
Himachal Pradesh	0.28
Jammu & Kashmir	0.33
Punjab	0.24
Uttarakhand	0.31

2) Higher RMSE

i) *Himachal Pradesh (0.28) and Jammu & Kashmir (0.33):* These states receive higher rainfall compared to the national average. The climate in these regions supports diverse vegetation and sustains local agriculture.
ii) *Uttarakhand (0.31):* Uttarakhand's ratio leans toward higher rainfall compared to the national average. Despite this, it faces challenges such as soil erosion and landslides due to heavy precipitation in certain regions.

The RMSE values for rainfall prediction using LSTM across various North Indian states highlight distinct climatic conditions and their implications. East and West Rajasthan exhibit the lowest RMSE of 0.21, reflecting their arid or semi-arid climates with below-average rainfall, necessitating efficient water management strategies. Haryana, Delhi, and Chandigarh also show low RMSE values of 0.22, indicating similar dry conditions and potential water scarcity issues. In contrast, East and West Uttar Pradesh, with RMSE values of 0.23, and Punjab with 0.24, have relatively wetter climates that support more extensive agricultural activities but still require effective water management. Higher RMSE values are observed in Himachal Pradesh (0.28), Jammu & Kashmir (0.33), and Uttarakhand (0.31), regions known for higher rainfall and significant variability due to their hilly terrains. These variations underscore the importance of tailored water resource management, agricultural planning, and infrastructure development to address the diverse climatic challenges across North India.

3.4 Interpretation

The study analyzes rainfall data of Northern India, revealing climate changes and varying annual precipitation forecasts. Results indicate diverse rain patterns across the considered subdivisions. Rajasthan has significantly lower mean and median rainfall compared to other regions, with West Rajasthan having the lowest values. Uttarakhand consistently shows higher values across mean, median, maximum, and minimum, indicating higher rainfall patterns. Himachal Pradesh and Jammu & Kashmir show significant differences between maximum and minimum rainfall, indicating potential variability as these come under hilly areas. East and West Uttar Pradesh have a moderate range between maximum and minimum rainfall. Overall patterns show disparities, with Uttarakhand experiencing higher rainfall and Rajasthan receiving the lowest rainfall. Understanding these disparities is crucial for water resource management, agriculture, and infrastructure planning.

4 Conclusion

The study examines the precipitation trends in northern India by utilizing historical data, statistical methodologies, and the LSTM technique. The data highlights substantial discrepancies in yearly precipitation, with Rajasthan experiencing the least amount and Uttarakhand receiving the most amount. The analysis emphasizes the significance of comprehending rainfall patterns concerning agriculture, as fluctuations affect soil moisture, crop yields, and irrigation requirements. Efficient solutions for managing water resources are essential for implementing sustainable practices and adapting to climate change. The study provides useful information that can be used for agricultural planning, water resource management, and disaster preparedness.

References

1. Jain, S.K., Kumar, V., Saharia, M.: Analysis of rainfall and temperature trends in northeast India. Int. J. Climatol. **33**(4), 968–978 (2013)
2. Kumar, V., Jain, S.K., Singh, Y.: Analysis of long-term rainfall trends in India. Hydrol. Sci. J.-Journal des Sciences Hydrologiques **55**(4), 484–496 (2010)
3. Saxena, R., Mathur, P.: Recent trends in rainfall and temperature over North West India during 1871–2016. Theoret. Appl. Climatol. **135**(3–4), 1323–1338 (2019)
4. Jain, S.K., Kumar, V.: Trend analysis of rainfall and temperature data for India. Curr. Sci. 37–49 (2012)
5. Mahanta, R., Sarma, D., Choudhury, A.: Heavy rainfall occurrences in northeast India. Int. J. Climatol. **33**(6), 1456–1469 (2013)
6. Dash, S.K., Kulkarni, M.A., Mohanty, U.C., Prasad, K.: Changes in the characteristics of rain events in India. J. Geophys. Res. Atmos. **114**(D10) (2009)
7. Xavier, A., Manoj, M.G., Mohankumar, K.: On the dynamics of an extreme rainfall event in northern India in 2013. J. Earth Syst. Sci. **127**, 1–13 (2018)
8. Parthasarathy, B., Dhar, O.N.: Secular variations of regional rainfall over India. Q. J. R. Meteorol. Soc. **100**(424), 245–257 (1974)
9. Yadav, R.K., Roxy, M.K.: On the relationship between north India summer monsoon rainfall and east equatorial Indian Ocean warming. Glob. Planet. Change **179**, 23–32 (2019)
10. Singh, A., Thakur, S., Adhikary, N.C.: Analysis of spatial and temporal rainfall characteristics of the North East region of India. Arab. J. Geosci. **14**, 1–16 (2021)
11. Dash, Y., Mishra, S.K., Sahany, S., Panigrahi, B.K.: Indian summer monsoon rainfall prediction: a comparison of iterative and non-iterative approaches. Appl. Soft Comput. J. ASOC-4445, 70, 1122–1134 (2018)
12. Dash, Y., Mishra, S.K., Panigrahi, B.K.: Rainfall prediction for Kerala state of India using artificial intelligence approaches. Comput. Electr. Eng. **70**, 66–73 (2018)
13. Dash, Y., Mishra, S.K., Panigrahi, B.K.: Predictability assessment of northeast monsoon rainfall in india using sea surface temperature anomaly through statistical and machine learning techniques. Environmetrics e2533, 1–10 (2018)
14. Dash, Y., Mishra, S.K., Panigrahi, B.K.: Neural network based approaches for prediction of the Indian summer monsoon rainfall. In: 2nd ICICICT, Kannur, Kerala, India, IEEE, pp. 550–554 (2019)
15. Dash, Y., Mishra, S.K., Panigrahi, B.K.: Prediction of southwest monsoon rainfall for Kerala, India using ε-SVR model. In: 2nd ICICICT, Kannur, Kerala, India, IEEE, pp. 746–748 (2019)
16. Dash, Y., Mishra, S.K., Panigrahi, B.K.: Rainfall prediction of a Maritime State (Kerala), India using SLFN and ELM techniques. In: ICICICT, Kannur, India, IEEE, pp. 1714–1718 (2017)
17. Dash, Y., Mishra, S.K., Panigrahi, B.K.: NEMR Predictability Assessment over Indian Peninsula using ELM. CI 2017, NCAR Technical Note NCAR/TN-536+PROC, 77–80 (2017)
18. Dash, Y., Mishra, S.K., Panigrahi, B.K.: Indian Summer Monsoon Prediction by Deep Long Short-Term Memory Neural Network using Time Series Based Approach. AGU Fall Meeting Abstracts, A33M-2968 (2019)
19. Dash, Y., Abraham, A.: Integrating empirical analysis and deep learning for accurate monsoon prediction in Kerala, India. Appl. Comput. Geosci., 14 (2024). https://doi.org/10.1016/j.acags.2024.100211

20. Dash, Y., Mishra, S.K., Panigrahi, B.K.: Indian Summer Monsoon Prediction by Deep Long Short-Term Memory Neural Network using Time Series Based Approach. AGU Fall Meeting Abstracts, pp. A33M-2968 (2019)
21. Dash, Y., Abraham, A., Kumar, N., Yadav, D.K., Singhal, N.: Climate predictors in Indian summer monsoon forecasting: a novel de-correlated RVFL ensemble strategy. Earth Sci. Inform. **18**, 134 (2025). https://doi.org/10.1007/s12145-024-01532-y
22. Dash, Y., Kumar, N., Raj, M., Abraham, A.: Deep learning approach for predicting monsoon dynamics of regional climate zones of India. Appl. Comput. Geosci. **23**, 100176 (2024)
23. Kaggle datasets. https://www.kaggle.com/datasets/kkhandekar/statewise-rainfall-1901-to-2015/data

Canal Water Data Monitoring System and Controlling System

Vaishali Niranjane(✉), Shivam Patil, Trushant Dharpure, Manish Bhosle, and Rutik Pawar

Yeshwantrao Chavhan College of Engineering, Nagpur, India
niranjane_vaishali@yahoo.com

Abstract. The development of canal automation systems has been made easier by the advancement of science and technology in the modern era. Canal irrigation is a typical method of obtaining water for agricultural use. In this research work, we suggest a flow control system for canal gates based on a microcontroller. The flow control system is composed of an intelligent Microcontroller based Remote Terminal Unit (RTU). The RTU can interface with a variety of sensors, a free local host cloud server, GSM, and numerous modules. A microcontroller-based system is very adaptable and is simple to change locally. Additionally, it can interface with different modules and is less expensive than PLC. The proposed system aims to improve the efficiency of irrigation operations, power consumption, measurement precision, water distribution, and imbalance responses. The remote terminal unit keeps an eye on security, gate opening, solar power status, and water level. Each component of the system is built to run on solar power. In this research, we put more an emphasis on the direct pipe outputs (DPOs), laterals, and distributaries flow monitoring activities. The security, gate opening, and water level are all tracked by the remote terminal unit. The suggested system has the ability to continuously control the flow at the gate location, reducing labor needs and water waste. The flow control system is made up of smaller systems that can operate on their own while yet working as a whole. The technology is very effective and efficient at regulating the water flow in canals, enhancing irrigation methods, and minimizing water waste.

Keywords: Remote Terminal Unit (RTU) · flow control system · water level measurement system · flow measurement system · gate actuator system · solar power supply · water distribution

1 Introduction

The management of water resources, water delivery to communities and agricultural areas, flood prevention, and river traffic are all made possible by extensive canal networks. Traditional canal water management techniques, however, are frequently labor-intensive and ineffective, which causes delays in decision- making and response times. [8] This research suggests the use of an Internet of Things (IoT) enabled canal water monitoring and controlling system that enables real-time monitoring and controlling

V. Sakalauskas et al. (Eds.): IBICA 2023, LNNS 1231, pp. 384–392, 2025.
https://doi.org/10.1007/978-3-031-78946-5_37

management in order to address these difficulties. The designed "Canal Water Data Monitoring and Controlling System" utilizes various sensors, including ultrasonic sensors to collect data on water levels and flow rates accurately. The data collected by these sensors are then processed and routed to a free local host IoT server (thinkspeak.com), where they can be monitored in real-time by an operator at remote locations through a user-friendly app (Thingshow). This real-time monitoring provides decision-makers with accurate information, enabling them to make quick and informed decisions regarding water management. The system also enables the remote control of canals gate related operations such as opening and closing gates, facilitating effective water management. This automation eliminates the drawbacks and inefficiencies of traditional canal monitoring systems, enabling quicker response times and more efficient water management [9].

The objectives includes developing a system that eliminates the drawbacks of traditional canal water management systems, improves water management efficient cy, and facilitates effective water distribution. By implementing this system, decision-makers can better manage water resources, ensuring a sustainable and efficient water management system [7].Overall, the designed Canal Water Data Monitoring and Controlling System promises to be a significant improvement over traditional canal water management systems with real-time monitoring, accurate data collection, and automated canal gate's controls. The system will enable quick and informed decision-making, leading to more efficient and sustainable water management [11].

2 Related Works

Several studies have investigated the usability of IoT for water management, including the work of author [1], who proposed the development of an IoT-based system for dam safety and water management. The proposed system uses sensors to monitor water levels and provides real-time data for monitoring and controlling purposes. A wireless sensor network (WSN)-based automated monitoring and reporting system for water quality was proposed, employing various routing strategies. They monitored the water quality using a ZigBee-based WSN and sent data to the monitoring station using various routing methods. [2] A real-time water quality monitoring network and water quality indices for the River Nile were developed. The system used sensors to gather information about the water's quality, and it used water quality indices to assess the water's condition. [3].

Several studies have also focused on water level sensing and controlling, such as a microcontroller- based automated water level sensing and controlling system [4]. The system used ultrasonic sensors to detect the water level and controlled the water flow using a microcontroller. Researcher Proposed solutions for advanced sensing of water infrastructures in urban environments, including the use of low-cost sensors and wireless communication technologies. They discussed the challenges of implementing advanced sensing technologies and proposed solutions to overcome them. [5] A low-power wireless liquid monitoring system using ultrasonic sensor system used a low-power wireless communication module to send the data to the monitoring station [6].An automatic water level control system using sensors to detect the water level and control the water flow used a microcontroller to control the water flow and maintain a constant water level [7].

3 Methodology

The methodology employed for the implementation of the proposed canal water data monitoring and controlling system is divided into three phases, each with its own set of sub-phases. These phases are Structural Implementation, Hardware Implementation, and Software and Network Implementation.

3.1 Structural Implementation

Structural Implementation, involves replicating the dam and canal structures at the project scale using acrylic and plastic materials. The dimensions of the structures are obtained using laser cutting technology, and a 3D model is created to ensure accuracy. The structures are then mounted onto a plywood base to ensure stability.

3.2 Hardware Implementation

The hardware implementation phase involves the installation of all the required electronic hardware necessary for yielding all the promised functionalities.

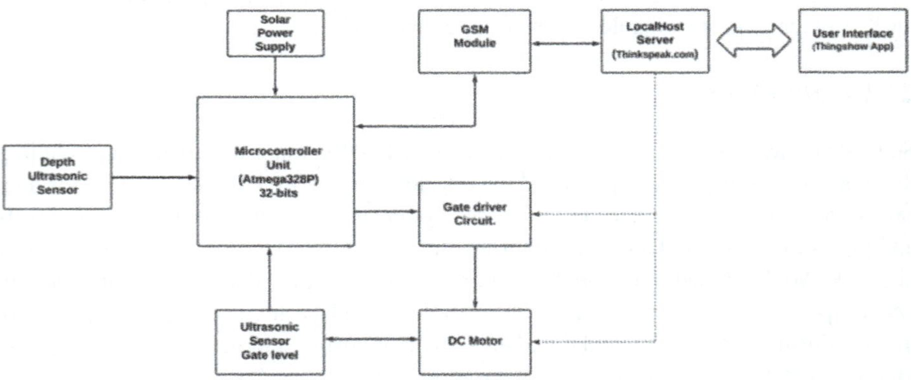

Fig. 1. Block Diagram of Canal water data monitoring and controlling system

The Fig. 1 shows block diagram of Canal water data monitoring and controlling system. The ultrasonic sensors (HC-SR04) are interfaced to the microcontroller through the GPIO (General Purpose Input/Output) ports. The gate driven circuitry is designed with a DC motor to control the canal gate and is connected to a microcontroller. [12] The GSM Module SIM TOOL 800 is used to transmit data to the free cloud- based server, which is ThingSpeak.com, with enabled ThingShow App (User interface) that will plot all the necessary graphs of the system. [13] Additionally, the entire system will be powered by a solar power supply. A scale model of the dam and canal structure was created using acrylic and plastic materials.

3.3 Software and Network Implementation

In the Third phase, Software and Network Implementation, the micro-controller is programmed to process water-related data fetched from the canal structures by the installed sensors. The free cloud-based server, ThingSpeak, is used to receive and store the data, which is then sent to the ThingShow app every 60 s. The GSM Module is used to receive and transmit command to microcontroller to perform task like opening the gate of canal. The micro-controller is programmed using Arduino IDE, and an effective algorithm is developed to efficiently implement the project. A GSM module is successfully integrated to transmit data to the cloud-based server. An LCD display was also installed to show the depth of water and the status of the canal gate. [10] The DC motor with attachments was connected to the microcontroller for user control of the gate opening and closing.

4 Experimental Result and Discussions

The designed Canal Water Data Monitoring and Controlling System is able to monitor and record the water level and flow rates of the canal using ultrasonic sensors and display the data on an LCD display. The system also allows remote monitoring of the water level, flow rates and canal gate status through a cloud-based server using GSM communication.

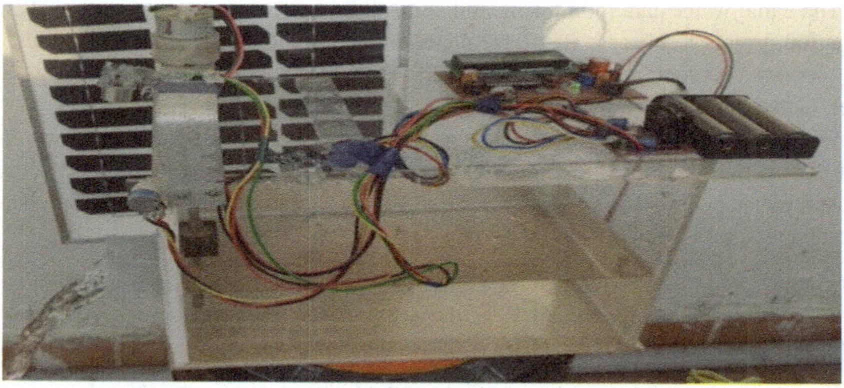

Fig. 2. Demonstration of Canal water data monitoring and controlling system

Figure 2 Shows Demonstration of Canal water data monitoring and controlling system In addition to monitoring, the system is also capable of controlling the canal gate using a DC motor. The motor is controlled by a motor driver module (L298) which is interfaced with the microcontroller. The system allows the user to remotely control the opening and closing of the canal gate using a mobile phone.

The GSM module and a set of buttons that are used to open or close the gate of the canal. When a call is made to the SIM number inserted in the GSM module, the microcontroller detects the call and waits for the user to enter a specific button code. If the user presses “0” button, the gate will close completely. If the user presses “1” button, the gate will open 25%. If the user presses “2” button, the gate will open 50%. If the user presses "3" button, the gate will open 75%. If the user presses "4" button, the gate will open 100%.The microcontroller receives the button code and sends a corresponding signal to the L298 motor driver module, which controls the DC motor that opens and closes the gate. The system is designed to ensure that the gate opens and closes smoothly and accurately, based on the user's inputs.

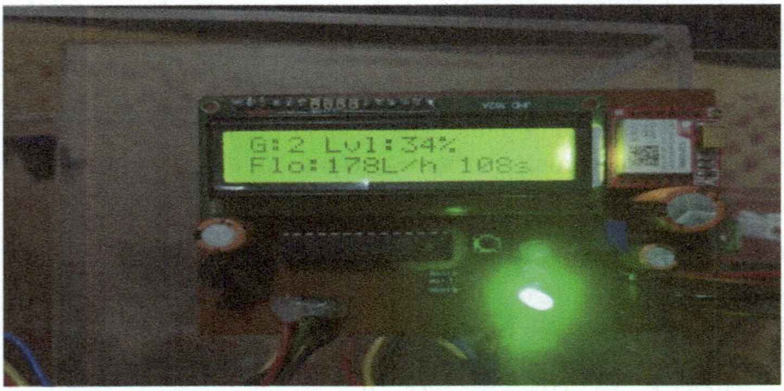

Fig. 3. Water level, Gate & Flow displayed on LCD

The canal water data monitoring and controlling system was tested in a small scale irrigation project to observe its performance in measuring water depth, flow rates, and gate opening percentages. Figure 3 shows the LCD display which is showing Water level, Gate opening & Flow of water.

Water depth measurements were taken using two HC- SR04 ultrasonic sensors placed at different points along the canal. The sensors were calibrated to provide accurate measurements within a range of 2 cm to 400 cm. Rough values of water depth were obtained as follows:

- At gate opening percentage of 25%, the water depth was approximately 10 cm.
- At gate opening percentage of 50%, the water depth was approximately 13cm.
- At gate opening percentage of 75%, the water depth was approximately 15cm.

Flow rate measurements were taken by calculating the water level in the canal and the percentage of the gate that is open. The gate is the structure that controls the flow of water in and out of the canal. Rough values of flow rates were obtained as follows:

- At gate opening percentage of 25%, the flow rate was approximately 178 L per hour.
- At gate opening percentage of 50%, the flow rate was approximately 256 L per hour.
- At gate opening percentage of 75%, the flow rate was approximately 356 L per hour.

Gate opening percentages were controlled using the L298 motor driver module and were observed visually. Rough values of gate opening percentages were obtained as follows:

- At a gate opening percentage of 25%, the gate was observed to be partially open.
- At a gate opening percentage of 50%, the gate was observed to be half open.
- At a gate opening percentage of 75%, the gate was observed to be almost fully open.

Table 1. Experimental observations of the performance of canal water data monitoring and controlling system

Sr. No	Water Level	Gate Opening	Water Flow
1	0%	0%	0
2	10%	0%	0
3	20%	0%	0
4	30%	25%	178 litres/hour
5	40%	0%	0
6	50%	0%	0
7	60%	50%	256 litres/hour
8	70%	0%	0
9	80%	75%	356 litres/hour
10	90%	0%	0

Experimental observation of the performance of canal water data monitoring and controlling system is shown in Table 1. These experimental observations provide a preliminary understanding of the performance of the canal water data monitoring and controlling system in measuring water depth, flow rates, and gate opening percentages. Further testing and calibration will be required to obtain more accurate measurements and to determine the system's suitability for large-scale irrigation projects.

Thingshow app provided detailed graphs and historical data about the water levels, Gate opening and flow rates as shown in Figs. 4.

All the processed data of canal water level, gate opening and Flow rates with time and date. The system was powered using a solar panel and battery, which proved to be a reliable and efficient source of power. Overall, the system provides an efficient and reliable solution for monitoring the water level and flow rate in a canal, while also allowing for easy control of the gate. The data collected can be analyzed and used to make informed decisions regarding water management, which can have significant impacts on agriculture, irrigation, and other industries that rely on water resources [13].

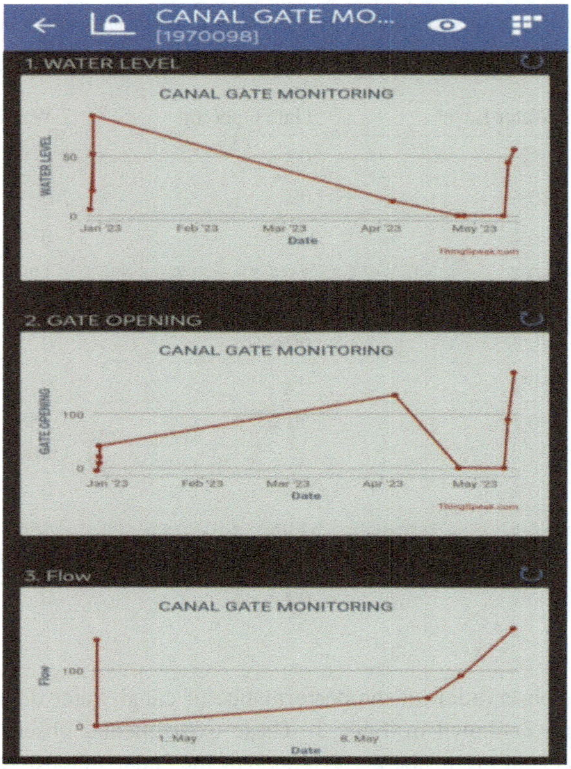

Fig. 4. Graphs and historical data about the water levels, Gate opening and flow rates on Thingshow app

5 Conclusion

Efficient management of water resources is essential for sustainable agriculture and economic growth in rural areas. The canal water data monitoring and controlling system provides a solution to the challenges faced by farmers and water resource managers in monitoring and controlling canal water levels and flow rates. The system's real-time monitoring and control features enable farmers to make informed decisions about when to irrigate their crops and how much water to use, thereby reducing water waste and improving crop yields. Moreover, the system's remote monitoring and control features make it an attractive option for large-scale water resource projects. The centralized platform can help reduce water waste and improve the efficiency of irrigation systems, which is crucial in regions where water is scarce. One of the significant benefits of the system is that it can detect and prevent malpractices done by workers in the canal. A vital phase toward developing efficient and durable irrigation systems is the canal water data monitoring and controlling system.

References

1. Mohod, N.: Usability of Internet of Things [IoT] for dam safety and water management. Int. J. Res. Advent Technol. **5**(1) (2017)
2. Khetre, A.C., Hate, S.G.: Automatic monitoring & Reporting of water quality by using WSN technology and different routing methods. IJARCET **2**(12), 3255–3260 (2013)
3. Madhavi, K.R., Jarugula, J., Karuna, G., Kaleru, S., Raju, K.S., Sunitha, G.: Bird species recognition using deep transfer learning. In: Reddy, K.A., Devi, B.R., George, B., Raju, K.S., Sellathurai, M. (eds.) Proceedings of Fourth International Conference on Computer and Communication Technologies. LNNS, vol. 606. Springer, Singapore (2023). https://doi.org/10.1007/978-981-19-8563-8_46
4. Narasimharao, J., Laxmaiah, B., Arumalla, R., Unnisa, R., Fatima, T., Nazare, S.S.: Restoration and deblurring the images by using blind convolution method. In: Reddy, K.A., Devi, B.R., George, B., Raju, K.S., Sellathurai, M. (eds.) Proceedings of Fourth International Conference on Computer and Communication Technologies. LNNS, vol. 606. Springer, Singapore (2023). https://doi.org/10.1007/978-981-19-8563-8_32
5. Kumar, V.N., et al.: Anomaly-based hierarchical intrusion detection for black hole attack detection and prevention in WSN. In: Reddy, K.A., Devi, B.R., George, B., Raju, K.S., Sellathurai, M. (eds.) Proceedings of Fourth International Conference on Computer and Communication Technologies. LNNS, vol. 606. Springer, Singapore (2023). https://doi.org/10.1007/978-981-19-8563-8_30
6. Dimlo, U.M.F., et al.: An improved blind deconvolution for restoration of blurred images using ringing removal processing. In: Reddy, K.A., Devi, B.R., George, B., Raju, K.S., Sellathurai, M. (eds.) Proceedings of Fourth International Conference on Computer and Communication Technologies. LNNS, vol. 606. Springer, Singapore (2023). https://doi.org/10.1007/978-981-19-8563-8_34
7. Munawar, S., Geetha, A., Srinivas, K.: Arrhythmia classification using BiLSTM with DTCWT and MFCC Features. In: Reddy, K.A., Devi, B.R., George, B., Raju, K.S., Sellathurai, M. (eds.) Proceedings of Fourth International Conference on Computer and Communication Technologies. Lecture Notes in Networks and Systems, vol. 606. Springer, Singapore (2023). https://doi.org/10.1007/978-981-19-8563-8_29
8. Subburaj, T., et al.: Discover Crypto-Jacker from blockchain using AFS method. In: Reddy, K.A., Devi, B.R., George, B., Raju, K.S., Sellathurai, M. (eds.) Proceedings of Fourth International Conference on Computer and Communication Technologies. LNNS, vol. 606. Springer, Singapore (2023). https://doi.org/10.1007/978-981-19-8563-8_15
9. Jagadeesh, S., Shrimali, M., Prasad, P.S., Maheswari, K., Narendrakumar, A., Koteswara, R.G.: Automated arrhythmia classification using Harris hawks optimization with deep learning model on IoT environment. In: 2023 Third International Conference on Artificial Intelligence and Smart Energy (ICAIS), Coimbatore, India, pp. 30–35 (2023). https://doi.org/10.1109/ICAIS56108.2023.10073736
10. Thaseen, A., Unnisa, R., Sultana, N., Madhavi, K.R., NagaJyothi, G., Kirubakaran, S.: Breast cancer detection using deep learning model. In: Reddy, A.B., Nagini, S., Balas, V.E., Raju, K.S. (eds.) Proceedings of Third International Conference on Advances in Computer Engineering and Communication Systems. LNNS, vol. 612. Springer, Singapore (2023). https://doi.org/10.1007/978-981-19-9228-5_57

11. Madhavi, K.R., Vooradi, S., Mounika, P., Yedlla, S., Tangudu, N.: Emotion analysis using convolutional neural network. In: Reddy, A.B., Nagini, S., Balas, V.E., Raju, K.S. (eds.) Proceedings of Third International Conference on Advances in Computer Engineering and Communication Systems. LNNS, vol. 612. Springer, Singapore (2023). https://doi.org/10.1007/978-981-19-9228-5_44
12. Nagajyothi, G., Kumar, G.P., Kumar, B.S., Kumar, B.P.D., Damodaram, A.K.: High-speed low area 2D FIR filter using Vedic multiplier. In: Reddy, A.B., Nagini, S., Balas, V.E., Raju, K.S. (eds.) Proceedings of Third International Conference on Advances in Computer Engineering and Communication Systems. LNNS, vol. 612. Springer, Singapore (2023). https://doi.org/10.1007/978-981-19-9228-5_48

Early Detection and Prediction of Water Quality Using Machine Learning Techniques

N. Pooja kumara, H. S. Vandana, and Bobby Lukose(✉)

Department of Computer Applications, Krupanidhi College of Management, Bengaluru, Karnataka, India

mcakric@krupanidhi.edu.in

Abstract. Forecasting the availability of clean water is becoming increasingly difficult. This study aims to identify the numerous contaminants present in water using various criteria and then use a machine learning classification algorithm to determine whether the water is clean. Each of the 1043 samples contains a unique set of 28 columns of data, including biochemical oxygen demand, temperature, PH, conductivity, dissolved oxygen, feces-associated coli form, total coli form, and nitrate. This study used Adaboost, RF, KNN, logistic regression, DT, SVM, and other machine learning techniques to construct a robust indicator and classification system for water quality. More accurate prediction models can be developed by examining more reliable data sources, highlighting the importance of ensuring that drinking water is safe. The correlation between predicted and actual values indicates the model or system's accuracy when predicting water quality. Based on the obtained accuracy, the model predicts which method (RF, DT, KNN, or LR) would be the most effective at purifying a specific range of water and, consequently, The KNN algorithm is most effective at detecting whether water is safe to consume (94%). In contrast, the SVM algorithm is the least effective (66%).

Keywords: Government Data · Random Forest · Linear Regression · Decision Tree · SVM · KNN · Adaboost

1 Introduction

Humans require water to survive, and the quality of that water directly impacts their ability to carry out daily tasks. Before water can be designated potable or suitable for human consumption, it must meet a minimum quality standard. Non-potable water is defined as water that is not suitable for human ingestion. Even after extensive purification, non-potable water is unsuitable for human consumption. In 2020, over 74% of the world's population, or 5.8 billion people, will have access to water that has been appropriately managed, indicating that it has been treated to at least the WHO's standard for safety and quality. However, more than two billion people worldwide consume contaminated or hazardous water. The drinking water individuals consume has a direct effect on their health. If their presence can be predicted, potentially harmful pollutants to public health can be eliminated before causing damage. People should be given early warnings

V. Sakalauskas et al. (Eds.): IBICA 2023, LNNS 1231, pp. 393–400, 2025.
https://doi.org/10.1007/978-3-031-78946-5_38

and notifications to prevent consuming contaminated water, which would benefit public health. The initiative will develop prediction models that accurately estimate water quality measurements using the provided criteria. Statistical analysis, machine learning, and other modeling techniques may be required. Validation against vast datasets is required to evaluate the constructed models' quality. Comparison of the model's predictions to independent data can shed light on the model's validity. The algorithms can analyze complex data connections and patterns to learn how to make accurate predictions.

2 Literature Review

SitiNurMahfuzah et al. [1] contrasted the efficacy of Random Tree and Random Forest for classifying river water quality. According to the data, Random Forest is more effective than Random Forest at classifying data. Umair Ahmed et al. [2] presented supervised learning models to evaluate water quality prediction using turbidity, PH, total dissolved solids, and temperature as water component attributes. Using minimal features, the proposed models accurately predicted the WQI in real-time with low error rates. To learn more about the country's water quality issue, the researcher conducted investigations in Pakistan and analyzed the data collected in the lab. Daud et al. [3] discovered an abundance of E. coli and fecal coli form in water samples collected from various locations in Pakistan and manually analyzed for numerous parameters. This was attributed to effluent and industrial refuse. Alamgir et al. [4] discovered an abundance of sulfates and total fecal coliforms in 46 samples from Orangi town, Karachi, through manual laboratory analysis. Theyazn et al. [5] classified water quality. Network, non-linear auto regression neural networks, support vector machine, and the model are a few examples of these models. They utilized the SVM machine learning technique and attained a 97.01 percent accuracy rate. Hasan and Alhammadi [6] analyzed the physical and chemical information required by the department to confirm the safety of Abu Dhabi's potable water. Several Algorithms, like logistic regression, decision trees, k-nearest neighbor, and support vector machines, were employed. Compared to other classifiers, the DT predicts the water quality level with the highest accuracy (97.70%).

Haq et al. [7] made water quality predictions. This study contrasted two distinct styles of acting. Through K-fold cross-validation, the author put the machine learning model through its trials in this research. The DT classifier was the most precise at 97.23 percent. Hassan et al. [8] addressed the water quality prediction problem using MLR, BTM, MN, RF, and SVM. The proposed solution comprises five steps: data preprocessing, managing missing values, correlating features, employing machine learning techniques, and modeling feature significance. 99.83% was the maximum accuracy of the MLR classifier. Machine learning techniques, fold cross-validation. The authors employed various statistical techniques, such as ridge regression, Lasso regression, least-squares support vector machines, XGBoost, and a hybrid model (XGBoost+Lasso regression). The results supported the utility of the hybrid framework proposed.

Wang et al. [9] The final prediction is then made using a machine learning model that combines the five most common methods (Bayesian network, random forest, partial least square, sparse partial least square, and AK hand linear regression). Three beaches on the eastern shore of Lake Erie in New York were evaluated by layering models and compared to five baseline models. The outcomes demonstrated that the approach of model stacking outperformed the distinct foundation models. The layering model consistently ranked first in accuracy year after year, averaging 78%, 81%, and 82.3% across the three examined coastlines.

Ding et al. [10] employed Genetic Algorithm, Back Propagation Neural Network, and Principal Component Analysis to develop a composite intelligent system for forecasting river water quality. This study examined 23 distinct water quality indices, each of which affects water quality in a subtle, non-linear manner. In addition to optimizing the BPNN parameters, principal component analysis (PCA) significantly sped up the training time of subsequent algorithms. The study's results indicated an overall prediction rate of nearly 91%, with an average prediction rate of 88.9% for pure water quality and a prediction rate of 93.1% for polluted water quality. Abobakr Saeed et al. [11] designed an efficient machine-learning algorithm based on the SVM model to predict the WQI of the Langat River Basin by analyzing six variables (ammonia nitrogen, dissolved oxygen, PH, chemical oxygen demand, suspended solids, and biochemical oxygen demand) of dual reservoirs in the catchment. Using this model, the investigator could estimate WQI values with a small mean absolute error, as demonstrated by the experimental findings.

3 Methodology

This data set contains 1043 samples analyzed for nine properties: biochemical oxygen demand, mobility, temperature, nitrate, conductivity, fecal coliforms, total coliforms, and PH. The relationship between pH, an indicator of water acidity/alkalinity (6.5–8.5), and other water parameters. Conductivity (1501U): indicates total dissolved solids; estimates salinity; identifies pollution and contamination.

Biological Oxygen Demand (BOD) > 5.0. Nitrate levels in potable water (10 mg/L) indicate nutrient pollution, water quality, and agricultural and land management. Fecal coliforms (2500) can indicate fecal contamination, a measurement of waterborne pathogens. Total coliforms per one hundred milliliters can be a health risk indicator.

Above 100 degrees Celsius, chemical processes, stratification, mingling, and the production of dissolved oxygen are triggered, along with biological activity and terminal pollution.

3.1 Algorithm Explanation

Support Vector Machines: SVM and other supervised learning algorithms [12] can be used for regression and classification. It performs reasonably well for regression, but its strength is categorization. The magnitude of the hyperplane is proportional to the feature variables in the dataset. The hyperplane is a straight line when just two input characteristics exist. All three conditions must be satisfied to compress the hyperplane into two dimensions. However, displaying the hyperplane graphically becomes more accessible when more than three feature variables are involved.

Decision Tree: classifier is the most well-known data classification technique [13]. Both theoretically and practically, the decision tree classifier is simple to understand.

It requires minimal computer processing time. However, this classifier may need to be more resistant to overfitting.

Ada Boosting: Using Ada boosting, numerous unsuccessful workbooks are merged and individually trained using data from incorrectly classified instances. Several manual adjustments were made to enhance the Ada boosting technique.

Logistic Regression, a supervised learning technique, is a shared machine learning (ML) technique. LR can predict a categorical dependent variable given an independent set of parameters. Consequently, the answer must be an absolute number or a fixed integer. For instance, the final result could be: Please provide a (Yes/No/0/1) response. True/False. The logistic regression method also yields a probability value between 0 and 1.

K-nearest neighbor: The KNN technique presupposes that new and existing data are comparable. KNN makes no assertions about the data it analyzes, unlike parametric algorithms. In addition, the algorithm is a slow learner because acquiring new information from the training set takes time. There is no need for a training phase because the dataset is saved, and an operation is performed on it during classification. It can be mastered briefly, as shown in Fig. 1.

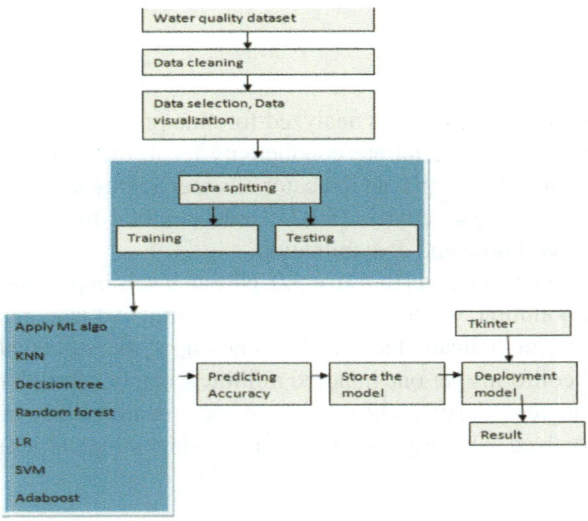

Fig. 1. Architecture Diagram

3.2 Architecture Diagram

4 Result

4.1 Confusion Matrix of Different Classifiers

The effectiveness of a taxonomy model can be evaluated using a confusion matrix, n-by-n matrix. Several correctly predicted test data points determine the model's accuracy. The efficacy of the classification algorithms (Adaboost, DT, KNN, SVM, Logistic

Regression, and Random Forest) was evaluated by calculating the misclassification rate, sensitivity, specificity, and prediction accuracy. It is essential that the test accurately determines whether the water is suitable for human consumption, as shown in Table 1.

$$\frac{TP + TN}{TP + TN + FP + FN}$$

Table 1. Comparison of confusion matric with different classifiers

Confusion Matrix	SVM (%)	DT (%)	LR (%)	Adaboost (%)	RF (%)	KNN (%)
True Positive	63.16	60.29	57.89	59.33	59.81	60.75
True Negative	0.48	31.10	34.45	33.01	33.01	33.01
False Positive	33.49	3.35	5.74	4.31	3.83	2.87
False Negative	2.87	5.26	1.91	3.35	3.35	3.35

The sensitivity of a test is its ability to precisely predict whether water is suitable for human consumption (also known as recall).

$$\frac{TP}{TP + FN}$$

The F1 scale ranges from 0 to 1, representing the harmonic mean of memory and precision ratings.

$$\frac{2 * Sensitivity * Precision}{Precision + Sensitivity}$$

The proportion of potentially potable water samples (TP minus FP) among all tested water samples can indicate the test's reliability. Here, the researcher presents the results of the water transportability prediction algorithms. The KNN algorithm yields superior results to extensively employed techniques such as Decision Tree is 91% accurate, random Forest, logistic regression, Adaboost, and support vector machine. On this dataset, SVM is 66% accurate, K-Nearest Neighbors is 93% accurate, logistic regression is 92% accurate, Random Forest is 92% accurate, and Adaboost is 92% accurate. These findings prove that the KNN algorithm's ability to discern whether a given water source is safe for human consumption has been enhanced.

4.2 Comparisons of the Accuracy of Various Classifiers

The proportion of correct estimates to total predictions measures categorization accuracy. Because it is easy to calculate and comprehend, it has become the standard for classifier model evaluation as shown in Table 2.

Table 2. Comparison results of classification analysis to six classifiers

Classifier	Accuracy %
SVM	66.02%
Decision Tree	91.38%
Logistic Regression	92.34%
Adaboost	92.34%
Random Forest	92.28%
KNN	93.77%

4.3 Describe the Classification Model

The output of water quality prediction depends on the parameter, and the model generates results based on the quality of water of the dataset values, as shown in Fig. 2. Each parameter has a specific format and range that varies the output quality. The prediction model or system to measure the accuracy output of water quality prediction. It quantifies the level of agreement or discrepancy between predicted and actual values. Specific accuracy tells the model's reliability and effectiveness and how purified the water is for drinking, as shown in Fig. 3. Based on the accuracy generated, the model predicts which algorithm is best for purifying that range of water.

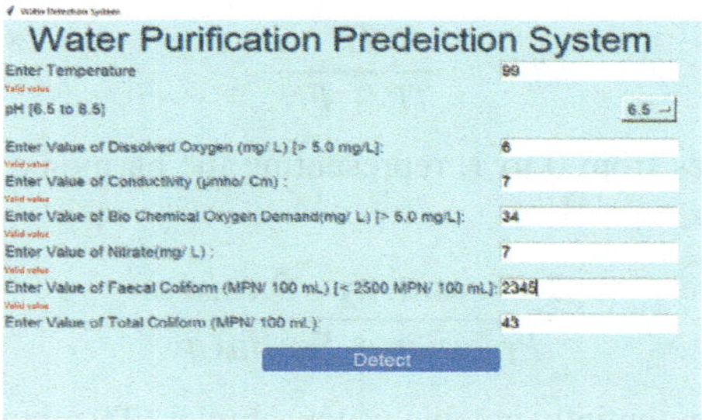

Fig. 2. User Inputs

5 Discussion

Our findings reveal the model's ability to predict water quality with an overall accuracy of 93.7%. Despite implementing imputation strategies, the residual impact of missing data on model performance can only partially be ruled out. This highlights the importance of more comprehensive and consistently collected data for even more robust predictions. While our model achieved promising results, its generalizability to other regions remains a consideration. Comparison with existing water quality prediction methods indicates that our approach offers competitive performance. The model's strong performance and potential real-world applications signify a step forward in environmental data science.

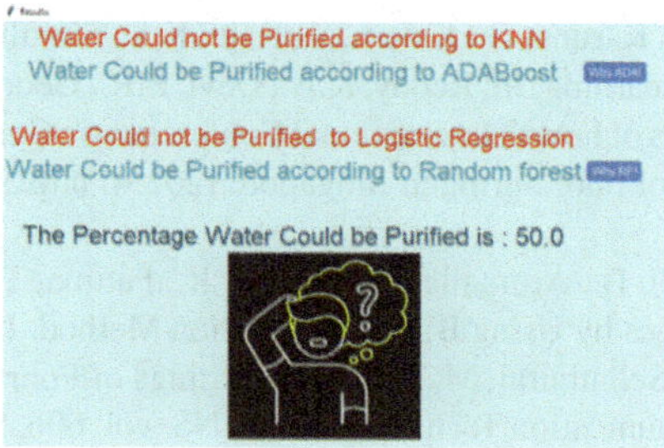

Fig. 3. Predicting the result

6 Conclusion

This study implements a machine learning system to classify water samples and estimate their quality score. The classification layer of the proposed architecture consists of AdaBoost, a K-Nearest Neighbors, a Support Vector Machine, and a Decision Tree classifier. A benchmark data set consisting of 1043 water samples and nine attributes was used to evaluate the efficiency of these models. With the collected data, the excellent classification results of one model were clarified with average precision—93.7%. Even if the K-Nearest Neighbors algorithm had the highest accuracy, this was evident from the data. These results support that machine learning algorithms may aid water quality analysis and forecasting. These findings also have important implications for developing machine learning models that accurately predict water quality for human consumption, agricultural irrigation, and other purposes. And comparison. Those looking for a new home may also rely on our trustworthy recommendations. Therefore, individuals will be better equipped to make informed decisions.

References

Nafi, S.N.M.M., Mustapha, A., Mostafa, S.A., Khaleefah, S.H., Razali, M.N.: Experimenting two machine learning methods in classifying river water quality. In: International Conference on Applied Computing to Support Industry: Innovation and Technology, pp. 213–222. Springer International Publishing, Cham (2019)

Ahmed, U., Mumtaz, R., Anwar, H., Shah, A.A., Irfan, R., García-Nieto, J.: Efficient water quality prediction using supervised machine learning. Water **11**(11), 2210 (2019)

Daud, M.K., et al.: Drinking water quality status and contamination in Pakistan. Biomed. Res. Int. **2017**, 7908183 (2017)

Alamgir, A., Khan, M.A., Hany, O.E., Shaukat, S.S., Mehmood, K., Ahmed, A., Ali, S., Riaz, K., Abidi, H., Ahmed, S., Ghori, M.: Public health quality of drinking water supply in Orangi town, Karachi, Pakistan. Bull. Environ. Pharmacol. Life Sci. **4**(11), 88–94 (2015)

Aldhyani, T.H., Al-Yaari, M., Alkahtani, H., Maashi, M.: Water quality prediction using artificial intelligence algorithms. Appl. Bionics Biomech. **2020**, 1–12 (2020)

Hasan, A.N., Alhammadi, K.M.: Quality monitoring of abu dhabi drinking water using machine learning classifiers. In: 2021 14th International Conference on Developments in eSystems Engineering (DeSE), pp. 1–6. IEEE (2021)

Madhavi, K.R., Jarugula, J., Karuna, G., Kaleru, S., Raju, K.S., Sunitha, G.: Bird species recognition using deep transfer learning. In: Reddy, K.A., Devi, B.R., George, B., Raju, K.S., Sellathurai, M. (eds.) Proceedings of Fourth International Conference on Computer and Communication Technologies. LNNS, vol 606. Springer, Singapore (2023). https://doi.org/10.1007/978-981-19-8563-8_46

Narasimharao, J., Laxmaiah, B., Arumalla, R., Unnisa, R., Fatima, T., Nazare, S.S.: Restoration and Deblurring the Images by Using Blind Convolution Method. In: Reddy, K.A., Devi, B.R., George, B., Raju, K.S., Sellathurai, M. (eds.) Proceedings of Fourth International Conference on Computer and Communication Technologies. LNNS, vol. 606. Springer, Singapore (2023). https://doi.org/10.1007/978-981-19-8563-8_32

Kumar, V.N., et al.: Anomaly-Based Hierarchical Intrusion Detection for Black Hole Attack Detection and Prevention in WSN. In: Reddy, K.A., Devi, B.R., George, B., Raju, K.S., Sellathurai, M. (eds.) Proceedings of Fourth International Conference on Computer and Communication Technologies. LNNS, vol. 606. Springer, Singapore (2023). https://doi.org/10.1007/978-981-19-8563-8_30

Dimlo, U.M.F., et al.: An Improved Blind Deconvolution for Restoration of Blurred Images Using Ringing Removal Processing. In: Reddy, K.A., Devi, B.R., George, B., Raju, K.S., Sellathurai, M. (eds.) Proceedings of Fourth International Conference on Computer and Communication Technologies. LNNS, vol. 606. Springer, Singapore (2023). https://doi.org/10.1007/978-981-19-8563-8_34.

Munawar, S., Geetha, A., Srinivas, K.: Arrhythmia Classification Using BiLSTM with DTCWT and MFCC Features. In: Reddy, K.A., Devi, B.R., George, B., Raju, K.S., Sellathurai, M. (eds.) Proceedings of Fourth International Conference on Computer and Communication Technologies. LNNS, vol. 606. Springer, Singapore (2023). https://doi.org/10.1007/978-981-19-8563-8_29

Subburaj, T., et al.: Discover Crypto-Jacker from Blockchain using AFS method. In: Reddy, K.A., Devi, B.R., George, B., Raju, K.S., Sellathurai, M. (eds.) Proceedings of Fourth International Conference on Computer and Communication Technologies. LNNS, vol. 606. Springer, Singapore (2023). https://doi.org/10.1007/978-981-19-8563-8_15

S. Jagadeesh, M. Shrimali, P.S. Prasad, K. Maheswari, A. Narendrakumar, A., Reddy, K.: Automated arrhythmia classification using Harris Hawks optimization with deep learning model on IoT environment. In: 2023 Third International Conference on Artificial Intelligence and Smart Energy (ICAIS), Coimbatore, India, pp. 30–35 (2023). https://doi.org/10.1109/ICAIS56108.2023.10073736.

Thaseen, A., Unnisa, R., Sultana, N., Madhavi, K.R., NagaJyothi, G., Kirubakaran, S. (2023). Breast Cancer Detection Using Deep Learning Model. In: Reddy, A.B., Nagini, S., Balas, V.E., Raju, K.S. (eds) Proceedings of Third International Conference on Advances in Computer Engineering and Communication Systems. Lecture Notes in Networks and Systems, vol 612. Springer, Singapore. https://doi.org/10.1007/978-981-19-9228-5_57

Madhavi, K.R., Vooradi, S., Mounika, P., Yedlla, S., Tangudu, N.: Emotion analysis using convolutional neural network. In: Reddy, A.B., Nagini, S., Balas, V.E., Raju, K.S. (eds.) Proceedings of Third International Conference on Advances in Computer Engineering and Communication Systems. LNNS, vol. 612. Springer, Singapore (2023). https://doi.org/10.1007/978-981-19-9228-5_44.

Nagajyothi, G., Kumar, G.P., Kumar, B.S., Kumar, B.P.D., Damodaram, A.K.: High-speed low area 2D FIR filter using vedic multiplier. In: Reddy, A.B., Nagini, S., Balas, V.E., Raju, K.S. (eds.) Proceedings of Third International Conference on Advances in Computer Engineering and Communication Systems. LNNS, vol. 612. Springer, Singapore (2023). https://doi.org/10.1007/978-981-19-9228-5_48

Bird Species Recognition and Detection Using Neural Network

Reshma Mathew, Sreekrishnadasan, and N. Rajkumar(✉)

Department of Computer Applications, Krupanidhi College of Management, Bengaluru, Karnataka, India
mcakric@krupanidhi.edu.in

Abstract. There is a wide variety of birds worldwide, and identifying them by name and call requires time and effort. To get around this problem, there is a need for a robust model that can reliably identify the bird species from its name and call. The methodology is to create a robust and efficient system for identifying 510 unique bird species by preprocessing a dataset of bird images from eBird, iNaturalist, and a Convolutional Neural Network (CNN) to classify these images using the EfficientNetV2 architecture in TensorFlow. This approach includes collecting and processing raw data, building and training a model and delivering it using Flask. This research utilized 80% of the data for training and 20% for evaluating its performance. Deploying the model has led to an increase of 86% in the ability to recognize and forecast the class name of birds and the corresponding bird sound. The result displays the use of the algorithm to identify bird species from photographs and improve the accuracy and efficiency of monitoring bird populations and their habitats.

Keywords: Data · CNN · Efficient Net · Fine-tuning · Transfer Learning

1 Introduction

It is necessary to know its species to identify a bird from a photograph. Identifying bird species can be difficult due to differences in appearance, background, and environmental conditions. Recently, alterations in the size and conduct of bird populations have become a pressing concern. Time and resources are required for the capture and study of animals. Deep learning has seen a stratospheric rise in popularity in recent years, outperforming earlier state-of-the-art methods in various fields, primarily due to the availability of high-performance computation [1]. A neural model is created for identification purposes, and the user can upload a picture to be analyzed by the model. The model then outputs the bird species to the user [2]. Convolutional Neural Network (CNN) is used because it is appropriate for imposing superior algorithms and offers accurate numerical precision accuracy. CNN algorithm can be used for visual processing that could help recognize bird species that researchers have studied as a non-intrusive and extraordinarily effective technique [3]. Using training data, CNN and transfer learning can classify species sounds more accurately.

V. Sakalauskas et al. (Eds.): IBICA 2023, LNNS 1231, pp. 401–408, 2025.
https://doi.org/10.1007/978-3-031-78946-5_39

CNN has demonstrated success in image categorization and the potential to address various computer vision issues. The development of algorithms to classify birds has assisted authorities in their search for bird species with highly accurate classifications. The datasets for these studies are collected and cleansed from numerous databases and websites. Developing and training a self-learning model using images of diverse bird species is essential to predict species. This endeavor will utilize the pre-trained EfficientNetV2 model available on TensorFlow Hub as its foundation. Due to its construction on ImageNet, a massive image dataset, the pre-trained model can detect various features relevant to our bird photo classification problem.

2 Literature Review

To acquire features from bird regions of interest (ROIs), Borana et al. [1] developed an end-to-end deep learning-based strategy based on transfer learning. To extract the Bird ROIs from the photos and feed them to the neural network, which has been constructed using a transfer learning technique and fine-tuned using the available dataset, a pre-trained Mask-RCNN is employed.

Bird Species identification with CNN [2] paper examines the entire experimental configuration using CNN Python. It accomplishes this by extracting characteristics from images that can be used for recognition. The employed method is adequate for identifying valuable video content and effectively labeling it.

Deep convolutional neural networks for acoustic detection of locally scarce bird species [3], pre-trained models are used as feature extractors. At the same time, transfer learning and training from scratch are investigated to classify bird sounds. Deep cascade characteristics are taken explicitly from multiple layers of several pre-trained models to categorize bird noises and then combined. By simply repeating the spectrogram to make it acceptable for trained models, a multi-view spectrogram is created to characterize bird noises.

In the architecture of EfficientNet [4], features from a CNN are extracted across multiple convolutional layers after locating a component using the proposed method. The EfficientNet architecture has been proven to be an efficient system for classification. For local bird recognition, various CNN architectures and transfer learning are used [5]. This study put out the issue of using three-dimensional convolutional recurrent neural networks to recognize bird sounds. The results show that 3D convolution and recurrent layers can utilize redundancy in the long-term temporal modeling of bird cries. The proposed architecture performs better than the standard CNN+RNN method.

Identification of Bird [6] This paper provides evidence that neural network characteristics aid in identifying bird species from photographs. The proposed strategy has proven effective in situations with limited acoustic data. They achieved a classification accuracy of 27.03% using a simple visual classifier. If a perfect rejection criterion is employed to exclude only incorrectly classified images, audio alone can obtain an accuracy rate of 37.33%.

In [7], the paper has been proposed to determine the loss of triplets, allowing us to differentiate between bird species. This article highlights the Red-vented Bulbul, house sparrow, crow, magpie, common kingfisher, myna, and cuckoo as seven indigenous bird species of Bangladesh. To identify this creature, six transfer learning strategies were utilized. In this study, 2800 images are trained using the model, while 700 images (20%) are used to complete the analysis.

This study offers a method to localize and identify bird species in a user-submitted photograph using mask RCNN, transfer learning, and CNN. Transfer learning quickens the learning process by reusing previously learned knowledge and reduces the requirement for extensive computational resources [8].

The suggested approach starts with component detection and extracts CNN features from many convolutional layers. The classifier judges after evaluating the data. 80% of bird species have been identified with the proposed method [9].

The deep learning work on bird species identification [10] focused primarily on identifying 27 indigenous bird species. The suggested method had a training dataset accuracy of 98.70% for identifying and differentiating uploaded photos as birds. The final goal of this research was to create an autonomous system for distinguishing fine-grained objects among photos of birds with common essential traits but little changes in appearance.

Using deep learning techniques for recognizing birds [11], This study investigates how CNN may be used to extract features from photos and use them for deep learning for bird identification. The investigation's findings include but are not limited to, the following: Using PCA on deep features reduces dimensionality and, as a result, training and testing time but also enables increasing identification accuracy, especially when using ANN.

DFNN [12] in this paper describes classifying different bird species utilizing Python Tensor Flow and deep learning techniques.

Researchers offer deep learning methods for detecting auditory birds [13]. The detection of birds in audio recordings is accomplished by modifying and fine-tuning DCNNs, initially developed for image classification.

Using both visual features extracted from unconstrained bird photos and auditory data collected from bird vocalizations, [14] presents a novel method for identifying different species of birds.

3 Methodology

3.1 Data Collection

A tagged dataset containing images of 500 distinct bird species would be ideal for this image classification task. Five images of each of 500 distinct species have been used, and for testing, an average of 100 images per species has been used for training and evaluation. At least 50,000 or 100 images for each species in the collection are recommended. It is conceivable that images from multiple sources must be compiled for such a vast collection.

- eBird:
- iNaturalist:
- Flickr API

3.2 Data Preprocessing

Among the activities involved in preparation are:

Resizing Photos: To be processed by EfficientNetV2, images must adhere to specific requirements, such as a resolution of 224 × 224 pixels. To ensure compatibility with widely used image processing software such as OpenCV and PIL, all images within the dataset must be resized to this format.

Enhanced Data: Data augmentation methods like rotation, horizontal flipping, magnification, and brightness/contrast modifications can expand the dataset and improve the model's generalizability.

Classification of Labels: Numbering the names of the numerous bird species from 0 to 499 using the range provided. This translation is essential, as machine learning algorithms require numerical data.

Test/Train Validation: Divide the data into a training set (80%) and a testing set (20%) to determine the model's effectiveness.

3.3 Model Development and Training (Algorithm Used)

3.3.1 EfficientNetV2 Architecture

The EfficientNetV2 family of CNNs primarily concerns accurate and rapid image classification [4]. Its design and training procedure are preferable to Efficient Net's due, among other things, to including Fused-MBConv blocks and progressive learning. The method surpasses conventional CNNs in scope, depth, and resolution while requiring fewer computational resources. Users can choose from three various sizes of EfficientNetV2 depending on their requirements and available computing resources: small (S), medium (M), and large (L). The TensorFlow Hub-hosted pre-trained EfficientNetV2 model will serve as the foundation for this project. The pre-trained model can recognize various characteristics relevant to our bird picture classification challenge since it was trained on ImageNet, a large-scale image dataset.

3.3.2 Transfer Learning

Transfer learning enables a model explicitly trained for one task to perform another [5]. Transfer learning will be utilized to apply the EfficientNetV2 model to the bird photo categorization challenge. To achieve this, swap out the pre-trained model's last layer for a dense layer created explicitly for the task. This custom layer must have 510 output nodes (one for each bird species) and use a softmax activation function to calculate class probabilities.

3.3.3 Fine-Tuning

After removing the last layer, the model will be fine-tuned by being taught to identify birds in a predefined set of photographs. During fine-tuning, backpropagation is used to modify the model's weights to learn our dataset's features more effectively [5, 6]. The loss function will be based on the categorical cross-entropy, and the model will be trained using the Adam optimizer with a learning rate of 0.001. Tracking the validation loss throughout training prevents overfitting and halts early after ten iterations.

3.4 Deployment Using Flask

Flask Web Application

Flask is a Python web framework that facilitates the creation and distribution of websites. It is ideal for undertaking any scale due to its adaptable and user-friendly design. The Flask-based web application will provide users with a predicted species name and an audio recording of the submitted bird in exchange for a photograph.

3.5 Model Integration

TensorFlow'sload_model() has been incorporated into the pre-built EfficientNetV2 model into the Flask project. The subsequent step is to develop a function that accepts an image as input, applies some foundational preprocessing (scaling and normalization), and then sends it to the model for prediction. This function must return the anticipated bird species' name and an audio sampling of the bird. Utilize the established dictionary that translates numbers into the names of bird species and audio recordings of birds singing. This feature will output the model's numerical prediction as connections to audio files and the names of bird species (Fig. 1).

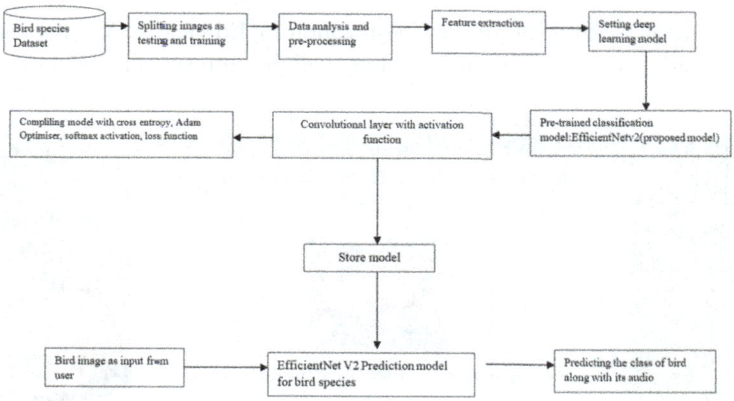

Fig. 1. Architecture diagram of bird species recognition using Neural Networks

3.6 Architecture Diagram

4 Results

The area under the accuracy curve graphically represents the model's dependability. In contrast, an epoch is merely one round of the algorithm applied to the entire training data set. It is observed that the model acquires knowledge over time by plotting accuracy versus epoch, as shown in Fig. 2. Directing and evaluating the accuracy graph demonstrates that 80 epochs were used to train and validate the dataset, indicating that the hyper-parameters were appropriately set. If the model genuinely learns from the data, its accuracy on the training and validation datasets should increase with each additional epoch. While the optimal model was discovered at 85 epochs (accuracy), the accompanying graph demonstrates that it reaches its limit after only 40 epochs. The minor difference between the two curves suggests that the model effectively works.

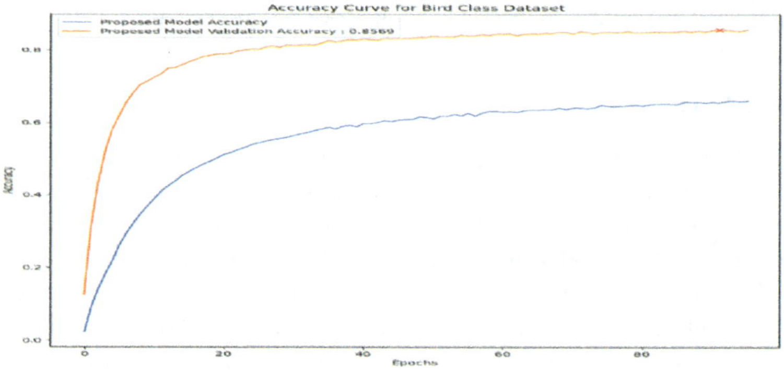

Fig. 2. Accuracy curve for the proposed model for Bird class Dataset

Figure 3 depicts the result of the classification by EfficientNetV2 that enables the user to determine the species of the bird along with its corresponding audio.

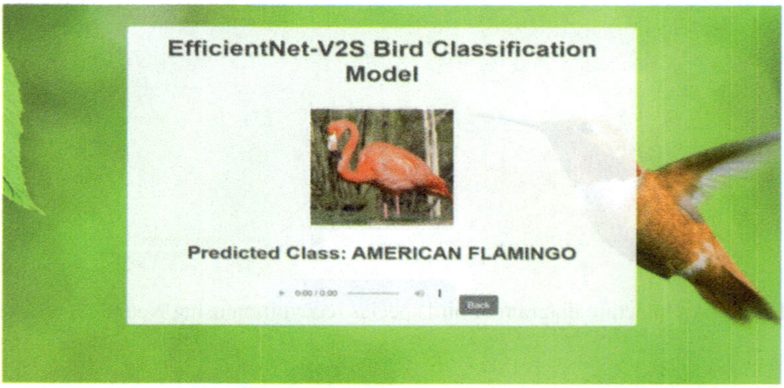

Fig. 3. Predicting the class of the species

5 Conclusion

This research aims to develop a robust and rapid photo classification system that reliably identifies 510 distinct bird species using the EfficientNetV2 architecture in TensorFlow. CNN algorithm with the EfficientNet architectural paradigm in this work model accurately predicted the species of birds. The system will be deployed using Flask, and the final interface will display the anticipated bird class name and predicted bird audio. The findings of this study indicate that the EfficientNetV2 pre-trained model and the CNN algorithm can effectively classify bird species. In evaluating bird image classification, the system significantly outperformed contemporary methods. This initiative achieves a higher level of comprehension and interactivity by displaying a bird's vocalizations alongside the identification result. Those interested in or enjoy animals could learn a great deal from this. The experiment's overall success demonstrates the potential for employing sophisticated machine learning algorithms to address real-world problems in disciplines such as ecology and conservation. As these technologies advance, they may assist us in better comprehending and protecting the natural world.

References

Borana, K., More, U., Sodha, R., Shirsath, V.: Bird Species identifier using convolutional neural network. In: International Journal of Engineering Research & Technology (IJERT) **9** (2021). ISSN: 2278-0181

Manasa Devi, P., Harshith Nikhil, V.S., Sai Nikhil V., Naga Venkata, D.G,, Golagani Abhishek S.Y.: Identification of bird species using convolutional neural network. **13**(04) (2022). ISSN: 0377-9254

Ming Zhong 2021 Acoustic detection of regionally rare bird species through deep convolutional neural networks Ecol. Inf. **64** 101333

Hoang, V.T., Jo, K.H.: Practical analysis on the architecture of EfficientNet. In: 14th International Conference on Human System Interaction (HSI), pp. 1–4. IEEE (2021)

Biswas, A., Rahman, M., Rajbongshi, A., Majumder, A.: Recognition of local birds using different CNN architectures with transfer learning. In: 2021 International Conference on Computer Communication and Informatics (ICCCI), pp. 1–6. IEEE (2021)

Anusri, M., Prashanth, M., Hema Devi, M., Raj Kumar, K.: Bird species identification using CNN. Neural Netw. **4** (2022). e-ISSN: 2582-5208

Madhavi, K.R., Jarugula, J., Karuna, G., Kaleru, S., Raju, K.S., Sunitha, G.: Bird species recognition using deep transfer learning. In: Reddy, K.A., Devi, B.R., George, B., Raju, K.S., Sellathurai, M. (eds.) Proceedings of Fourth International Conference on Computer and Communication Technologies. LNNS, vol. 606. Springer, Singapore (2023). https://doi.org/10.1007/978-981-19-8563-8_46.

Narasimharao, J., Laxmaiah, B., Arumalla, R., Unnisa, R., Fatima, T., Nazare, S.S.: Restoration and Deblurring the Images by Using Blind Convolution Method. In: Reddy, K.A., Devi, B.R., George, B., Raju, K.S., Sellathurai, M. (eds.) Proceedings of Fourth International Conference on Computer and Communication Technologies. LNNS, vol. 606. Springer, Singapore (2023). https://doi.org/10.1007/978-981-19-8563-8_32

Kumar, V.N., et al.: Anomaly-based hierarchical intrusion detection for black hole attack detection and prevention in WSN. In: Reddy, K.A., Devi, B.R., George, B., Raju, K.S., Sellathurai, M. (eds.) Proceedings of Fourth International Conference on Computer and Communication Technologies. LNNS, vol. 606. Springer, Singapore (2023). https://doi.org/10.1007/978-981-19-8563-8_30

Dimlo, U.M.F., et al.: An improved blind deconvolution for restoration of blurred images using ringing removal processing. In: Reddy, K.A., Devi, B.R., George, B., Raju, K.S., Sellathurai, M. (eds.) Proceedings of Fourth International Conference on Computer and Communication Technologies. LNNS, vol. 606. Springer, Singapore (2023). https://doi.org/10.1007/978-981-19-8563-8_34.

Munawar, S., Geetha, A., Srinivas, K.: Arrhythmia classification using BiLSTM with DTCWT and MFCC features. In: Reddy, K.A., Devi, B.R., George, B., Raju, K.S., Sellathurai, M. (eds.) Proceedings of Fourth International Conference on Computer and Communication Technologies. LNNS, vol. 606. Springer, Singapore (2023). https://doi.org/10.1007/978-981-19-8563-8_29

Subburaj, T., et al.: Discover crypto-jacker from blockchain using AFS method. In: Reddy, K.A., Devi, B.R., George, B., Raju, K.S., Sellathurai, M. (eds.) Proceedings of Fourth International Conference on Computer and Communication Technologies. LNNS, vol. 606. Springer, Singapore (2023). https://doi.org/10.1007/978-981-19-8563-8_15

Jagadeesh, S., Shrimali, M., Prasad, P.S., Maheswari, K., Narendrakumar, A., Koteswara Reddy, G.: Automated arrhythmia classification using Harris hawks optimization with deep learning model on IoT Environment. In: 2023 Third International Conference on Artificial Intelligence and Smart Energy (ICAIS), Coimbatore, India, pp. 30–35 (2023). https://doi.org/10.1109/ICAIS56108.2023.10073736.

Thaseen, A., Unnisa, R., Sultana, N., Madhavi, K.R., NagaJyothi, G., Kirubakaran, S.: Breast cancer detection using deep learning model. In: Reddy, A.B., Nagini, S., Balas, V.E., Raju, K.S. (eds.) Proceedings of Third International Conference on Advances in Computer Engineering and Communication Systems. LNNS, vol. 612. Springer, Singapore (2023). https://doi.org/10.1007/978-981-19-9228-5_57

Madhavi, K.R., Vooradi, S., Mounika, P., Yedlla, S., Tangudu, N.: Emotion analysis using convolutional neural network. In: Reddy, A.B., Nagini, S., Balas, V.E., Raju, K.S. (eds.) Proceedings of Third International Conference on Advances in Computer Engineering and Communication Systems. Lecture Notes in Networks and Systems, vol. 612. Springer, Singapore (2023). https://doi.org/10.1007/978-981-19-9228-5_44.

Nagajyothi, G., Kumar, G.P., Kumar, B.S., Kumar, B.P.D., Damodaram, A.K.: High-Speed Low Area 2D FIR Filter Using Vedic Multiplier. In: Reddy, A.B., Nagini, S., Balas, V.E., Raju, K.S. (eds.) Proceedings of Third International Conference on Advances in Computer Engineering and Communication Systems. Lecture Notes in Networks and Systems, vol. 612. Springer, Singapore (2023). https://doi.org/10.1007/978-981-19-9228-5_48

A Study on the Application of Digital Technology to the Dissemination of Intangible Cultural Heritage - Taking the Art of Paper Cutting as an Example

S. Venkateswara Rao[1], T. Rama Mohan[2], Md.Farooq Alfaz[3], M. Madhu Sudhan Rao[4(✉)], and M. Ravichand[5]

[1] Department of English, QIS College of Engineering and Technology (Autonomous), Ongole, AP, India

[2] Department of English Madanapalle Institute of Technology and Science (Autonomous), Madanapalle, AP, India

[3] Department of English Annamacharya Institute of Technology and Sciences (Autonomous), Rajampet, AP, India

[4] Department of English, Koneru Lakshmaiah Education Foundation, Vaddeswaram, A.P, India
madhu.css@kluniversity.in

[5] Department of English, VR Siddhartha Engineering College, Siddhartha Academy of Higher Education Deemed-to-be University, Vijayawada, Andhra Pradesh, India

Abstract. As a world conservation culture, how to spread and inherit intangible cultural heritage in the digital technology development environment, this study on the paper-cut art as a sample to examine the transmission of intangible cultural heritage under digital technology; This paper makes a comprehensive analysis and research on the comparison of the specific and cultural transmission achievements of paper-cut art under different technical environments and the comparison of coupling degree; Through analysis, it is concluded that under the support of digital technology, As an important asset of our civilization, intangible cultural heritage provides an effective means to strengthen the protection and propagation of intangible cultural heritage. The application of digital media technology can improve the understanding of the paper-cut art of intangible cultural heritage and further lay the foundation for the inheritance of intangible cultural heritage, advanced science and technology and digital technology have positive significance for the modeling and teaching of paper-cut art and the production of works in the development of network; It can better spread culture and art, expand influence and protect intangible cultural heritage, and to a certain extent, it also promotes the country's economic development and social progress.

Keyword: digital technology; Intangible cultural heritage; Cultural communication; Paper cut art Intangible cultural heritage protection

V. Sakalauskas et al. (Eds.): IBICA 2023, LNNS 1231, pp. 409–419, 2025.
https://doi.org/10.1007/978-3-031-78946-5_40

1 Introduction

With the continuous development of technology, many hand-made intangible cultural heritage cultures are gradually disappearing. Intangible cultural heritage has been relying on people themselves. It is the most vulnerable part of living culture and its tradition. Because of its fragility, it is imperative to protect it. It is a hand-made skill that is also a heritage that gradually disappears. As an important wealth of mankind, intangible cultural heritage is related to the future of a country and nation. If we are careless, it will be irretrievable. Therefore, it is our most important task to strive for effective protection means for the protection of culture. Zhang Chenglong (2022) pointed out that the development of Chinese paper cutting art mainly comes from two aspects: Intangible Cultural Heritage Inheritance and contemporary communication. Analyze the cultural value of paper-cut art in intangible cultural heritage from the development dilemma, sort out the situation of paper-cut art in Intangible Cultural Heritage Inheritance, and summarize that the communication subject and new media communication mode should be established at present [1]. Zhang Quan et al. (2022) pointed out that taking advantage of the characteristics and protection status of paper-cut art as well as the advantages of digitalization, actively paying attention to the paper-cut art by the government, protection institutions, non-governmental organizations, the public, etc., and implementing it in the aspects of digitalization, such as storage, display, dissemination, application, support, etc., is enough to protect the inheritance and development of culture [2]. Liu Xinyu (2022) pointed out that due to the introduction of many intangible cultural heritage protection lists, Promote many traditional folk cultures to be well preserved. In the process of carrying forward the art of paper cutting, find the market positioning of paper cutting, promote its products, vigorously promote the input of Fujian school through the Internet, and expand marketing channels. Paper cutting clips can also be added to film and television dramas to integrate local characteristics, so as to effectively promote publicity [3]. Qiu Yu et al. (2022) This paper analyzes the development of paper cutting art in the digital age. We should take Internet technology as an auxiliary, strengthen the promotion of it with the power of digital effect, and vigorously publicize it in combination with current hot spots, so as to find a new development direction for paper cutting art [4]. Liu Jiaquan (2020) pointed out that from the perspective of inheriting the traditional process of intangible cultural heritage, we should study the development performance of intangible cultural heritage paper cutting in the integrated media environment, analyze the cultural connotation of intangible cultural heritage paper cutting with artistic characteristics, and combine paper cutting skills with integrated media, so as to make the dissemination and innovation of intangible cultural heritage paper cutting in the integrated media environment [5]. Zhou Yucheng (2020) intangible cultural heritage culture is facing timely development opportunities under the continuous development of the new era. We need to strengthen the training of inheritors and make efforts in the development of literature and art, digital protection and financial support. With the help of new technology, science and technology give art a sense of the times and make it develop into an innovative change [6]. Li jiarou (2018) pointed out that social media uses the function of gathering people to spread and expand intangible cultural heritage, effectively attract forums, Weibo, wechat and other platforms, perfectly integrate into daily life, implement strategies according to the audience goals, spread technical resources in a variety of ways, and reflect the value of intangible cultural heritage [7]. Jiang Kai (2022) proposed that the emergence of short videos brings a glimmer of hope to the transmission of intangible cultural heritage at the boundary of multiple modes of transmission. The short-term screen uses fragment time to attract the attention of young strong groups, set up a transmission mechanism, and establish a live path to achieve the ideal effect of transmission [8]. This study analyzes the application of intangible cultural heritage culture communication through digital technology. Under the current background, the culture with national imprint is facing great challenges. Based on the application of modern digitization, Through internet television and various digital communication means, the maximum range of communication is expanded, and the human–computer interaction is more obvious, breaking the traditional restrictions on the time and place of cultural heritage visits, it adopts various ways to transmit, transmit and integrate into the lives of all living beings at the same time, so that more people can understand cultural skills, and more and more people will join in the protection and inheritance of intangible cultural heritage.

In the digital era, the information processing and communication methods of new media have shown many advantages and have become the main source of information for people. With the all-round development of the times, traditional TV media gather together in many fields of technical functions and communication, trying to create a new platform. Digital media uses electronic information technology to store large amounts of content, uses retrieval software to complete retrieval quickly, and the Internet has developed rapidly. It has many incomparable advantages over traditional media in many aspects. Compared with traditional language and writing, digital media has more diverse performance, higher media production efficiency, and can create more artistic styles. And by using digital technology, pictures, sound effects, special effects and colors can be spread out in the form of digital media, making creation more efficient. Network technology has replaced many functions such as telegram, instant message, e-mail and so on. With the changes of the times, it has reduced costs and brought convenience to people, including traditional newspapers. With the introduction of e-books and periodicals, the content of the network is becoming richer and richer, and newspapers are gradually introduced into the public eye. The emergence of digital media can transmit better and better things in communication and attract more people's attention. At present, people are far from enough to create content. Therefore, in the future, we need to continue to export to the development of content and make use of the advantages of your digital media to promote the rapid development of digitalization in China.

2 Digitalization to Intangible Cultural Heritage Communication

Intangible cultural heritage represents the spiritual height of human cultural heritage and is the spiritual home of the Chinese nation. It is not only an important part of excellent traditional culture, but also an important symbol of a country and national history and culture. With the development of the times, people's long-standing life, expectation and yearning for a better life in the nation is based on intangible cultural heritage. Therefore, this has important practical significance for the inheritance and protection of intangible cultural heritage. The protection and dissemination of intangible cultural heritage is a great project to inherit the national spirit, which requires the active participation of the whole people. In addition to cultivating intangible cultural heritage research professionals, intangible cultural heritage should be displayed in people's lives in more forms, so that more people can understand our national intangible cultural heritage, understand history, spontaneously join the favorite national culture, and spread and inherit in combination with daily life. The methods of information dissemination are becoming more and more extensive through the advancements in the digital age. For paper-cut, in order to improve the audience's understanding and love of paper-cut art, through a wide range of communication channels, the use of electronic magazines, new media, microblogging and other network resources to improve people's understanding and love of paper-cut art, the paper cutting skills can be displayed in many ways, and the materials and characters of paper cutting can be made into videos with the narrative content of the story, It is suggested that an exchange platform for paper-cut art lovers can also effectively utilize resources, establish characteristic brands, and design products with paper-cut art logo. In combination Under the influence of current national art elements and communication, we can also use the digital derivatives of paper-cut works as intangible cultural heritage to improve the value preservation and appreciation effect of the digital industry, so that more people can feel the process of intangible cultural heritage through the media, participate in it, and buy their favorite cultural products. According to the characteristics of paper-cut art, we should find the right time to spread great value at the lowest cost, give play to its important economic value in protecting the value of culture and art, and promote the synchronous development of culture and art and economy.

3 Paper Cutting Art of Intangible Cultural Heritage of Digital Technology

3.1 Comprehensive Value Analysis of Paper Cutting Art Under Different Technical Environments

China's paper-cut art has been developed for 1500 years. From the Northern Dynasty to modern times, it has been quite proficient in skills. As one of the intangible cultural heritage, Chinese paper-cut art is a diversified cultural heritage integrating culture, history, art and sculpture. As a kind of intangible cultural heritage, by comparing the value of paper-cut art in different environments, the following table is obtained:

As shown in Table 1, the value analysis data of domestic paper-cut arts such as Nanjing paper-cut, Yangzhou paper-cut and Ansai paper-cut are counted. It can be found that with the application of digital technology, the value of paper-cut art is more prominent. The data in the table are visually processed to get Fig. 1:

Table 1. value analysis of paper cutting art under different technical environments

Group	Cultural value	Artistic value	Historical value
Under the traditional technology environment	23.26	33.25	26.54
Under the environment of digital technology	46.25	49.26	52.03

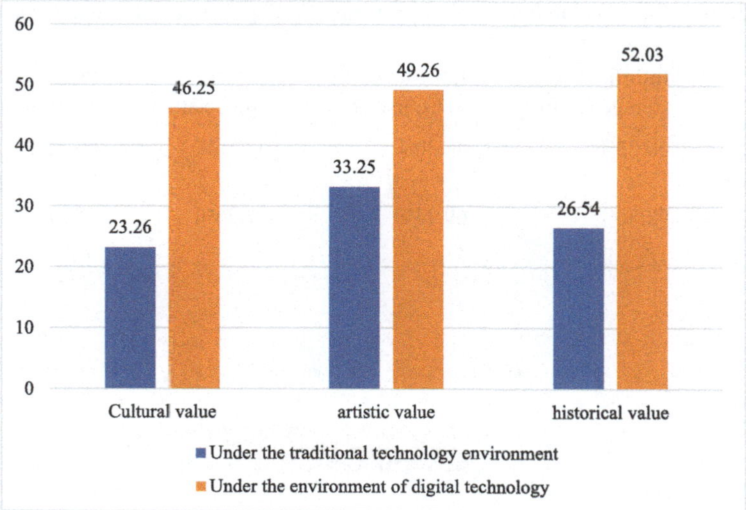

Fig. 1. value analysis of paper cutting art under different technical environments

As shown in Fig. 1, it can be seen from the figure that the impact and role of digital technology on paper-cut art is very obvious, which is different from the conservative and limitations of traditional paper-cut inheritance methods. Supported by digital technologies such as the Internet, big data, VR technology, etc., it plays a good role in promoting the development of paper-cut art and the transformation of related commodities and industries, It can better meet the characteristics of the inheritance of skills in the information age, so as to realize the comprehensive protection and promotion of paper-cut art based on modern information technology.

3.2 Comparative Analysis of Intangible Cultural Heritage Communication Under Different Technologies

Taking advantage of the convenience of the network and the popularity of digital technology, it is more convenient than manual work in data collection and analysis and operation of paper-cut data, especially using virtual reality technology to realize the communication mode of human–computer interaction for teaching and experience, realizing immersive cutting and design of paper-cut works in the network environment, and realizing good simulation and modeling design for colors, graphics and required element data, It is a new technological model that plays a positive role in the dissemination of intangible cultural heritage culture; By comparing the cultural transmission characteristics of paper-cut art under the two technologies, the following table is obtained through data analysis:

As shown in Table 2, security is stored according to information from the table; The scale of inheritance and dissemination; Artistic appreciation of paper cutting; It is suitable for analyzing the effect in product design; The effect of paper cutting propagation and security under digital technology is higher than that under traditional technology. The visual processing of the data in the table shows Fig. 2:

Table 2. Analysis of cultural transmission of paper cutting art under different technologies

Group	Information storage security	Inheritance and transmission scale	Artistic appreciation	Product design application
Traditional technology	56.28%	46.30%	52.36%	46.23%
Digital technique	72.36%	69.24%	72.26%	76.21%

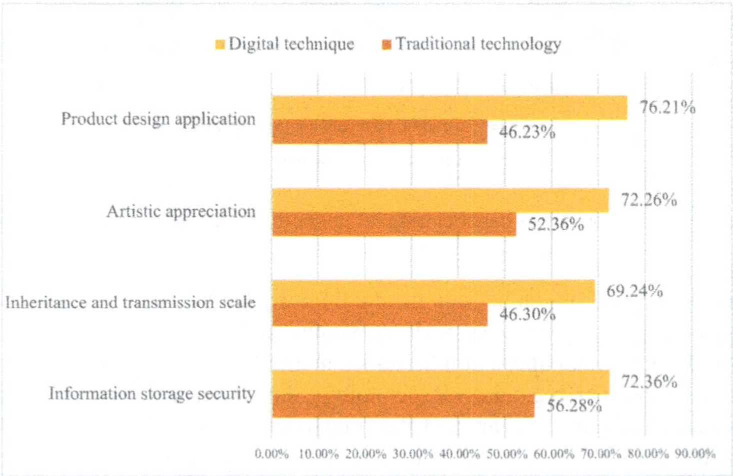

Fig. 2. Analysis of cultural transmission of paper cutting art under different technologies

As shown in Fig. 2, it can be seen from the figure that the application of digital technology in art communication and social development is more advantageous; In the current art field and Intangible Cultural Heritage Inheritance, the use of high-tech technology to optimize and improve the paper-cut technology not only ensures the inheritance of art and culture, but also achieves results in many fields of social development, providing help for economic development.

3.3 Analysis of Intangible Cultural Characteristics of Paper Cutting Art

As a kind of folk art, paper cutting mainly takes paper as the processing object and uses scissors for artistic creation. It used to be used for daily decoration, and slowly applied to various hollowed out materials to show the artistic techniques. Although the creation is simple, it contains the connotation of folk customs, It also has different artistic styles for artistic paper cutting; The following table is obtained by analyzing the characteristics of paper cutting in terms of creative techniques, structural patterns, image styles, colors and so on:

Table 3 presents the artistic traits of paper cutting. The traits of paper cutting under digital technology have been improved nearly one third higher than traditional data. The data in the table is represented through a bar chart below.

Table 3. Analysis of artistic characteristics of paper cutting under different technologies

Group	Line to line connection line to line disconnection	Patterning of composition and modeling	The image is exaggerated, concise, beautiful and full of rhythm	Simple and bright color	Knife technique should be "stable, accurate and skillful"
Traditional technology	56.23%	52.21%	56.11%	46.25%	48.26%
Digital technique	71.26%	73.25%	80.22%	70.11%	69.15%

As shown in Fig. 3, under digital technology, the size, angle and shape structure of paper cutting can be accurately calculated by using science and technology and accurate algorithms. In virtual reality technology, digital technology can also be used in teaching and modeling to innovate and modify paper cutting works, which greatly saves resource waste and loss. In terms of cultural and artistic inheritance, it can also expand the scope of influence in the network development environment.

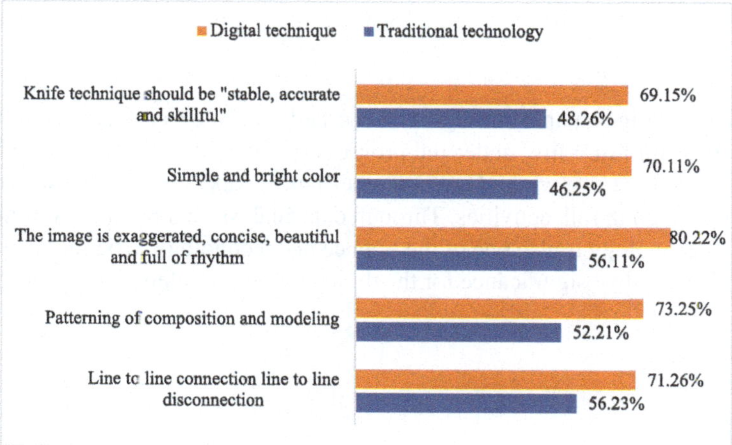

Fig. 3. Analysis of paper cutting art characteristics under different technologies

3.4 Analysis on the Coupling Degree of Paper Cutting Art and Intangible Cultural Heritage Communication Under Different Technologies

In different development environments, the impact on art is also different. As a historical and cultural art, paper-cut art has a better inheritance environment and development environment in the application of digitalization; It has not only made achievements in intangible cultural heritage in the art world, but also integrated into our living environment as a kind of material for economic development and the embellishment and cultural essence in people's lives in the market economy and surrounding development; By comparing the coupling data under the two technologies, the following table is obtained:

As shown in Table 4, after analyzing the effectiveness and stability of paper-cut art as intangible cultural heritage, we found that the coupling degree of digital technology is higher and more coupling than that of traditional technology before and after coupling; Where $t < 10$; $P < 0.05$, the data was statistically significant; Visualize the data and get Fig. 4:

Table 4. Analysis of the coupling degree of intangible cultural heritage culture and paper cutting art under different technologies

Group	Effectiveness of paper cutting art communication		Stability of paper cutting art communication	
	Before coupling	After coupling	Before coupling	After coupling
Traditional technology	36.41%	40.26%	40.12%	44.56%
Digital technique	56.23%	65.11%	68.39%	66.23%
t	8.235	8.056	7.956	7.235
P	0.035	0.032	0.039	0.037

As shown in Fig. 4, paper cutting art can be said to exist everywhere in our life. From decorations to furniture life, under the progress of social development, integrating art into life is one of the best ways of cultural inheritance; Paper cutting art also occupies an important position in folk activities. Through data analysis and research, it is also found that digital technology is also one of the science and technology to better inherit culture and art; It has positive significance for the dissemination of intangible cultural heritage.

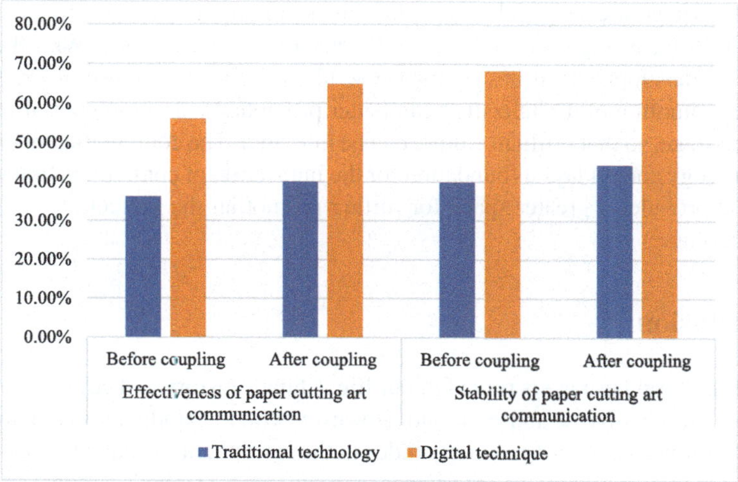

Fig. 4. Analysis of the coupling degree of intangible cultural heritage culture and paper cutting art under different technologies

4 Discussion

China has a long history and culture. In the past, China has inherited many excellent traditional cultures and traditional crafts. Paper cutting is the representative of traditional handicrafts. It has certain historical significance. Paper cutting is a very important intangible cultural heritage in China. With the changes of the times, the development process of paper cutting began to face a crisis. He Jia (2022) pointed out that for the dissemination and innovation of intangible cultural heritage, we should expand publicity in the form of handicraft art, so that more people can participate in the experience, integrate aesthetics from entertainment, and participate in craft production while playing, Thus, it is displayed in the public view with a new situation [9]. Zhou Kai (2022) pointed out that joining the dissemination of intangible cultural heritage in the cultural tourism industry, describing skills in a flexible way, and professional personnel talking about and understanding our national cultural symbols, the combination of the two makes intangible cultural heritage tourism a unique dissemination activity, reflecting a sense of ceremony and participation, and allowing national culture to show value in exchanges [10]. Yang Yanjun (2022) proposed to spread intangible cultural heritage in the form of documentaries, learn from previous communication experience, explore the development mode theory and communication mode of intangible cultural heritage television records, and systematically sort out documentaries [11]. Yin Yali (2022) has played a great role in the promotion and sales of intangible cultural heritage products with the help of digital technology and has a good understanding of intangible cultural heritage technology. According to the habits of the daily public, digital content and communication should be served to the local shopping platform people, and private capital and vocational schools should be guided to mine digital content products to realize the digital transformation of intangible cultural heritage technology communication [12]. Zhang Di (2021) pointed out that throughout the historical evolution of paper-cut art, its cultural connotation and historical heritage are very rich, and it has cross era significance in life and art. Traditional skills have stagnated under the influence of new technology. With the help of information development, it has brought new ideas to the inheritance of paper-cut art [13]. The promotion of digitalization can better promote the development of intangible cultural heritage, so that brilliant culture can be inherited. The combination of digitalization technology and art lays a foundation for the innovation of communication ways and means and provides a greater space for further promoting the protection of intangible cultural heritage.

5 Conclusion

Intangible cultural heritage is rooted in our life. China's history and culture are brilliant and long-standing, diverse and profound. However, with the gradual loss of many skills, many cultures have experienced vicissitudes in the current era of change. In view of this situation, we should explore the inheritance path suitable for China's national conditions. Cultural heritage is people's intangible capital for China and the cohesion of the cultural nation, It is also a responsibility to strengthen its protection. With the help of digitalization under the change of the times, take paper cutting as an example, integrate skills into daily life through the power of new media, record it by means of digital multimedia, research and use it, and make use of the video communication of network media, As well as the identification and certification of brand effect, the experience of virtual paper cutting system, mastering skills in interesting paper cutting, becoming the inheritor of paper cutting skills, attracting the attention and participation of the public, so that more young people can understand our culture, participate more, and transfer it to more people through online and offline platforms, With AR network platform as the media, people from different regions and different cultural backgrounds will widely participate in the protection and inheritance of paper cutting technology, integrate different regional cultures into paper cutting, and constantly improve the diversity of cultural heritage. Let people all over the country feel intangible cultural heritage anytime and anywhere, so as to realize the inheritance and promotion of intangible cultural heritage.

References

Zhang, C.: Intangible cultural heritage inheritance and contemporary communication of paper cutting art. Paper Making Inf. (06), 70–71 (2022)

Zhang, Q., Liu, Z., Zouchengdong.: Research on the digital protection and development path of intangible cultural heritage -- Taking fuyang paper cutting as an example. J. Fuyang Normal Univ. (Soc. Sci. Edn.) (03), 10–17 (2022)

Liu, X.: Research on Meishan paper cutting inheritance and innovative design from the perspective of intangible cultural heritage brand communication. Brand and Standardization (S1), 1–3 (2022)

Qiu, Y., Yu, Z.: Innovative development and inheritance of Nanjing paper cutting in the era of digital empowerment. Design **35**(10), 152–154 (2022)

Liu, J.: Communication innovation of intangible cultural heritage paper cutting in the integrated media environment. Audiovisual (11), 196–197 (2020)

Zhou, Y., Chen, L.: Research on the inheritance and development of intangible cultural heritage promoted by science and technology -- Taking Xiaogan carved paper cutting as an example. J. Yangtze Univ. (Soc. Sci. Edn.) **43**(04), 33–37 (2020)

Li, J.: Using social media platforms to help the dissemination of intangible cultural heritage projects -- Taking Fuyang paper cutting as an example. Today Media **26**(03), 101–103 (2018)

Jiang, K.: Production and consumption of intangible cultural heritage space under short video ecology. Media Observ. (7), 90–95 (2022)

He, J., Zhang, Z.: Analysis on the feasibility of hand-made art to help the dissemination of intangible cultural heritage. J. Guizhou Univ. (Art Edn.) **36**(04), 80–85 (2022)

Zhou, K., Zhang, Y.: The function and path of intangible cultural heritage tourism culture communication from the perspective of ritual view. J. Shandong Univ. (Philos. Soc. Sci. Edn.) (04), 40–46 (2022)

Yang, Y., Su, H.: Challenges and countermeasures of intangible cultural heritage TV documentaries -- Communication Research Based on Lasswell's "five W" model. Media Forum **5**(13), 71–74 (2022)

Yin, Y.: Research on the digital communication of intangible cultural heritage of traditional skills in Jiangsu. Jiangsu Shang Lun (7), 37–39 (2022)

Zhang, D.: Discussion on the inheritance of paper cutting art and design from the perspective of intangible cultural heritage protection. Paper Making Information (10), 77–78 (2021)

Detailed Investigation of the Internet of Things in Education

C. S. Srinivas[1], Dasam Sridevi[2], V. Sunitha[3], M. Ravichand[2](✉), and M. Madhu Sudhan Rao[4]

[1] Department of English, Department of Mathematics and Humanities, Mahatma Gandhi Institute of Technology, Hyderabad, India
cssrinivas_eng@mgit.ac.in
[2] Department of English, VR Siddhartha Engineering College, Siddhartha Academy of Higher Education Deemed-to-be University, Vijayawada, Andhra Pradesh, India
dsridevi@vrsiddhartha.ac.in, ravichandenglish@gmail.com
[3] Department of English, School of Social Sciences and Languages, Vellore Institute of Technology, Vellore, Tamilnadu, India
sunitha.venugopal@vit.ac.in
[4] Department of Engineering English, Koneru Lakshmaiah Education Foundation, Vaddeswaram, A.P, India
madhu.css@kluniversity.in

Abstract. People in today's culture are often fairly ready to accept newly developed types of technology. We now link a vast assortment of other control systems, including those found in autos and home appliances of all kinds, to the internet. The educational sector is not an exception and is among the many other industries that are actively adopting device interconnectivity to increase both the speed at which they complete their work and the quality of the work that they produce. This is being done to increase both the speed at which they complete their work as well as the quality of the work that they produce. This is being done as part of an attempt to enhance the quality of their output as well as the amount of work they generate. Establishments that provide higher education are faced with a diverse set of concerns, some of which include ensuring the safety of their student populations, improving the effectiveness of their administrative systems, and addressing other topics connected to these issues. It is only to be expected that the number of customers looking for intelligent gadgets will continue to climb. This is something that can only be predicted. It was projected that the market for the Internet of Things (IoT) in education would reach $19.5 billion by 2027, which would represent an increase in the compound annual growth rate of 17.7% from 2020 to 2027. In 2019, it was estimated that the market for IoT in education would be worth $5.3 billion. By 2027, it is projected that it will reach $19.5 billion.

Keywords: Digital Optimization · Artificial Intelligence · Internet Connectivity · Education · Internet of Things

V. Sakalauskas et al. (Eds.): IBICA 2023, LNNS 1231, pp. 420–426, 2025.
https://doi.org/10.1007/978-3-031-78946-5_41

1 Introduction

The Internet of Things (IoT) in education has the potential to greatly help students with special needs by creating an accessible learning environment for them. For pupils with visual impairments, teachers may use voice assistants to provide audio information, while teachers for those with hearing impairments can translate lectures into texts. By using interconnected microphones, tablets, and VR gadgets, students with unique needs may learn at their own pace while still achieving excellent results in the classroom. It's possible to feel uplifted and at peace here. The potential for IoT in the academic world is enormous. However, incorporating IoT into routine classrooms will take some time. Making these technologies easily accessible and intuitive for users should be your priority. The views of teachers, parents, and students must be altered. They must also begin to support new educational policies. (Kalaiselvi, 2021) A growing number of educational institutions have made tentative moves toward implementing IoT platform integration solutions. In addition, they have a favorable reaction. It is hoped that as technology continues to advance, more and more kids will be able to make use of its many benefits in the classroom.

The research aimed to fulfill the following objectives:

- To study the internet of things in education
- Why the education sector absolutely must have access to the internet of things
- To study Edtech IoT
- The influence of the internet of things on learning

Education, when taken in its traditional definition, refers to the act of receiving or imparting systematic teaching, most often in an educational setting such as a school or university; however, education via the Internet of Things is a more enjoyable process. Traditional education refers to the act of receiving or imparting systematic teaching, most often in an educational setting such as a school or university. To put it in more straightforward terms, we may say that it is an eye-opening trip. Even though it's possible that all that was required for traditional teaching was a notebook, some pencils, a board, and some chalk, the Internet of Things (IoT) is playing a significant part in the educational sector in today's globe, especially in 2020 with the issue involving Covid-19. This is especially the case in terms of the global educational landscape. Before we can learn more about the Internet of Things in the educational sector, we need to be well-versed in the obstacles that the educational sector experienced previous to the adoption of IoT. This is because the Internet of Things was very recently deployed in the educational sector.

2 Internet of Things in Education

Because of the positive impact that technology has had on teaching strategies, educational institutions have enthusiastically welcomed this development. The phenomenal possibility that presented itself to the education industry as a result of the proliferation of mobile technology is still being realized today.The mobile app development services seek to design educational applications that will help students more effectively manage and advance their education following the shifting educational standards. Software developers are putting in a lot of effort to offer individualized solutions for the education sector in the hopes of improving worker performance and the level of satisfaction felt by both students and teachers. (Quezada-Sarmiento et al., 2018). The Internet of Things (IoT) is the most recent piece of technology that is affecting the instructional sphere (obviously in a positive way). The school system has been drastically altered as a result of its implementation, which was a game-changer at the time. A significant number of educational establishments are now in the process of putting in place linked smart devices to support their e-learning and smartboard infrastructure.

Many factors are contributing to the widespread adoption, including the following:

- The rising popularity of Internet of Things-enabled devices may be attributed, in large part, to the fact that cloud services are becoming easier and more affordable to use.
- Rapid development is possible for customized app packages that are intended for use by educational establishments.

The Internet of Things can make life easier for those in management positions, classroom teachers, students, and their parents. It is possible to efficiently optimize the use of the resources, and while maintaining the same level of work, a lucrative outcome may be achieved. (Arras et al., 2022).The following facets of the educational sector are targeted through the development of applications for the internet of things:

- Using app management solutions to get the best possible pricing and feasibility.
- Providing helpful solutions for classroom management to enhance the learning experience.

Since this is one of the most important obligations of the industry that deals with education, the educational process must be continuously developed and enhanced. The problem is that the tactics utilized by the educational system in the United States to enhance the overall quality of education are not only inefficient but also somewhat dated. This makes it difficult to make progress. Schools and colleges can broaden access to education when new teaching posts are filled; however, this does not always correspond to an enhancement in the quality of the educational experience as a whole.

The same thing is happening with a wide variety of instructional modes and kinds of information. Even though printed textbooks quickly grow out of date and do not correspond with contemporary teaching practices, educational institutions of higher learning allot budgetary resources toward the procurement of printed textbooks. In addition to this, the infrastructure for the adoption of new policies inside the educational system is intricate and complex. As a consequence of this, educational institutions need solutions that are adaptable enough to account for the realities that are currently in place. (Trabelsi, 2021)

The Potential of Connected Devices in the Classroom

The "Internet of Things" is a network of computerized devices that can communicate with one another and share data over the internet. This network is referred to by the phrase "Internet of Things". Internet of Things (IoT) technology is currently being implemented in elementary, secondary, and tertiary educational institutions with the goals of tracking students' attendance, ensuring their safety, including them in the learning process, and adapting course material to meet the requirements of students who have specific requirements (Fig. 1).

Smart boards, door locks, fire alarms, security systems, mobile applications, and educational software are just some examples of devices that may be linked to the internet. (Senthil et al., 2022)

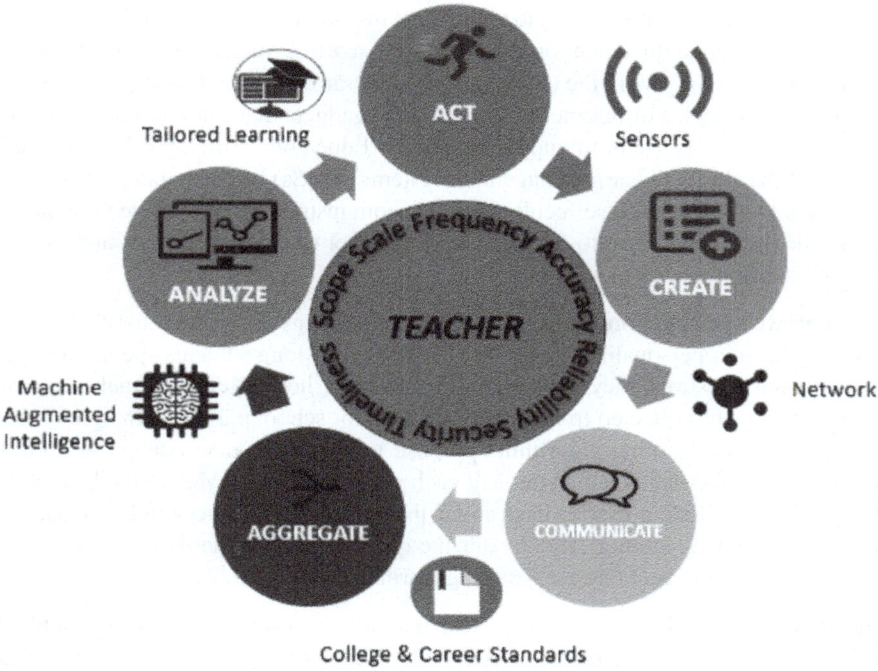

Fig. 1. Edtech IoT

The influence of the Internet of Things on Learning

A variety of instructional activities are made easier with the help of linked gadgets. With the help of smart classroom technology, security systems, and management software, teachers now have the resources they need to streamline the educational process for their students. Personalized learning is made possible by technological advancements, and it's this kind of learning that helps pupils develop marketable abilities like critical thinking, technical literacy, creativity, and adaptable problem-solving. Let's look at how the IoT may benefit classrooms.

Enhanced Student Participation: Ever since the Internet was made available to almost everyone, there has been a dramatic growth in the amount of content being consumed. Young people nowadays are used to communicating with video and visuals, and they take in a great deal of information via social media. Teachers nowadays need more than just a textbook to keep their pupils engaged in the classroom.Through the use of IoT devices, educators can provide students with interactive learning materials, hence increasing student engagement. Think about the use of interactive tables. Using gamified software on such a gadget, learning may become an exciting adventure across space and time. Traveling in a student-chosen path, with students making historical predictions and completing assignments along the way. (Wijaya et al., 2021). A recent survey found that the majority of pupils (67%) preferred gamified learning to more conventional methods of instruction. IoT gadgets make them much more interesting to use.

Improved Resources for Teachers: Many hours of a teacher's day are spent on administrative chores that have nothing to do with actual instruction. On the contrary, computers can handle tasks like document organization, test scoring, student attendance tracking, resource monitoring, etc. in the classroom and laboratory.Wearables and smart sensors are only two examples of Internet of Things (IoT) gadgets that can monitor where children are and alert teachers to supply shortages. Educators' time may be better used because of the ability to learn management systems (LMSs) to offer students immediate feedback on their work. The effect is that classroom instructors have more time on their hands, which they may use to experiment with novel teaching strategies and improve students' academic outcomes.

Differentiated Instruction: Individualized learning plans are created for each student using Big Data in personalized education. Data on how long students spend on assignments, which questions they choose to skip on tests, how often they make common errors, etc., may be collected from any device used for school purposes. In light of these findings, machine learning algorithms propose modifying course content and implementing new functionalities to better suit each student's needs. (Mohammadian, 2020). Sharing the information they collect about their users, IoT devices help advance the field of customized learning. The greater the quantity of data that is processed by the algorithms, the more accurate the resulting learning routes will be.

Upgraded Security: Every school has a responsibility to ensure the safety of its students. Each student's arrival may be recorded by the school's security system thanks to the Internet of Things. Each student and staff member may have a smart card programmed to act as their building ID.A system like this might track student attendance and stop intruders from entering the building. As a bonus, a school's laboratories, libraries, and other restricted areas may all be protected by an access control system.Cashless payment solutions for schools are another valuable IoT appliance. Some American educational institutions have used cashless payment systems to better control the flow of funds. This prevents students' funds from being misplaced or stolen. (Raghavan et al., 2018)

3 EdTech IoT

IoT is a network of internet-connected devices. These items can send data wirelessly without human involvement. IoT technologies utilized in education are:

- Smart Boards replaced blackboards. Smartboards let instructors write, educate, and illustrate using photos, sounds, and videos. This makes conversation entertaining, fascinating, and engaging. This eliminates any misunderstanding caused by a chalkboard and textbook where pupils can't grasp the images. Teachers mayquickly exchange notes/illustrations and lessons.

- Tablets are replacing textbooks in many locations. Tablets replace textbooks, examinations, and homework. This transition minimizes paper use, which helps the environment. Tablets improve student-teacher interactions. Addiction and screen time are the two downsides.
- Almost every youngster nowadays has a cellphone. IoT specialists turned game-filled phones into instructional ones. This helps bring together individuals with similar interests from across the world. This increases student-teacher-student interactions.
- In many institutions, instructors and students use attendance monitors. This captures attendance and lets students monitor per-subject/overall attendance. No fraudulent attendance or proxy may be entered, ensuring students attend class. This offers the institution a formal record. (Papaefstathiou, 2016)
- AV or audiovisual technology is old. Pro AV has just been released and linked with IoT. AV rooms are popular in schools. Teachers utilize AVs to educate or analyze student work while presenting. IoT and AV go hand-in-hand in schools and outdoors, even if AV doesn't appear like IoT.
- AV is used in schools, universities, corporations, retail businesses, and security. IoT's biggest function in AV is interacting with the building. Multiple IoT tools can control AV. Some are as easy as using a remote, while others need a touchpad.
- Internet, however beneficial, constantly threatens the protection of our data. IoT provides emergency indications, audio augmentation, Wi-Fi clocks, and hearing-challenged warnings. This gives users a sense of security. It contains emergency tones, live bulletins, bell schedules, and pre-recorded educational messages (Fig. 2).
- C-pen is a student-friendly gadget. It's a portable gadget for scanning and sharing textual or smartphone material. This is straightforward to use, so no notes will be lost or forgotten. This also defines a highlighted term, produces notes, and translates almost 40 languages with proper pronunciation. (Jogdand et al., 2018)

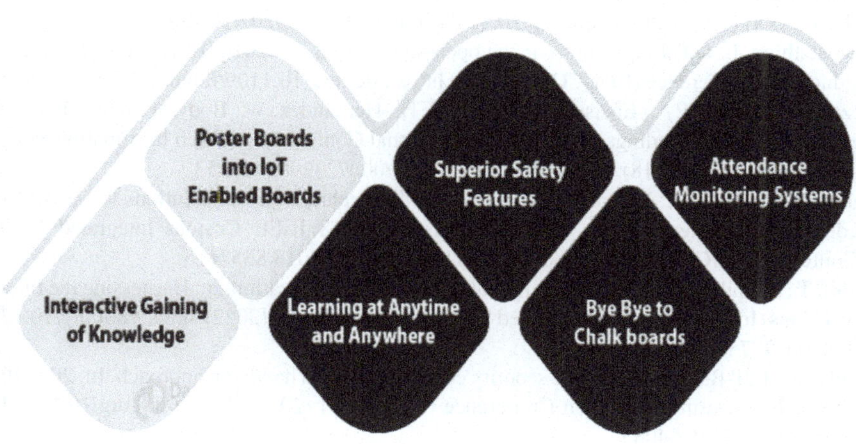

Fig. 2. Application of Iot In Education

4 Conclusion

The Internet of Things (IoT) is becoming an increasingly important component of the educational system. Because the implementation of the Internet of Things in the educational sector is still in its infancy, it is essential to have an in-depth comprehension of the challenges that were confronted by educators before the IoT was made available. The process of digitalization inside the educational system is both rapid and unavoidable in the modern world. Because of this, developing software for the Internet of Things can turn out to be profitable. These types of hybrid methods can fulfill the immediate and practical requirements of the educational system. When it comes to learning, the Internet of Things opens up a vast number of hitherto unimaginable avenues. It helps to improve the learning process by making it more exciting and personalized for each student. The Internet of Things (IoT) is beneficial to educators because it makes their work simpler by automating routine tasks. This makes their jobs easier. A system composed of devices that can communicate with one another paves the way for enhanced surveillance and control of limited areas, in addition to the creation of new educational opportunities.

References

Arras, P., Van Merode, D., Tabunshchyk, G.: Project-oriented teaching approach for IoT education. In: Dependable IoT for Human and Industry, pp. 515–533 (2022). https://doi.org/10.1201/9781003337843-30

Jogdand, G., Kadam, S., Patil, K., Mate, G.: IoT transaction security. J. Adv. Scholarly Res. Allied Educ. **15**(2), 711–716 (2018). https://doi.org/10.29070/15/57056

Kalaiselvi, K.: IoT application in education. securing IoT. In: Industry 4.0 Applications with Blockchain, pp. 29–50 (2021). https://doi.org/10.1201/9781003175872-2

Mohammadian, H.D.: IOT-education technologies as solutions towards SMEs' educational challenges and 14.0 readiness. In: 2020 IEEE Global Engineering Education Conference (EDUCON) (2020). https://doi.org/10.1109/Educon45650.2020.9125248

Papaefstathiou, I.: IoT design course using open-source tools. In: 2016 IEEE Global Engineering Education Conference (EDUCON) (2016). https://doi.org/10.1109/Educon.2016.7474540

Quezada-Sarmiento, P.A., Enciso, L., Washizaki, H., Hernandez, W.: Body of knowledge on IoT education. In: Proceedings of the 14th International Conference on Web Information Systems and Technologies (2018). https://doi.org/10.5220/0007232904490453

Raghavan, K., Kang, J., Natarajan, A.: IOT-education technologies as solutions towards SMEs' educational challenges and 14.0 Readiness. In: 2018 IEEE Custom Integrated Circuits Conference (CICC) (2018).https://doi.org/10.1109/Cicc.2018.8357116

Senthil, T., Deepika, J., Rajan, C.: Internet of things (IoT) in education. In: Harnessing the Internet of Things (IoT) for a Hyper-Connected Smart World, pp. 25–45 (2022). https://doi.org/10.1201/9781003277347-2

Trabelsi, Z.: IOT-Based smart home security education using a hands-on approach. In: 2021 IEEE Global Engineering Education Conference (EDUCON) (2021). https://doi.org/10.1109/Educon46332.2021.9454085

Wijaya, I.D., Suprianto, D., Hadid Firdaus, V.A.: Design of IoT Infrastructure as support empirical learning. In: IoT Laboratory. 2021 Universitas Riau International Conference on Education Technology (URICET) (2021). https://doi.org/10.1109/Uricet53378.2021.9865890

Implementation of Machine Learning-Assisted Group Discussions for Improving Spoken English Proficiency

N. Balamurugan[1], C. Raghavendra Reddy[2(✉)], V. V. Parthu[2], G. Chenna Reddy[3], R . Retna Mony[3], and P. Maruthi Kumar[4]

[1] Department of Humanities and Sciences, Nalla Narasimha Reddy Group of Institutions, Hyderabad, India

[2] Mohan Babu University (Erstwhile Sree Vidyanikethan Engineering College), Tirupati, India
raghavendrareddy4323@gmail.com

[3] School of Liberal Arts and Sciences, Mohan Babu University, Tirupati, India

[4] Department of English, Mohan Babu University, Tirupati, India

Abstract. This paper is an attempt to implement Deep Learning and AI using Logical Regression Algorithm to investigate the impact of group discussions on the improvement of spoken English proficiency among non-native English speakers. The study involved 100 participants from Mohan Babu University, Tirupati, Andhra Pradesh, who were enrolled in an English language course, in I year B.Tech. The participants were divided into two groups: the experimental group and the control group. The experimental group participated in group discussions while the control group did not. The participants were evaluated before and after the intervention using a speaking test. The results of the study showed that the experimental group had a significant improvement in their speaking scores compared to the control group. The experimental group participants demonstrated better-speaking fluency, pronunciation, grammar, and vocabulary. Additionally, the experimental group participants reported feeling more confident and comfortable speaking in English. The findings of this study suggest that group discussions are an effective method for improving spoken English proficiency among non-native English speakers.

Keywords: Natural language processing · better-speaking · fluency · pronunciation · grammar · vocabulary · monologue

1 Introduction

English is a globally recognized language that is widely used in various fields such as education, business, and communication. Proficiency in English is considered a valuable asset, particularly for non-native speakers who aspire to succeed in an international setting. However, mastering spoken English can be a daunting task for many non-native English speakers due to various challenges such as limited exposure to the language, lack of practice, and fear of making mistakes. One such method is group discussions,

V. Sakalauskas et al. (Eds.): IBICA 2023, LNNS 1231, pp. 427–435, 2025.
https://doi.org/10.1007/978-3-031-78946-5_42

which provide an interactive learning environment for learners to practice their speaking skills, receive feedback, and learn from their peers. Group discussions allow students to participate in meaningful discussions, express their views, and improve their critical thinking skills.

2 Literature Review

The traditional approach to teaching spoken English has been focused on providing students with grammatical rules and vocabulary lists, which is often ineffective in improving their speaking skills. Therefore, it is essential to explore innovative teaching methods that can enhance the development of spoken English proficiency among non-native English speakers. Group discussions provide opportunities for learners to practice their speaking skills, receive feedback from their peers, and learn from the mistakes of others. Therefore, group discussions should be incorporated into English language courses to enhance the speaking skills of non-native English speakers.

3 Methodology

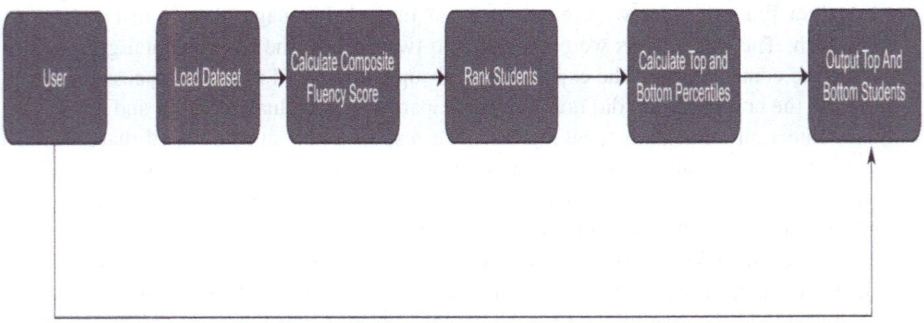

Fig. 1. Architecture Diagram

The study involved 100 participants from Mohan Babu University, Tirupati, AP who were enrolled in an English language course, in I year B.Tech. The participants were divided into two groups: the experimental group and the control group. The experimental group participated in group discussions, while the control group did not. Both groups were given the following questionnaire as a pretest to assess their initial standards. On a scale of 1 to 10, how confident do you feel when speaking in public? How do you rate your ability to articulate your thoughts clearly and concisely when speaking to others? (Fig. 1).

When you speak, do you feel like you use appropriate vocabulary and grammar? How comfortable are you with improvising or speaking spontaneously without prior preparation? How often do you practice your speaking skills, such as rehearsing a speech or recording yourself speaking?

How well do you feel you can adapt your speaking style to different audiences or situations? Have you received any feedback from others on your speaking skills? If so, how did you respond to that feedback?

Do you feel that your speaking skills have improved over time? If so, what have you done to improve them? When speaking to someone, do you feel like you can engage them and maintain their interest throughout the conversation?How do you perceive your overall effectiveness as a communicator, both in terms of speaking and listening skills? How often do you speak English in your daily life? Can you describe a recent conversation you had in English? Do you find it easy to express yourself in English?

a) Pronunciation: Can you pronounce the word "schedule"? What is the correct stress in the word "photography"? Can you say the following sentence with correct intonation: "I'm going to the store."
b) Grammar: Which sentence is correct: "I have been studying English for three years" or "I am studying English for three years"? What is the correct form of the verb "to be" in the present tense: "am", "is", or "are"? Which sentence is grammatically incorrect: "She have a cat" or "She has a cat"?
c) Vocabulary: What does the word "ambivalent" mean? Can you give an example of a phrasal verb? What is the difference between "affect" and "effect"?

The participants were evaluated before and after the intervention using a speaking test. The speaking test comprised three tasks: a monologue, a dialogue, and a group discussion. The monologue task required the participants to talk about a topic of their choice for two minutes. The dialogue task required the participants to engage in a conversation with the examiner for two minutes. The group discussion task required the participants to participate in a discussion with their peers for five minutes.

d) Monologue:

Content: Does the speaker stay on topic and provide relevant information? Organization: Is the monologue well-structured and easy to follow?

Vocabulary: Does the speaker use a range of appropriate vocabulary and phrases? Pronunciation: How clear and accurate is the speaker's pronunciation?

Fluency: Is the speaker able to speak smoothly and without hesitations or pauses?

e) Dialogue:

Interaction: Does the speaker engage in a natural and effective conversation with their partner? Listening: Does the speaker listen actively and respond appropriately to their partner's contributions?

Turn-taking: Does the speaker take turns appropriately and avoid interrupting their partner? Initiative: Does the speaker initiate and maintain the conversation effectively? Accuracy: Is the speaker's grammar and vocabulary use accurate and appropriate for the conversation?

f) Group discussion:

Collaboration: Does the speaker work well with others in the group?

4 Contribution:

Does the speaker actively contribute to the discussion and provide meaningful input? Respect: Does the speaker listen to others' opinions and show respect for their viewpoints? Leadership: Does the speaker take initiative in guiding the discussion and keeping it on track? Communication: Does the speaker communicate effectively and clearly with others in the group? (Figs. 2, 3, 4, 5, 6, 7, 8, 9 and 10 and Table 1).

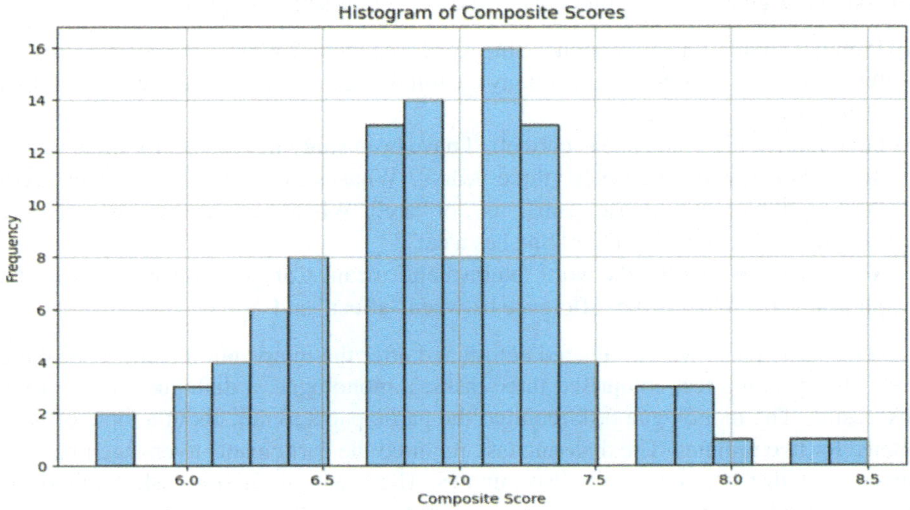

Fig. 2. Histogram of Composite Scores

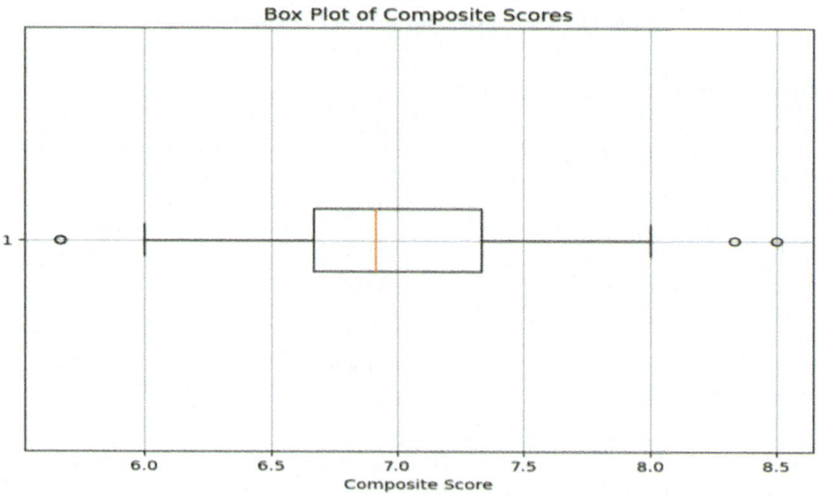

Fig. 3. Boxplot of Composite Scores

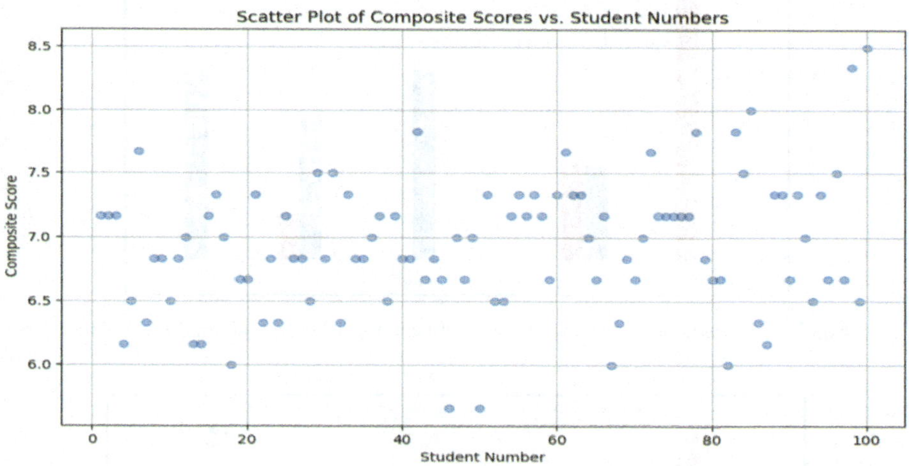

Fig. 4. Scatter plot of Composite Scores VS Student Numbers

Table 1. Descriptive Statistics For Composite Scores

Descriptive Statistics for Composite Scores:	
Statistic	Value
Count	100
Mean	6.936667
Std	0.519961
Min	5.666667
25%	6.666667
50%	6.916667
75%	7.333333
Max	8.5

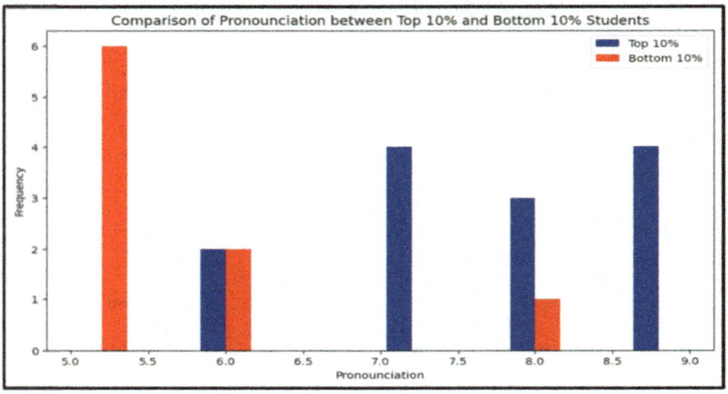

Fig. 5. Comparision of Pronounciation between top 10% and button 10% students

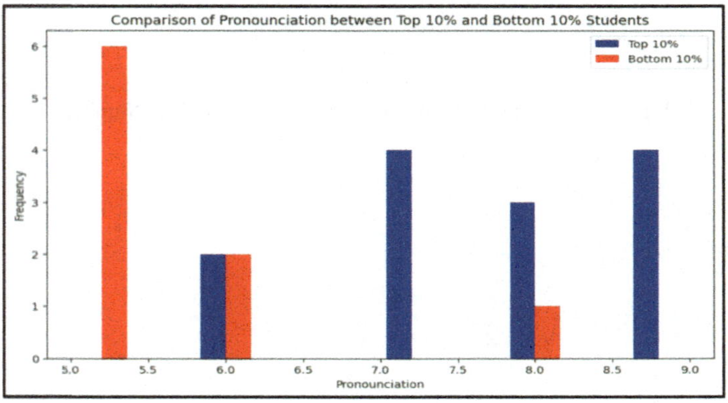

Fig. 6. Comparision of Grammer between top 10% and button 10% students

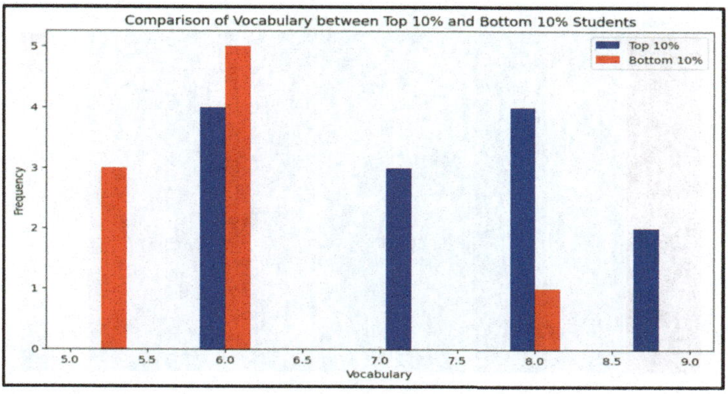

Fig. 7. Comparison of Vocabulary between top 10% and bottom 10% students

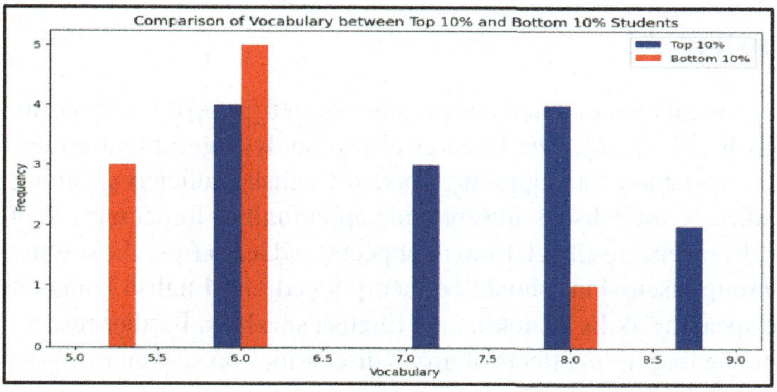

Fig. 8. Comparison of Vocabulary between Top 10% and bottom 10% students

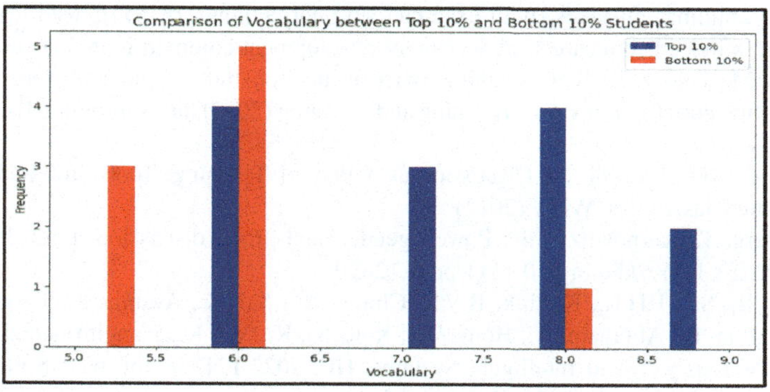

Fig. 9. Comparison of Vocabulary between top 10% and bottom 10%

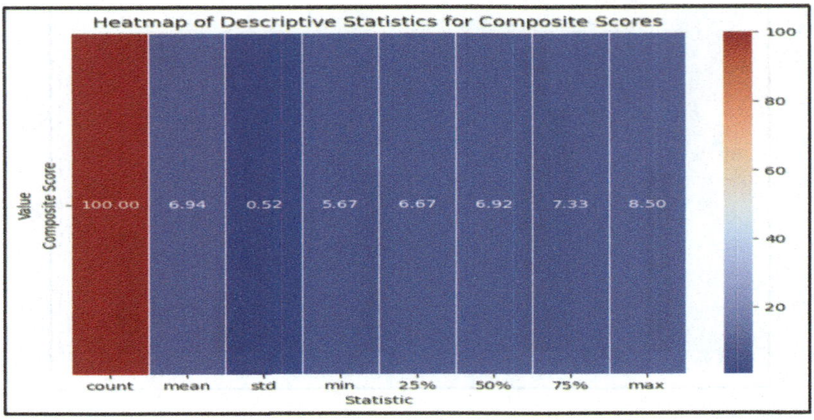

Fig. 10. Heatmap of Descriptive Statistics for composite scores

5 Conclusion

After the implementation of machine learning-assisted group discussions for improving spoken English proficiency, The findings of this study suggest that group discussions are an effective method for improving spoken English proficiency among non-native English speakers. Group discussions provide opportunities for learners to practice their speaking skills, receive feedback from their peers, and learn from the mistakes of others. Therefore, group discussions should be incorporated into English language courses to enhance the speaking skills of non-native English speakers. Further research is needed to investigate the long-term effects of group discussions on spoken English proficiency.

References

1. Common grammar mistakes (with examples and corrections). (n.d.). Indeed, career guide. https://www.indeed.com/career-advice/career-development/common-grammarmistakes
2. Asebedo, M., Jagosz, T.: Team teaching: twice as nice for students and educators. In: International Symposium for Innovative Teaching and Learning (2017). https://doi.org/10.4148/2573-4911.1014
3. Brookfield, S.D., Preskill, S.: Discussion as a Way of Teaching: Tools and Techniques for Democratic Classrooms. Wiley (2012)
4. Mendelberg, T., Karpowitz, C.F.: Power, gender, and group discussion. Polit. Psychol. **37**, 23–60 (2016). https://doi.org/10.1111/pops.12320
5. Naseeba, B., Sai, P.H.R., Karthik, B.V.P., Chitteti, C., Sai, K., Avanija, J.: Toxic Comment Classification. In: Abraham, A., Hong, TP., Kotecha, K., Ma, K., Manghirmalani Mishra, P., Gandhi, N. (eds.) Hybrid Intelligent Systems. HIS 2022 LNNS, vol. 647. Springer, Cham (2023). https://doi.org/10.1007/978-3-031-27409-1_80
6. Hundiwala, S.: A Complete Kit for Group Discussion. Arihant Publications India limited (2019)
7. Challa, N.P., Madhavi, K.R., Naseeba, B., Bhanu, B.B., Naresh, C.: Sentiment Analysis from TWITTER Using NLTK. In: Abraham, A., Hong, TP., Kotecha, K., Ma, K., Manghirmalani Mishra, P., Gandhi, N. (eds.) Hybrid Intelligent Systems. HIS 2022. LNNS, vol 647. Springer, Cham (2023). https://doi.org/10.1007/978-3-031-27409-1_78

8. VM, R.: Seven simple ways to improve your speaking skills | British Council. Britishcouncil.in (2021). https://www.britishcouncil.in/blog/seven-simple-waysimprove-your-speaking-skills
9. Abbagalla, S., Rupa Devi, B., Anjaiah, P., Reddy Madhavi, K.: Analysis of COVID-19-impacted zone using machine learning algorithms. In: Reddy, K.A., Devi, B.R., George, B., Raju, K.S. (eds.) Data Engineering and Communication Technology, vol. 63, pp. 621–627 (2021). https://doi.org/10.1007/978-981-16-0081-4_62

Execution of Natural Language Processing via Squabble to Enhance Students' Essay Writing Cognition

V. V. Parthu[1], C. Raghavendra Reddy[1(✉)], N. Balamurugan[2], Guda Chenna Reddy[3], R. Retna Mony[1], and P. Maruthi Kumar[4]

[1] Mohan Babu University (Erstwhile Sree Vidyanikethan Engineering College), Tirupati, India
raghavendrareddy4323@gmail.com
[2] Department of Humanities and Sciences, Nalla Narasimha Reddy Group of Institution, Hyderabad, India
[3] School of Liberal Arts and Sciences, Mohan Babu University (Erstwhile Sree Vidyanikethan Engineering College), Tirupati, India
[4] Department of English, Mohan Babu University, Tirupati, India

Abstract. This paper is an attempt to execute and demonstrate AI using Google Gemini Developer's Access. Squabble is an important requirement for essays and should be written by students in the most common genres. In many cases, this has not been fully communicated to students. This paper presents research on the concept of Squabble when undergraduates arrive at university, the difficulties they encounter in developing Squabble in their essays, and the nature and quality of the instruction they receive. As a framework, a three-part definition to explain Debating in terms of what students need to learn is used. The results show that students have only partial or incorrect conceptions of Squabble. Many of the problems they encounter are caused by their lack of knowledge of what an argumentative essay requires, especially the need to develop one's position in academic Squabble. It is related to inconsistent and vague reasoning, rather than clarifying to improve writing instruction, an "essay writing framework" based on a three-part definition has been proposed. This approach puts reasoning at the centre of the lesson, explaining other aspects of writing according to its function in reasoning development.

Keywords: Natural language processing · academic writing · Squabble · evidence · persuasion · structure

1 Introduction

Much has been written about the rhetorical and linguistic structure of Squabble, and academic writing in general, but less attention has been paid to the teaching and learning of reasoning. Given the important role of argumentation in academic essays, this is surprising. The research described in this article examined learning needs and teaching delivery in a case study of his first-year student in Engineering Programs at Mohan Babu University, Tirupati. In this program, as in most of the field, the primary form of writing and assessment is the argumentative essay. The purpose of this research is to:

V. Sakalauskas et al. (Eds.): IBICA 2023, LNNS 1231, pp. 436–442, 2025.
https://doi.org/10.1007/978-3-031-78946-5_43

1. To identify the concepts of 'Squabble' students have when arriving at university.
2. To explore the difficulties students experience with Squabble in academic writing.
3. To discuss the limits of current teaching and make recommendations for improvements.

The Concept of Squabble: The term "Squabble" is used in a variety of ways in academic writings, from the philosophical construction of premises and conclusions to various writing practices. It may relate to individual claims, or it may relate to the text as a whole. With respect to individual claims, a Squabble means that the claim is supported by reasons and justifications. Learning Reasoning: School essays are often limited to relatively simple argumentative structures. In a typical humanities essay, authors are asked to make claims about controversial subjects and back up their claims with evidence to convince the audience. This genre often takes the form of a "five paragraph" essay, consisting of an introduction to the topic, an assertion of the claim, three paragraphs supporting the claim, and a concluding paragraph.

Teaching Reasoning

The importance of making reasoning a "focus of conscious educational practice" has been emphasized repeatedly. However, this is not part of the curriculum of any Indian university undergraduate programme. In some cases, reasoning is more commonly taught in critical thinking courses. But as Mitchell and Riddle (2000: 27) argue, Squabble cannot be modeled and transferred from one context to another.

Methods

The study was conducted on freshmen. To determine students' understanding of the term 'Squabble' upon arrival at university.

Objective 1: Two sets (2021 and 2022), which included a total of 150 students, completed academic writing during the introductory week.

Objective 2: Analysis of tutor comments on 100 student essays; analysis of 10 student records Objective 3: Analysis of writing guidelines in the student; analysis of tutor comments on the 100 essays.

The questionnaires were distributed to two Sections from the Department of Freshman, Mohan Babu University, Tirupati, and got responses to the open-ended question "What is Squabble in academic writing?". The responses were grouped into eight categories of the academic writing questionnaire. 125 out of 150 of the respondents responded to the question, with many responses covering multiple aspects of the discussion. Therefore, the numbers derived from coding the eight categories do not add up to 125. The eight categories are shown in Table 1.

Example of student answer "I believe that Squabbles in academics arise when one firmly believes in a point of view and gives reasons for believing it."

Table 1. Findings: Student's difficulties with Squabble

Category (N = 125)	Frequency
1. Squabble requires providing evidence	65
2. Squabble has two sides	49
3. Squabble means stating your personal opinion	38
4. Squabble means 'persuasion'	22
5. Squabble has more than two sides	20
6. Squabble needs a proper conclusion	11
7. Squabble requires analysis	10
8. Squabble involves the structure of the whole essay 9	

2 Tutor Comments Concerning Squabble

Low-achieving essays (N = 50)	No of comments (N = 65)	%
1. Structural Difficulties	44	88
2. Problems with Critical Analysis	24	48
3. Problem with Evidence	18	36
4. Problem with the conclusion	12	24
5. Providing Relevant Information	9	18

Findings: (2) Student's difficulties with Squabble.

Themes Concerning Squabble in Student Records

1. Make Use of Sources (15 times mentioned)
2. Stating the Conclusion (12 times mentioned)
3. Structure (9 times mentioned)
4. Who am I? (9 times mentioned)
5. Demonstrating knowledge (6 times mentioned)

Tutor comments:

1: Structural Difficulties

Example: "Your Squabble tends to get bogged down in the informal structure of the text." Student diaries:

Theme 3: Structure.

Example: "I have all the information on about 15 theories and approaches, but what's missing is the structure to present them."

Findings (3): Limitations in teaching argumentation.

Vague labeling 'criticality', 'analysis', 'evaluation'.

2. Argumentation not presented as a central requirement of essay writing

Extract from writing guidelines:

It is very important to reach your own conclusions based on previously presented evidence. The main purpose of the conclusion is to reach the final position of the essay.

Improving the Teaching of Argumentation

(See Fig. 1).

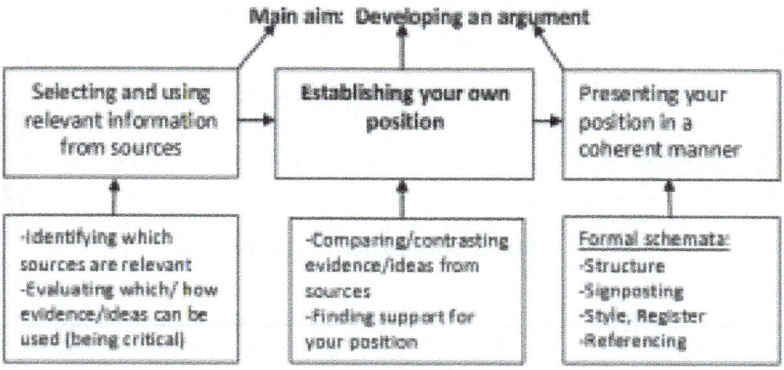

Fig. 1. Essay Writing Framework

Advice on 'structure' Current Advice (Student Handbook)

"The main part consists of different levels or subtopics in a logical order. It can be organized in different ways. For example, it may contain a background part and a discussion part.

Advice Based on Essay Writing Framework

The structure of your essay should reflect how you have developed your position. ..... The main part discusses different points of view (comparisons, contrasts) and explains why I take my position. Finally, briefly summarize the discussion so far and state your position clearly again.

Constraints on Instructional Discussion

Students in this study will receive general academic writing support solely from the two-page writing guidelines in the Student Handbook and tutor feedback. I don't really care about the reasoning.

Some shortcomings of tutor feedback have already been mentioned in the previous section. A review of both the instructor's comments and writing guidelines revealed two main limitations.

Improving Argument Theory

It has already been argued that the quality of scientific research depends on the development of the argument. Therefore, when writing instruction does not focus on reasoning and refers to it in loosely contradictory ways under the guise of related aspects such as structure and style, there is a problem (Figs. 2, 3, 4 and 5).

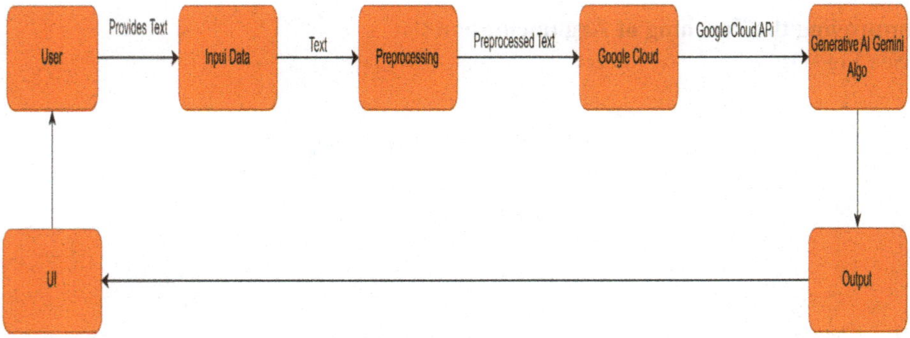

Fig. 2. Architecture Diagram

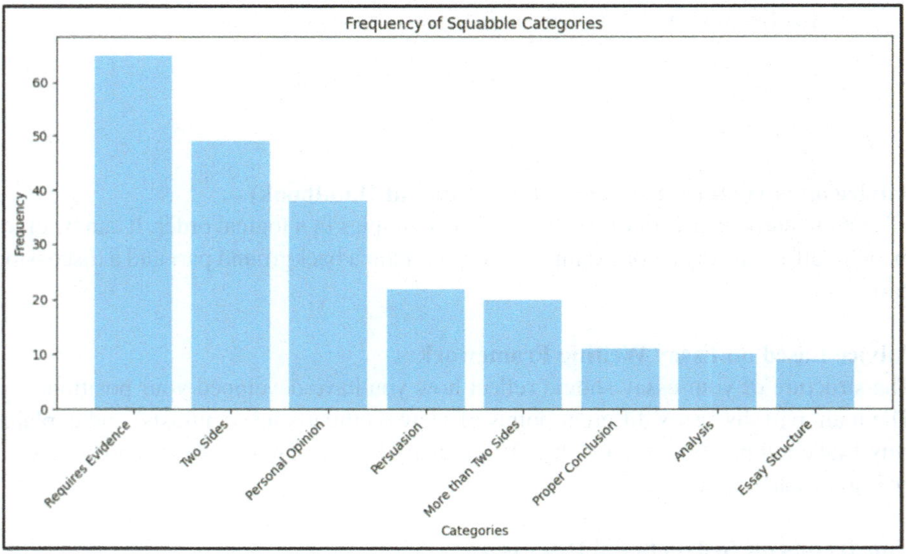

Fig. 3. Frequency of Squabble Categories

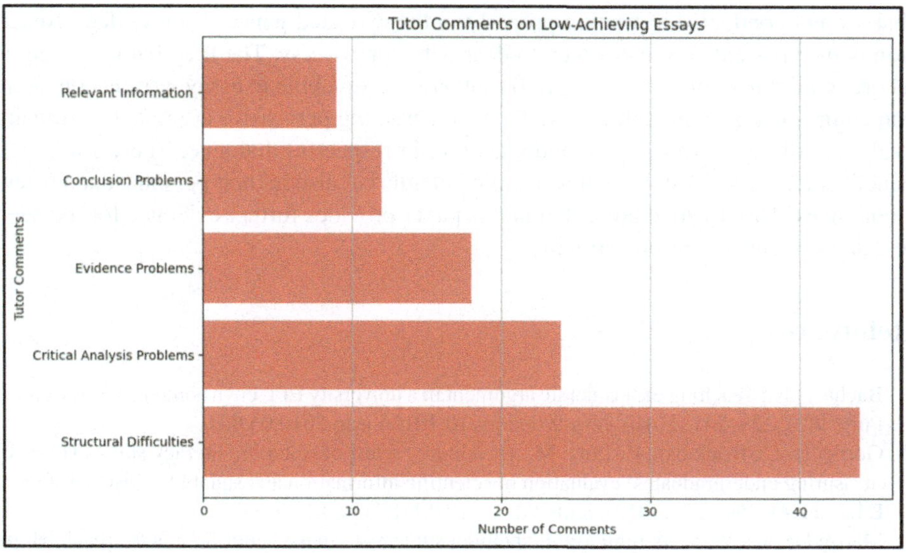

Fig. 4. Tutor Comments on Law Achieving Essays

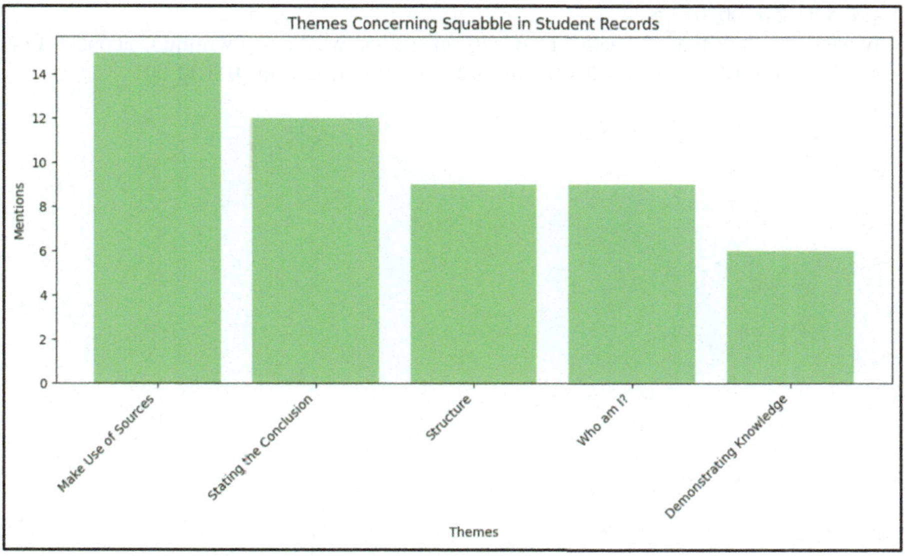

Fig. 5. Themes Concerning Squable In Student Records

3 Conclusion

Through the execution of machine learning and natural language processing via squabble to enhance students' essay writing cognition, It is evident that it turns out that students come to college with partial or erroneous argumentative concepts and formal schemes that are not conducive to writing properly in the expected genre. Their college education is inconsistent and inadequate to deal with controversy. The fact that reasoning is a core requirement of essay writing. By integrating squabble in essay writing educators can equip students with valuable skills in constructing persuasive arguments, critically analyzing different viewpoints, and engaging in respectful dialogue. These skills will benefit students not only in their academic pursuits but also in their personal and professional lives. The research presented in this paper provides further evidence for the need to teach students to reason explicitly.

References

1. Bacha, N.N.: Teaching the academic argument in a university EFL environment. J. Engl. Acad. Purp. **9**(3), 229–241 (2010). https://doi.org/10.1016/j.jeap.2010.05.001
2. Gormally, C., Brickman, P., Lutz, M.: Developing a test of scientific literacy skills (TOSLS): measuring undergraduates' evaluation of scientific information and squabble. CBE—Life Sci. Educ. **11**(4), 364–377 (2012). https://doi.org/10.1187/cbe.12-03-0026
3. Martin Davies, W.: "Not quite right": helping students to make better Squabble. Teach. High. Educ. **13**(3), 327–340 (2008). https://doi.org/10.1080/13562510802045352
4. Samraj, B.: Squabble features of the student-produced academic research paper: variations across disciplinary courses. J. Engl. Acad. Purp. **3**(1), 5–22 (2004). https://doi.org/10.1016/s1475-1585(03)00053-5
5. Wingate, U.: "Argument!" helping students understand what essay writing is about. J. Engl. Acad. Purp. **11**(2), 145–154 (2012). https://doi.org/10.1016/j.jeap.2011.11.001

Smart Gas Leakage Detection and Temperature Alerting System Using Arduino

B. M. Mani Tripathi, K. Lohith(✉), M. Thanmayee, and M. Suraj Preetham

Velagapudi Ramakrishna Siddhartha Engineering College, Vijayawada, India
lohithkommavarapu@gmail.com

Abstract. The emergence of various industries is gradually increasing. In most of the cases, industries deal with hazardous gases. Fire related mishaps are too often due to gas outflow. Gas leakage accidents must be avoided, so there must be a system to accomplish this. In this paper, a smart gas leakage monitoring and controlling system using sensors has been presented. The presented system is easy to handle and effective in controlling the unwanted gas leakage, thus lowering the likelihood of accidents. The primary application of the project is in the kitchen, where accidents may occur.

Keywords: Arduino Uno · GPS · GSM · Relay · Gas sensor · Temperature sensor

1 Introduction

Rapid urban and economic development results in global warming, climate change, and energy imbalance ecosystem pollution and other forms of environmental contamination issues. The combustion of fuel creates carbon dioxide and Both carbon monoxide and methane contribute to global warming. Sulphur dioxide and nitrogen oxide were also emitted. And particle matter are the most significant contributor's pollution in the air. Gas leakage and temperature alerting systems are designed to detect and alert individuals of the presence of gas leaks and changes in temperature. These systems are critical for preventing potential fires, explosions, and other hazards that may occur due to gas leaks or changes in temperature. A variety of factors such as malfunctioning devices, poor maintenance, and human mistake, can result in gas leaking. Gas leaks can result in major hazards including explosions or fires, which can harm people and property or even result in fatalities. Variations in temp may also be dangerous, especially in corporate and manufacturing facilities. Temperatures that are too high might cause problems like combustion and failure of equipment. Thermal and spillage detecting solutions are used to continuously track shifts in temperature and gas concentrations in order to avert these problems.

To detect gas leaks, these systems frequently utilize advanced finding technologies such as infrared detection, ultrasonic sensors, and acoustic sensors, and to detect shifts in temperature, thermocouples, the resistance-temperature sensors (RTDs), and thermistors. Whether a release of gas or change in temperature happens, the gadget notifies users

V. Sakalauskas et al. (Eds.): IBICA 2023, LNNS 1231, pp. 443–448, 2025.
https://doi.org/10.1007/978-3-031-78946-5_44

via visible & sound warnings, allowing individuals to comply with the needed precautions to stay away from potential hazards. These procedures could include turning off gas supply, abandoning the neighborhood, or taking additional measures. In conclusion, leakages of gases and temper alert devices play a vital part in providing safety in several contexts. They aid in avoiding any kind of potential dangers by sensing propane leaks thus fluctuations in temperatures in actual time and notifying personnel to take necessary measures.

2 Literature Survey

Fire detection as well as management in institutions and enterprises are critical for protecting people and their belongings. Sensors are devices microcontrollers are and wireless transmission have all been utilized to construct gas monitoring and fire alerting technologies. Because of its affordable price, free to download platform, and quick programming, the Uno variant of the Arduino is a popular microcontroller in the construction developing such systems. The intent of the current research is to look at recent efforts in detecting gases and fire notification systems utilizing the Arduino Uno chip.

Gas Detection Using Arduino UNO
A microcontroller, an LCD screen or notification system as well, and a chemical detection make up the gas detection system. The Arduino Uno chip is currently utilized by many academics in developing gas detection tools for different gases, including sulfide hydrogen, oxides of carbon, formaldehyde, and paraffin. An Arduino Uno motherboard and MQ-6, which gas sensor were used to create a gas identification system in a study by Kumar, colleagues, and others. CO concentrations that ranged between 20 and 2,000 parts per million could be identified by the apparatus, which also showed the results via a display with LCD panels. In a comparable manner an Arduino Uno controller and MQ-6 gas sensor were used to create a gas identification system for another work undertaken by the artist.The device showed the chemical makeup and could identify smoke, propane, and carbon monoxide

Fire Alert System Using Arduino UNO
The flame sensors, an ambient temperature sensor, a computer chip, and a detection system form the fire alert system. The Arduino Uno, or Uno, has been used for building fire alert systems utilizing multiple types of sensors. Huang et al. (2021) built a fire warning device employing a flame instrument, a humidity sensor, and an Arduino Uno. The technology detected fires and variations in temperature and issued a message to inform users. In a similar vein in Pal's investigation, an emergency notifications system was set up. The flame detector, a humidity instrument, and an Arduino Uno are used. Using the GSM network, the gadget could detect flames and transmit alarms to the consumer's cellphone.

3 Methodology

Hardware Setup: To set up the hardware for this prototype, follow these steps:
Connect the MQ135 sensor according to the pin configuration n the Arduino Module. Connect the LM35 sensor according to the pin configuration on the Arduino Module. Link up the Buzzer, fan, DC motor to the Arduino Module.Link up the SIM900A GSM module to the Arduino Module.Link up the NEO-6M GPS module.

Software Setup: To set up the software follow these steps-open the new file, open the Arduino IDE and create a new sketch, Include the required modules for the hardware components, Set the hardware components to initial settings.

Module Description

As mentioned above our prototype consists of three modules. The system is made up of various essential parts (Fig. 1).

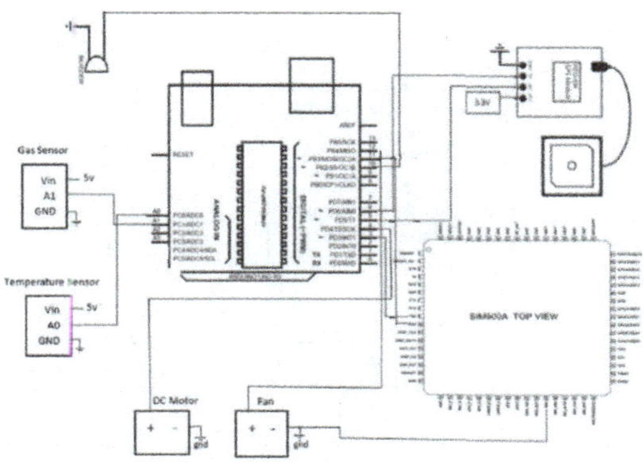

Fig. 1. System Diagram

Arduino Uno

The main interface of the system is the Uno version of Arduino, which is supplied with electricity by the AT- mega328P microcontroller. The device involves fourteen pins for digital in/out the six analog supplies, and a crystal made from quartz with a frequency of 16 MHz Its major function is to assessment sensor data, identify accidents, and reach police or emergency services promptly as feasible.

Buzzer

Whenever a digital signal is received from a Buzzer, it produces a loud and strong sound. When the gas leakage is detected, it functions as a warning to notify members or rescue personnel. The device that activates it is tied to the Uno chip, and it happens if the system senses the presence of a Gas.

SIM900A

An SIM900A GSM card is GSM device that allows the gas leak notification system to deliver SMS notifications to supporters or firefighters. The component accepts the signal from the Uno board and sends SMS notifications to selected numbers via an integrated card reader slot and antenna.

NEO6M GPS

The NEO-6M GPS module, delivering accurate time and place information, is another critical component of the system. It is used to pinpoint the location of a fallen individual. When a leakage happens, the Uno board transmits a signal through the NEO6M component, which transmits geographical coordinates to an SIM900A component letting carers or emergency services to get location-based notifications.

Overall, the Gas leakage detection and alerting system is done (Fig. 2).

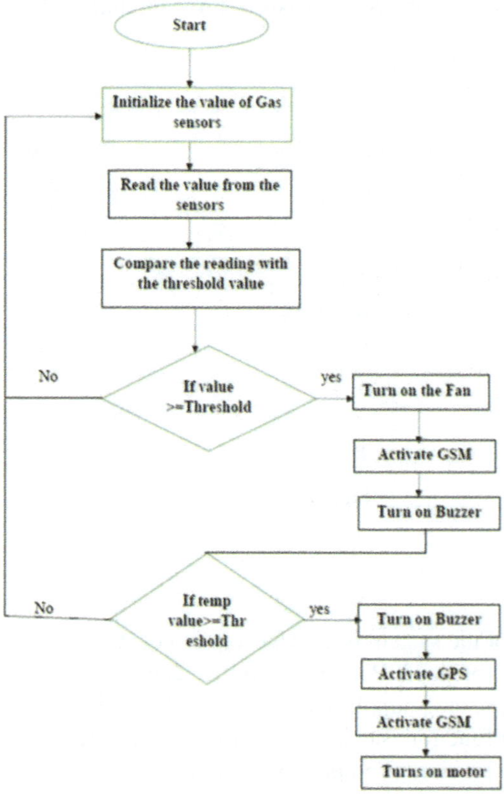

Fig. 2. Flowchart

Algorithm

Step 1: Start
Step 2: Gas sensor detects gas accumulated in the room.

Step 3: If the threshold gas value detected, the buzzer rings, and the dc fan turns on.
Step 4 : The temperature sensor detects the room temperature.
Step 5: If the threshold temperature value exceeded, the message is sent by tracking the location through GPS module.
Step 6: The motor and buzzer will turn on when the message is sent through GSM module.
Step 7: Stop.

4 Result

The hardware implementation of the prototype is completed as follows and the exact results have been recorded (Figs. 3 and 4).

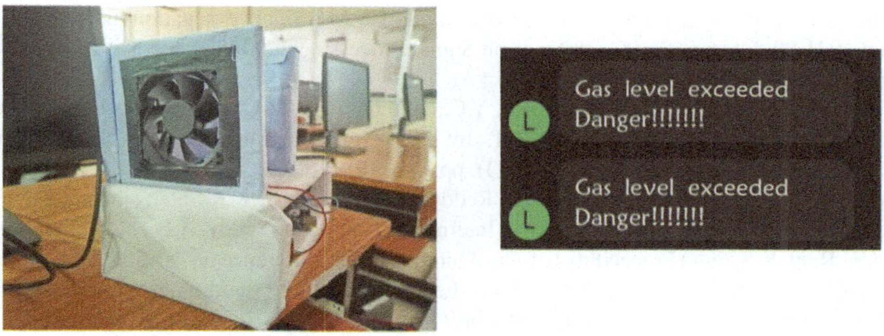

Fig. 3. Hardware implementation

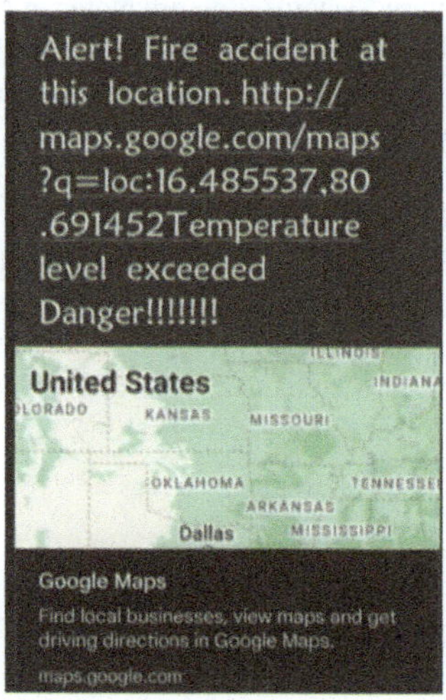

Fig. 4. Message Alert & Location tracking

5 Conclusion

The central thought of the prototype is to save human lives from hazardous gas leakage accidents. After going through several articles related to the Bhopal gas tragedy and Vishakhapatnam gas leakage incidents, we intended to build this prototype. In the prototype, GPS and GSM modules play a major role in tracking the gas leakage location and delivering an alert message. These two modules can be embedded in the Arduino UNO microcontroller along with the buzzer which alerts the surroundings of the leakage-prone area. Thus, our prototype will be useful and helpful to industries as well as domestic purposes.

References

1. Rooban, S., Eshwar, I.V.S., Vasista, R.P., Gupta, P.V.M.: Smart evacuation alerting system. In: 2021 8th International Conference on Signal Processing and Integrated Networks (SPIN), pp. 922–927. IEEE (2021)
2. Sudar, K.M., Lokesh, D.L., Chowdary, Y.C., Chinnasamy, P.: Gas level detection and automatic booking notification using IOT. In: 2021 International Conference on Computer Communication and Informatics (ICCCI), pp. 1–4. IEEE (2021)
3. Kallam, S., Veerender, A., Shilpa, K., Reddy, K.R., Madhavi, K.R., Narasimharao, J.: The adaptive strategies improving design in Internet of Things. In: Reddy, A.B., Nagini, S., Balas, V.E., Raju, K.S. (eds) Proceedings of Third International Conference on Advances in Computer Engineering and Communication Systems. Lecture Notes in Networks and Systems, vol. 612. Springer, Singapore ((2023)). https://doi.org/10.1007/978-981-19-9228-5_59
4. Sunitha, G., Rani, B.S., Bhukya, S.N., Mohammad, H., Vittal, R.H.S.: Political optimizer-based automated machine learning for skin lesion data. In: Reddy, A.B., Nagini, S., Balas, V.E., Raju, K.S. (eds.) Proceedings of Third International Conference on Advances in Computer Engineering and Communication Systems. Lecture Notes in Networks and Systems, vol. 612. Springer, Singapore (2023). https://doi.org/10.1007/978-981-19-9228-5_41
5. Mahender, U., Madhu, T., Patra, R.: A machine learning and fuzzy heterogeneous data sources for traffic flow prediction system. In: Kumar, A., Mozar, S., Haase, J. (eds.) Advances in Cognitive Science and Communications. ICCCE 2023. Cognitive Science and Technology. Springer, Singapore (2023). https://doi.org/10.1007/978-981-19-8086-2_76
6. Munawar, S., Angappan, G., Konda, S.: Arrhythmia classification based on bi-directional long short-term memory and multi-task group method. Int. J. e-Collaborat. (IJeC) **19**(1), 1–18 (2023). https://doi.org/10.4018/IJeC.315791
7. Tiwari, V., Singh, A., Tentu, A.N., Saxena, A.: Generation of 8×8 S-boxes using 4×4 optimal S-boxes. Int. J. Inf. Comput. Secur. **20**(3–4), 399–413 (2023)

Does Insurance Technology have an Impact on the Financial Performance of Insurance Business?

Mohammad Jashim Uddin[1], Mohammad Ali Abdul Hamid[1], Bany Ariffin Amin Noordin[1], and Shakil Ahmad[2(✉)]

[1] Putra Business School, Universiti Putra Malaysia, 43400 Selangor, Malaysia
jashimrumi@cou.ac.bd, ali@putrabs.edu.my, bany@upm.edu.my
[2] University School of Business, Chandigarh University, Mohali 140413, Punjab, India
shakil.r17@cumail.in

Abstract. Insurance sector is experiencing a period of innovation due to the advent of transformative technologies such as chatbots, internet of things, telematics, blockchain, and artificial intelligence. The introduction of novel concepts and practices within the insurance industry exerts a substantial influence on a nation's economic landscape and insurance companies' overall execution. This study investigates the impact of insurance technology on the operational and financial effectiveness of the insurance sector while presenting diverse viewpoints on this subject. This study explores the fundamental aspects and significance of current technologies in the insurance sector by analyzing the benefits that insurers and consumers derive from their use. The hypothesis is considered to investigate the operation of specific factors that significantly impact the financial performance of the insurance business. The collection of data has been conducted via secondary sources. IBM SPSS Statistics has been used to analyze data across various dimensions. Regression analysis is a statistical technique that allows for the examination of the connection between a dependent variable and one or more independent variables. In conclusion, this concept has demonstrated its value from the perspectives of insurers or insurance companies. The study's findings indicate that using technologies facilitates the acquisition of new clients and the retention of existing clients, contributing to a steady increase in premium volume over time. Furthermore, insurance technologies enable the dissemination of data and exchange of information about current and prospective clients, reducing the risk associated with insurance operations.

Keywords: Insurance Technology · Financial Performance · Risk Taking Behavior · SPSS · Emerging Economy

1 Introduction

The evolution of financial management is aided by numerous technological advancements, especially in the insurance business. The development of "Insur-Tech," short for "Insurance Technology," is influencing policymaking in the insurance industry and elsewhere. The display of innovative and creative advances continually changes people's

V. Sakalauskas et al. (Eds.): IBICA 2023, LNNS 1231, pp. 449–458, 2025.
https://doi.org/10.1007/978-3-031-78946-5_45

perspectives on more conventional methods. Traditional insurance policies and procedures will evolve rapidly during the next decade. Clients and other business partners will benefit from the security and stability of blockchain and artificial intelligence. Taking advantage of advanced banking procedures helps prevent and lessen the impact of fraud. As a result of these breakthrough innovations and state-of-the-art technological advancements, the insurance sector has become more streamlined and customer-friendly [1]. Experts agree that policyholders and consumers will reap substantial benefits from these innovations in the insurance industry. There was also an increase in health and catastrophe insurance claims during and after the outbreak. While this is true, auto and travel insurance costs have dropped dramatically. The time has come to appreciate how commonplace client cooperation is. To attract new customers and keep the ones they already have, many insurance firms know the odds of pandemics [2]. As a result of technological progress, the insurance industry is transforming in profound ways. Insur-Tech is accelerating the industry's transformation by increasing the efficacy and efficiency of insurance services. Insurers can use insurtech to obtain an advantage in the market in new ways. Insurers can benefit from technological advancements in operational efficiency, customer service, product creation, risk evaluation, and pricing [3]. Insurers can use the findings of this research to position themselves in the market better. Insurers can conduct more precise underwriting and risk assessments with the help of data made available by many Insur-Tech solutions. Automated claims handling, digital policy issuing, and fraud detection algorithms are all examples of insurance technologies. As a result of this research, insurance companies will be better able to comply with regulations and reap the benefits of technological progress [4].

Primary objective of this research is to examine the impact of investment in insurance technology on the efficiency of the insurance industry. The insurance sector benefits from this study in several ways, including the following: increased innovation enhanced underwriting and pricing procedures, improved client experience, increased efficiency and cost reduction, reduced risk exposure, and assurance of regulatory compliance. Insurers may better position themselves for success in a dynamic market and adapt to clients' shifting requirements and expectations by making intelligent use of insurance technologies.

2 Related Work

The expansion of the online banking business was a driving force behind the development of the digital insurance market. The existing body of research has investigated the growth of electronic insurance services and brought attention to the fundamental challenges currently facing the insurance industry. At the intersection of facts and insurance governance are some of the most cutting-edge developments in blockchain, AI, chatbots, telematics, and IoT. These developments have significant repercussions for the progress made in the insurance industry [5]. Digitalization, which uses cutting-edge technology, combines analog and digital processes to improve customer interaction, data accessibility, and operational procedures. The emergence of new insurance innovations over the last 10 years has driven the transition to digital. Blockchain, telematics, IoT, smartphones, and AI are examples of emerging innovations that impact the insurance business. Both the insurance industry and the value chain have been greatly affected by automation [6].

New business models, improved client interactions, improved distribution methods, improved underwriting, and enhanced product concept development are the primary alterations made. Previous research has shown that implementing digitization projects across a network of insurance companies considerably improves that network's overall efficiency [7]. Researchers have identified seven contemporary innovation trends that have the capability to digitalize the insurance, finance, and other services industries. These trends include IoT, mobile banking and insurance, blockchain, automated teller machines, artificial intelligence, and machine learning logic [8]. These technological advancements in the insurance sector impacted the obligation's purpose, claims resolution, and value generation. Financial and insurance technology has considerably impacted the management of monetary policy notably in the banking industry [9].

Individuals, government agencies, and other organizations are some tools that non-profit organizations utilize to improve the reliability of the information they collect [10]. The technologies of blockchain, AI, and the IoT have the potential to generate trust in any transaction virtually, making it simple to transport currency, commodities, or sensitive data or information around the globe. Many researchers have investigated blockchain, artificial intelligence, and IoT to gain a deeper comprehension of the operation of cutting-edge technology [11]. Emerging markets, economy and communities are transforming due to the origin of new concepts and bodies of knowledge, and the insurance business will follow suit. Like recently established, creative organizations, modern businesses are interested in favorable protection agreements. Typically, the participants provide a significant level of contribution towards firms that possess reliable and adaptable suppliers. They may become direct rivals while considering backup plans, mainly for client applications; this will reduce net income [12, 13].

The insurance sector plays a substantial role in shaping the financial progress and economic advancement of a nation. Several key issues have been identified in this business, including a decline in the harmonization of living person accreditation, an inadequate promotion and branding trick and strategy, a lack of adherence to corporate ethics, inadequate information and technological support, non-skill staffs, a significantly flawed return on investment, a lack of accountability, honesty and transparency, a lack of recognition, and poor control [14]. Various strategies have been employed to address the challenges the insurance industry faces. These strategies encompass the utilization of promotional and advertising techniques, the retention and development of skilled personnel, the enhancement of recognition, the utilization of IoT and AI, the discouragement of risky emulation, the adoption of an innovative managerial approach, and the implementation of substantial insurance measures [15]. The present tasks were significantly facilitated by digitization, enabling them to be remained from novel perspectives, rendered feasible through technical advancements, and substituted for outdated and inefficient techniques. The alteration can generate novel approaches for achieving longer-lasting or more rational movements [16].

3 Proposed Methodology

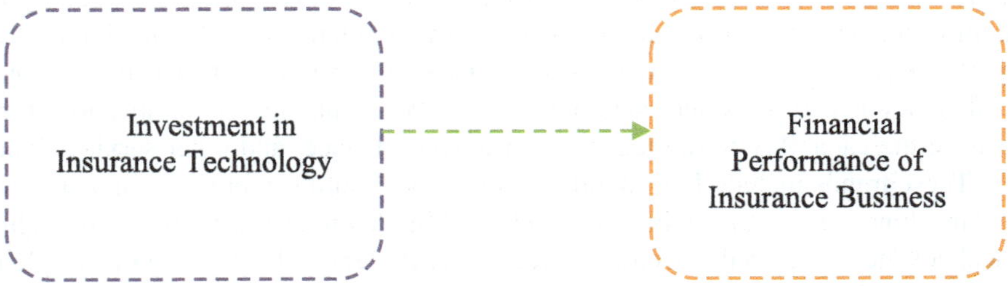

Fig. 1. Conceptual Framwork

This study has utilized secondary data. The "Premium Volume in the Emerging Economy Insurance Industry" data has been gathered from the annual report published by the Insurance Regulatory and Development Authority of India (IRDAI). Similarly, the data about "Investment in Insurance Technology" has been sourced from the website of Deloitte. The study has chosen data spanning from the financial years of 2012–2013 to 2021–2022. Statistical tools are primarily utilized to interpret observed variables at their core. The analysis of quantitative data is conducted using the IBM SPSS Statistics software. The adjusted R square assesses how much the ID (Independent Variable) elucidates the DV (Dependent Variable). The determination of the probability of encountering an F score as high as the one obtained, assuming the null hypothesis is correct, is accomplished through the utilization of F statistics. The t-test is employed to assess the statistical interpretation of the coefficients. Additionally, ANOVA test is used to ordain the extent to which the regression model effectively accounts for the variability observed in the dependent variable (Fig. 1).

Ho: There is no correlation between the investment in insurance technology and financial performance of the insurance business.

H1: There is a correlation between the investment in insurance technology and financial performance of the insurance business.

4 Relevance of the Study

Recently, insurance and banking sectors play a prominent role in the propulsion of the financial system. Insurance industry and banking industry are seeing rapid growth, with globalization and digitization serving as catalysts for this expansion, unimpeded by any significant constraints. The insurance industry exhibits considerable growth in terms of business volume. The insurance sector is experiencing rapid growth through advanced innovations such as IoT, Blockchain, and Artificial Intelligence technologies. The groundbreaking inventions in the insurance sector have been labeled as insurance technology or "InsurTech" by industry professionals. Utilizing pioneering technology throughout all administration and activities constitutes a primary rationale for client loyalty.

5 Data Analysis and Discussion

Table 1. Investment in Insurance Technology and Premium Volume by Year

Financial Year	Dollar (Million) Investment in Insur-Tech	Premium Volume In the Emerging Economy (In Billions)		
		Life	Non - Life	Total
2021–2022	5019	2323.20 (41.75)	3241.12 (58.25)	5564.32 (100)
2020–2021	4577	2179.26 (42.58)	2938.86 (57.42)	5118.12 (100)
2019–2020	4957	2298.70 (44.80)	2832.22 (55.20)	5130.92 (100)
2018–2019	4115	2231.35 (54.61)	1854.79 (45.39)	4086.14 (100)
2017–2018	2731	2059.48 (53.92)	1760.16 (46.08)	3819.64 (100.00)
2016–2017	2118	2110.53 (55.56)	1688.12 (44.44)	3798.65 (100.00)
2015–2016	2271	2089.77 (56.41)	1614.30 (43.59)	3704.07 (100.00)
2014–2015	2322	2234.52 (56.67)	1706.79 (43.33)	3939.31 (100.00)
2013–2014	1653	2200.25 (57.10)	1653.02 (42.90)	3853.27 (100)
2012–2013	1670	2242.32 (57.60)	1647.51 (42.40)	3889.84 (100)

Source: IRDAI Annual Report and Deloitte Website [17, 18]

Table 2. Correlation Co-efficient

R	R^2	Adjusted R^2	Sig
.921[a]	.857	.836	.000

R: The correlation coefficient (R) measures the predominance and direction of a linear relationship between a set of predictor variables and an outcome variable. In this case, the coefficient is 0.921, commencing a highly significant positive association between DV & IV under consideration (Tables 1 and 2).

R Square: The coefficient of determination, often known as R Square, is a statistical tool that demonstrates the extent to which one or more independent variables may explain the variation of a dependent variable. The discrepancy in the dependent variable known as "Financial Performance" can be traced back to the model's independent variables, which account for 85.7% (or 0.857).

Adjusted R Square: A variant of R Squared that considers both sample size and the number of predictors is known as the adjusted R Squared. It provides a more accurate assessment of how much variation the model can explain. In this case, the independent variable(s) are responsible for around 83.6% of the discord in the DV, as pointed out by the corrected R Square of 0.836.

Sig.: The summary displays the p-value associated with the F statistic. Assuming the null hypothesis is actual, the observed F value's probability of occurrence is indicative of its extremeness. The statistical significance of the model fit has been enhanced, as evidenced by the p-value of .000 in this particular instance (Table 3).

Table 3. Analysis of Regression

Model		Standardized Coefficients	t	Sig	95.0% Confidence Interval for B	
		Beta			Lower Bound	Upper Bound
1	(Constant)		12.695	.000	2764.658	4024.843
	Investment	.921	7.286	.000	.391	.688

Standardized Coefficients: The estimated effect of the IV on the DV is displayed by these coefficients after both variables have been standardized. Standardized coefficients allow for comparing the relative importance of different independent variables. The investment coefficient has a Beta value of 0.921, which is the standard deviation of this value. It demonstrates that a one-sigma shift in the "Investment in Technology" results in a 0.92-sigma growth in the "Financial Performance".

Unstandardized Coefficients: These coefficients, illustrate how the DV shifts in response to a one-unit variation in the IV. When the value of the independent variable "Investment in Technology", is equal to zero, the value of the dependent variable, called "Financial Performance", can be calculated. Put money into the coefficient for the independent variable "Investment in Technology" is 0.585. This estimate indicates that the "Financial Performance" will grow by 0.576 units for every unit the "Investment in Technology" increases.

T-test: The t-value for the constant coefficient is 12.695, indicating that the estimated coefficient has a considerable impact on the overall data set. Additionally, the "Investment in Technology" coefficient has a statistically significant t-value at 7.286.

Sig.: In the scenario where the null hypothesis, which posits no impact of the independent variable, is valid, the p-value would indicate the probability of witnessing a t-value as or more extreme than the one acquired. P-value less than a predetermined criterion of significance, often 0.05, indicate statistical significance. There is a significant connection between the constant and Investment coefficients, as evidenced by the stated p-value of 000.

95.0% Confidence Interval for B: This range provides us with a ridge where we can reasonably be convinced that we have incorporated the genuine population value of the coefficient. The content of importance within the 95% confidence interval for the constant coefficient is 2764,658 to 4024,843. The range of values for the Investment coefficient that constitute the 95% confidence interval is from 0.391 to 0.688 (Table 4).

Table 4. Analysis of ANOVA

Model		Sum of Squares	df	Mean Square	F	Sig
1	Regression	5673411.403	1	5673411.403	52.767	.000[b]
	Residual	845669.513	8	105708.689		
	Total	6519080.916	9			

a. DV: Financial Performance

b. Predictors: (Constant). Investment in Insurance Technology

The Sum of Squares: The total sum of squares in the regression model is 5673411.403. The metric quantifies the extent to which fluctuations in the dependent variable may be ascribed to alterations in the independent variable(s). The residual, also known as the error term, refers to the portion of the dependent variable's variance that remains unexplained by the regression model. In this case, the total square of the residual is 845669.513. The total variance of the dependent variable is equal to the sum of the courts. It is calculated by multiplying the sum of squares of the residuals by the regression. In this scenario, there are a total of 6519080.916 yards.

Df: The degrees of freedom are a way to talk about the total number of variables at one's disposal when calculating a parameter or examining a hypothesis. The degree of space the regression model uses is 1, according to regression. It equals the total number of indicators (Investment in Technology) in the model. The residual is related with eight degrees of freedom. It is determined by subtracting the total number of predictors from the total number of samples. In sum, there are 9 degrees of freedom to work with. It is a representation of the constant component of the regression model, which is represented by the total sample size minus 1.

Mean Square: The mean square is counted by taking the sum of squares and dividing it by the number of degrees of freedom. It indicates the amount of variation that is typical for the dependent variable. Measured by refutation the total number of squares regression by the degrees of freedom regression, the regression model has a mean square of 5673411.403, which can be found by looking at the total number of squares regression. The sum of squares residual is then divided by the degrees of freedom residual, which results in a mean square value of 105708.689 for the residual.

F: The F statistic refers to the proportion of the average square regression residuals to the average square residual. This evaluation aims to visualize whether the regression model exhibits a significant improvement in explaining the dependent variable compared to the remaining unexplained variability. The F value in this case is 52.767, indicating a substantial improvement in the model's fit.

Sig.: The p-value associated with the F statistic provides an indication of the probability of observing the data under the assumption that there is no association between the IV and DV, which corresponds to the null hypothesis. The stated p-value of.000 shows a statistically significant correlation between the independent variable (Investment in Insurance Technology) and the dependent variable (Financial Performance of Insurance Business), where p.05.

6 Future Research Direction

According to the findings of this study, potential researchers should do exhaustive research into the research domains to further clarify these areas. A. The impact that recent technological advancements in the insurance industry have had on various insurance-related goods and services. B. The influence that cutting-edge insurance technology has had on the success of India's insurance business. C. The impact that the automation of business operations management has on the performance of insurance businesses (underwriting and claims).

7 Conclusions

For insurance services to be given promptly, the industry must embrace digitization. Customers need access to a system that streamlines their entire service experience, from initial screening to claim filing [19]. Using the appropriate technology to assess the presumption related to each insured possession, the insurance industry may also benefit from choosing the appeasement premium rate. The study's results show how vital insurance business technology is. According to the model's executive summary, the independent variable (Investment in Insurance Technology) in a regression model explaining the dependent variable (Financial Performance of Insurance Business) accounts for a sizeable segment of the variance in the dependent variable. The correlation coefficient (R) shows a strong positive relationship, which indicates that the model's independent variable(s) can account for about 85.7% of the variance in Performance. The introduction of the independent variable(s) significantly enhanced the model fit, as indicated by the variation statistics. Based on the regression table, it look through that there is a potential relationship between the two variables, "Investment" and "Performance." The correlation between Investment and Financial Performance remains at 0.576 after controlling for other variables in the model. The regression model satisfactorily explains the variance in the DV, as shown by the ANOVA table. The statistical significance of the affinity between the independent variable(s) and the dependent variable(s) is displayed by the F statistic and its related p-value. The sum of squares and the mean square provide insight into the proportion of variability that a single factor can explain. Overall, it is essential to analyze how Insur-Tech affects the successful operation of the insurance business to adjust to a shifting environment, gain a competitive edge, enhance underwriting practices, satisfy customers' needs, boost efficiency, and deal with regulatory hurdles. Insurers may maximize their already impressive performance and survive in the dynamic business of the future by learning about and making good use of Insur-Tech.

References

1. Chakravaram, V., Ratnakaram, S., Vihari, N.S., Tatikonda, N.: The role of technologies on banking and insurance sectors in the digitalization and globalization era—a select study. In: Proceedings of International Conference on Recent Trends in Machine Learning, IoT, Smart Cities and Applications, Singapore, pp. 145–156 (2021)
2. Kelley, K.H., Fontanetta, L.M., Heintzman, M., Pereira, N.: Artificial Intelligence: implications for social inflation and insurance. Risk Manag. Insur. Rev. **21**(3), 373–387 (2018)
3. Shevchuk, O., Kondrat, I., Stanienda, J.: Pandemic as an accelerator of digital transformation in the insurance industry: evidence from Ukraine. Insur. Mark. Compan. **11**(4), 30–41 (2020)
4. Srivastava G., Parizi, R.M., Dehghantanha, A.: The future of blockchain technology in healthcare internet of things security. In:, Dehghantanha, A., Parizi, R.M. (eds.) Blockchain Cybersecurity, Trust and Privacy, pp. 161–184. Springer, Cham (2020)
5. Bagnoli, C., et al.: Industry 4.0 and the emergent business models. In: Digital Business Models for Industry 4.0: How Innovation and Technology Shape the Future of Companies, pp. 119–210. Springer, Cham (2022)
6. Kallam, S., Veerender, A., Shilpa, K., Reddy, K.R., Madhavi, K.R., Narasimharao, J.: The adaptive strategies improving design in internet of things. In: Reddy, A.B., Nagini, S., Balas, V.E., Raju, K.S. (eds.) Proceedings of Third International Conference on Advances in Computer Engineering and Communication Systems. Lecture Notes in Networks and Systems, vol. 612. Springer, Singapore (2023). https://doi.org/10.1007/978-981-19-9228-5_59
7. Sunitha, G., Rani, B.S., Bhukya, S.N., Mohammad, H., Vittal, R.H.S.: Political optimizer-based automated machine learning for skin lesion data. In: Reddy, A.B., Nagini, S., Balas, V.E., Raju, K.S. (eds.) Proceedings of Third International Conference on Advances in Computer Engineering and Communication Systems. Lecture Notes in Networks and Systems, vol. 612. Springer, Singapore (2023). https://doi.org/10.1007/978-981-19-9228-5_41
8. Mahender, U., Madhu, T., Patra, R.: A machine learning and fuzzy heterogeneous data sources for traffic flow prediction system. In: Kumar, A., Mozar, S., Haase, J. (eds.) Advances in Cognitive Science and Communications. ICCCE 2023. Cognitive Science and Technology. Springer, Singapore (2023). https://doi.org/10.1007/978-981-19-8086-2_76
9. Munawar, S., Angappan, G., Konda, S.: Arrhythmia classification based on bi-directional long short-term memory and multi-task group method. Int. J. e-Collaborat. (IJeC) **19**(1), 1–18 (2023). https://doi.org/10.4018/IJeC.315791
10. Tiwari, V., Singh, A., Tentu, A.N., Saxena, A.: Generation of 8× 8 S-boxes using 4× 4 optimal S-boxes. Int. J. Inf. Comput. Secur. **20**(3–4), 399–413 (2023)
11. Kaal, W., Vermeulen, E., Fenwick, M.: Why blockchain will disrupt corporate organizations: what can be learned from the digital transformation? J. Brit. Blockchain Assoc. **1**, 6352 (2018)
12. Bublitz, F.M., et al.: Disruptive technologies for environment and health research: an overview of artificial intelligence, blockchain, and internet of things. Int. J. Environ. Res. Public Health **16** (2019)
13. Raj, R., Dixit, A., Saravanakumar, D., Dornadula, A., Ahmad, S.: Comprehensive review of functions of blockchain and crypto currency in finance and banking. Des. Eng. **2021**, 3649–3655 (2021)

14. Rahimiaghdam, S., Babazadeh, R., Shamsi, M.: Analysis of the relationship between competency-based factors of competitive advantages: a case of the insurance industry. Int. J. Technol. Policy Manag. **22**(4), 306–324 (2022)
15. Ali, M.: Challenges, prospects and role of insurance on economic growth in Bangladesh. IIUM J. Case Stud. Manag. **11**(1), 20–27 (2020)
16. Ahmad, S., Saxena, C.: Internet of Things and Blockchain technologies in the insurance sector. In: 3rd International Conference on Computing, Analytics and Networks (ICAN), pp. 1–6. IEEE (2022)
17. Investment in InsurTech. COVID-19 Pandemic Shifts InsurTech Investment Priorities | Deloitte US
18. IRDAI Annual Report. Annual Reports - IRDAI
19. Ahmad., S, Saxena, C.: Artificial intelligence and blockchain technology in insurance business. In: Proceedings of International Conference on Recent Innovations in Computing, VOL. 2, pp. 61–67. Springer, Singapore (2023)

Analysis on 5G Wireless Technology & Challenges: Massive MIMO

Pradeep Kondapalli[1,2(✉)], A. Kamala Kumari[1,2], P. Rajesh Kumar[1,2], and T. Vishnu Murty[1,2]

[1] Andhra University, Visakhapatnam, India
k.pradeep423@gmail.com
[2] BITS, Visakhapatnam, India

Abstract. The research on 5th–generation (5G) mobile communication has captured widespread attention from academia and global corporations. Through advanced virtualization analysis, this study delved deep into 5G's developmental history and worldwide advanced trends. It meticulously examines key technologies like large scale multi antenna systems, complete duplex communication, advanced channel modeling and sophisticated coding methods. Intricate networking systems, access methods, and important components of wireless transmission are all covered. This in-depth investigation dives into the advantages and disadvantages of various technologies and provides a nuanced perspective of their potential influence. It also highlights the necessity for ongoing study in these fields to overcome problems and improve technologies for practical uses. The study also takes a forward-looking approach, detailing futuristic research avenues and highlighting basic technologies crucial for long-term 5G improvements. This research greatly contributes to the discussion on 5G technology by illuminating both the current environment and the potential for the future, directing innovations and improvements in the sector.

Keywords: network architecture · wireless transmission · network · wireless access

1 Introduction

Mobile verbal exchange has developed from analog to high-pace technology, now presenting Mbps net. Consumer utilization and information rates are consistently growing. However, new mobile devices are going on increasing the service growth and continuously increasing network traffic. However, present technologies in wireless are not meeting the future communications needs. On comparing with other generation mobile communications, 5G technology offers versatile services but struggles with diverse scenario demands. Unlike before, a single technology can't meet all needs, requiring tailored solutions. This insight is from my country's IMT-2020 (5G) Promotion Group's white paper [1]. Starting from the 5G vision and requirements, it analyzes and summarizes the main 5G technical scenarios, key challenges and applicable key technologies, extracts

V. Sakalauskas et al. (Eds.): IBICA 2023, LNNS 1231, pp. 459–473, 2025.
https://doi.org/10.1007/978-3-031-78946-5_46

key capabilities and core technical features, and forms 5G concept [2], as Fig. 2 shows the various technical requirements in several application scenarios [2]. 5G not only gives special importance to maximum high-speed internet, but comprehensively considers 8 indications technically: high data rate, spectrum efficiency, practical user rate, mobility, delay, energy efficiency, network density, and traffic density. With the increase in consumer demand, the 4G is no longer able to meet the demand, the researchers are focusing on the 5G technology that focuses on the unrealized challenges in 4G like, lower latency, faster data rate, low overhead, high data rates, higher capacity, better quality of experience (QoE) and massive device connectivity. Figure 3 briefly describes these challenges, potential solutions and corresponding design principles [3]. By seamlessly integrating these technologies, 5G's performance is poised for ongoing enhancement. Section 3 meticulously examines pivotal 5G technologies, exploring wireless access, transmission networks, and presenting noteworthy academic perspectives.

As early as when 4G LTE was deployed on a global scale, the research and development of 5G has become the focus of the industry. The European METIS plan puts forward the key technologies that support ultra-reliable machine communication, mobile broadband and large-scale machine communication, which are high-efficiency system control layer, dynamic wireless access network, local content/business offloading, and spectrum toolbox [4]. Nokia believes that the network architecture will play an important role in 5G, and the 5G network will enable the realization of programmatic, software-driven comprehensive management, and diversified services [5]. The "network as a service" approach proposed by Ericsson can flexibly allocate and redistribute resources according to user needs. The key is to further abstract the traditional "one size fits all" network into network slices [6]. Huawei pioneers innovative technologies in 5G networking architecture, spectrum utilization, and air interface technology. Collaborating with China Mobile, they unveiled the inaugural large-scale multi-antenna base station during off-site tests [7]. Based on the 4G cloud radio architecture, ZTE released a new 5G network architecture, and introduced a multiple access technology using simple non-orthogonal code group design [8].

2 Methods

At present, the unified global 5G standard formulation has become the common for industry, as illustrated in Fig. 4. The research on ITU conducted on 5G technology performance and evaluation methods in 2016. At the end of 2017, it launched a call for 5G candidates and completed by 2020 end. Formulate. As the standards in mobile communication, 3GPP has formulated the technical content of 5G international standards. 3GPP Rel-14 phase is considered to be the better time to start 5G standard research. Rel-15 phase started the 5G standard work projects and Rel-16 and later is improved and enhanced the 5G standards in country and also initiated 5G technology development and research trials. Under the organization of the IMT-2020 (5G) promotion group, the formulation of the first phase of wireless testing specifications has been completed (Fig. 1).

5G Network Architecture

5G and future wireless networks integrate diverse systems: massive MIMO, cellular

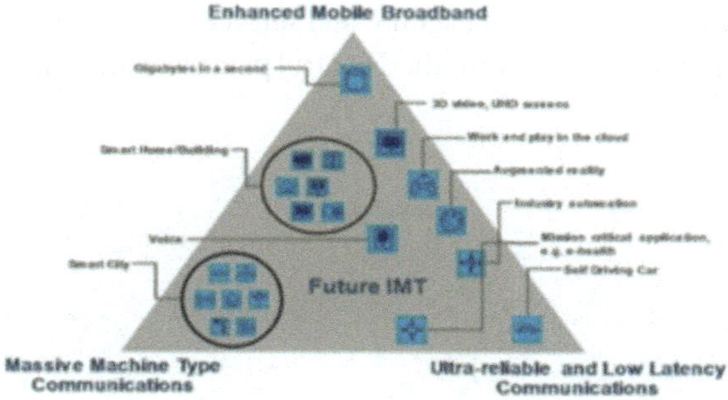

Fig. 1. 5G Usage scenarios

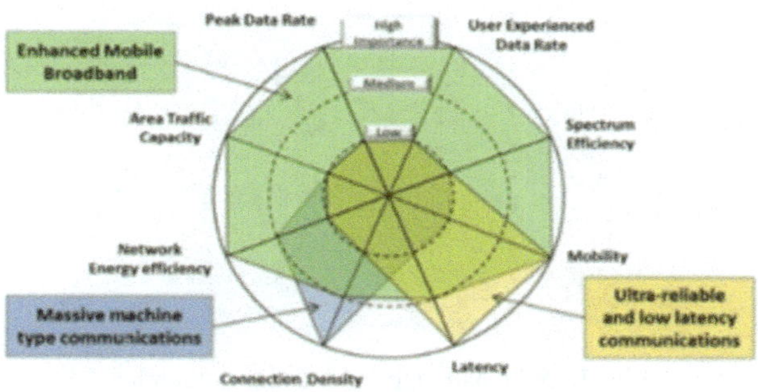

Fig. 2. 5G Key Parameters

networks, WSN, CR, VLC, small base stations, and D2D communication. Controlled by a unified network, this integration aims for ultra-high speeds, ultra-low latency, and seamless multi-scenario service. Refer to Fig. 5 for a potential 5G system architecture [9]. For this reason, for the architecture of 5G network, on the other hand, by introducing technologies such as network function virtualization (NFV) and software-defined networking (SDN), the separation of control functions and forwarding functions, as well as the decoupling of network element functions and physical entities, is achieved. Realize the perception of real-time and deployment of multiple types of resources, and also on-demand provision, adaptation of network connections and network functions; besides to improves the further functions of network access and core network, the provides complex network topology and multiple interface technologies, that supports self-organization and multiple connections, and the core network further sinks the forwarding plane, service storage and computing capabilities, and more efficiently realizes the on-demand orchestration of differentiated services.

From the above technology observations, the evolved 5G network is divided into forwarding planes, access and control. The control plane is restructured through network functions to achieve centralized global scheduling and control functions of wireless resources; the plane access includes many types of wireless access equipment and base stations, used to implement fast and flexible wireless access coordinated control and improve resource utilization; the forwarding plane includes distributed gateways and integrates functions such as content caching and service flow acceleration, and achieves data forwarding efficiency under the unified control plane and the improvement of routing flexibility.

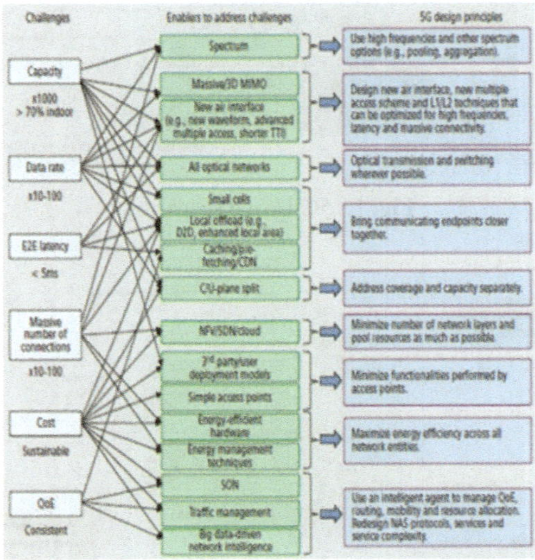

Fig. 3. 5G Challenges and design principles

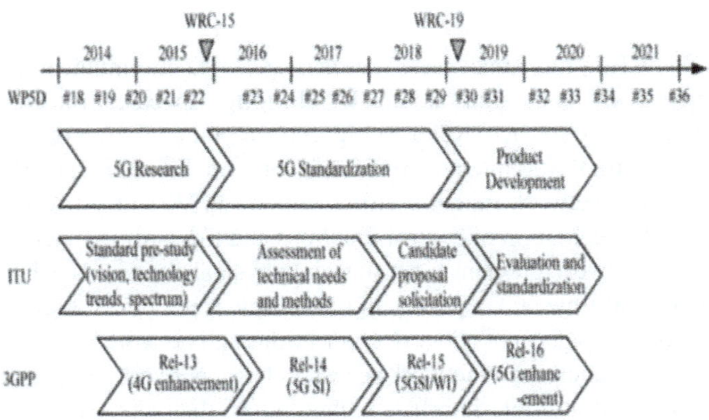

Fig. 4. 5G Work plan

5G Key Technologies

Wireless Transmission Technologies Large Scale Multi-antenna: In 2010, Bell Labs' Thomas introduced the large-scale multi-antenna concept in 5G wireless communication [10]. This technology, also termed massive multiple-input multiple-output or massive antenna system, involves a communication system where the base station's antennas significantly outnumber those of the terminals. This configuration achieves robust high-speed signal transmission by maximizing channels through numerous antennas. Moreover, it simplifies MAC layer design, ensuring low-latency transmission. Figure 6 illustrates the envisioned application scenario of massive multi-antenna in the next generation. In the 5G large-scale antenna scenario, there are two types of cells co-existing; macro cells and micro cells, and the network can be classified as homogeneous or heterogeneous. The scenarios are divided into outdoor and indoor scenarios. According to related test documents, 70% of mobile communication system comes from indoors. In this way, huge scope receiving wires channels can be partitioned into full scale cell base stations to outside clients and indoor clients, and miniature cell base stations to open air clients and indoor clients. At the same time, micro-cells are also available. As a relay base station for transmission, the channel also includes from the macro cell base station to the micro cell base station.

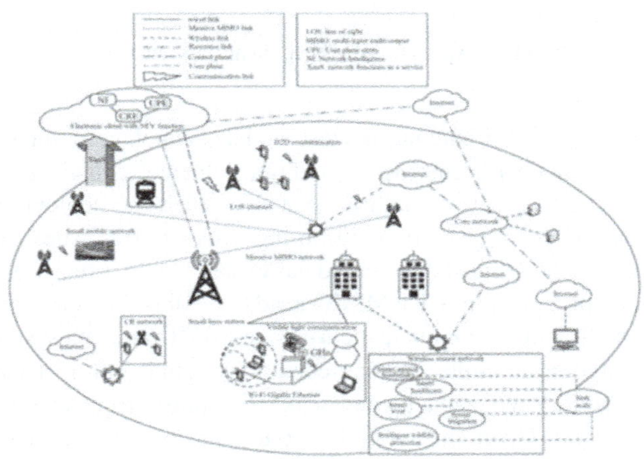

Fig. 5. 5G network architecture

As the number of base station antennas tends to infinite, the user antennas also increases. Massive MIMO technology will bring the following series of advantages in the entire 5G systems. Massive MIMO system spatial resolution is greatly compared with the conventional system. Massive MIMO is realized with deep mining without base station splitting and space resources. Beam forming technology precisely focuses beams over a small area with minimal energy, significantly reducing interference. This method is integrated with cell splitting and small cell clusters, particularly in short-distance wireless transmissions using millimeter wave frequency bands. By concentrating signal strength in specific directions and user groups, it ensures high-speed and reliable signal transmission. Contrasted with a solitary reception apparatus, gigantic MIMO innovation can improve range and energy and proficiency over various measurements (space, polarization, recurrence, time area). Because of these achievable advantages, massive MIMO technology is considered a key viable technology in 5G.

For large-scale multi-antennas, the current industry implementation focuses on TDD, and the development of FDD requires good consideration of channel estimation, information return and interference control. At the same time to make full use of technical

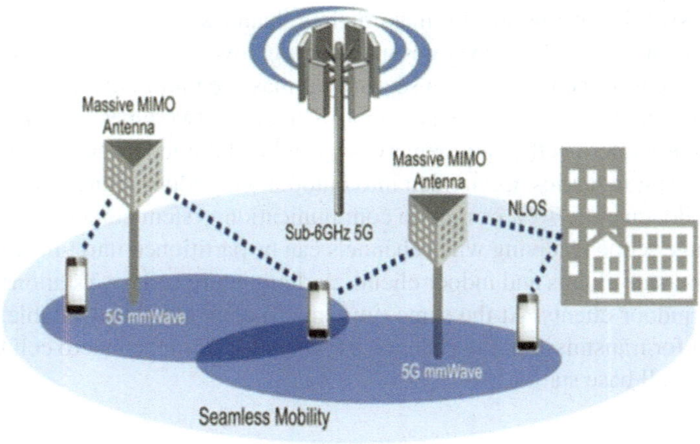

Fig. 6. Massive MIMO application scenarios

advantages, large-scale multi-antenna and test model channel measurement work is also needed in special practical scenarios. In addition, channel equalization and channel estimation had received great research interest. However, the deployment of massive MIMO technology over base station could bring a lot of cost issues, such as greatly increased overhead processing and increased hardware [11]. In actual scenarios, because the design and completion of massive MIMO need to flexibly adapt to the complex radio environment, it is difficult to complete the Massive MIMO antenna system.

In essence, large-scale multi-antenna technology stands out as a viable key solution. It enhances both system capacity and peak rate while curbing energy consumption and transmission delay. Given the escalating communication demands, especially with antennas ranging from 4 to 128, traditional multiple-input multiple-output solutions in experimental networks fall short. In 5G's massive MIMO technology, base station antennas will significantly increase, potentially forming arrays from 10×10 to 100×100 or even larger. However, the design and engineering of these extensive multi-antenna systems face challenges due to crucial technical issues.

Channel Modeling and Channel Encoding

A couple of parameters is used for Channel model characterization through the abstract description of the wireless behavior, and then accurately describe the propagation mechanism of signals in wireless communication. It is the most effective methods to evaluate wireless technology characteristics. Many research institutions or organizations have done a lot of channel measurement and modeling work, as shown in Table 1.

Polar and Low Density Parity Check (LDPC) are key for 5G channel coding. At the beginning of the discovery of LDPC codes, due to the limitations of hardware technology and the complexity of LDPC codes at that time, LDPC codes have not been paid attention. The evolution of integrated technology in recent years, the LDPC codes in actual communication systems has gradually become feasible. LDPC codes have re-entered people's vision is used in optical fiber communications, deep space communications, digital satellite video, digital watermarking, mobile communications. It is widely used in fixed wireless communication and digital subscriber line (DSL, digital subscriber line). Although the LDPC code has good anti-interference performance, the complexity of the encoding and decoding of LDPC codes has always been the biggest obstacle to the application of LDPC codes in practical communication systems. At present, research on LDPC codes is mainly focused on finding encoding and decoding algorithms with low-complexity and the application of LDPC codes in actual communication. Literature [12, 14] proposed several low-complexity decoding schemes for LDPC codes. Literature [15] studies the application of multiple LDPC codes in Rayleigh fading channels, literature [16] studies the application of multiple LDPC codes in frequency selective channels, and literature [17, 19] studies multiple LDPC codes in MIMO channels Applications. The new channel coding method i.e., polar codes is proposed by Arikan based on the phenomenon of channel polarization [20, 21]. It is the only one that has been theoretically proven. It is used in binary discrete memoryless channel (BDMC, binary Discrete memoryless channel) The capacity of the communication system can reach Shannon's limit.

Compared with the Turbo code widely used in 4G, LDPC code is decoded based on a parallel iterative algorithm of sparse matrix, and due to the parallel structure, it is easier for hardware implementation. Therefore, more advantages are associated with LDPC codes for applications in massive-capacity communication, which also cater to the development trend of 5G. The Polar code can be realized by a simple encoder and decoder, and the encoding and decoding complexity is only *(nlogn)*. Considering its excellent performance, Polar code has obvious advantages compared to Turbo code.

The 5Gnetwork requires the end-to-end delay to be reduced to less than 5 times that of the 4G network, the number of connected devices reaches 10–100 times that of the 4G network, and the data rate reaches 10–100 times that of the 4G network. Therefore, this brings severe challenges to the 5Gnetwork in terms of network delay, energy consumption, and frequency band utilization. In order to meet these high-performance requirements of 5G networks, this article hopes to choose the coding scheme that not only has good anti-interference performance, but also has high energy utilization, low system delay and high spectrum utilization, and can be applied to multiple applications. In the terminal scenario, LDPC and Polar codes, as candidate codes for 5G network channel coding, have their own advantages and disadvantages when applied to 5G networks: 1) Compared with LDPC codes, Polar codes can gradually achieve arbitrary binary symmetrical discreteness in terms of coding complexity. The channel limit of the memory less channel has low coding and deciphering multifaceted nature, and the unraveling calculation doesn't need complex iterative computations and is applied to frameworks with multi-terminal; 2) In terms of frequency band utilization, Polar codes are far inferior to multi-element LDPC In addition, the performance of Polar codes under medium and short code lengths is not as good as that of multiple LDPC codes. Since these codes have their own disadvantages and advantages when considered for 5G networks, some scholars have proposed a cascading scheme of LDPC and Polar codes [22], uses LDPC codes as outside codes and Polar codes as inside codes. Experiments prove that LDPC-Polar cascade scheme can well integrate the respective advantages of the two coding schemes at different code lengths.

Table 1. Major research institutions and channel models

Research structure or organization	Standard model	Brief introduction
COST	COST259 [12] COST273 [13] COST2100 [14]	Compared with COST 259,COST273improvesthe modeling of angle features and can be applied to more scenes and higher transmission frequency bands. COST 2100 mainly proposes the concept of CVR (cluster visibility region) to describe the changes in the number of scattering clusters experienced when the mobile platform is in different positions, and further accurately describes the channel characteristics
3GPP	TR 25.996 [15] 36.814 [16] TR 36.873 [17]	TR 25.996 gave the first space-basedmodeling (SCM, spatial channel model). This channel model is suitable for a center frequency of 2 GHz and a bandwidth of 5 MHz, including urban macro cells, urban microcells, and urban suburbs. Scene, and any antenna array can be configured at the same time. TR 36.814 adds the modeling of relay, pico cell and femto cell scenarios. Then, TR 36.873 gave 3D MIMO channel modeling, focusing on modeling the elevation angle
WINNER	SCME [18] WINNER I/II/ + [19]	SCME,WINNERI/II, and WINNER are expanded based on SCM, support bandwidth100MHz,carrier

(*continued*)

Table 1. *(continued)*

Research structure or organization	Standard model	Brief introduction
ITU	M.2135 [20]	In order achievements of 3GPP,IST-WINNERandother research institutions, M.2135 is proposed, which is suitable for bandwidth 20–100 MHz and center frequency 450 MHz–5 GHz
METIS	D1.2 [21]	Based on WINNER II/extended, of 5G channel modeling have been improved. The main aspects include spectrum range, spatial continuity, 3D, spherical waves, etc. The applicable center frequency is 380 MHz–86 GHz

Full Duplex

Self-interference antenna systems mainly rely on increasing the loss between transmitting and receiving antennas, including separation of transmitting and receiving signal, isolation of transmitting and receiving antennas, antenna cross-polarization, antenna nulling method, etc.; analog circuit automatically includes the circular isolation and interference cancellation, reconstructing the self-interference signal through analog circuit design and directly subtracting the reconstructed self-interference signal from the received signal, etc.; The digital domain self-interference cancellation method primarily depends on parameter estimation and reconstruction. Once reconstructed, the self-interference is subtracted from the received signal, effectively eliminating any residual self-interference; the principle of the full-duplex terminal self-interference elimination method is shown in Fig. 7.

Altogether, for full-duplex innovation to be broadly utilized in future remote organizations, there is still a lot of work to be accomplished for full-duplex examination. It isn't simply important to persistently and profoundly study the self-obstruction disposal issue of full-duplex innovation, yet additionally consider more fully the opportunities and challenges faced by full-duplex technology, including designing low-power, low-cost, and miniaturized antennas to eliminate self-impedance; unraveling the physical layer encoding, modulation, power distribution, and beam forming of full-duplex systems shape, channel evaluation, modulation, decoding, and so forth; design media access layer and higher-level protocols, determine interference coordination strategies, network resource management, and full-duplex frame structure in full-duplex systems; full-duplex technology and large-scale multiple antennas effective blend of technology and system performance analysis, and so on.

Wireless Access Technology for Multi-access

Multiple access technology is pivotal in modern mobile communication systems and has been a key aspect across various generations of mobile technologies. In addition to supporting traditional OFDMA, 5G introduces several new multiple access technologies like SCMA, NOMA, PDMA, and MUSA. These innovations not only increase user connections through multi-user superimposed transmission but also enhance the system's spectrum efficiency significantly. Through free scheduling and contention access, they notably reduce latency.

Non-orthogonal multiple access (NOMA) employs power domain multiplexing, offering higher spectrum efficiency at the cost of increased receiver complexity. Future advancements in computing power make this approach viable. NOMA relies on continuous interference cancellation technology at the receiver end for multi-user detection and power domain multiplexing at the transmitter end based on specific algorithms. However, challenges exist, including intricate non-orthogonal transmission receivers and ongoing research on power domain multiplexing technology.

Sparse Code Multiple Access (SCMA) utilizes code domain multiplexing, integrating QAM modulation and signature transmission. It maps input bit streams to specific codes from a codebook, spreading multi-dimensional SCMA codewords across sparse physical resource elements. SCMA achieves an overload rate of up to 300% due to the excess number of codewords compared to resource elements, and research focuses on optimal codebook design, low-complexity receiver algorithms, efficiency studies, and integration with other wireless technologies.

Pattern Division Multiple Access (PDMA) is a non-orthogonal multiple access technology based on joint transmitter and receiver design It encodes and transmits multiple user signals in power, space, and coding domains, distinguishing them by characteristic patterns for interference cancellation. At the receiving end, a low-complexity, high-performance serial interference cancellation receiver algorithm detects multiple users, significantly boosting system capacity and downlink spectrum efficiency. Implementing PDMA requires solving challenges related to pattern design, receiver simplification, and integrating PDMA with MIMO designs for spatial domain coding patterns. Multi-user Shared Access (MUSA) is an uplink non-orthogonal multiple access technology based on multiple codes in the complex domain. It is suitable for scheduling-free multi-user shared access solutions and is conducive to low-cost, low-power 5G massive connections (Internet of Everything). Each access user uses the SIC receiver-based, complex-domain multiplex code sequence with low correlation to expand its modulation symbols, and the expanded symbols can be sent in the same time-frequency resources. The receiving side uses linear processing and block-level SIC to separate the information of each user. The extended sequence will directly affect the performance and receiver complexity of MUSA, and is a key part of MUSA. MUSA is a potential multi-access technology in 5G, but its implementation is still challenging. For example, the mapping method of the propagation sequence, the standard of pattern selection, and how the system capacity changes as the number of users increases.

The following is a comparison of the characteristics of the above four multiple access technologies, as shown in Table 2. In addition to the above multiple access technologies, for 5G communications, researchers have also proposed Bit Division Multiple Access

(BMA), Software Defined Multiple Access (SoDeMA) and so on. In short, the reform and innovation of multiple access technologies in the 5G era will enable future wireless access technologies for mobile communications to reach a new level.

Dynamic TDD

The 5G networks key feature deployment of small cells (cell radius less than a few meters) and different requirements from gigabit speeds (i.e., ultra-low latency). The air interface based on TDD is proposed for the deployment of small cell signal propagation with small delay, and flexible allocation of uplink and downlink transmission resources in each subframe. This kind of TDD with flexible choice of downlink and uplink configuration is called dynamic TDD. In case of adaptive TDD downlink and uplink configuration, distributed cells can adapt to service requirements more flexibly, which also has a certain effect on reducing base station energy consumption. Adaptive TDD technology is mainly used in node cells with low-power with small coverage, while dynamic TDD technology is generally not used in macro base station cells with large coverage. Ultra-dense small cell networking and massive applications become the basic content of 5G wireless communication systems. The deployment of adaptive TDD may cause downlink and uplink subframe interleaving interference and reduce performance of the system. The major challenges of dynamic 5G TDD contain faster UL/DL, shorter TTI switching, and the combination of MIMO. In order to cope with these challenges, four solutions are currently being considered: small cluster interference mitigation (CCIM), power control, FeICIC/eICIC, and the use of MIMO technology.

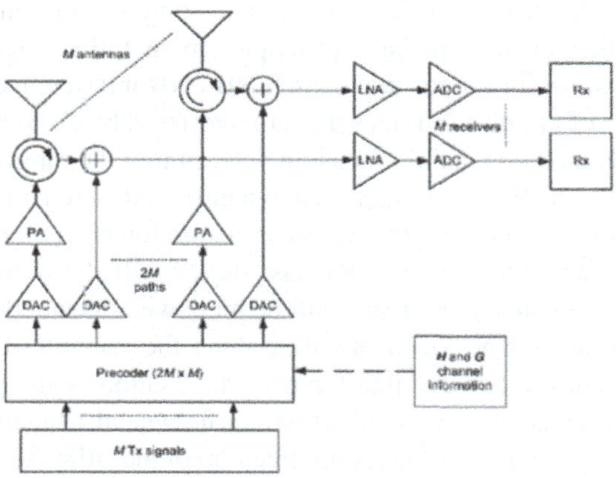

Fig. 7. C-RAN architecture

CCIM is a method to classify small clusters according to a certain threshold's in between cells. Only one or more cells is contained in each cluster. All cells in each cluster actively transmit any subframe or a subset of all subframes, either all uplinks or downlinks, so that the interference between base stations in the same cluster and user-user interaction interference between them is alleviated. Coordination between multiple cells belonging to the same cluster is necessary. The transmission directions of cells belonging to different clusters can be different in a subframe. By freely selecting different TDD configurations, the benefits of the reconfiguration of TDD uplink and downlink based on service adaptation can be obtained. CCIM essentially includes two functions, namely the formation of cell clusters and cooperative transmission in each cell cluster. Literature In order to form a cell cluster reasonably, base station measurement is necessary, and the purpose of base station measurement is to evaluate the interference level from another base station. In addition, the signals and procedures related to base station measurements must be supported. Concerning the development states of coordinated effort inside the cell group, further examination is required. eICIC relies on almost blank subframes (ABS) to coordinate the inter-layer interference between the macro cell and the small cell. However, the eICIC solution does not solve the interference control on the cell-specific reference signal (CRS). To ensure backward compatibility, the CSR cannot be a blank subframe. FeICIC takes CRS interference into account and uses reduced power almost blank subframes (RP-ABS) to increase system capacity. Learn from the idea of inter-cell interference coordination (ICIC) and enhanced inter-cell interference coordination (eICIC) that orthogonalize resource allocation in the time or frequency domain to solve the interference in adjacent cells. These ICIC-based solutions may cause unnecessary waste of resources in interference suppression. The main challenge for FeICIC is the intelligent scheduling and coordination between the macro cell and the small cell and how to reduce power. eICIC and FeICIC were originally designed to solve the downlink interference problem in heterogeneous networks. In this project, interference coordination based on interference suppression (IM) methods will be used in interference suppression between small cells.

In a dynamic TDD system, the performance of the uplink will decrease significantly. In order to improve the performance of the uplink, some power control schemes have been proposed in the industry. The basic principles are as follows: reduce the transmission power of downlink subframes that cause eNB-eNB interference, and increase the transmission power of uplink subframes that are interfered by eNB-eNB. Current interference suppression methods mainly focus on determining the range of power changes and control strategies, such as some static and dynamic control schemes. However, in the method based on power control, increasing the transmission power may cause additional interference, and reducing the power of the base station will also reduce the coverage of the cell. In the above methods, the focus will be on power enhancement and reduction. Although there are not many documents describing the method of combining large-scale multiple antennas to achieve interference cancellation, some potential ideas are still worth considering, such as the method of interference alignment. The interference alignment technology maximizes the spatial freedom of the MIMO system, thereby significantly improving the system capacity. The core idea of the interference alignment method is to adjust the interference from different sources to a constrained signal subspace, then the target signal can be received in the null space of this subspace. Create a joint interference alignment beamforming problem, then a single beamforming solution can handle beamforming problems in several different interference scenarios.

In order to be able to apply D2D communication technology well, people need to focus on solving the potential technical difficulties of D2D communication. First, the D2D discovery technology needs to detect and identify neighboring D2D end users, and then establish a D2D communication link. Since the D2D communication technology in the cellular network will inevitably bring additional interference to cellular communication, efficient wireless resource allocation and interference management schemes are essential, through efficient scheduling and management of wireless resources and control of the transmission power of D2D users, etc. Method to reduce the interference caused by D2D communication to the cell. Finally, communication mode switching is also one of the research points that people pay special attention to, because it will determine whether the spectrum efficiency of the system can be improved, and affect the degree of interference between cellular users and D2D users. Now people have considered factors such as interference between D2D users, path loss, channel quality, and distance, and formulated user communication mode switching criteria. Because D2D communication technology has the preferences of improving organization execution and advancing future organization design, it has pulled in wide consideration from analysts, what's more, has completed exploration on some central questions and accomplished a few outcomes. However, for the research of D2D communication technology, there are still a few issues and difficulties that have not been explained. For instance, regarding correspondence mode exchanging, most archives don't consider the client's versatility, yet by and by, the client is in a portable state, which wills greater affect correspondence mode exchanging. As far as asset assignment and obstruction the executives, individuals watch out for multiplex the uplink or downlink on the D2D correspondence connect, rather than considering the dynamic choice of the D2D correspondence interface dependent on the phone downlink and uplink conditions. What kind of cellular communication link is multiplexed, in addition, further design is needed for potential network scenarios based on D2D communication technology, and new network scenarios will introduce new resource allocation problems and interference problems. Therefore, new the resource allocation and interference management schemes of China are also worthy of further study.

3 Conclusion

This paper presents and sums up the advancement history of 5G and the examination progress at home and abroad in detail, and analyzes the 5G cellular network architecture based on virtualization. From the three viewpoints of wireless transmission, wireless access, and network, it thoroughly presents the likely key innovations and most recent advancements of 5G, and examines the advantages and disadvantages of some key advances and the directions for future research. At long last, the key undertakings and fundamental directions of future development of 5G are prospected. The future 5G network will be more open, smart, and flexible. To fulfill the development and need of mobile communication, 5G is expected to become the main mobile communication system in 2020. Compared with the widely popular 4G mobile communications, 5G has clear advantages in resource use, transmission rate, and spectrum usage. It will also gain significant advantages in terms of user experience, transmission delay, and wireless network coverage. All in all, 5G researches are currently at a critical phase of development, and the research on various technologies is advancing rapidly. In the future, there is still a lot of work to be done urgently for normalization and technical assessment around user needs, scale, and cost and energy consumption.

4 Future Outlook

As of now, all through the world a great deal of interest in technical feasibility studies, normalization, and item improvement for future 5G portable communication networks. The advancement of 5G should be coordinated on a worldwide scale under basic structure. Be that as it may, in the 5G communication framework, continuous research and development, such as massive MIMO and channel modeling not only effectively improve the efficiency of wireless spectrum usage, but also accelerate the wireless data transmission rate and support the access of more terminals. The support of 5G far surpasses the scenarios, data volume, and device access that 3G and 4G networks meet. The realization of this network requires continuous technological development and innovation. Likewise, 5G will also give more attention to client experience in all perspectives, and will be sent astutely as per the individual necessities of various clients, so clients can get to it conveniently and quickly at any time and anywhere. Simultaneously, the future of 5G innovation lies not only in the further improvement of data transmission speed, but also in that it is an extension of human capabilities. All holistic items are in an ongoing arranged state and can detect and interact with one another.

References

1. Agyapong, P., Iwamura, M., Staehle, D., et al.: Design considerations for a 5G network architecture. IEEE Commun. Mag. **52**(11), 65–75 (2014)
2. Tullberg, H., Popovski, P., Gozalvez-Serrano, D., et al.: METIS system concept: the shape of 5G to come. IEEE Commun. Mag. (2015)
3. Koumaras, H., et al.: 5GENESIS: the genesis of a flexible 5G facility. In: 2018 IEEE 23rd International Workshop on Computer Aided Modeling and Design of Communication Links and Networks (CAMAD). IEEE (2018)
4. Gupta, A., Jha, R.K.: A survey of 5G network: architecture and emerging technologies. IEEE Access **3**, 1206–1232 (2015)
5. Marzetta, T.L.: Non-cooperative cellular wireless with unlimited numbers of base station antennas. IEEE Trans. Wireless Commun. **9**(11), 3590–3600 (2010)
6. Zeng, Y., Zhang, R., Chen, Z.N.: Electromagnetic lens-focusing antenna enabled massive MIMO: performance improvement and cost reduction. In: IEEE/CIC International Conference on Communication in China, pp. 454–459 (2014)
7. Bai, T.Y., Desai, V., Robert, W., et al.: Millimeter wave cellular channel models for system evaluation. In: IEEE International Conference on Computing, Networking and Communications (ICNC), pp. 178–182 (2014)
8. Fossorier, M.C., Miodrag, M., et al.: Reduced complexity iterative decoding of low-density parity check codes based on belief propagation. IEEE Trans. Commun. **47**(5), 673–680 (1999)
9. Yazdani, M.R., Saied, H., et al.: Improving belief propagation on graphs with cycles. IEEE Commun. Lett. **8**(1), 57–59 (2004)
10. Apat, S.K., Mishra, J., Srujan Raju, K., Padhy, N.: State of the art of ensemble learning approach for crop prediction. In: Kumar, R., Pattnaik, P.K., R. S. Tavares, J.M. (eds.) Next Generation of Internet of Things. Lecture Notes in Networks and Systems, vol. 445. Springer, Singapore (2023). https://doi.org/10.1007/978-981-19-1412-6_58
11. Raju, V.R., Reddy, D.A., Narsimha, D., Srinivas, K., Rani, B.K.: Adaptive closed-loop deep brain stimulator coding techniques for target detections in Parkinson's. IETE J. Res. **69**(8), 4945–4960 (2023). https://doi.org/10.1080/03772063.2021.1962742

12. Fayaz, R., et al.: An intelligent Harris hawks optimization (IHHO) based pivotal decision tree (PDT) machine learning model for diabetes prediction. Int. J. Intell. Syst. Appl. Eng. **10**(4), 415–423 (2022)
13. Priya Remamany, K., et al.: A localized bloom filter-based CP-ABE in smart healthcare. Appl. Sci. **12**(24), 12720 (2022). https://doi.org/10.3390/app122412720
14. Rani, B.K., Rao, M.V., Patra, R.K., et al.: Vehicle type classification using graph ant colony optimizer based stack autoencoder model. Multimed Tools Appl **81**, 42163–42182 (2022). https://doi.org/10.1007/s11042-021-11508-5
15. Munawar, S., Geetha, A., Srinivas, K.: Squirrel search-based optimal feature extraction with Bi-LSTM for the arrhythmia classification using ECG. J. Theor. Appl. Inf. Technol. **100** (2022)
16. Pothalaiah, S., Lakshmaiah, D., Doss, B., Sairam, N., Srikanth, K.: Design of CMOS base band analog. Cogn. Comput. Models Commun. Syst. **123** (2022)
17. Lakshmaiah, D., et al.: A novel low-power frequency-modulated continuous wave radar based on low-noise mixer. Cogn. Comput. Models Commun. Syst. **165** (2022)
18. Hemanand, D., Vinoda Reddy, G., Sathees Babu, S., Balmuri, K.R., Chitra, T., Gopalakrishnan, S.: An intelligent intrusion detection and classification system using CSGO-LSVM model for wireless sensor networks (WSNs). Int. J. Intell. Syst. Appl. Eng. **10**(3), 285–293 (2022)
19. Natesan, G., Konda, S., de Prado, R.P., Wozniak, M.: A hybrid mayfly-aquila optimization algorithm based energy-efficient clustering routing protocol for wireless sensor networks. Sensors **22**(17), 6405 (2022). https://doi.org/10.3390/s22176405
20. Peng, R., Chen, R.R.: Design of non-binary LDPC codes over GF(q) for multiple-antenna transmission. In: IEEE Military Communications Conference (MILCOM 2006), pp. 1–7 (2006)
21. Arikan, E.: Channel polarization: a method for constructing capacity-achieving codes for symmetric binary-input memoryless channels. IEEE Trans. Inf. Theory **55**(7), 3051–3073 (2009)

Development and Construction of System-On-Chip Implementing Discrete Cosine Transform for Image Processing Applications

Sunil Kumar Singh[1], Mannem. Saimanasa[1], B. Premalatha[2], T. D. Bhatt[2], M. Kishore Kumar[3], and Voruganti Naresh Kumar[3](✉)

[1] CMR Engineering College, Hyderabad, Telangana, India
kumarsunil@cmrec.ac.in
[2] CMR College of Engineering & Technology, Hyderabad, Telangana, India
{bpremalatha,dr.td_bhatt}@cmrcet.ac.in
[3] CMR Technical Campus, Hyderabad, Telangana, India
nareshkumar99890@gmail.com

Abstract. The Discrete Cosine Transform (DCT) is a crucial component in numerous image processing applications. The execution of the Discrete Cosine Transform (DCT) is frequently the arithmetic unit that consumes the most time in the field of image processing. The overall efficiency of real-time image applications is affected by the substantial number of Discrete Cosine Transform (DCT) computations involved in image compression. The modification of images using Discrete Cosine Transform (DCT) is frequently challenging to fulfill the demands of real-time processing as a result of the substantial computational burden and software implementation involved. While the Digital Signal Processor (DSP) is indeed robust in its capabilities for digital signal processing, it presents challenges in terms of integrating external peripherals because to its complicated and diverse design, as well as its excessive use of power. On the other hand, the utilization of field programmable gate array (FPGA) is a viable approach to address these limitations, hence facilitating a more streamlined and compact design of the discrete cosine transform (DCT) module, as well as enabling seamless system integration. The majority of image Discrete Cosine Transform (DCT) transformations are frequently carried out on hardware platforms utilizing Field-Programmable Gate Arrays (FPGAs), which may not be considered a cost-effective solution. This study focuses on the implementation of the image Discrete Cosine Transform (DCT) transformation within a System on Chip (SoC) architecture. The utilization of SoC offers several advantages, including enhanced adaptability, simplified peripheral integration, reduced power consumption, and improved communication bandwidth across numerous processors. This functionality becomes beneficial for integration inside a real-time image processing system.

Keywords: Image compression · DCT · Coding Redundancy · FPGA · System on Chip

V. Sakalauskas et al. (Eds.): IBICA 2023, LNNS 1231, pp. 474–482, 2025.
https://doi.org/10.1007/978-3-031-78946-5_47

1 Introduction

Compression is the process of transforming visual data into a more compact format, aiming to minimize the overall number of bits required for its representation. The fundamental concept behind all compression approaches is to express commonly occurring components using shorter codes and less frequently occurring elements using longer codes [1–3]. Consequently, the data block necessitates a reduced memory capacity compared to the scenario when all parts were encoded with identical lengths.

The capacity to efficiently and effectively compress photos is widely recognized as a highly significant attribute of picture storage [4, 5]. In the absence of compression, it can be observed that an image possessing a resolution of 1024 pixels by 1024 pixels and an 8-bit color depth would necessitate a storage capacity of 3 gigabytes. Furthermore, the transmission of such image across a high-speed ISDN connection, operating at a rate of 64 kilobits per second, would require around 2.35 min. When the image is compressed with a compression ratio of 10:1, the storage space needed is reduced to 300 kilobytes, and the transmission time is reduced to less than two seconds. The presence of large image files remains a prominent obstacle within distributed systems in many situations. Compression is a fundamental component of available solutions for reducing the size of dimensions in a manner that is both manageable and capable of transmission. It is well recognized as one of the most commonly employed types of compression. One further option is to enhance the bandwidth; however, the cost associated with this approach often renders it less desirable.

The feature of interconnectedness among neighboring pixels, leading to the presence of redundant information, is a characteristic commonly observed in the majority of photos. The primary goal is to identify an image representation with a lower degree of association. The problem of representing a digital image with a reduced amount of data is addressed by picture compression, which aims to decrease the data required for image representation [6–9]. The deletion of redundant data acts as the primary mechanism upon which the process of reduction is based. From a mathematical perspective, this process involves transforming a two-dimensional array of pixels into a statistically uncorrelated dataset. The modification is performed prior to the image being stored or sent. Subsequently, the compressed image undergoes a process of decompression, with the aim of reconstructing either the exact original image or an image that closely approximates the original [10].

2 Literature Survey

In their study, Singh et al. [11] put forth a method for image compression and decompression on FPGA utilizing the Discrete Cosine Transform (DCT) and Inverse Discrete Cosine Transform (IDCT). The individual conducted an analysis and devised various methodologies for the Discrete Cosine Transform (DCT). Subsequently, they proceeded to develop a one-dimensional, eight-word DCT core that may be employed in numerous compression techniques. The methodology employed in this study involves utilizing input data in the form of 8-bit words. The processing of these 8-bit words necessitates a total of 512-clock cycles.

In their study [12] put out a proposal for the FPGA implementation of the 2D transform in future video coding. The FVC 2D transform hardware unit carries out the DCT-2, DCT-5, DCT-8, DST-1, and DST-7 operations on 4*4 and 8*8 transform units. In order to perform 1D transforms, it is necessary to have two reconfigurable data routes and incorporate multiplications by constants. In the event of the most unfavorable scenario, this system has the capability to efficiently handle ultra-high-definition video frames per second while exhibiting reduced energy consumption.

In their study [13] presented a novel FPGA-based architecture for future video coding, specifically targeting the adaptive multiple core transform. This approach incorporates the utilization of adaptive multiple core transform for enhancement. Specifically, it employs five varieties of DS/DCT, including DCT-2, DCT-5, DST-1, and DST-7, together with transform unit sizes of 64*64. The utilization of this approach leads to a rise in the computational complexity of both the encoder and decoder. The successful implementation of a deeply pipelined high-performance architecture necessitates the development of an initial architecture for the FVC adaptive multiple core transform. This architecture should be capable of supporting various block sizes, including 4*4, 8*8, 16*16, and 32*32. The VHDL architecture was created for various FPGA chips in order to efficiently process high frame rates for varying transform sizes.

The Efficient design of variable size High Efficiency Video Coding (HEVC) 2D-DCT for Field-Programmable Gate Array (FPGA) platforms was proposed by Chen et al. [14]. The suggested design supports several block sizes, including 4×4, 8×8, 16×16, and 32×32. The developed architecture utilizes a four-stage Discrete Cosine Transform (DCT) and employs System Verilog and FPGA technologies. In comparison to existing approaches, the suggested system demonstrates improved performance and decreased costs. Notably, the system offers the advantage of achieving ultra-high quality (4K@30fps) in television displays, while concurrently reducing hardware expenses.

3 Proposed Method

Discrete Transform Cosine

The technique under consideration is extensively employed in the field of image compression. This transformation exhibits a favorable attribute of energy compaction. The dependency of the Discrete Cosine Transform (DCT) lies solely on the chosen order of transformation, rather than being influenced by the statistical characteristics of the input data. The utilization of the Discrete Cosine Transform (DCT) has witnessed a growing trend in its application for image and video coding, specifically in widely adopted techniques like the JPEG and MPEG standards. The Discrete Cosine Transform (DCT) is utilized to partition the image into sub-bands by considering the block size, which is determined according to its significance in relation to the spatial quality of the image. The Discrete Cosine Transform (DCT) shares similarities with the Discrete Fourier Transform (DFT) in that it converts an image or signal from the spatial domain to the frequency domain. The discrete cosine transform (DCT) possesses a primary benefit over the discrete Fourier transform (DFT) in that it exclusively employs real multiplications. This characteristic results in a reduction in the overall quantity of necessary multiplications, which is not observed in the DFT.

This study presents a novel design for a low-complexity Time Sharing Medium Access Control (MAC) system based on the Discrete Cosine Transform (DCT). The determination of the cost of a transformation matrix is often based on the quantity of arithmetic operations necessary for its computation. Let us consider a sequence R (n) consisting of N points. The N-point Discrete Cosine Transform (DCT) is formally defined.

Entropy coding is the concluding stage of discrete cosine transform (DCT)-based coding. The next phase in the process results in further lossless compression by encoding quantified DCT coefficients in a more condensed manner, taking into account the statistical properties of the coefficients. The JPEG proposal outlined two distinct approaches for entropy encoding:

3.1 Block Diagram

Figure 1 illustrates the process of Discrete Cosine Transform (DCT) image compression. Interpixel redundancy arises from the inherent association between adjacent pixels within an image. This observation suggests that the adjacent pixels exhibit a lack of statistical independence.

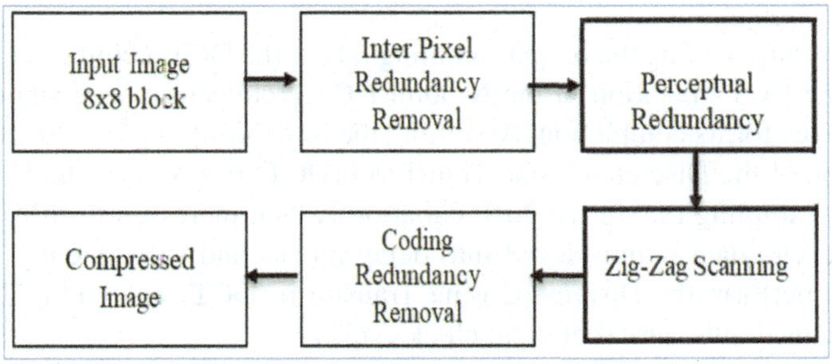

Fig. 1. DCT image compression process

Interpixel redundancy arises from the inherent connection among adjacent pixels in an image. This observation suggests that the adjacent pixels exhibit a lack of statistical independence. The probabilities of the grey levels are not uniformly distributed. The prediction of a pixel's value can be achieved by leveraging the strong interconnections between the values of neighboring pixels.

The display of information is intricately linked to the presence of coding redundancy. The data is encoded in a manner that facilitates comprehension by computational systems. According to the assertion, a picture exhibits coding redundancy when the coding scheme employed to represent the grey levels of the image utilizes an excessive number of code symbols to encode each grey level, surpassing the minimum requirement for representation.

The primary factor influencing the intricacy of the DCT architecture is the presence of several multipliers. To overcome these challenges, a proposed architecture is introduced that utilizes a singular multiplier and adder for the operation of each X(0) to X(7). This

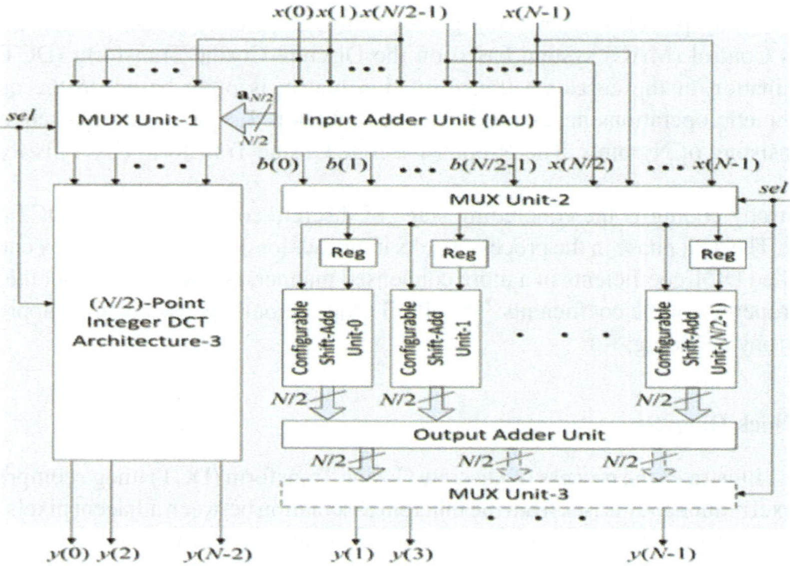

Fig. 2. Time sharing MAC based DCT architecture

is achieved by augmenting the output sampling rate of the DCT architecture, as depicted in Fig. 2. The DCT operation in the N-point DCT architecture necessitates a total of 2N clock cycles for its completion. Assuming the input sampling rate for accessing the X(0) equation of the Discrete Cosine Transform (DCT) is 1 Mega samples per second (MSPS), the sampling rate of the X(0) equation is then increased to 8 MSPS. During each clock cycle, data is introduced into delay blocks and subsequently fed into the multiplier to perform the Discrete Cosine Transform (DCT) operation. The resulting output is obtained after a total of eight clock cycles.

3.2 FPGA – SoC Implementation

Embedded applications are a rapidly growing domain within the discipline. Nevertheless, these entities are progressively encountering formidable functional limitations, such as the capacity for complex computations and streamlined communication. In contrast, advancements in technology have facilitated the capacity to include hundreds of millions of transistors onto a solitary silicon chip. The advancement of technology has facilitated more efficient resource allocation by enabling the grouping of resources within a minimal circuit of the entire system. The architectural framework that is linked to a streamlined design process effectively harnesses the capabilities of technology to enhance performance and capacity. This capability enables the optimization of parallelism inside the application and the customization of the architecture. Parallelism refers to the utilization of several processors or calculation components, commonly known as multiprocessors. Locality, on the other hand, pertains to the minimization of circuits, sometimes even limited to a single circuit. Specialization involves dedicating these processors or components to a specific application or a group of applications, such as image processing for measurement purposes. Regarding the design flow, effectively managing a substantial quantity of resources entails elevating the degree of abstraction and presenting a systematic approach for transitioning from this level to an optimal implementation.

The term "system on a chip" (SoC), sometimes known as a single chip system, pertains to the consolidation of multiple intricate and diverse components into a solitary silicon circuit or chip, resulting in a fully integrated system. This word has gained significant traction within the industrial sector. Despite the absence of a universally accepted or widely recognized definition. There is a viewpoint held by certain individuals that posits the notion that any circuit possessing a high level of intricacy can be classified as a System-on-a-Chip (SoC). However, adopting such a perspective would likely result in the inclusion of virtually every circuit already in existence.

A more suitable characterization of a single chip system can be provided as follows: it refers to a comprehensive system that is integrated onto a solitary silicon substrate, wherein numerous intricate functions coexist, including processors, Digital Signal Processors (DSPs), various types of memory buses or communication networks, converters, analog blocks, and other relevant components. The single-chip system is required to incorporate a software processing unit, commonly referred to as a CPU, and its functionality should be independent of external components, with the exception of any essential ones that may be necessary for its operation. The system in question should possess quasi-autonomous capabilities, allowing it to operate independently from external components and solely rely on the chip's internal functionalities. Consequently, the implementation of an on-board system necessitates the integration of both hardware and software components.

4 Results and Discussion

Figure 4 illustrates the exemplar input photos utilized for the compression process. Initially, the image is transformed into a binary format, after which the compression procedure commences.

(a) (b)

Fig. 4. Examples of input images

Upon the conversion of the input image, it is transformed into binary bits. In this process, each individual pixel will be transformed into binary representation and subsequently subjected to analysis using the Discrete Cosine Transform (DCT) algorithm. This analysis will determine whether pixels are suitable for compression.

Figure 5 illustrates the outcome of compressing the input image into binary bits. The Discrete Cosine Transform (DCT) initiates the compression process by considering the Region of Interest (RoI). In cases where the RoI exhibits a high degree of significance, the compression level is reduced. Conversely, when the RoI is less significant, the compression level is increased.

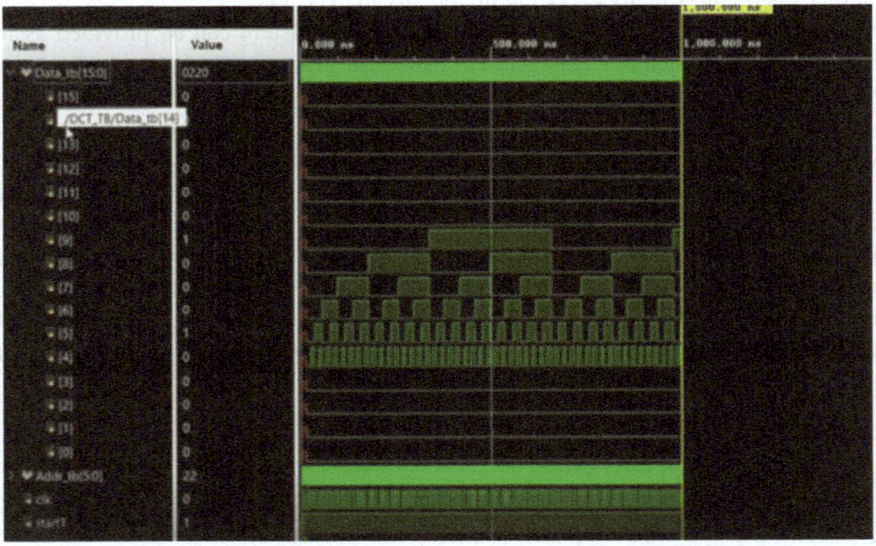

Fig. 5. Binary bit representation input image after compression

Table 1. The suggested strategy is being evaluated to the current technique using an input image. (a)

Compression method	Quality	CR	PSNR	MSE
JPEG format	41%	21.22	30.6	44.82
	82%	10.00	33.5	23.54
Proposed Model	Block Size	CR	PSNR	MSE
	16	21.10	29.88	78.45
	8	10.60	34.10	39.20

Table 1 presents an elucidation of the extent to which the hibiscus image is reduced at various stages. The explanation of the contrast was provided with respect to the compression factor, peak signal-to-noise ratio, and mean square error.

Table 2. The new strategy was assessed to the current technique using an input image. (b)

Compression method	Quality	CR	PSNR	MSE
JPEG format	41%	21.22	30.6	44.82
	82%	10.00	33.5	23.54
Proposed Model	Block Size	CR	PSNR	MSE
	16	21.10	29.88	78.45
	8	10.60	34.10	39.20

Table 2 provides an explanation of the extent to which the poppy image is compressed at various phases. The comparison was elucidated with respect to Compression Ratio, Peak Signal to Noise Ratio, and Mean Square Error.

5 Conclusion

This study examines the research and application of the Field-Programmable Gate Array (FPGA)-based Discrete Cosine Transform (DCT) algorithm. Beginning with an examination of the inherent attributes of the algorithm, it is initially seen that the algorithm plays a crucial part in the investigation of image processing applications. The research on the algorithm has yielded positive outcomes, which will significantly contribute to the advancement of image processing applications in research. In order to investigate this subject matter, the initial step entails doing an analysis of the algorithm, which encompasses several concepts and procedural stages in its execution. Software implementation refers to the process of implementing a high-level programming language without taking into account the specific low-level data flow and fundamental execution and control mechanisms. This involves examining the algorithm and identifying the most fundamental operations, such as addition, logical AND, circular shift, and so on. The implementation of these fundamental processes can be achieved by utilizing logic resources. It is crucial to analyze the input data stream efficiently, and the analysis algorithm exhibits a particular level of parallelism that aligns well with the FPGA's properties. Field-Programmable Gate Array (FPGA) refers to a configurable integrated circuit that combines programmable logic gates. This device enables the implementation of crucial algorithms and algorithmic stages through on-site programming. By allocating logic resources, the FPGA enhances computational speed while simultaneously reducing computational complexity.

References

1. Ahmed, N., Natarajan, T., Rao, K.R.: Discrete cosine transform. IEEE Trans. Comput. **C-23**(1), 90–93 (1974)
2. Rao, K.R., Yip, P.: Discrete Cosine Transform: Algorithms, Advantages, Applications. Academic, San Diego (1990)

3. Britanak, V., Yip, P., Rao, K.R.: Discrete Cosine and Sine Transforms. Academic, New York (2007)
4. Magli, E., Taubman, D.: Image compression practices and standards for geospatial information systems. In: Proceedings of IEEE International Geoscience and Remote Sensing Symposium (IGARSS), vol. 1, pp. 654–656 (2003)
5. Clarke, R.J.: Relation between the Karhunen-Loève and cosine transforms. IEE Proc. F Commun. Radar Signal Process. **128**(6), 359–360 (1981)
6. Madanayake, H.L.P.A., Cintra, R.J., Onen, D., Dimitrov, V.S., Bruton, L.T.: Algebraic integer based 8 8 2-D DCT architecture for digital video processing. In: Proceedings of ISCAS, pp. 1247–1250 (2011)
7. Madanayake, A., et al.: A row-parallel 2-D DCT architecture using algebraic integer-based exact computation. IEEE Trans. Circuits Syst. Video Technol. **22**(6), 915–929 (2012)
8. Potluri, U.S., Madanayake, A., Cintra, R.J.: Improved 8-Point approximate DCT for image and video compression requiring only 14 additions. IEEE Trans. Circ. Syst. I Reg. Pap. **61**(6) (2014)
9. Bouguezel, S., Ahmad, M.O., Swamy, M.N.S.: A multiplication free transform for image compression. In: the 2nd International Conference on Signals, Circuits and Systems, pp. 1–4 (2008)
10. Bouguezel, S., Ahmad, M.O., Swamy, M.N.S.: Low-complexity 8x8 transform for image compression. Electron. Lett. **44**, 1249–1250 (2008). https://doi.org/10.1049/el:20082239
11. Singh, K.K., Pandey, D.: Implementation of DCT and IDCT based image compression and decompression on FPGA. In: 2017 International Conference on Inventive Systems and Control (ICISC), pp. 1–4. IEEE (2017)
12. Mert, A.C., Kalali, E., Hamzaoglu, I.: An FPGA implementation of future video coding 2D transform. In: 2017 IEEE 7th International Conference on Consumer Electronics-Berlin (ICCE-Berlin), pp. 31–36. IEEE (2017)
13. Garrido, M.J., Pescador, F., Chavarrías, M., Lobo, P.J., Sanz, C.: A high-performance FPGA-based architecture for the future video coding adaptive multiple core transform. IEEE Trans. Consum. Electron. **64**(1), 53–60 (2018)
14. Chen, M., Zhang, Y., Chao, L.: Efficient architecture of variable size HEVC 2D-DCT for FPGA platforms. AEU-Int. J. Electron. Commun. **73**, 1–8 (2017)

Monitoring of Greenhouse for Improved Medicinal Plants on Pharmacological Demand via IoT

Shiva Sai[1], Anima Nanda[2](✉), B. K. Nayak[3], and Aakanshya Samantaray[2]

[1] Department of Electronics and Communication Engineering, Puducherry Technological University, Puducherry 605014, India

[2] Department of Biomedical Engineering, Sathyabama Institute of Science and Technology, Deemed to be University, Chennai 600119, India
dranimananda@gmail.com

[3] Department of Botany, Kanchi Mamunivar Government Institute for Postgraduate Studies and Research (Autonomous), Pondicherry 605008, India

Abstract. The development of greenhouse control systems is a result of the interaction of several environmental elements that have an impact on growth and production. It is obvious that a precise understanding of the factors restricting the primary production among medicinal plants is required in order to match the supply of food and pharmaceuticals for the need of the world's population. The objective of the work is to monitor remotely and automate greenhouses using IoT technology to promote the growth of the bioactive substances among medicinal plants. An Arduino-based greenhouse system collects the environmental data, which is then transmitted in both offline and online modes to an Android smartphone. A GSM modem is used to transmit SMS messages that show the state of the environmental parameters at any given time. The user receives an SMS when the sensor value exceeds a set threshold. All farmers can manage their greenhouses from anywhere as long as they are constantly aware of the conditions inside. A control actuators unit is connected for maintaining the cooling fan, exhaust fan, water pump, artificial light, and motor pump to change environmental conditions and send SMS. Additionally, data parameters transmitted over Ethernet to mobile devices, doing away with the requirement for SMS and enable the database to record all environmental parameters. Therefore, the user can use Android mobile to monitor and control parameters. The farmers will be benefitted from the increased production and economic growth if the internet of things is successfully implemented in managing the greenhouse environment.

Keywords: Greenhouse · Android mobile phone · GSM modem · SMS · control actuators unit

V. Sakalauskas et al. (Eds.): IBICA 2023, LNNS 1231, pp. 483–489, 2025.
https://doi.org/10.1007/978-3-031-78946-5_48

1 Introduction

The World Health Organization claims that pharmaceutical products must grow rapidly in order to feed the world's expanding population [1]. Farms today may span hundreds of acres. Increasing fertiliser and electricity costs, as well as constraints on irrigation, are placing pressure on farmers to use their resources more effectively [2].

2 Relevant Review

Agriculture has long been a leader in automation; many commercial farms utilise harvesters that are GPS-guided. Agriculture, which focuses mostly on medicinal plants, has long been a trailblazer in automation. As the concept of the "Internet of Things" gains more and more traction, numerous technologies are being created to make it possible to collect, analyse, and control devices across wireless data networks. Greenhouse crop growing is more impacted by the environment [3, 4]. The quality and productivity of medicinal plant development in greenhouses are significantly influenced by temperature, relative humidity, illumination, and soil moisture [5, 6]. The precise contributions of the various variables to getting the best crop output In order to work, IoT systems typically collect and exchange data in real-time. The core elements of an IoT system are smart devices, IoT applications, and graphical user interfaces. IoT has the potential to have an infinite future.

3 Methodology

There are four sensors in the greenhouse-style setup. These sensors serve as the microcontroller system's input. Analog data is the input feed that is given to the microcontroller. The controller transforms this data into a digital format. The embedded system, which is created using code, can control the parameter values once they have been tracked. The user controls the android phone. In order to control, the Android application is utilized in accordance with the user's knowledge and the desired result. Figure 1 displays the fundamental block diagram of an Arduino-based Internet of Things-based greenhouse monitoring system. The amount of water in the soil was measured using a soil moisture sensor that used the electrical potential difference between two sensor probes [4]. The sensor output is high if the soil has an adequate amount of water. The motor runs because the electrical conductivity is lower and the output is lower when the land is dry. To accurately detect humidity and temperature, we inserted temperature and humidity sensors within the smart greenhouse. If the temperature exceeds the stated limit, a fan will be turned on immediately as a coolant to lower the temperature. As soon as the temperature reached the predetermined level, the fan turned off automatically. However, a heater bulb would be turned on to boost the temperature to the desired level if it dropped below the ideal level. The air vapours were detected using a humidity sensor. The environment's change in relative humidity (RH) would cause values to be shown. The ambient air was measured using a thermistor and a capacitive humidity sensor.

Sensor LDR Light Dependent Resistor can be spelled LDR. The sensor transmitted a signal to the Arduino if there was no sunlight in the greenhouse, and the Arduino

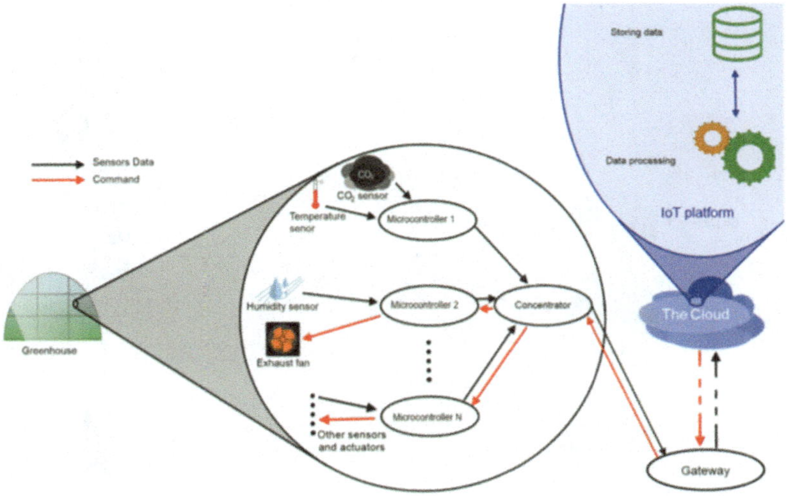

Fig. 1. Basic block diagram of a greenhouse monitoring system powered by Arduino and the Internet of Things

then delivered the corresponding signal to the interface unit to turn on the lighting. Based on the sensor output, the Arduino turns off the bulbs when the rate of sunlight reaches the required level. As a result, the Arduino program to turn on and off the lights in the greenhouse was published in accordance with the strength of the sun. Simply said, a photo-resistor is a variable resistor that responds to light. Wireless Module Information was transmitted to the cell phone using a Wi-Fi module. This was specifically accomplished via the Wi-Fi Module's Serial Port. Along with the network, the gateway node was powered on and started; the sensor node would use that one. The sensor node looked for and delivered the application to all available networks. The gateway recognized the request, connected the node, and gave the network the node's address. Data was gathered by the sensor node and transmitted to Arduino by the order gateway. An app could be hosted on the Wi-Fi module or downloaded from another application processor. The microcontroller's Wi-Fi connection was made possible by the app. The Green House content is given in Fig. 1.

A gadget could transmit sensor data to a distant location using an internet of things gateway. As a result of IoT cloud platforms using messages count as a payment unit (1 message instead of N messages, where N is the number of greenhouses count), this gateway collects data from all the greenhouses, filters it based on message priority, and periodically sends it to the cloud. This results in a low cost. These greenhouses formed a star network since they were all linked to the same gateway. Final output of greenhouse effect to mobile via web are displayed in Figs. 2, 3, 4, 5 and 6.

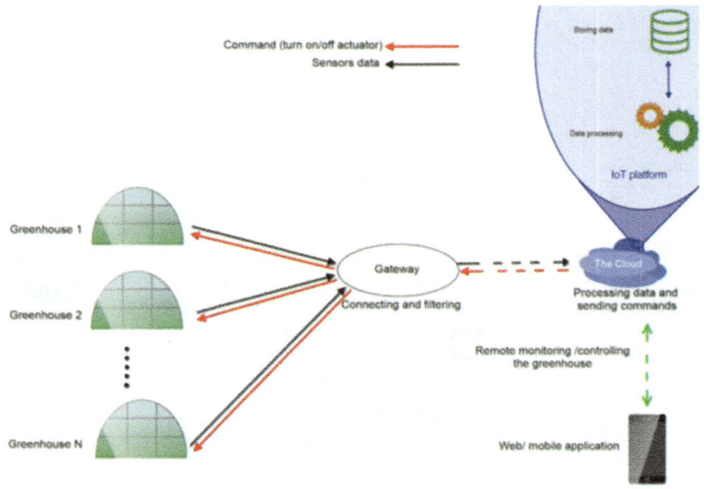

Fig. 2. Final output of greenhouse effect to mobile via web

4 Results and Discussions

Every greenhouse contained three microcontrollers, two of which were used just for measurements and were individually coupled to a similar set of sensors, while the third was linked to actuators. The described Greenhouse Monitoring System (GMS) leverages a range of sensors to continually measure various environmental parameters critical to greenhouse operations. Each parameter is monitored by a designated sensor, ensuring precise readings [7, 8]. The system's mobile alert feature serves as an essential communication channel, updating relevant parties about the greenhouse's status whenever necessary. This is particularly significant when sensor readings cross predetermined thresholds, signifying potential risks or issues.

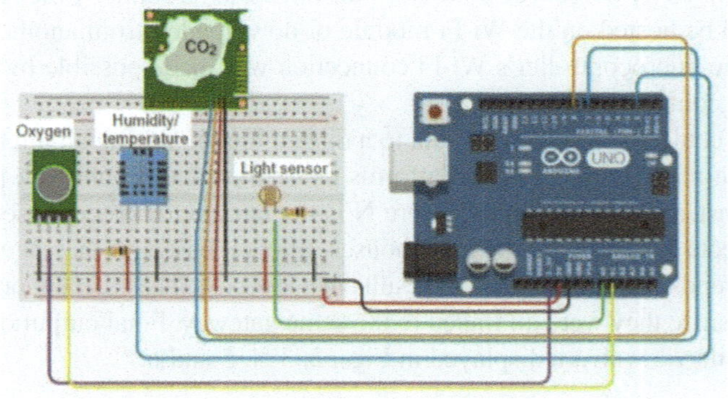

Fig. 3. Connection of all the sensors in the circuit

The GMS not only monitors but also takes corrective measures to stabilize the environment using actuators. These actuators, guided by sensor input, adjust the parameters

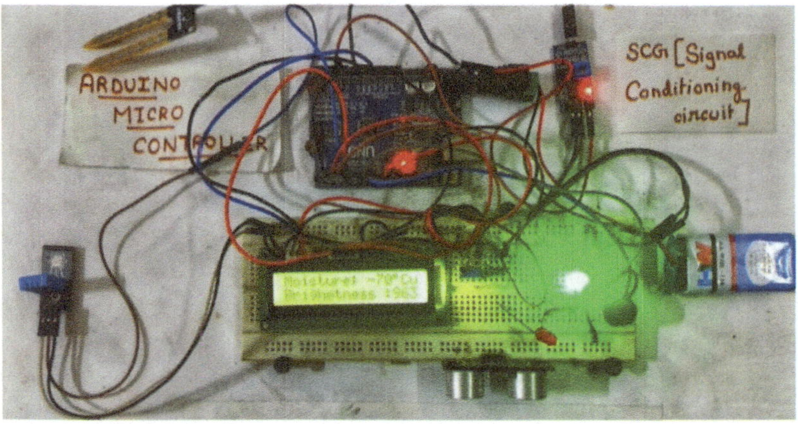

Fig. 4. Fixation of Arduino micro controller with Signa conditioning circuit.

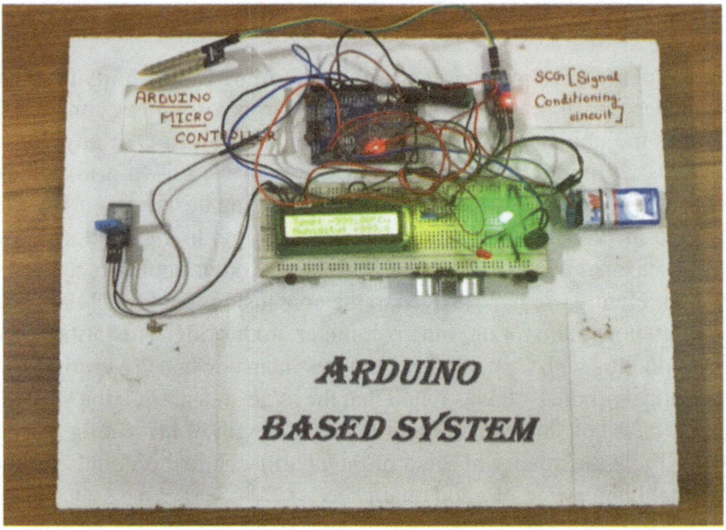

Fig. 5. Design of the all the sensors and display with Arduino

to ensure optimal conditions are maintained within the greenhouse. There's a notable flexibility in operation modes, with both automatic and manual options. In manual mode, users can control actuators based on SMS notifications, allowing for human intervention when needed. However, this manual mode was not utilized in the described work.

Prototyping

We were able to create a prototype in order to test how the system functions in actual use. Dimensions: 115 cm in height, 77 cm in width, and 102 cm in length. Materials: Made of wood, with a 1mm nylon sheet covering it.

Fig. 6. Prototype with the automation of parameter reader in Green house.

5 Conclusion

The benefit of Smart Greenhouse over traditional farming is our ability to produce crops free of pesticides and insecticides and to create an environment conducive to plant growth. Due to the ability of this type of greenhouse to preserve any climatic state, medicinal plants with a high pharmaceutical value can be grown primarily there. The Internet of Things technology was effectively used during the current project to monitor and regulate the medicinal plants in a greenhouse setting. In a greenhouse with certain actuators, the sensors and actuators were tested, and other actuators and greenhouses were simulated. Finally, we have arrived to the conclusion that farmers are not required to visit the green house site for ongoing parameter monitoring. The farmers also can get Controlled production with Cost effective and Less man power and Minimal time yield of the pharmaceutical products. It was found that the large area production with monitoring the parameters based on demand and minimize the length of harvesting period. Further it was observed that the quality of yield of the pharmaceutical products from the green house medicinal plants were of good quality.

References

1. Stipanicev, D., Marasovic, J.: Network embedded greenhouse monitoring and control. In: Proceedings of 2003 IEEE Conference on Control Applications, vol. 2, pp. 1350–1355 (2003)
2. Sahu, K., Mazumdar, S.G.: Digitally greenhouse monitoring and controlling of system based on embedded system. Int. J. Sci. Eng. Res. **3**(1) (2012)
3. Rangan, K., Vigneswaran, T.: An Embedded Systems Approach to Monitor Green House, pp. 61–65. IEEE (2014)
4. Turnell, D.J. de Fatima, Q.V., Turnell, M., Deep, G.S., Freire, R.C.S.: Farm Web-an integrated, modular farm automation system. In: Proceedings of IEEE International Conference on Systems, Man, and Cybernetics, vol. 2, pp. 1184–1189 (1998)
5. Reddy Madhavi, K., Vijaya Sambhavi, Y., Sudhakara, M.,, Srujan Kotagiri Raju: COVID-19 Isolation Monitoring System. Lecture Notes on Data Engineering and Communication Technology, vol. 63, pp. 601–609 (2021)

6. Mazidi, M.A., Mazidi, J.G., Mc Kinlay, R.D.: The 8051 Microcontroller & Embedded Systems, 2nd edn. Pearson Education Inc. (2008)
7. Khandelwal, S.A.: Automated green house management using GMS modem. Int. J. Comput. Sci. Inf. Technol. **3**(1) (2014)
8. Waykole, U.A.: Greenhouse automation system. In: 1st International Conference on Recent Trends in Engineering & Technology, Special Issue of International Journal of electronics, Communication & Soft Computing Science & Engineering, pp. 161–166 (2014)

A Fog-Based Security Framework for Smart Agriculture

Ch. Rajyalakshmi[1], R. Rajeswara Rao[2], V. Mahalakshmi[3](✉), Suresh Babu Jugunta[4], K. Reddy Madhavi[4], and S. Shanthi[5]

[1] B V Raju Institute of Technology, Narsapur, Medak, India
lakshmi.chatragadda@bvrit.ac.in
[2] CSE, JNTU-GV, Gurajada, Vizianagaram, India
rrraao.cse@jntukucev.ac.in
[3] Department of Computer Science, College of Computer Science and Information Technology, Jazan University, Jazan, Saudi Arabia
mlakshmi@jazanu.edu.sa
[4] School of Computing, Mohan Babu University, Tirupati, India
[5] Department of Civil Engineering with Computer Application, School of Engineering, Avinashilingam Institute for Home Science and Higher Education for Women, Coimbatore, India

Abstract. Smart agriculture, a phase of Agriculture 4.0, integrates IoT and smart computing technologies, addressing intelligence and automation in agriculture. However, security challenges persist. This study proposes a security architectural framework combining IoT, blockchain technology, fog computing, and SDN. Tested under DDoS attacks, the framework performs well, demonstrating the potential of these technologies in enhancing security in agriculture.

Keywords: IoT · Blockchain · SDN cyber security · fog computing

1 Introduction

The Agriculture Organization (FAO) of the United Nations emphasizes the need for a 70% increase in food production by 2050 to meet global demand, as 550 million people suffer from nutrition deficiencies and 820 million lack food daily. The world's population is expected to grow to over two billion. Agriculture 4.0 addresses climatic alteration and extreme application of resources, enhancing efficiency and safety through cutting-edge ICTs and addressing modern agriculture challenges. Wireless Sensor Networks and IoT technologies in Agriculture 4.0 offer benefits like environmental monitoring, disease detection, yield prediction, and reduced labor costs, but may also pose security risks due to outdated firmware or software. The top priority in IDS systems is ensuring network security, as disruptions can lead to dreadful outcomes.

To stop attacks and identify malicious network activity, network traffic must be monitored and categorized. Agricultural technology revolutionizes agriculture by utilizing sensors to collect data, enabling decision-making and enhancing crop quality.

V. Sakalauskas et al. (Eds.): IBICA 2023, LNNS 1231, pp. 490–498, 2025.
https://doi.org/10.1007/978-3-031-78946-5_49

Cloud computing, physical hardware, and IoT enable automation and data sharing, enabling faster and more effective agricultural tasks compared to human labor. The implementation of a secure mechanism in agriculture 4.0 is crucial to mitigate security risks, ensuring the integrity of data and ensuring the scalability of secure systems, despite the potential for incorporating traditional technologies. Agroterrorism, a danger since the sixth century B.C., uses agricultural crises and internet resources to injure crops and in turn the animals, causing a crucial economic loss and garnering public attention. Damage sensors in agriculture can result in social discontent and financial harm. The agriculture industry has undergone numerous revolutionary movements, enhancing its performance and effectiveness. The Green Revolution revolutionized the farming sector, utilizing newly discovered crops and agrochemicals. From 1900 to 1930, machines increased production, while biotechnology improved yield and pest resistance. In the twenty-first century, digitalization has the ability to support humanity. Agriculture development is transforming the industry, promoting economic growth and social stability. Smart agriculture aims to balance population growth with constrained food output, driven by advancements in science and technology. This revolution addresses both output and time limitations.

The main problem with agriculture throughout the traditional agricultural age (1784–1870), which was subjugated by animals resources and man power, was that it was not operating as efficiently as it could. Resource waste was a big problem during the automation of agriculture during the era of the 20th century. The main problem with automation of agriculture during its prompt progress (1992–2017) was its absence of astuteness. The phrase "smart agriculture," which was first used in 2017, refers to the use of cutting-edge information technology to improve and optimize agricultural operations. It transforms conventional farming practices into intelligent ones.

A new approach called Smart-agriculture for agriculture that participates with information technologies like the bigdata, the internet and IoT, enhancing services as well as the productivity. However, it also presents security and privacy challenges and presents significant obstacles. Agriculture 4.0 is a transformative shift in the sector, leveraging big data and new technology to enhance crop performance and productivity, transforming the sector into an intelligent, effective, and environmentally responsible one.

2 Related Work

Smart agriculture utilizes smart technology to enhance production quality, thereby boosting production volume and improving overall wellness and nutrition for all socioeconomic strata, ultimately contributing to a country's economic contribution. The effectiveness of agricultural resources is enhanced through the application of smart technologies, promoting efficient use of resources and achieving an ideal cost of cultivation.

The agricultural sector, a key economic sector, is significantly impacted by the waste of secondary resources. Smart technology can help reduce this waste, promoting environmental sustainability and efficient use of time, thereby enhancing agricultural output. A moment a while, when implementing agricultural 4.0 technology, the following factors could pose a threat to the agricultural sector:

- Data theft involving businesses and clients.

- Using devices and sensors to control resources.
- Destroying the controls for the objects.
- Reputational harm in the event of a data breach.

Infrastructure damage, sensor malfunctions that effect cow and poultry breeding, and control system hacks that impair greenhouse farming are all potential threats to agriculture 4.0. All of these could disrupt agricultural operations by damaging the hardware and software of the IoT infrastructure. In addition, malicious attacks, unauthorized access, privacy violations, and other problems present difficulties for data-collection technologies. The environment has changed as agriculture-based contexts, thrift management of water, cattle and farmlands are now the focus of AI and ML research. The irrigation industry has tried numerous times to conserve water and increase output through monitoring, control, and decision-making options [1]. The research was also particular on diseases that cover on the plant leaf [2], horticulture [3], vineyards [4], hydroponics [5], alarm facilities [6], technological innovations and integrations of IoT [7], resource trackability [8] and cloud based command control [9] (Table 1).

Table 1. Applications of IoT in Smart Agriculture: Agriculture 4.0

S.No.	Area of Applications	Period of Inception
1	Intelligent soil cultivation system	(2018–2023)
2	Efficient irrigation mechanisms	(2019–2023)
3	Smart fertilizer systems	(2018–2023)
4	Intelligent pest detection and treatment systems	(2018–2021)
5	Intelligent livestock agriculture	(2021–2022)
6	Smart harvesting system	(2019–2020)
7	Smart farm management system	(2018–2021)
8	Intelligent groundwater quality management system	(2018–2022)

Despite the abundance of intelligent agricultural options, they are still in their infancy and offer a low level of intelligence. The IoT devices communicates to base station, although many efforts have limited automated mechanisms [10]. on many instances, there is also no implementation with the Internet, with the exception of a few instances where local systems store data on the cloud.

2.1 Threats and Adversaries in Agriculture 4.0

In agriculture 4.0, there are a number of security concerns, starting with the storage, processing, and transmission of data via Internet access. In a multi-layered paradigm, Fig. 1 depicts security vulnerabilities Agriculture 4.0. Security flaws may either be inadvertent or occasionally discovered. Agriculture surroundings are easily accessed by animals, farm workers, and machine tools, all of which might generate issues [8–10]. The majority of security risks are very widespread, but some are specific to those who work in tough conditions, like Intelligent Agriculture.

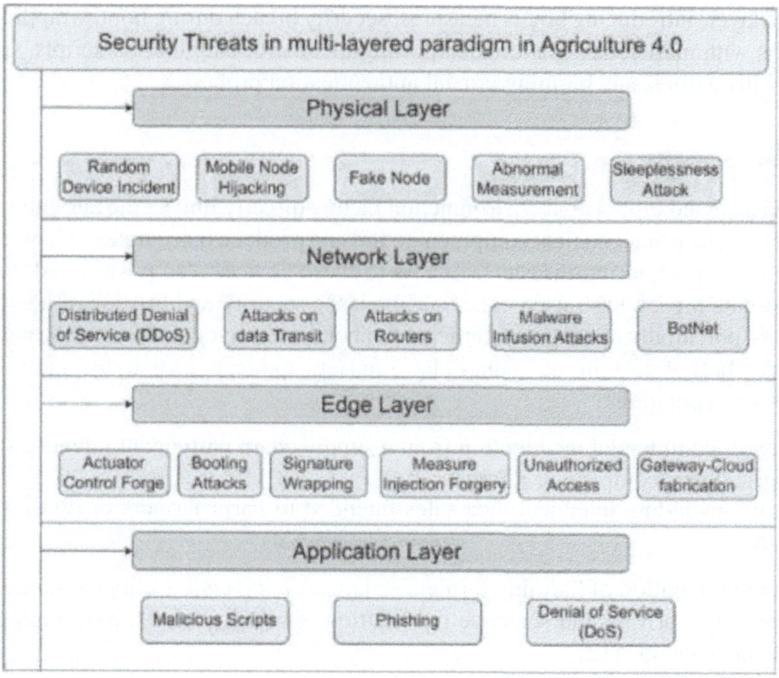

Fig. 1. Security Vulnerabilities in IoT, Fog-based Agriculture 4.0

2.1.1 Security

This mainly relates to tangible devices like sensors and actuators. Intentional or accidental human action, viruses, spyware, or cyberattacks can all cause physical equipment to malfunction. Numerous sensors and technologies are used in smart farming applications, and this variety creates a number of security vulnerabilities, including, incidents from randomized devices, hijacking of nodes, generate spurious nodes, superfluous measurements and attacks that causes unrest.

2.1.2 Network Layer

This layer's goal is to transmit the data produced by the physical layer's sensors to the cloud, which is the most reliable computing facility. The assaults that target the network layer's resources the most frequently are concentrated on [8, 9, 11]. The network layer security dangers are: DDoS attack, attacks on transit of data, attacks at routers, injecting malwares, spread of bots.

2.1.3 Edge Layer

The edge has essential elements that interact with all layers while monitoring, controlling, and accessing resources. Instead of being centralized in the cloud, large amounts of data from the physical layer can be processed locally. These are the key security issues in the edge layer: introducing bogus measures, security breach during boot, wrapping original texts with malicious signatures, importing incorrect configuration scripts, spurious gateway diversions, and harming crucial authentication processes.

2.2 Data Security

Extreme heat and other harsh environmental factors directly impact the agricultural sector, causing electromechanical equipment to fail and produce inaccurate data, potentially leading to manufacturing disasters. Access to these devices also poses a risk to farmlands. A data breach has raised concerns about the security and privacy of agricultural machines, potentially leading to significant financial and personal losses for farmers. The deliberate theft of data through smart platforms and applications that disregard security and privacy regulations;

- Thefts of confidential information from a supply chain participant intended to harm an agricultural company or a farmer;
- Threats, including unethical data sales intended to harm farmers or diminish their income.

With the intention of harming farmers or jeopardizing civic wellbeing, unrestricted alien contact to private and sensitive information is being carried out through drones, sensors, and cameras [12].

2.3 Vulnerabilities of IoT

Communication between nodes is made possible by four key systems that make up the Internet of Things. An IoT system is made up of essential sensor technology, using IoT for building gateways, cloud-centric servers, data vaults, and applications in mobile devices for the remote control. Current and emerging threats and vulnerabilities are present in the architectures of IoT systems used in agriculture. Hardware and software issues of IoT devices, networking protocols for communication, as well as data manipulation and storage can all be related to these defects.

IoT vulnerabilities, risks, and threats can be categorized using a variety of criteria. All IoT devices and applications, even in the agricultural framework, are subject to these risks. First, the two primary classification categories of security problems in IoT systems in the agriculture sector are internal vs. external risks and passive vs. active risks [13].

By using a layer-based system, IoT dangers are characterized in another exciting way. Each of the tiers below incorporates numerous technologies, according to [12], which may result in a varied kind of security threats, weaknesses. The application layer, middleware layer, internet layer, access gateway layer, and edge technology layer are the five main layers that make up the IoT architecture.

3 Existing Works

Blockchain technology is increasingly being utilized in various sectors, including the agriculture and medical sectors. It is being applied in food safety, trackability and monitoring of food items at production stages. A system employing sensor-based equipment is being developed for sustainable recording and verification. A blockchain-based security framework is being proposed for smart greenfield farming, ensuring confidentiality, integrity, and availability. Recent research on security issues in smart farming is also being conducted.

Gupta et al. [14] discuss the use of clouds in vehicles and present an E-ACO architecture relevant to IoV. They discuss access control models, authorization frameworks, and scenarios for cloud-assisted connected cars and vehicular IoT. Gupta et al. [15] propose a secure, trustworthy edge infrastructure-based approach for V2V and V2I communication, using trusted cloudlets for legitimacy, anonymity, and authenticity. They present AB-ITS, a formal attribute-based model, and demonstrate its implementation on the AWS IoT platform for safety applications.

Edge computing and cloud computing face security lapses, posing significant threats to the market. While edge computing speeds up tasks, it's crucial for clients to understand the potential risks. The hasty development of edge computing has led to a disregard for safety risks in its early stages.

4 Proposed Work

In this section, the suggested security framework for agriculture 4.0 is explained. Figure 2 shows the block diagram of the suggested security framework.

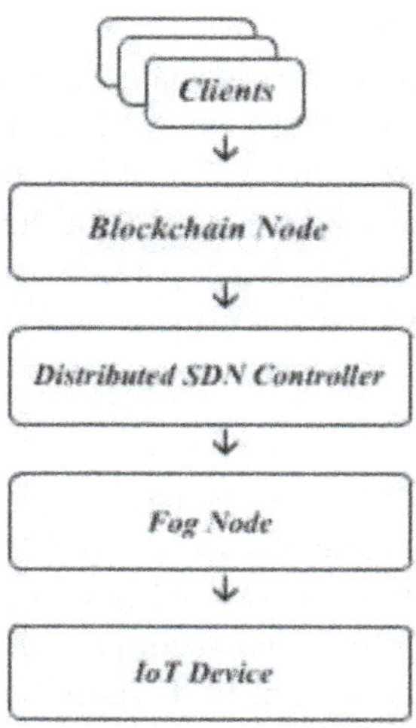

Fig. 2. Framework for Security in Agriculture 4.0

The interconnected Blockchain nodes and modules distribute, share and update the latest versions of information. When certain conditions are met, smart contracts will use the data provided to trigger and enforce the necessary actions, enabling verifiable transactions in the Blockchain without the need for a third party to get involved [16]. The modular design of the Ethereum Blockchain platform allows for the establishment of business rules specific to each application domain, enhancing its flexibility and adaptability.

The NOS, a Networking Operating System, is a crucial component in the SDN network, integrating autonomous operations with centralized control across geographically scattered controllers [17]. By controlling all transactions between network and application devices, this can aid in network monitoring since it enables modify and control the flow in network more effectively responding to the changing needs [17]. It keeps a steady view of the network from a distance, sending data to fog nodes via one API and to applications via different APIs (MQTT, CoAP, etc.) [17]. Additionally, the SDN controller synchronizes with a Block-chain contractor is employed appropriately protect the SDN flow tables in the Block-chain, preventing rule forgery.

The layer for fog computing is bundled with numerous fog nodes, the small pieces of the kit used for processing, information communication and storage to complete tasks that a typical Fog node at the end is unable to complete. Not every bit of collected data is processed on the cloud. Instead, all genuine analytic and latency-sensitive applications are hosted by this layer.

4.1 Smart Agriculture 4.0

The 5G network is a new wireless technology that can connect to millions of devices in geographically vast space and has upload and download speeds that are 100 times faster than 4G and LTE. The million smart devices that make up the precision agriculture networks are connected to each other over across the geographically vast space in order for data to be exchanged properly.

The Agriculture 4.0 scenario is thus especially well suited to 5G technology [12, 14, 15]. Meng and Cheng [18] presented a framework for the digitalization of agriculture networks and function modules for an intelligent irrigation system. An ardent research insight is also about the potential applications of 5G technology for agricultural planting, livestock farming, and traditional farming expansion using machine learning and deep learning is very well foreseen with the new innovations [19–21].

5 Conclusion

Precision agriculture utilizes automated resources and advanced agricultural methods to enhance productivity, resource efficiency, and minimize expenses. IoT technologies, edge computing, big data handling, and security features are essential for securing data throughout the agricultural system. In the initial phases of developing technologies, security measures were limited due to limited computation power. As precision agriculture introduces new complexities, incorporating high-quality, valid technologies with security features is crucial for ensuring the security of these advancements and advancing agricultural development. Software-defined networking is being integrated into the agricultural internet, enhancing communication security and system accuracy, and incorporating blockchain technology, fog computing, and machine learning methods for intrusion detection. Future work could incorporate an intrusion detection system using deep learning algorithms to support learning the sensor data in the Agriculture 4.0.

References

1. Dash, S., Panda, R., Padhy, S.: Blockchain-based intelligent medical IoT healthcare system. SPAST Abstr.2021, 1, 2190. https://spast.org/techrep/article/view/2190. Accessed 22 Apr 2022
2. Charatsari, C., Lioutas, E.D., Papadaki-Klavdianou, A., Michailidis, A., Partalidou, M.: Farm advisors amid the transition to agriculture 4.0: professional identity, conceptions of the future and future-specific competencies. Sociol. Rural. **62**, 335–362 (2022)
3. Singh, K.J., De, T.: Mathematical modelling of DDoS attack and detection using correlation. J. Cyber Secur. Technol. **1**, 175–186 (2017)
4. Raj, M., et al.: A survey on the role of Internet of Things for adopting and promoting agriculture 4.0. J. Netw. Comput. Appl. **187**, 103107 (2021)
5. Benavides, E., Fuertes, W., Sanchez, S., Sanchez, M.: Classification of phishing attack solutions by employing deep learning techniques: a systematic literature review. Dev. Adv. Def. Secur. **152**, 51–64 (2020)
6. Vankdothu, R., Hameed, M.A., Ameen, A., Unnisa, R.: Brain image identification and classification on internet of medical things in healthcare system using support value based deep neural network. Comput. Electr. Eng. **102**, 108196 (2022)
7. David, L.G., Patra, R.K., Falkowski-Gilski, P., Divakarachari, P.B., Antony Marcilin, L.J.: Tool wear monitoring using improved dragonfly optimization algorithm and deep belief network. Appl. Sci. **12**(16), 8130 (2022). https://doi.org/10.3390/app12168130
8. Mehbodniya, A., Rao, M.V., David, L.G., Nigel, K.G.J., Vennam, P.: Online product sentiment analysis using random evolutionary whale optimization algorithm and deep belief network. Pattern Recognit. Lett. **159** (2022)
9. Sunitha, D., Patra, R.K., Babu, N.V., Suresh, A., Gupta, S.C.: Twitter sentiment analysis using ensemble based deep learning model towards COVID-19 in India and European countries. Pattern Recognit. Lett. **158**, 164–170 (2022). ISSN 0167-8655
10. Balmuri, K.R., Konda, S., Lai, W.C., Divakarachari, P.B., Gowda, K.M.V., Kivudujogappa Lingappa, H.: A long short-term memory network-based radio resource management for 5G network. Future Internet **14**(6), 184 (2022). https://doi.org/10.3390/fi14060184
11. Ahmad, S.J., Unissa, I., Ali, M.S., Kumar, A.: Enhanced security to MANETs using digital codes. J. Inf. Secur. Appl. **66**, 103147 (2022). ISSN 2214-2126. https://doi.org/10.1016/j.jisa.2022.103147
12. Soma, S., Sonth, M.V., Gowre, S.C.: Design of two-dimensional photonic crystal based ultra compact optical RS flip-flop. Photon Netw. Commun. **43**, 109–115 (2022). https://doi.org/10.1007/s11107-021-00955-7
13. Sudhamani, C.: Detection probability maximization through optimization of samples in cognitive radio networks. Multimed Tools Appl **81**, 12275–12285 (2022). https://doi.org/10.1007/s11042-021-11089-3
14. Chinna Rao, R., Lakshmi, K.M., Raja, C., Varma, P.B.S., Rao, G.R.K., Patibandla, A.: Real-time implementation and testing of VoIP vocoders with asterisk PBX using wireshark packet analyzer. J. Interconnection Netw. **22**(Supp. 01), 2141030 (2022)
15. Rao, M.V., Reddy, D.A., Ampavathi, A., Munawar, S.: Data mining for cyber-physical systems. Data Mining Mach. Learn. Appl. 235–280. https://doi.org/10.1002/9781119792529.ch10
16. Vasques, A.T., Gondim, J.J.C.: Amplified reflection DDoS attacks over IoT mirrors: a saturation analysis. In: Proceedings of the 2019 Workshop on Communication Networks and Power Systems (WCNPS), Brasília, Brazil, 3–4 October 2019, Toulouse, France, pp. 1–6. IEEE (2019)

17. Nunes, B.A.A., Mendonca, M., Nguyen, X.-N., Obraczka, K., Turletti, T.: A survey of software-defined networking: past, present, and future of programmable networks. IEEE Commun. Surv. Tutor. **16**, 1617–1634 (2014)
18. Meng, H., Cheng, Y.: Research on key technologies of intelligent agriculture under 5G environment. J. Physics Conf. Ser. **1345**, 042057 (2019)
19. Martini, B., et al.: IndoorPlant: a model for intelligent services in indoor agriculture based on context histories. Sensors **21**, 1631 (2021)
20. Sriharsha, A.V.: Detection of holes on Indian roads using information and communication technologies. In: 2023 International Conference on Advances in Computing, Communication and Applied Informatics (ACCAI), Chennai, India, pp. 1–7 (2023). https://doi.org/10.1109/ACCAI58221.2023.10199826
21. Madhavi, R.K., Madhavi, G., Devi, B.R., Kora, P.: Detection of pneumonia using deep transfer learning architectures. Int. J. Adv. Trends Comput. Sci. Eng. **9**(5), pp. 8934–8937 (2020). ISSN 2278-3091. https://doi.org/10.30534/ijatcse/2020/292952020

A Systematic Review of Blockchain Application in Smart and Sustainable Agriculture

Gotte Ranjith Kumar[1], Ravi Kumar Poluru[2], Seetha Durgadevi[3], P. Viswanatha Reddy[4], K. Dinesh Kumar[5], and A. V. Sriharsha[6](✉)

[1] School of Computer Science and Artificial Intelligence, SR University, Warangal, India
[2] Information Technology, Institute of Aeronautical Engineering, Hyderabad, India
p.ravikumar@iare.ac.in
[3] Department of CSE (Data Science), B. V. Raju Institute of Technology, Narsapur, India
durgadevi.s@bvrit.ac.in
[4] Department of CSE, Viswam Engineering College, Madanapalli, India
[5] School of Computing, Amrita Vishwa Vidyapeetham University, Bengaluru, India
kk_dinesh@blr.amrita.edu
[6] Department of CSE (DS), School of Computing, Mohan Babu University, Tirupati, India
avsreeharsha@gmail.com

Abstract. Sustainable practices in agriculture require a reliable environment for intermediaries. Blockchain Technology (BT) can help achieve this by enhancing customer trust and democratization. However, there are barriers to adoption, due to insufficient global contingent standards and benchmark practices. This comprehensive review examines the obstacles and facilitators, enablers in blockchain adoption for the development of sustainable agriculture as smart systems, offering suggestions for successful implementation and filling knowledge gaps.

Keywords: Blockchain technology (BT) · smart systems · sustainable agriculture

1 Introduction

A distributed ledger that maintains information in progressive chains with all the transactional information in a distributed platform is BT. It is a secure [1, 2] method of recording information, ensuring it cannot be altered, hacked, or manipulated. Each transaction in the digital ledger is authorized by a digital signature, safeguarding the information.Reliable data is crucial for informed decision-making in sustainable agriculture, enabling farmers, producers, and suppliers to make informed choices. The existing systems hasshort-commonssuch as non-adaptability, inadequate data collection and biased data, making them unsuitable for the multi-actor agriculture market.

Smart and sustainable agriculture aims to reduce ecological footprints by utilizing ICT and IoT technologies in the farm supply chain. This involves processes from production to distribution and trading. However, issues are posed [3–5] by the disproportionate knowledge and distorted information in supply chains. Customers demand

V. Sakalauskas et al. (Eds.): IBICA 2023, LNNS 1231, pp. 499–507, 2025.
https://doi.org/10.1007/978-3-031-78946-5_50

information about food origin, requiring a system to verify products for sustainability. This pressure on food suppliers is crucial for maintaining a sustainable system.BT is being explored for its potential in the agriculture sector, addressing issues like data security, redundant spending, traceability, information asymmetry, but [6, 7]limited research on successful adoption and full-stack development is available. Hence, blockchain adoption issues must be examined for smart and sustainable agriculture. The current study presents blockchain adoption enablers, barriers, and management actions for smart and sustainable agriculture. "*What factors help or hinder the use of BT in sustainable, smart agriculture?*" is the question to ponder in this study.

The scope of the research question explores the adoption of smart agriculture, its enablers barriers to blockchain technologies in agriculture, which is the epitome of SLR applying the PRISMA (Preferred Report Items for Systematic reviews and Meta Analysis) technique.The study aims about status of the projectfor the adoption and identifying the enablers for sustainable and smart agriculture. According to the review, there are several barriers to the acceptance of BT in sustainable-smart agriculture, including: (a) insufficient government regulatories, (b) scanty capital investments, (c) unavailability of benchmark standards, (d) privacy, (e) infidelity, (f) expandability, (g) education, (h) feasibility, (i) security. Additionally, a number of enablers, namely like(a) collaboration with stakeholders, (b) shared responsibilities with partners, (c) enhanced trustworthiness and QoS, (d) sustainability and flexible infrastructure, (e) usability and acceptability, (f) enhanced supply chain efficiencycan help BT be successfully adopted in smart systems of sustainable agriculture. This study helps researchers and practitioners understand BT adoption issues and implement it for smart and sustainable agriculture.

2 Related Works

BT, a decentralized ledger, securely stores digital assets, revolutionizing industries like payments, cybersecurity, and healthcare by preventing modification of data stored on blockchain.

2.1 Purpose of Blockchain

A basic blockchain is a system with distributed ledger that enables transactions to be securely, decentralized, and fraud-proof. It timestamps information, [8–10] preventing tampering and mitigating with third parties like banks and brokers. As a decentralized shared database system, blockchain is a published distributed ledger system, providing a vast network of transactions without intermediaries. Its purpose is to mitigate tampering and ensure reliable transactions.

2.2 Blockchain as Effective Solution

BT offers a solution to the challenges faced by agriculture supply chains, like unavailability of essential factors such as transparency, traceability and trust with regard to the provenances of the food procuring systems.The systems lack many vital characteristics such as centralization, asymmetricity, incompatibility, lack of reliable information about the food provenance medium. Consumers are increasingly concerned about food quality and safety, and the support of BTthat assures the provenance of reliable and trustworthy solution to alleviate issues.

2.3 Integration of Blockchain

BT is a decentralized and distributed system that ensures data integrity and consistency. It is based on smart contracts, which are pre-written programs that execute financial transactions, records, and queries of data. The network is public, allowing seeing transactions on subscription. Blockchains can be permissioned or permission-less, deliver security and authentication. The ledger of transactions is to publish, validate and representin a sequential collection of blocks. Independent multiple agents connect in the network with no central authority in a decentralized ecosystem. This technology ensures that transactions are seen, assimilated and checked with correctness and then represented in the storage structures on the storage devices which ensures the integrity and intactness of the data.

3 Applications

The immutable ledgering methodology imposed on a decentralized networking system which is immutable and allowing the end user browse through the historical contents of transactions with due subscriptions and verifications is a BT. The distributed knowledge system, features with a consensus algorithm, concepts of smart contracts, transparencies and proofs-of-ownership, ensures reliable infrastructure for sustainable smart agriculture applications, preventing transaction alteration or removal.

Smart system in agriculture are in vogue by modern farmers of 20th century, towards the leading objectives of increased production, optimize input usage, saving of labor, access to markets [11]. They use smart agriculture systems for data collection, production management, and equipment operation. Some farmers use free software online, like the Bureau of Meteorclogy. Free software from private organisations like GrainCorpand AgWorldis used by some of them [12, 13]. Agriculture systems have increased productivity and food supply chains [14, 15]. Many researchers and organisations consider Agriculture 4.0, which is composed of precision agriculture, BT,IoT, and BD for sustainable agriculture. Measuring, monitoring and responding to the field metrics in order to quantify the crops or the cultivated live stock from low-put to high-production levels and facilitating the sustainability in the agriculture are the important traits of Precision agriculture. [16, 17], where a complete visualization and analysis can be made from the variables used in the system for ascertaining the sustainability in the crop growth.

Complexity, knowledge gaps, and inconsistent information cause trust and reliability issues in farm-to-consumer [18–20]. Our research shows that BT can provide various types of derivation methodologies for delivery of food, secure financial transactions, and transparent traceability in the smart systems for agricultural supply chains. Several security issues are identified, and global standards and industry best practises are sought. The concepts of the sustainable and smart agriculture are not properly studied and put them into the gamut of understanding in developing a system aligned with the block chain technology, hitherto technologies were into the developing the models of learning and handling big data. The insight into the factors of adoptions, enablers conjoining them into a systematic structure is a novel thought in the development of smart systems of sustainable agriculture also allows to identify the barriers, adoption levels and gaps in the system development.

4 Methodology

Qualitative research is a systematic investigation conducted in a natural setting [21], focusing on the behavior and interactions of individuals and groups.Sometimes quantitative methods like surveys cannot answer the research question. Numerical data collection can measure patterns like consumer shopping behaviour, according to Busetto et al. [9]. However, qualitative methods are better for finding hidden patterns.

A qualitative study is an approach of conducting research that allows for flexibility, openness, and context responsiveness, enabling researchers to investigate previously unexplored research issues. The researcher can find relevant topics and perform detailed walk-through of investigation of the current research topic by conducting reviews, questionnaires and collecting the opinions of the specialist.

Criteria for Inclusion/Exclusion

A systematic review is a crucial tool for a research study to achieve its purpose. It involves identifying literature with specific attributes to ensure reliable results. In order to reduce bias and avoid incorrect conclusions, a defined inclusion/exclusion criteria sets the parameters for the review. Time window, topics in the study, languages and outcome are among the criteria for selecting publications. By defining these criteria, the research study aims to produce reliable and reproducible results (Fig. 1).

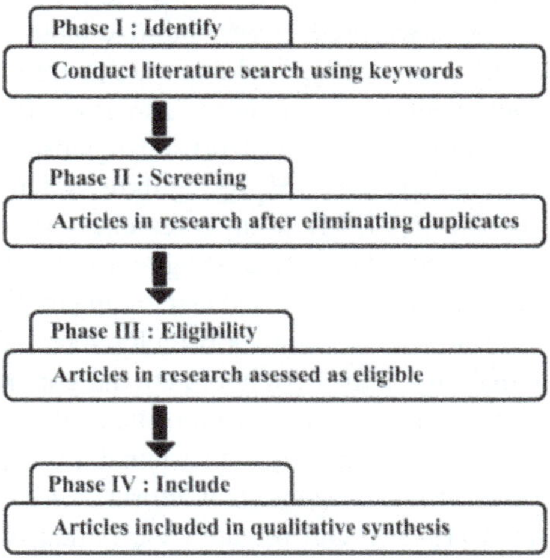

Fig. 1. PRISMA flow diagram for qualitative systematic literature review

The outcomes are presented based on criteria such as published literatures in various databases, progress in publications over time beginning in the year 2015. In terms of publications that prevail with the concepts of BT for smart systems of agriculture, Hindawi has published 19, MDPI has published 12, IEEE has published 28, SpringerNature has published 44, Springer has published 12, Elsevier has published 19, ScienceDirect has published 11 putting into the total of 145 (Table 1).

Table 1. Criteria for selecting publications in the review

Parameters	Inclusion	Exclusion
Date Range (Jan 2015 to Sep 2023)	Publications of Research consensus within this period	Publications of Research consensus not in this period
Topic of Study	Blockchain, Smart Agriculture, Farming	other than topics of inclusion
Keywords	blockchain, "agriculture", "cloud", "computing", "enablers", "provenance", "traceability", "security"	all other synonymous and generic search words that lead to bias
Redundancy	represent unique titles of publications	scrutinize the titles, authornames, locations of publications
Qualified	all publications that contribute with relevance to the smart agriculture	all other contributions that does not reflect ideas of agriculture
Language	english	non-english
Criteria	read thoroughly the full text of the article and certify the importance of the literatyre suitable to the keywords	publications with individual presence of keywords

5 Discussions

Since 2015, there has been a discernible rise in the number of articles published on BT pertaining to the agriculture industry. Figure 2 depicts how articles on blockchain and smart agriculture were distributed throughout several databases between 2015 and 2023.

In 2021, a trend is emerging to adopt BT for agricultural supply chains, payments, sustainability, and security, focusing on smart agriculture and sustainable practices (Fig. 3).

The study targeted on identifying the relevant articles matching with the themes and keywords and develop a classification model, discriminating the context relevant publications. Thematic analysis of data from 145 articles identified five themes as adoption enablers: blockchain for food supply chains, payments in agriculture, sustainability, smart agriculture, data security, and barriers of blockchain adoption. The iterative process assessed the content of each article, contributing to the reliability of the study.The

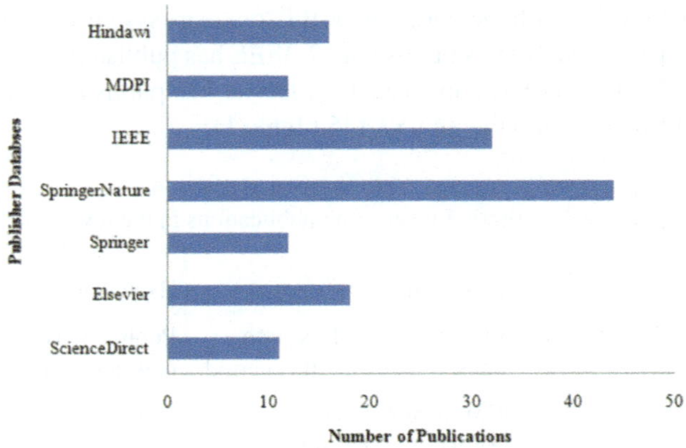

Fig. 2. Number articles in the publisher databases during 2015–2023.

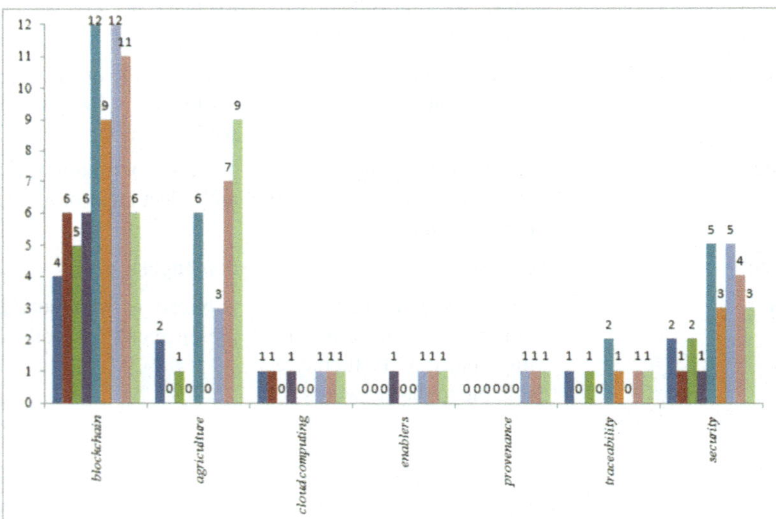

Fig. 3. Thematic – Keyword wise distribution of publications during 2015–2023.

articles reviewed discusses about the acceptance of enablers of BT in conceiving the forms of smart systems for sustainable agriculture, while also addressing the barriers to its adoption. These enablers-barriers interactions are specific to the development of smart systems for sustainable agricultural practices, providing valuable insights to the industry. A good example of BT integration with other technologies is an AI-based transparent and traceability system for food supply chains created by Qian et al. [14]. A few other examples show that the use of BT in the agriculture sector is feasible, including the consortium blockchain system [15], farm management system [16], and food trackability system [17]. Fast and secure financial transactionsfor agriculture by BT [18, 19]. The permissioned blockchain model created by Pranto et al. [19] andplatform created by Rijanto [18] both facilitate the development of the quick payments enabler.Because it encompasses numerous channels and stages in the food supply chains, issuing a sustainability certification is a laborious process. Nevertheless, blockchain can improve sustainable supply chains by resolving anomalies in the wine industry's certification process [11, 20]. More examples of sustainability adoption facilitators include an architecture for sustainable agriculture [10, 21], an information system with supply chain [12], and a double chain structure assuring sustainability [13].

BT is being increasingly applied in the development of smart systems for sustainable agriculture, enhancing aesthetics of collaborations with the stakeholders, shared tasks, customer trust, QoS, sustainability in value chains, infrastructure, data security, efficiency and agricultural democratization. The Ethereum blockchain network system enables robust transactions, with smart contracts enabling farmers to publish agricultural assets and customers to purchase them.

The BT systems can enhance trust and transparency and trackability in agricultural supply chains, promoting more sustainable supply chains. Examples include consortium blockchain-based trustworthy systems for agricultural supply chains, and functional blockchain-based traceability systems for dairy supply chains. BT enhances transparency in transactions within a blockchain network, reducing tampering and fraud. It allows for secure storage of data, such as farm management systems for food safety and traceability, ensuring all participants can access the same information. Hyperledger Fabric is being used to develop a permission-based blockchain application for fish farm management, enhancing data security and integrity. This system records trust values on a blockchain, preventing DDOS attacks. The system also enhances network security by securely storing data collected from IoT sensor devices in a private blockchain, thereby enhancing data security and usability.BT adoption in smart and sustainable agriculture faces barriers like lack of government regulations, resource capital requirements, security concerns, standards, trust, scalability issues, awareness, and ease of use.

6 Conclusions

Blockchain adoption in agriculture can enhance stakeholder collaboration, shared responsibilities, customer value, quality of service, trust, sustainable supply chains, democratization, and data useability. However, barriers like unclear government regulations, technology immaturity, scalability issues, privacy policies, power consumption, high establishment costs, and security can slow adoption. Technical barriers can be resolved with blockchain service providers' support, but defining global standards and centralized authorities may take more time. The discussions on the study about the highlights the potential of blockchain in shaping the future, suggesting that future research should include more databases, blockchain-related technologies, and applications to enhance the understanding of this technology.

References

1. Agricultural Productivity. https://www.agriculture.gov.au/abares/research-topics/productivity. Accessed 27 Aug 2022

2. Dufty, N., Jackson, T.: Information communication technology use in Australian agriculture: a survey of broadacre, dairy and vegetable farms. ABARES **18**(15), 1–22 (2018)
3. Sriharsha, A.V., et al.: Adaptable fog computing framework for healthcare 4.0. In: Proceedings of the 14th International Conference on Soft Computing and Pattern Recognition (SoCPaR 2022).SoCPaR 2022. Lecture Notes in Networks and Systems, vol. 648. Springer, Cham (2023)
4. Eitzinger, A., et al.: GeoFarmer: a monitoring and feedback system for agricultural development projects. Comput. Electron. Agric. **158**, 109–121 (2019)
5. Boursianis, A.D., et al.: Smart irrigation system for precision agriculture—the AREThOU5A IoT platform. IEEE Sens. J. **21**, 17539–17547 (2021)
6. Zhang, D.: The innovation research of contract farming financing mode under the blockchain technology. J. Clean. Prod. **270**, 122–194 (2020)
7. Awan, S., et al.: IoT with blockchain: a futuristic approach in agriculture and food supply chain. Wirel. Commun. Mob. Comput. **2021**, 14 (2021)
8. Teherani, A., Martimianakis, T., Stenfors-Hayes, T., Wadhwa, A., Varpio, L.: Choosing a qualitative research approach. J. Grad. Med. Educ. **7**, 669–670 (2015)
9. Busetto, L., Wick, W., Gumbinger, C.: How to use and assess qualitative research methods. Neurol. Res. Pract. **2**, 14 (2020)
10. Konda, S., Balmuri, K., Mamidala, K.K.: Classification and mining behavior of data. Data Min. Mach. Learn. Appl. 21–55. https://doi.org/10.1002/9781119792529.ch2\
11. Patra, R.K., Mahendar, A., Madhukar, G.: Inductive learning including decision tree and rule induction learning. Data Min. Mach. Learn. Appl. 209–234 (2022).https://doi.org/10.1002/9781119792529.ch9
12. Bhavani, R., Muni, T.V., Tata, R.K., Narasimharao, J., Murali, K., Kaur, H.: Deep learning techniques for speech emotion recognition. In: 2022 International Conference on Futuristic Technologies (INCOFT). Belgaum, pp. 1–5. (2022). https://doi.org/10.1109/INCOFT55651.2022.10094534
13. Hanumanthakari, S., Pullela, S.K., Bhukya, S.N., Vijayalakshmi, K., Ahmad, S.R., Kumar, N.: IoT based patients monitoring system in healthcare service. In: 2022 International Conference on Automation, Computing and Renewable Systems (ICACRS), Pudukkottai, India, pp. 1324–1329 (2022). https://doi.org/10.1109/ICACRS55517.2022.10029295
14. Gulati, K., Nayak, K.M., Priya, B.S., Venkatesh, B., Satyam, Y., Chahal, D.: An examination of how robots, artificial intelligence, and machinery learning are being applied in the medical and healthcare industries. Int. J. Recent Innovation Trends Comput. Commun. **10**(2s), 298–305 (2022). https://doi.org/10.17762/ijritcc.v10i2s.5947
15. Reddy, G., et al.: An intrusion detection using machine learning algorithm multi-layer perceptron (MlP): a classification enhancement in wireless sensor network (WSN). Int. J. Recent Innovation Trends Comput. Commun. **10**(2s), 139–145 (2022). https://doi.org/10.17762/ijritcc.v10i2s.5920
16. Mubeen, S., Kulkarni, N., Tanpoco, M.R., Kumar, R.D., Naidu, M.L., Dhope, T.: Linguistic based emotion detection from live social media data classification using metaheuristic deep learning techniques. Int. J. Commun. Netw. Inf. Secur. (IJCNIS) **14**(3), 176–186 (2022). https://doi.org/10.17762/ijcnis.v14i3.5604
17. Bhaskar, T., Shiney, S.A., Rani, S.B., Maheswari, K., Ray, S., Mohanavel, V.: Usage of ensemble regression technique for product price prediction. In: 2022 4th International Conference on Inventive Research in Computing Applications (ICIRCA), Coimbatore, India, pp. 1439–1445 (2022). https://doi.org/10.1109/ICIRCA54612.2022.9985521
18. Patil, S., Balmuri, K.R., Frnda, J., Parameshachari, B.D., Konda, S., Nedoma, J.: Identification of triple negative breast cancer genes using rough set based feature selection algorithm & ensemble classifier. Hum. Centric Comput. Inf. Sci. **12** (2022)

19. Kesana, M., Rikhari, S.: Multi-Modal Medical Image Fusion Using Laplacian Re-Decomposition (2022). https://doi.org/10.1201/9781003240310-7
20. Reddy Madhavi, K., Madhavi, G., Rupa Devi, B., Kora, P.: Detection of pneumonia using deep transfer learning architectures. Int. J. Adv. Trends Comput. Sci. Eng. 9(5), 8934–8937 (2020). https://doi.org/10.30534/ijatcse/2020/292952020. ISSN 2278–3091
21. Reddy Madhavi, K., Kora, P., Venkateswara Reddy, L., Avanija, J., Soujanya, K.L.S., Prabhakar, T.: Cardiac arrhythmia detection using dual-tree wavelet transform and convolutional neural network, soft computing. Impact Factor **3**(64) (2021)

An Evaluation of a Refined Cryptosystem for Safe Image Sharing on Internet of Things Devices

Jonnadula Narasimharao[1(✉)], Y. Lakshman Kumar[1], Abdul Subhani Shaik[2], B. Suresh Ram[2], Shyam Sundar Yella[3], and Amit Kumar Singh[3]

[1] CMR Technical Campus, Hyderabad, Telangana, India
jonnadula.narasimharao@gmail.com

[2] CMR College of Engineering and Technology, Hyderabad, Telangana, India
bsureshram@cmrcet.ac.in

[3] CMR Engineering College, Hyderabad, Telangana, India
{shyamsundar.y,amitkumarsingh}@cmrec.ac.in

Abstract. Image encryption can be achieved by utilizing several spatial domain techniques to manipulate the placements of image pixels. The image scrambling method based on recursive sequences is one example. Several recursive sequences, including the Fibonacci sequence, cellular automata, and chaotic maps, are used to jumble images. Another method for encrypting images is to jumble blocks or coefficient matrices in the transform domain. However, because there aren't any security keys or there isn't much key space, these methods have very poor security levels. Moreover, it is well known that encryption methods based solely on permutations are susceptible to plaintext assaults. Altering the values of image pixels using a combination of logic operations and image bit plane decomposition is another method for encrypting images. Because the outcomes of its logic operations and deconstruction process are predictable, this technique has a much lower security level. Attacks using plaintext are not unheard of. One way to attain higher security levels is to use different approaches to jumble the placements of image blocks or pixels while altering their values. This study presents a new edge mapped combined key generation (EM-CKG) algorithm for grayscale and RGB picture encryption and decryption. EM-CKG generates a binary image of the same size as the original image that needs to be encrypted.

Keywords: Encryption · decryption · edge mapped combined key generation

1 Introduction

Network technologies and media services offer widespread advantages for individuals and organizations to gather, trade, and distribute photographs and videos inside multimedia networks and wireless or mobile public channels [1]. Ensuring the security of images is a significant difficulty in applications pertaining to storage and transmission [2]. Video surveillance systems, for example, are used for homeland security to monitor

V. Sakalauskas et al. (Eds.): IBICA 2023, LNNS 1231, pp. 508–516, 2025.
https://doi.org/10.1007/978-3-031-78946-5_51

several important areas, such as business and financial districts and public transportation. Every day, enormous volumes of private information-containing movies and photos are created, sent, or recovered. Furthermore, physicians in various branches of a health service organization may exchange medical pictures and patient records via networks for many clinical purposes. Two instances of this method are the Advanced Encryption Standard (AES) and the Data Encryption Standard (DES). These encryption techniques, however, are very computationally expensive and have a low error resilience [3]. In the modern era of computers and quick communication, information must be protected from unauthorized users when transferred by any electronic device, whether it is a message or plain text. For this reason, security of visual data is a crucial consideration in communication system design. When transmitting visual data across unsafe channels, confidentiality and security are added using data masking techniques and visual cryptography. Techniques for data concealing aim to send data in an undetectable manner by embedding it in digital media [4]. The two well-known cryptosystems are the private key and the public key. By employing them, we may protect the secret data safely and prevent an intruder from deciphering its meaning. Three exemplary techniques are the Advanced Encryption Standard (AES), Rivest, Shamir, Adleman (RSA), and Data Encryption Standard (DES) [5].

One method of data hiding is steganography, which allows for secret communications that mask the message's very existence [6]. Another method where information is hidden but still intimately linked to the object in which it is contained is watermarking [7]. Conversely, in visual cryptography, also known as visual secret sharing (VSS), a dealer (holder of the secret image) shares the original input image with a group of players P.

2 Literature Review

When it comes to coded caching, Behrouz Zolfaghar suggested using a conventional, straightforward network architecture with one server, one hub, a shared link joining the server and hub, and private links joining the users and hub [1]. Jawad Ahmad offers a secure real-time scheme for Internet of Things (IoT) systems that cleverly combines chaos-based lightweight picture encryption with occupancy monitoring. First, the single overhead camera was utilized to extract video frames from the real-time footage. The intelligent system counts people in and out, encrypts the current frame, and sends occupancy count data to the cloud computing platform when individuals are spotted in a frame [2]. Muhammad Seyed Self-adaptive color picture encryption is suggested by a design of the compound one-dimensional chaotic function that combines nearest-neighboring coupled-map lattices (NCML) with piecewise nonlinear chaotic maps [3]. ZHENG LEI suggested The multicast performance of the coded caching strategy will be negatively impacted in wireless multicast networks due to the possibility of disparate consumers having very varied link quality to the server. The wireless multicast physical channels are divided into numerous logic parallel partially shared connections by a parallel delivery technique, which is presented to address the asymmetric link characteristics and satisfy the demand for fairness delivery by a different user [4]. In the field of chaotic image encryption, Takeshi Koshiba suggested a thorough assessment that might highlight current trends and throw light on less-explored subjects [5, 6]. Zhang Li-bo

An analysis of TDBMP's limitations has led to the proposal of a specific plaintext attack as a means of breaking the method. Subsequently, they put up an improved approach to address the inadequacies in the original scheme mentioned above [7]. In contrast to two baseline techniques based on conventional coded caching, Shirin Saeedi Bidokhti's proposal improves either on rate or decoding latency [8]. Janan Adeyad suggested Information sharing across public networks, such as the internet, has rapidly risen in recent years. This data may come in the form of text, music, pictures, or videos. One crucial concern is data transmission [9] security. In [10], we offer a time-domain, low-power temperature-sensing circuit designed for Internet of Things applications. In this work, we explore the design and implementation of a real-time machine vision system for medical imaging. Even if the RX's various sub-blocks have had a lot of work done to improve their power efficiency [11], it becomes harder to reduce the power allocation of the device as several important RF blocks [12].

3 Proposed Method

More importantly, as Fig. 1 and 2 demonstrate, we have implemented a new combined key generation (CKG) scheme that uses two key images for increased security.

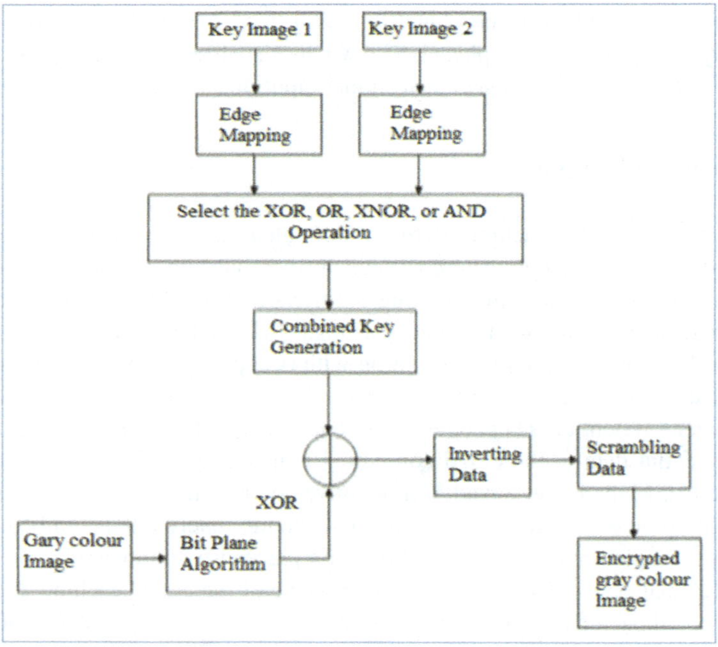

Fig. 1. Block diagram of proposed EM-CKG encryption

This method enables the new image encryption and decryption algorithm to be implemented in a way that allows for the encryption and decryption of both 3D images, i.e., Key image and Original image.

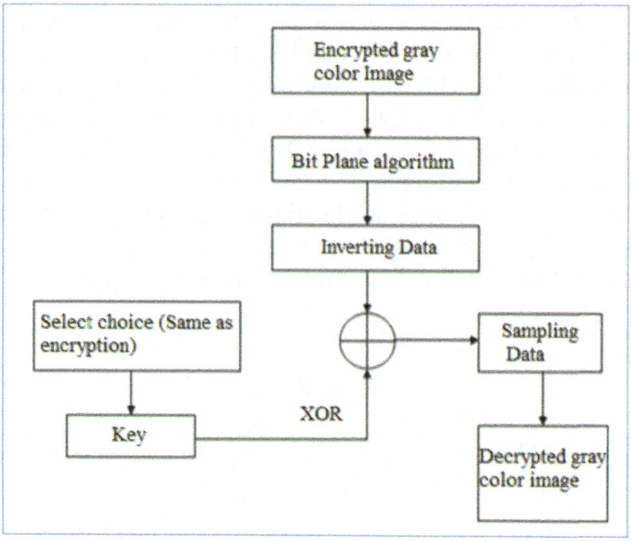

Fig. 2. Proposed decryption algorithm.

Encryption Algorithm

Step 1: Choose and open an RGB (gray/color) picture that has to be encrypted.
Step 2: Next, utilize the bit plane algorithm to convert the image into a number of bit planes, such as "8" bit planes for grayscale images and "24" bit planes for RGB images.
Step 3: Choose and read the two key images that match the input image's size.
Step 4: To obtain the binary data, use edge mapping.
Step 5: Apply combined key generation to the newly created key image, taking into account the user's selected XOR, AND, XNOR, and OR encryption operations.
Step 6: After that, XOR the new key image's bit planes with those of the original grayscale or RGB image.
Step 7: To further security, the XORed image will next be inverted, or its bit planes will be shuffled.
Step 8: Next, we used binary to decimal and numeric to string operations to do a scrambling process for further security concerns.
Step 9: At last, the CKG algorithm produced an entirely encrypted image.

BIT Plane Algorithm

Each of the binary numbers used to define a digital discrete signal has a corresponding bit position in the bit plane, as shown in Table 1. For instance, there are 16-bit planes for a 16-bit data representation: the most significant bit set is included in the first-bit plane, while the least significant bit is found in the 16th bit plane [13]. It is evident that the first bit plane provides the most accurate but imprecise estimate of a medium's values, and that the bit plane's contribution to the final stage decreases with increasing bit plane number [14]. A bit plane is thus added to provide a more accurate approximation. An m-bit dataset's bits on the nth bit plane provide a value of 2(m-n) if they are set to 1, and nothing else [15]. Consequently, bit planes have the potential to provide fifty percent of the numerical value of the preceding bit plane. Bitmap and bit plane are sometimes used interchangeably, although bitmap refers to data, whereas bit plane refers to memory location. One aspect of using bit-planes is determining if they contain random noise or valuable information. Comparing each pixel (X, Y) to its three neighboring pixels (X-1, Y), (X, Y-1), and (X-1, Y-1) is one way to compute this. It is not noise if the pixel is identical to at least two of the three neighboring pixels. Between 49% and 51% of the pixels in a noisy bit-plane will be noise.

For instance, the bit planes function as follows in the 8-bit number 10110101 i.e.,181 in decimal.

Table 1. Bit plane

Bit Plane	Value	Contribution	Running Total
1	1	1*2^7 = 128	128
2	0	0*2^6 = 0	128
3	1	1*2^5 = 32	160
4	1	1*2^4 = 16	176
5	0	0*2^3 = 0	176
6	1	1*2^2 = 4	180
7	0	0*2^1 = 0	180
8	1	1*2^0 = 1	181

Inverting

To perform an inversion, use the command "fliplr," which flips an array from left to right. After applying logical operations, this procedure is used to provide an encrypted image greater security.

Here, binary to decimal and numeric to string operations are used to accomplish the scrambling.

4 Results and Analysis

As shown the Fig. 4 represented the image encryption gray scale and Fig. 5 represented the image encryption with color images.

As shown the Fig. 6 represented the gray color image, first picture is original image and next is key image as input image and third picture as encrypted image of gray scale image and final picture is decrypted gray scale image.

As shown the Fig. 7 represented the color image, first picture is original image and next is key image as input image and third picture as encrypted image of color image and final picture is decrypted color image.

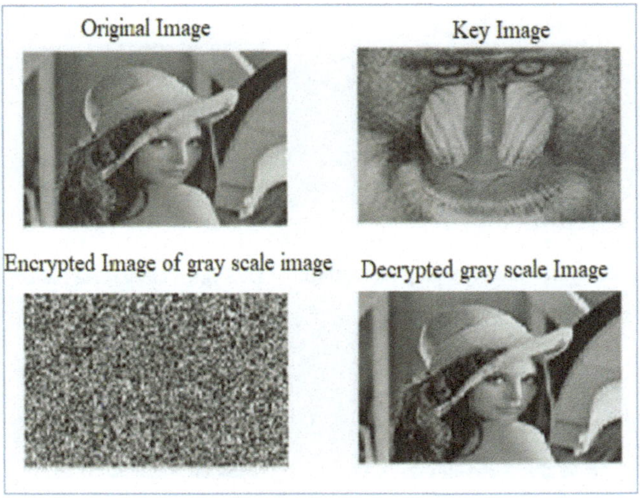

Fig. 4. Image Encryption with Gray Scale.

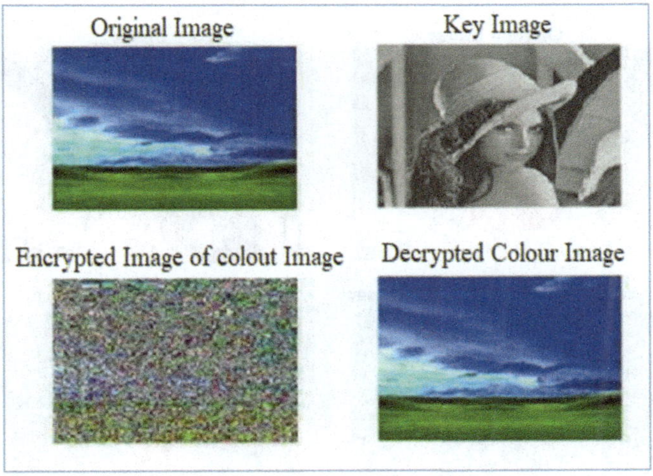

Fig. 5. Image Encryption with color images

As shown the Fig. 8 represented the color image, first picture is original image and next is key image1 as input image and third picture is key image 2 and next colour picture is encrypted image of color image and final picture is decrypted color image.

As shown the Fig. 9 represented to compare the both the images, picture one is histogram of original image and second picture is histogram of decrypted image.

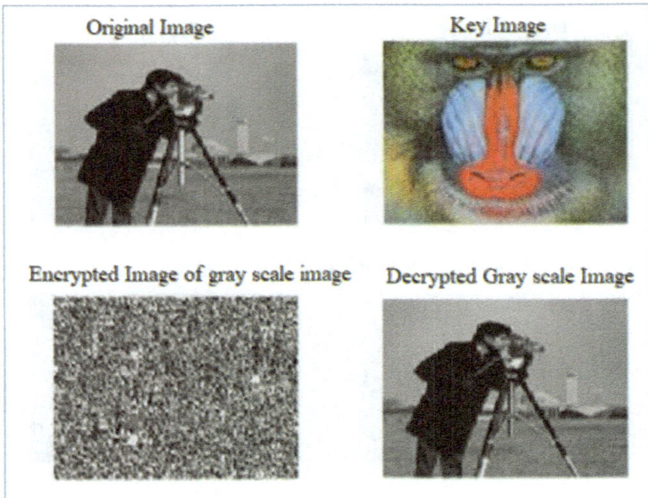

Fig. 6. Image Encryption with Secret key image for Gray scale.

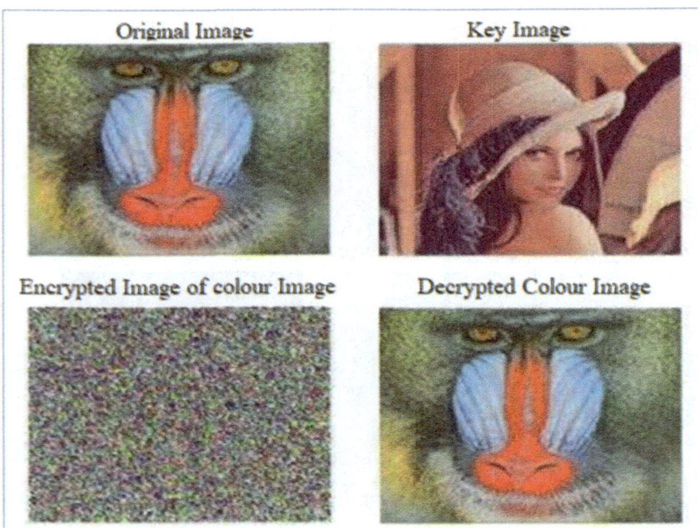

Fig. 7. Image Encryption with Secret key image for Color images

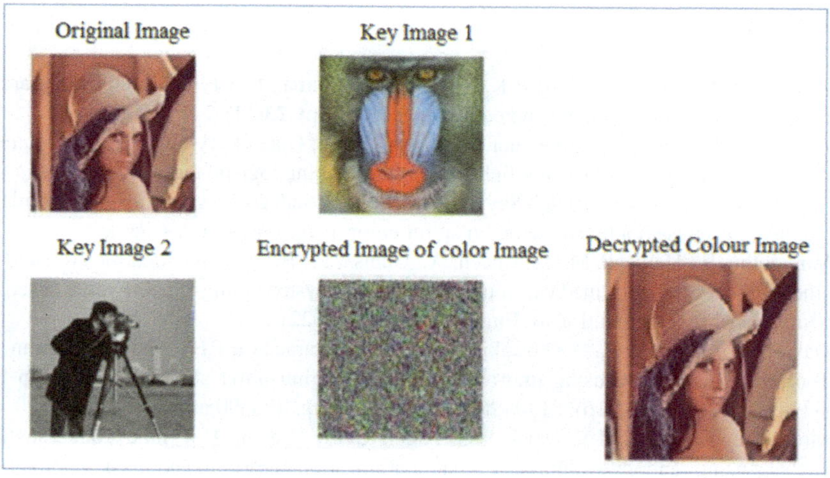

Fig. 8. Image Encryption with two Secret Keys

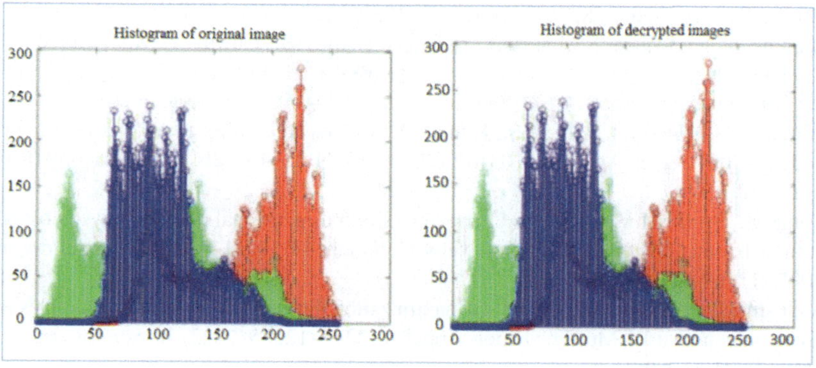

Fig. 9. Histogram of original image and Decrypted image.

5 Conclusion

Presents a novel edge-mapped combined key generation (EM-CKG) algorithm for grayscale and RGB image encryption and decryption in IOT, combining logical operations and scrambling techniques. Compared to other single key image methods like bit plane crypt and edge map crypt, the suggested algorithm offers a number of advantages. The security issue has been successfully resolved. Given that they function at the binary levels, hardware implementation is incredibly simple. Additionally, they function well for multimedia protection in real-time applications like mobile phone services and wireless networks.

References

1. Zolfaghari, B., Singh, V., Rai, B.K., Bibak, K., Koshiba, T.: Cryptography in hierarchical coded caching: system model and cost analysis. Entropy **23**(11), 1459 (2021)
2. Ahmad, J., Larijani, H., Emmanuel, R., Mannion, M., Qureshi, A-U-H.: Secure occupancy monitoring system for IoT using lightweight intertwining logistic map.
3. Seyed Mohammad seyedzadeh,Seyyed Mohammad Sadegh moosavi Using self-adaptive coupled piecewise nonlinear chaotic map for color image encryption scheme.:
4. Vankdothu, R., Hameed, M.A., Ameen, A., Unnisa, R.: Brain image identification and classification on internet of medical things in healthcare system using support value based deep neural network. Comput. Electr. Eng. **102**, 108196 (2022)
5. David, L.G., Patra, R.K., Falkowski-Gilski, P., Divakarachari, P.B., Antony Marcilin, L.J.: Tool wear monitoring using improved dragonfly optimization algorithm and deep belief network. Appl. Sci. **12**(16), 8130 (2022). https://doi.org/10.3390/app12168130
6. Mehbodniya, A., Rao, M.V., David, L.G., Nigel, K.G.J., Vennam, P.: Online product sentiment analysis using random evolutionary whale optimization algorithm and deep belief network. Pattern Recogn. Lett. **159**, 1–8 (2022)
7. Sunitha, D., Patra, R.K., Babu, N.V., Suresh, A., Gupta, S.C.: Twitter sentiment analysis using ensemble based deep learning model towards COVID-19 in India and European countries. Pattern Recogn. Lett. 158, 164–170 (2022). ISSN 0167–8655
8. Balmuri, K.R., Konda, S., Lai, W.C., Divakarachari, P.B., Gowda, K.M.V., Kivudujogappa Lingappa, H.: A long short-term memory network-based radio resource management for 5G network. Future Internet **14**(6), 184 (2022). https://doi.org/10.3390/fi14060184
9. Ahmad, S.J., Unissa, I., Ali, M.S., Kumar, A.: Enhanced security to MANETs using digital codes. J. Inf. Secur. Appl. **66**, 103147 (2022). ISSN 2214–2126. https://doi.org/10.1016/j.jisa.2022.103147
10. Soma, S., Sonth, M.V., Gowre, S.C.: Design of two-dimensional photonic crystal based ultra compact optical RS flip-flop. Photon Netw. Commun. **43**, 109–115 (2022). https://doi.org/10.1007/s11107-021-00955-7
11. Sudhamani, C.: Detection probability maximization through optimization of samples in cognitive radio networks. Multimed Tools Appl **81**, 12275–12285 (2022). https://doi.org/10.1007/s11042-021-11089-3
12. Chinna Rao, R., Lakshmi, K.M., Raja, C., Varma, P.B.S., Rao, G.R.K., Patibandla, A.: Real-time implementation and testing of VoIP vocoders with asterisk PBX using Wireshark packet analyzer. J. Interconnection Netw. **22**(Supp01), 2141030 (2022)
13. Rao, M.V., Reddy, D.A., Ampavathi, A., Munawar, S.: Data mining for cyber physical systems. Data Min. Mach. Learn. Appl. 235–280 (2022).https://doi.org/10.1002/9781119792529.ch10
14. Suresh, M., Shaik, A.S., Premalatha, B., Narayana, V.A., Ghinea, G.: Intelligent & smart navigation system for visually impaired friends. In: Garg, D., Narayana, V.A., Suganthan, P.N., Anguera, J., Koppula, V.K., Gupta, S.K. (eds.) Advanced Computing. IACC 2022. Communications in Computer and Information Science, vol. 1781. Springer, Cham (2023). https://doi.org/10.1007/978-3-031-35641-4_30

Evaluation on Obtaining Advanced Features of CNN's Deep Learning VGG 16 Architecture for Breast Cancer Detection

G. Divya[1(✉)], Sudha Arvind[1], B. Premalatha[2], Y. Aruna Suhasini Devi[2], M. Kumara Swamy[3], and Lal Bahadur Pandey[3]

[1] CMR Technical Campus, Hyderabad, Telangana, India
cmrtc.paper@gmail.com, divyasep9@gmail.com, sudharvind.ece@cmrtc.ac.in

[2] CMR College of Engineering and Technology, Hyderabad, Telangana, India
{bpremalatha,yarunasuhasini}@cmrcet.ac.in

[3] CMR Engineering College, Hyderabad, Telangana, India
{m.kumaraswamy,lalbahadurpandey}@cmrec.ac.in

Abstract. According to this study, The biggest cause of death is breast cancer for women in the contemporary era. With 14% of cases predicted to result in death in 2018, breast cancer is anticipated to have the second-highest estimated mortality toll among women among all cancer types. With a 29% incidence rate across all cancer types, this has consequently suggested a serious worldwide health concern. In addition, 2,46,660 women had a diagnosis of breast cancer in 2016. Early detection has a major impact on the possibilities for diagnosis and treatment of breast cancer, which helps explain the disease's 97.5% five-year survival rate. Conversely, if the diagnosis is delayed and the disease has spread to other organs, the patient's five-year survival percentage drops to just 20.4%. In 2020, there were 6,84,996 deaths worldwide due to breast cancer (WHO, 2021). Early identification of breast cancer remains a significant global issue. Deep learning algorithms are rapidly gaining popularity due to their remarkable image classification performance, particularly for imaging applications linked to breast histopathology diagnosis. CNN are among the most often used deep learning model for health picture detection and analysis among all deep learning subcategories. Nevertheless, CNN requires regular parameter modifications and has a significant implementation computing cost. To address this issue, a number of previously trained models are being constructed using the provided network design. This work derives a transfer learning model from the 16-layer complex model architecture.

Keywords: CNN · Breast Cancer · deep learning · VGG 16

1 Introduction

Breast cancer remains a prominent public health issue, standing as the primary contributor to cancer-related fatalities in women on a global scale. Therefore, early detection of the illness is crucial to halting its effects and reducing the morbidity rates associated

V. Sakalauskas et al. (Eds.): IBICA 2023, LNNS 1231, pp. 517–524, 2025.
https://doi.org/10.1007/978-3-031-78946-5_52

with it in women. Different entities with distinct clinical and histological features are included in the category of breast cancer cells. Figure 1 shows two rows of photos, one row showing pictures of healthy cells and the other row showing pictures of malignant cells. Regrettably, the formation of abnormal breast cells is the cause of this disease, which can spread to nearby healthy organs.

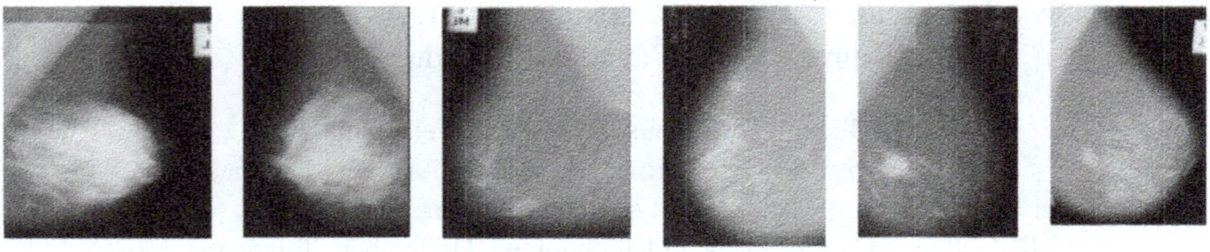

Fig. 1. Image of breast histopathology: bottom row, top row, wicked

Various techniques are employed in the examination of breast cancer. Using radiological images, such as magnetic resonance imaging (MRI) and mammography, the first method does clinical screening. But non-invasive imaging can't pinpoint the cancerous location. To do this, pathologists need to use Computer-Aided Diagnosis (CAD) to assess the histopathological images and identify the questionable area. Machine learning methods are commonly utilized to examine images of malignant pathology. Computer-aided design (CAD) systems are one of the most widely used models for analysis in digital histopathology image processing. In contrast to earlier CADs that depended on the features collected, DL is essential for numerous classification tasks and may provide high-level features and remarkable outcomes from histopathological images. Many academics have recently become aware of the shortcomings of traditional computer-aided design (CAD) methods, and as a result, they are thinking about using deep learning algorithms for task classification.

Researchers have focused much of their emphasis on CNN-based feature extraction models among DL models for the classification of histological breast images. The aims of this study are as follows: The already trained deep learning VGG16 will be examined as a characteristic extraction for breast cancer histopathology picture categorization in the first phase of the project. In the research section that follows, the capacity of three distinct classifiers to categorize the features retrieved from the breast histopathology images is assessed. The visual geometry group (VGG16), mobileNetV2, and ResNet101 were the three classifiers.

2 Literature Review

The authors of [1] presented a survey of various techniques for breast cancer classification, incorporating histological image analysis (HIA) with diverse artificial neural network (ANN) architectures. The study in Artificial Intelligence in Medicine structured its examination according to the relevant datasets, following a chronological progression that culminated in 2012 as the initial period when ANNs were employed in HIA. PNNs and ANNs are the two methods that are most widely used. However, morphological and textural traits have been the foundation of most feature extraction methods.

This section contains information on the work that is connected to the recently completed research. Essentially, there exist two approaches for the detection of breast cancer. Many different kinds of research initiatives make use of machine learning. Deep learning, however, provides solutions for certain problems utilizing machine learning techniques. This section contains information on deep learning and machine learning methods. To create a CNN system that uses characteristics from magnetic resonance imaging (MRI) to anticipate the molecular classification of breast cancer. 216 patients having available pre-treatment MRIs and immunohistochemically stained pathological data were included in an IRB-approved experiment. Using 3D slicer, the original post-contrast MRI images [1] were considered for 3D segmentation. A fourteen-layer CNN architecture was created.

Mengko T., Wisesty UN, and Purwarianti The research [2] was to identify breast cancer through gene mutation. When a gene's DNA sequence is altered to result in a different consequence, this is known as a genetic alteration. The DNA sequence of that gene is altered permanently.

An examination of the neural network-based cancer categorization system" The effectiveness of neural networks in classifying cancer diagnoses, particularly during the initial phases., was demonstrated by the authors in [3]. A neural network is a set of algorithms designed to mimic the structure and functions of the human brain, aiming to uncover important patterns within a dataset. Neural networks, as used here, are networks of neurons, which may have synthetic or organic origins. To evaluate several deep learning, machine learning, and data mining methods for predicting breast cancer. This was demonstrated in numerous machine learning tasks where the ensemble approaches were used in the effective solutions. The bulk of the publications used imaging, while only a small number of studies included genetics, according to the researchers.

The reviewers of [5] examined current studies that treat breast cancer by utilizing deep learning and a variety of imaging modalities. The dataset, architecture, application, and evaluation criteria were taken into consideration when organizing the experiments. They focused on deep learning architectures for breast imaging created using MRI, ultrasound, and mammography. They attempted to use DLR-based CAD systems to offer state-of-the-art research on breast cancer imaging in their papers. For their investigation, they employed confidential data and CNN categorization. After analysing these surveys, we will make a contribution by investigating the simultaneous use of genome sequencing and imaging to predict breast cancer and to collect further data that will support breast cancer treatment and early diagnosis. We will also offer guidance to researchers in academia interested in conducting studies in this field..

In this study, deep learning models—a subfield of artificial intelligence—are used to classify breast cancer with invasive ductal carcinoma. As a consequence, linear discriminant analysis was used to classify the data and the ridge regression method was used to identify the most effective features. 98.59% was the highest classification success rate attained. As a result, it is possible to acknowledge the suggested strategy as a successful model in the categorization [6].

Our findings demonstrate that one can learn predictive signals in tumor tissue images even in the absence of domain [7] experience. To target the patients who are not clustered, we propose a new ensemble classification stage that follows the ensemble clustering stage in our innovative classification framework, which [8] we provide in this paper [9]. The results show that applying the described framework leads to a more accurate classification. In the end, the findings were verified by statistical testing, visualization techniques, cluster quality assessment, and professional clinical interpretation [10].

This study provides a thorough rundown of the ANN-based BHIA methods. Firstly, for more thorough analysis, we divide the BHIA systems into two categories: deep neural networks and traditional neural networks [11]. Next, pertinent research utilizing BHIA systems is showcased [12]. Next, we examine the current models to identify the best algorithms. In conclusion, download links to publicly available datasets are included for the benefit of future researchers [13, 14].

3 Proposed System

CNN is regarded as the most successful deep learning method. It is being used in many data science applications, including the classification of medical photographs and the extraction of features from input training photos using CNN models that have already been trained. The core components of a CNN comprise the input stack, a series of hidden layers, and the output layers. The primary computations occur within the hidden layers, typically consisting of multiple fully connected layers (FC), convolutional layers (CL), and several pooling layers. Adding more layers to a model makes it more complex and allows the system to detect finer details for better decision-making. The basic architecture of CNN is shown in Fig. 2. Using a small, sliding filter (such a 3x3), The convolutional and pooling layers derive characteristics from the training dataset's input data.CL in CNN is utilized to separate a unique set of features from an image sequence. One CL, for instance, regulates shapes, edges, and areas that are bright or dark. On the other hand, others are responsible for the identifiable objects and forms in the picture. The next key layer, the pooling layer, is in charge of simplifying the feature map while maintaining the most important data, such average and maximum pooling. The categorization layer is also known as the FC layer. After the FC stages have finished their voyage, the output layer, also known as the last layer, generates probability for the input using the softmax activation function.

Because of its tremendous growth over the past few years, CNN has built a variety of distinct architectures. Among the well-known CNNs is VGG16. It has a simpler layer architecture and a smaller filter size than earlier systems.

VGG16

Among all, VGG16 stands out with one of the most visually prominent deep learning architectures. The visually impressive architecture of VGG16 is depicted in Fig. 3. Thirteen convolutional layers (Conv.), three FC layers, sixteen weight layers, and thirteen other strata totaling forty-one have been disrupted. Each Conv Layer in VGG16 uses a tiny 3x3 kernel, or filter, with stride 1. Conversion layers typically come before max pooling layers. The input for VGG16 consists of three 224x224 pixel fixed-size channels. The VGG16 model has three fully connected (FC) layers with different depths. While the first two FC layers have the same dimension of 4096, the last FC layer has a dimension of 1000, which correlates to the class label quantity in the image Net dataset. The likelihood of the input layer is ascertained by the outcome layer, also referred to as the soft-max layer.

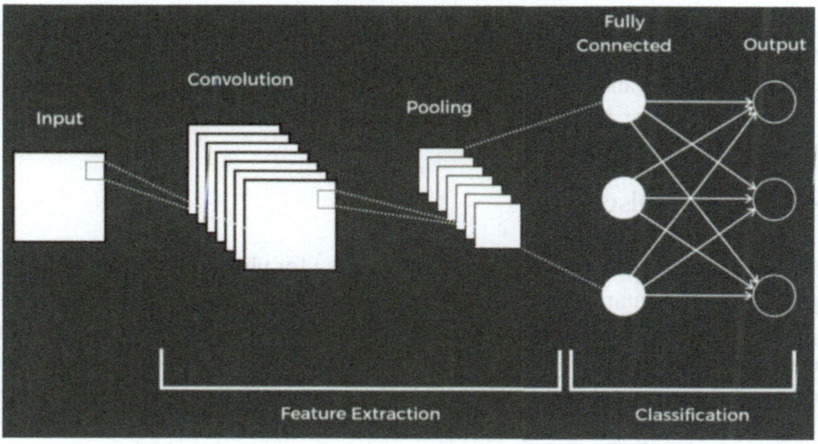

Fig. 2. Layers of convolution neural network

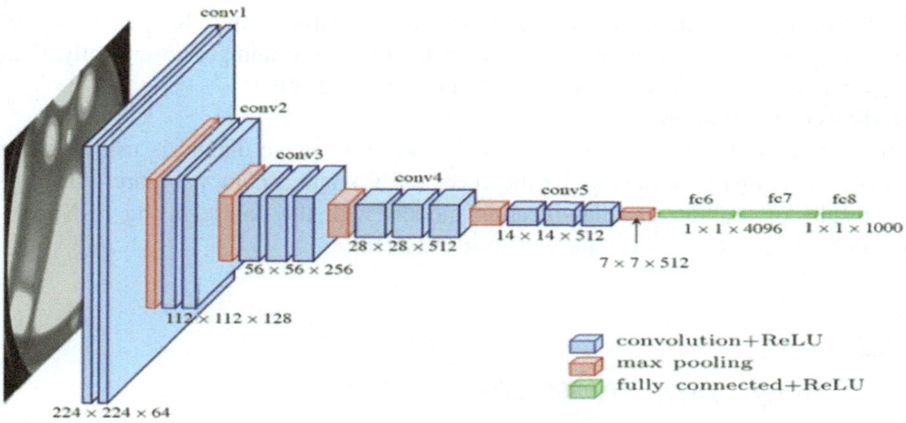

Fig. 3. Architecture of VGG16 model

The VGG16 needs extensive training, just like any other pre-trained model, if the number of weights is initialized at random. Consequently, transfer learning (TL) strategies are commonly used by CCN algorithms. TL is a technique that uses a model designed for one activity and applies it to another task in a way that makes sense. Stated simply, we use the same input to train a CNN algorithm on an identical problem, potentially yielding a different type of answer. Here, the VGG 6 models are trained using a large number of real-world object pictures from the ImageNet dataset. Photographs showing breast histopathology exhibit a range of textures, forms, and histology features, such as nuclei and cytoplasm. Using the supplied histopathology photos, the suggested method uses VGG16 to extract deep typical traits. The recommended method's primary steps are as follows: Initially, the original training pictures are resized to fit the VGG16 input layer's dimensions. Secondly, in the context of histopathological images, the final three layers (FC) at the top of the VGG16 stack, representing the categorization stage, are omitted. This is why the layers involved in convolution and pooling are termed as the retained layers. The method is used to the input image to extract 4096 features. We split the data into training and testing halves after removing the features from each of the input photos so that we may apply a variety of classification techniques.

4 Results and Discussion

The current study makes use of the smallest database for photos related to breast cancer histopathology. The dataset consists of 322 photos that have been classified as benign and malignant. The MIAS dataset includes the following features: the x, y image coordinates of the irregularity's center, backdrop tissue type, current aberration class, severity of deviation, reference number for the MIAS database, and an estimate of the circle's diameter in pixels.

Initial Experimental Configuration

Images from the MIAS dataset were used in the two main investigations of this study. For each image, 4096 features are extracted using the VGG16. The dataset is built using these properties in conjunction with the labels for the image classes. Ten percent is used for testing, while the remaining ninety percent is used for training. Subsequently, three classifiers were trained: VGG16, mobileNetV2, and resNet101.

Methods of Evaluation

This paper assesses the classification performance of image diagnosis using various metrics, including loss, area under the curve (AUC), and accuracy. Figure 4 displays the results of training the classification models with the dataset using resNet101, mobileNetV2, and Visual Geometry Group (VGG16).

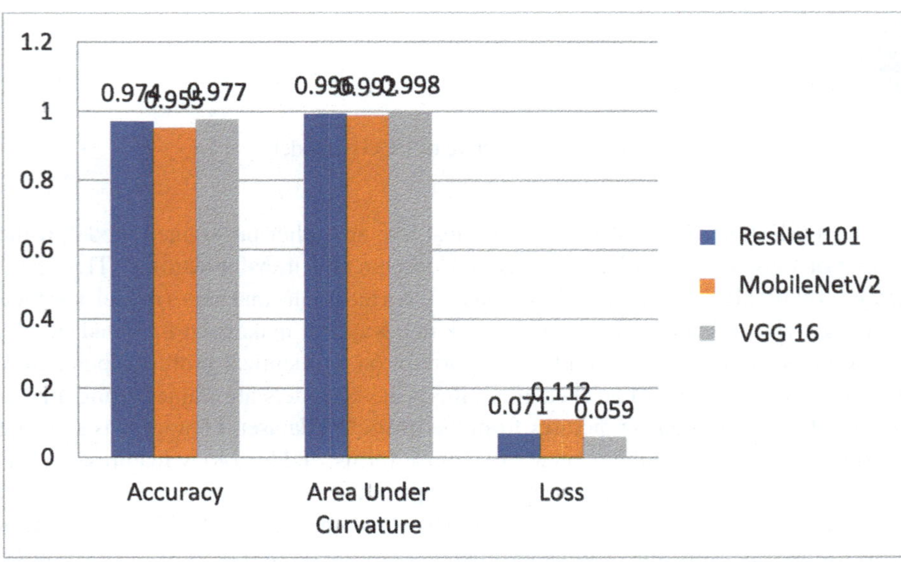

Fig. 4. Comparing the suggested model to earlier deep learning methods on the same dataset.

We examined the dataset against the VGG16 model since the model's training results were quite good. The values for accuracy, area of curvature, and loss are, respectively, 0.9356, 0.9730, and 0.2356. Testing and training accuracy differ greatly because there is a lack of testing data.

5 Conclusion and Future Enhancements

Using images of breast histopathology to detect high-level traits can improve the efficiency of diagnostic process. Therefore, the primary aim of this work is to extract high-level features from breast pictures using the trained CNN deep learning model VGG16. In order to accomplish that, we eliminated VGG16's final fully connected layer. Following that, a set of classifiers was used to categorize the features that had been collected. A collection of performance measures was established for the test datasetsection after thorough tests were conducted on the MIAS dataset (public dataset). The results of the experiment outperformed various state-of-the-art techniques. This demonstrates the potency of the features extracted from VGG16. Real-time applications and extensive screening: Future developments in hardware and processing power may allow for the use of VGG16 in real-time or large-scale screening programs for the diagnosis of breast cancer. This would entail creating effective systems that quickly handle enormous amounts of medical imaging data, enabling the earlier and more precise identification of breast cancer..

References

1. Ha, R., Mutasa, S., Karcich, J., Gupta, N., Pascual Van Sant, E., Nemer, J., Jambawalikar, S.: Predicting breast cancer molecular subtype with MRI dataset utilizing convolutional neural network algorithm. J. Digital Imaging **32**, 276–282 (2019). https://doi.org/10.1007/s10278-019-00179-2
2. Wisesty, U.N., Mengko, T., Purwarianti, A.: Gene mutation detection for breast cancer disease: a review. IOP Conf Ser Mater Sci Eng **830**, 32051 (2020). https://doi.org/10.1088/1757-899X/830/3/032051
3. Mahmood, M., Al-Khateeb, B., Alwash, W.M.: A review on neural networks approach on classifying cancers. IAES Int. J. Artif. Intell. **9**(2), 317 (2020). https://doi.org/10.11591/ijai.v9.i2.pp317-326
4. Fatima, N., Liu, L., Hong, S., Ahmed, H.: Prediction of breast cancer, comparative review of machine learning techniques, and their analysis. IEEE Access **8**, 150360–150376 (2020). https://doi.org/10.1109/ACCESS.2020.3016715
5. Pang, T., Wong, J.H.D., Ng, W.L., Chan, C.S.: Deep learning radiomics in breast cancer with different modalities: Overview and future. Expert Syst. Appl. **158**, 113501 (2020). https://doi.org/10.1016/j.eswa.2020.113501
6. Toğaçar, M., Ergen, B., Cömert, Z.: Application of breast cancer diagnosis based on a combination of convolutional neural networks, ridge regression and linear discriminant analysis using invasive breast cancer images processed with autoencoders. Med. Hypotheses **135**, 109503 (2020). https://doi.org/10.1016/j.mehy.2019.109503
7. Turkki, R., Byckhov, D., Lundin, M., Isola, J., Nordling, S., Kovanen, P.E., et al.: Breast cancer outcome prediction with tumour tissue images and machine learning. Breast Cancer Res. Treat. **177**, 41–52 (2019)

8. Vankdothu, R., Hameed, M.A., Ameen, A., Unnisa, R.: Brain image identification and classification on Internet of medical Things in healthcare system using support value based deep neural network. Comput. Electr. Eng. **102**, 108196 (2022)
9. David, L.G., Patra, R.K., Falkowski-Gilski, P., Divakarachari, P.B., Antony Marcilin, L.J.: Tool wear monitoring using improved dragonfly optimization algorithm and deep belief network. Appl. Sci. **12**(16), 8130 (2022). https://doi.org/10.3390/app12168130
10. Mehbodniya, A., Rao, M.V., David, L.G., Nigel, K.G.J., Vennam, P.: Online product sentiment analysis using random evolutionary whale optimization algorithm and deep belief network. Pattern Recogn. Lett. **159**, 1–8 (2022)
11. Sunitha, D., Patra, R.K., Babu, N.V., Suresh, A., Gupta, S.C.: Twitter sentiment analysis using ensemble based deep learning model towards COVID-19 in India and European countries. Pattern Recogn. Lett. **158**, 164–170 (2022). ISSN 0167–8655
12. Balmuri, K.R., Konda, S., Lai, W.C., Divakarachari, P.B., Gowda, K.M.V., Kivudujogappa Lingappa, H.A.: Long short-term memory network-based radio resource management for 5G Network. Future Internet 2022, **14**(6), 184 (2022)
13. Ahmad, S.J., Unissa, I., Ali, M.S., Kumar, A.: Enhanced security to MANETs using digital codes. J. Inf. Secur. Appl. **66**, 103147 (2022). https://doi.org/10.1016/j.jisa.2022.103147
14. Soma, S., Sonth, M.V., Gowre, S.C.: Design of two-dimensional photonic crystal based ultra compact optical RS flip-flop. Photon Netw. Commun. **43**, 109–115 (2022). https://doi.org/10.1007/s11107-021-00955-7
15. Sudhamani, C.: Detection probability maximization through optimization of samples in cognitive radio networks. Multimed Tools Appl **81**, 12275–12285 (2022). https://doi.org/10.1007/s11042-021-11089-3
16. Chinna Rao, R., Lakshmi, K.M., Raja, C., Varma, P.B.S., Rao, G.R.K., Patibandla, A.: Real-time implementation and testing of VoIP vocoders with asterisk PBX using wireshark packet analyzer. J. Interconn. Netw **22**(Supp01), 2141030 (2022)
17. Rao, M.V., Reddy, D.A., Ampavathi, A., Munawar, S.: Data mining for cyber physical systems. Data Min. Mach. Learn. Appl. 235–280 (2022).https://doi.org/10.1002/9781119792529.ch10
18. Thaseen, A., Unnisa, R., Sultana, N., Madhavi, K.R., NagaJyothi, G., Kirubakaran, S.: Breast cancer detection using deep learning model. In: Proceedings of Third International Conference on Advances in Computer Engineering and Communication Systems: ICACECS 2022, pp. 669–677. Springer, Singapore (2023)

Quality-Aware Node-Level Routing Protocol for Energy-Efficient Cognitive Wireless Sensor Networks

Rajashekar Reddy Eduru and Suraya Mubeen(✉)

Department of ECE, CMR Technical Campus, Hyderabad, Telangana, India
suraya418@gmail.com

Abstract. In the field of energy-efficient cognitive radio sensor networks (CRSN), a prominent challenge arises from the limited energy resources of sensor nodes and the multifaceted obligations associated with channel-dependent data transmission. Additionally, the need for inter-sensor node communication to share channel-dependent data poses another hurdle. To enhance network longevity, achieving energy homeostasis among cluster heads (CHs) is crucial, necessitating the use of an uneven clustering strategy. Furthermore, CH selection considers number of open channels to perform optimal channel allocation. Sensible channel distribution should factor in cluster members' energy consumption, and selecting an appropriate next-hop for relaying should involve evaluating connection stability to boost transmission efficiency. To effectively maintain cluster, CH rotation sensitive to both energy and spectrum dynamics is required. Energy and Spectrum-Aware Unequal Clustering (ESAUC) protocol addresses these challenges by jointly addressing energy and spectrum limitations, promoting network longevity by equalizing residual energy allocation and minimizing overall energy usage. The culmination of this effort introduces the Quality-Based Clustering Node Disjoint Routing Protocol (QCL-NDRP), featuring a source-destination edge-based route selection mechanism. This protocol considers Quality of Service (QoS) metrics, like distance, available energy, and spectrum availability, to evaluate path cost function and efficiently mitigate interference among paths. Results reveals that proposed approach performs better than ESAUC method considering Network Lifetime, throughput, Packet Delivery Ratio, and end-to-end latency.

Keywords: Cognitive Radio Sensor Networks (CRSN) · Uneven Clustering · Energy and Spectrum-Aware Protocol (ESAUC) · Quality-Based Clustering Node Disjoint Routing Protocol (QCL-NDRP) · Network Lifetime Optimization · Interference Mitigation

1 Introduction

The demand for efficient and sustainable communication networks has driven research efforts towards enhancing the performance of Cognitive Wireless Sensor Networks (CWSNs). In this context, the paper introduces a novel Energy-based Quality-Aware Node-Level Routing Protocol designed to optimize the routing mechanisms within

V. Sakalauskas et al. (Eds.): IBICA 2023, LNNS 1231, pp. 525–532, 2025.
https://doi.org/10.1007/978-3-031-78946-5_53

CWSNs [1]. The protocol uniquely integrates energy considerations with a quality-aware approach, aiming to extend the network's operational lifetime while ensuring reliable and high-quality data transmission. As the landscape of wireless sensor networks evolves, the proposed protocol seeks to address the challenges associated with energy constraints and the need for quality-driven routing strategies, thus contributing to the advancement of resource-efficient CWSNs. This paper focuses on the importance of Energy-based Quality-Aware Node-Level Routing Protocol, discussing its key features, mechanisms, and anticipated impact on overall performance of cognitive wireless sensor networks. By combining energy efficiency with a focus on data quality, proposed protocol is capable of providing a better solution that aligns with evolving requirements of modern wireless communication systems. Furthermore, proposed protocol places a significant emphasis on node-level routing, acknowledging pivotal role individual sensor nodes play in shaping network dynamics. By integrating energy-awareness at the node level and incorporating quality metrics into routing decisions, the proposed protocol aims to balance between prolonging the network's lifespan and ensuring the timely and dependable delivery of high-quality data.

2 Background Analysis

The development of better routing protocols for Cognitive Wireless Sensor Networks (CWSNs) is crucial to address unique challenges posed by dynamic and resource-constrained nature of wireless sensor networks. Wireless sensor nodes typically operate on limited energy resources, often relying on batteries. In the context of CWSNs, where nodes are equipped with cognitive capabilities to adapt to changing environments and network conditions, energy demands can be even more pronounced. Traditional routing protocols may not be well-suited to handle cognitive aspect and may overlook energy conservation measures that are essential for prolonging network lifetime [1–5]. The focus on node-level routing is critical due to the heterogeneous nature of sensor nodes within a network. Nodes may have varying energy levels, computing capabilities, and sensing capacities [6–10]. Routing decisions at the node level can enable the network to adapt dynamically to the changing energy states of individual nodes, optimizing the overall energy consumption and extending the network's operational lifetime [11–15]. While several routing protocols exist for WSNs, there is a need for specialized protocols that integrate energy awareness and QoS considerations at the node level in the context of cognitive capabilities [16–21]. Existing protocols might not adequately address the dynamic energy constraints and quality requirements of CWSNs. Motivation for developing an energy-based quality-aware node-level routing protocol lies in bridging the gap between energy efficiency and data quality. By incorporating cognitive capabilities into the routing decisions at node level, the protocol aims to intelligently adapt to the network's energy landscape while meeting the QoS demands of diverse applications. The background analysis specifies that the critical need for a specialized routing protocol in CWSNs that integrates energy considerations, leverages cognitive capabilities, and ensures the delivery of high-quality data.

3 Methodology

One method for clarifying the idea of a "disjoint path," which is sometimes referred to as a "disjoint route," is to determine the frequency with which numerous pathways from a source to a destination demand the recurrent participation of the same nodes more than twice. There are three distinct categories that can be applied to these configurations: non-disjoint, in which nodes participate in multiple paths sequentially; link disjoint, in which nodes participate in multiple paths without the consecutive participation of two or more nodes; and node-disjoint, also known as complete disjoint, in which each node contributes to a single path exclusively. An example of a Node-disjoint multipath network may be seen in Fig. 1.

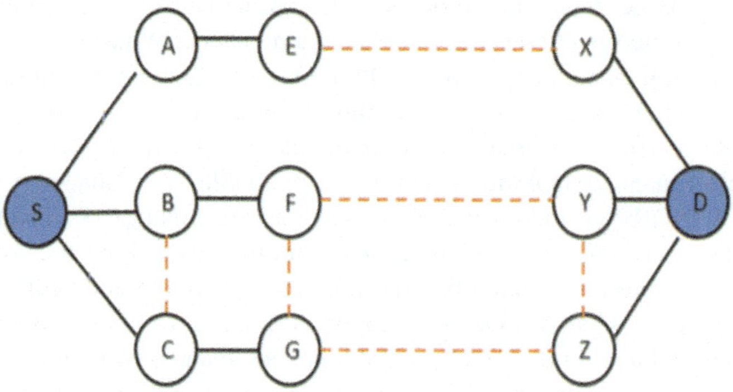

Fig. 1. Node Disjoint Multipath Network

After giving one-of-a-kind IDs to each and every node in the network, the likelihood of having k node-disjoint paths between nodes is figured out by applying Eq. 1.

$$p = \prod_{i=2}^{k} \frac{\left(n - 2 - \sum_{i=1}^{j} d_i\right)}{(n-2)} \tag{1}$$

Take into consideration the following scenario: a particular source node, denoted by the letter S, and a sink node, denoted by the letter D, are singled out. There are k unique pathways that pass through these nodes, and none of them connect to any of the others. The notation j is used to signify the jth node inside the ith path within these routes, where While the existing likelihood of a connection between nodes A and B is indicated as α 1 for the ith route, the probabilities of linkages between nodes B and C, nodes B and F, and between nodes d_i -1 and d_i as specified in Eq. 2.

$$\beta = \alpha^{d_i} + 1 \tag{2}$$

When interpreted as the probability of their being k different paths leading from S to D,p_k is the variable that is denoted as p_k in Eq. 3.

$$P_k = \prod_{i=1}^{k} [1 - (1 - \beta)] \tag{3}$$

The node-disjoint multipath stands out as a more robust technique of assuring reliable data delivery when compared to the other accessible choices. This is because it uses more than one path to transport the data. This step of node-disjoint multipath discovery is an important part of the multipath routing technique.Source routing is an alternate method for multipath discovery that involves defining a number of different routes in the Route Request (RREQ) and choosing the destination node based on the paths that have been identified. A different name for this approach is "source discovery." This easy strategy has the potential to uncover multipaths; however, it also has the potential to generate considerable routing load and overhead owing to the issuance of a large number of RREQ messages.

QCL-NDRP, which is a node-disjoint multipath routing protocol that places an emphasis on interference-free operations and quality-of-service awareness. The protocol makes use of an edge-based source-destination routing selection technique, providing specifics on the route selection process and its maintenance. While selecting best route, QCL-NDRP computes path cost function by taking into account a number of quality-of-service parameters, such as distance, amount of energy that is still available, and availability of spectrum. This helps to eliminate the possibility of interference between many possible routes. A maximum number of available paths that may be achieved using the QCL-NDRP protocol between a source and destination is shown graphically in Fig. 2. This is an illustration of the possible multipaths linking the source and the destination, which were computed based on features of source node edge and destination node edge. It is essential to keep in mind that number of node-disjoint multipaths must not be more than number of edges that link source and destination. During "route discovery" phase, major objective is to locate various node-disjoint routes that are unaffected by interference. Selection of a path takes into account quality-of-service criteria such as distance (Euclidean), available energy levels, and spectrum availability in an effort to reduce amount of interference that occurs between many possible routes.

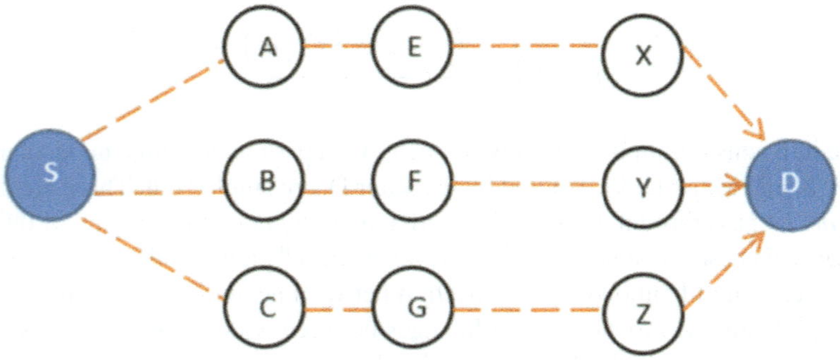

Fig. 2. QCL-NDRP Protocol

The major path was chosen from among the available disjoint paths after taking into account the total costs associated with all of the other possible routes that might be taken between the origin and the destination. When choosing the principal routing route, it is necessary to evaluate all possible routing pathways and compare the expenses associated with traveling to each node in the network. The price of each node is determined by taking into consideration a number of different criteria, such as the amount of energy, distance, and spectrum that is accessible. Because it has access to more spectrum resources and more residual energy, a node that has a higher node cost is able to handle a larger data traffic load than one that has a lower cost. If the cost of the jth node includes the highest possible node cost, then the total route cost, represented by the letter "P," will automatically take that amount. Following this, once all of the conceivable route costs have been computed (P = 1, 2,..., k), the main path is determined to be the one with the highest path cost. This identifies the primary path as the one with the capacity to handle the largest data traffic volume and establishes it as the way with the highest level of dependability among the node-disjoint paths. The expression for the principal route is shown by the Eq. 4.

$$Primaty_{PATH} = max\{PATH\_COST_i \ wherei \in P\} \tag{4}$$

The process of data transmission is kicked off by the source node located somewhere along the path that was found. This beginning takes place after the route has been mapped out and established. It is possible that the node energy levels may decline over time, which will have a negative effect on the forwarding connection. It is referred to as a "dead node" when the energy stored in a node runs out completely, which also causes the connection to the node to be severed. The source node will continue its data transmission attempts despite this condition, which will have an effect on performance. In these kinds of situations, a route alert packet is sent in the opposite direction, to the location of the node that originated it. After receiving this warning packet, the source node will continue to delete both the whole route and the data that has been transferred via the currently active pathways.

4 Results and Discussions

During experimentation, NS2 simulator program is used to simulate and evaluate not only the suggested protocol but also other protocols that already exist, such as ESAUC. The fundamental objective of the NS2 network simulator is to provide assistance for research in the field of networking.

During the process, the size of the network increased from fifty to two hundred fifty nodes inside a 1000 by 500 m region. Within the network, the nodes were dispersed in an unpredictable manner. The size of each data unit was 1024 bytes, and an initial energy allotment of 100 J was given to each node in the network. The effectiveness of the strategy that was suggested was evaluated using a number of different factors, including as end-to-end latency, routing overhead, packet delivery ratio, and energy usage as specified in Table 1.

The end-to-end delay specified in Fig. 3 is quantified as the amount of time it takes for a packet to travel across a network from its place of origin to its final destination. This delay is shown in the chart that has been supplied. This measure is used to evaluate the performance of the proposed approach in relation to the QoS routing protocols that are already in existence. The QCL-NDRP uses route selection factors to reduce path

Table 1. Network Simulation Parameters

Simulation Parameter	Value
Area of Network	1000 m × 500 m
Size of packet	1024 bytes
Duration of Simulation	200s
Nodes	50 to 250
Base station	1
Initial Energy of Node	100j

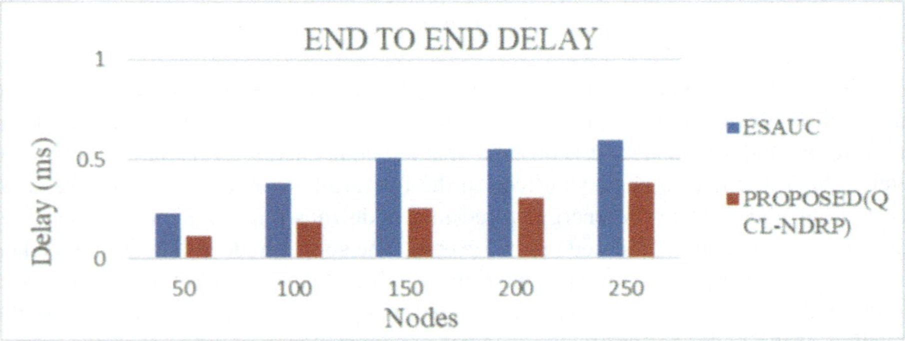

Fig. 3. Graphical representation of End to End Delay versus nodes

interference along the selected routing path. These criteria include distance, remaining energy, and spectrum availability.

During data transmission, the end-to-end latency will be lowered as a direct consequence of this improvement. In comparison to the ESAUC technique, the suggested network displayed a peak delay that was a considerable amount lower, clocking in at just 0.381 ms, while the ESAUC method revealed a peak delay that was much longer, clocking in at up to 0.597 ms. In conclusion, the suggested network shown significantly enhanced performance in terms of decreased latency.

The distribution of control packets is an integral part of the process of route discovery as well as route maintenance. During the course of the application's functioning, a substantial amount of control packets will be sent out into the network. The amount of control packets that are sent around between sensor nodes in the network has a direct bearing on the amount of overhead that is incurred as specified in Fig. 4.

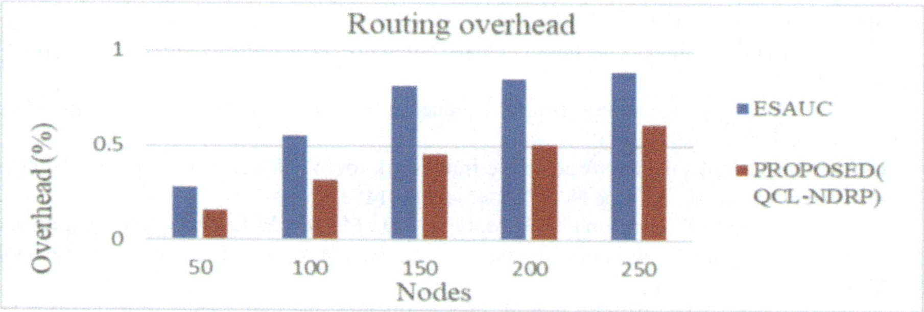

Fig. 4. Routing Overhead

5 Conclusion

In conclusion, a complete assessment of a variety of aspects is required in order to solve the issue of energy efficiency in Cognitive Radio Sensor Networks (CRSN). These factors include the reliable establishment of relay connections, the fair distribution of energy consumption among cluster heads (CHs), the effective allocation of channels, the equitable distribution of channels, the efficient allocation of spectrum, and effective cluster maintenance that takes energy and spectrum awareness into account. In order to meet these issues front on, the established procedure known as ESAUC is put into action. This protocol successfully lengthens the lifetime of the network by bringing the amount of energy used by the sensor nodes to an equilibrium while simultaneously lowering the amount of energy that is used overall. In addition, the findings provide a brand new protocol that has been given the name Quality-based Clustering Node Disjoint Routing Protocol (abbreviated as QCL-NDRP). It does this by using a route selection mechanism that is based on the source-destination edge, all the while taking into consideration Quality of Service (QoS) criteria like as distance, remaining energy, and spectrum availability. This strategy works to optimize the path cost functions, which in turn helps to reduce interference between different routes. According to the findings of the simulations, the recently suggested QCL-NDRP approach outperforms the ESAUC algorithm that is being used at the moment in terms of essential performance measures. These metrics include Network Lifetime, throughput, Packet Delivery Ratio, and end-to-end latency. Development of more energy-efficient CRSNs has been pushed forward because to this breakthrough.

References

1. Ejaz, W., Ul Hasan, N., Kim, H.S.: Distributed cooperative spectrum sensing in cognitive radio for ad hoc networks. Comput. Commun. **36**(12), 1341–1349 (2013)
2. Ahmad, A., Ahmad, S., Rehmani, M.H., Hassan, N.U.: A survey on radio resource allocation in cognitive radio sensor networks. IEEE Commun. Surv. Tutorials **17**(2), 888–917 (2015)
3. Yu, J., Qi, Y., Wang, G., Guo, Q., Gu, X.: An energy-aware distributed unequal clustering protocol for wireless sensor networks. Int. J. Distrib. Sens. Netw. **7**(1), 202145 (2011)

4. Stephan, T., Joseph, K.S.: PSO assisted OLSR routing for cognitive radio vehicular sensor networks. In Proceedings of the International Conference on Informatics and Analytics, pp. 1–8 (2016)
5. Lo, B.F.: A survey of common control channel design in cognitive radio networks. Phys. Commun. **4**(1), 26–39 (2011)
6. El-Mougy, A., Ibnkahla, M.: A cognitive framework for WSN based on weighted cognitive maps and Q-learning. Ad Hoc Netw. **16**, 46–69 (2014)
7. Ding, L., Melodia, T., Batalama, S.N., Matyjas, J.D., Medley, M.J.: Cross-layer routing and dynamic spectrum allocation in cognitive radio ad hoc networks. IEEE Trans. Veh. Technol. **59**(4), 1969–1979 (2010)
8. Younis, O., Fahmy, S.: HEED: a hybrid, energy-efficient, distributed clustering approach for ad hoc sensor networks. IEEE Trans. Mob. Comput. **3**(4), 366–379 (2004)
9. Huynh, V.V., Nguyen, H.S., Hoc, L.T.T., Nguyen, T.S., Voznak, M.: Optimization issues for data rate in energy harvesting relay-enabled cognitive sensor networks. Comput. Netw. **157**, 29–40 (2019)
10. Madhavi, K.R., Nawi, M.N.M., Reddy, B.B., Baboji, K., Kishore, K.H., Manikanthan, S.V.: Energy efficient target tracking in wireless sensor network using PF-SVM (particle filter-support vector machine) technique Measurement. Sensors **26**, 100667 (2023)
11. Avanija, J., Kumar, K.N., Kumari, C.U., Jyothi, G.N., Raju, K.S., Madhavi, K.R.: Enhancing network forensic and deep learning mechanism for internet of things networks. J. Sci. Ind. Res. **82**(05), 522–528 (2023)
12. Vankdothu, R., Hameed, M.A., Ameen, A., Unnisa, R.: Brain image identification and classification on internet of medical things in healthcare system using support value based deep neural network. Comput. Electr. Eng. **102**, 108196 (2022)
13. David, L.G., Patra, R.K., Falkowski-Gilski, P., Divakarachari, P.B., Antony Marcilin, L.J.: Tool WEAR MONITORING USING IMPROVED DRAGONFLY OPTIMIZATION ALGORITHM AND DEEP BELIEF NETWORK. Appl. Sci. **12**(16), 8130 (2022). https://doi.org/10.3390/app12168130
14. Mehbodniya, A., Rao, M.V., David, L.G., Nigel, K.G.J., Vennam, P.: Online product sentiment analysis using random evolutionary whale optimization algorithm and deep belief network. Pattern Recogn. Lett. **159**, 1–8 (2022)
15. Sunitha, D., Patra, R.K., Babu, N.V., Suresh, A., Gupta, S.C.: Twitter sentiment analysis using ensemble based deep learning model towards COVID-19 in India and European countries. Pattern Recogn. Lett. 158, 164–170 (2022). ISSN 0167–8655
16. Balmuri, K.R., Konda, S., Lai, W.C., Divakarachari, P.B., Gowda, K.M.V., Kivudujogappa Lingappa, H.: A long short-term memory network-based radio resource management for 5G network. Future Internet **14**(6), 184 (2022). https://doi.org/10.3390/fi14060184
17. Ahmad, S.J., Unissa, I., Ali, M.S., Kumar, A.: Enhanced security to MANETs using digital codes. J. Inf. Secur. Appl. **66**, 103147 (2022). ISSN 2214–2126 https://doi.org/10.1016/j.jisa.2022.103147
18. Soma, S., Sonth, M.V., Gowre, S.C.: Design of two-dimensional photonic crystal based ultra compact optical RS flip-flop. Photon Netw. Commun. **43**, 109–115 (2022). https://doi.org/10.1007/s11107-021-00955-7
19. Sudhamani, C.: Detection probability maximization through optimization of samples in cognitive radio networks. Multimed Tools Appl **81**, 12275–12285 (2022). https://doi.org/10.1007/s11042-021-11089-3
20. Chinna Rao, R., Lakshmi, K.M., Raja, C., Varma, P.B.S., Rao, G.R.K., Patibandla, A.: Real-time implementation and testing of VoIP vocoders with asterisk PBX using wireshark packet analyzer. J. Interconnection Netw. **22**(Supp01), 2141030 (2022)
21. Rao, M.V., Reddy, D.A., Ampavathi, A., Munawar, S.: Data mining for cyber physical systems. Data Min. Mach. Learn. Appl. 235–280 (2022).https://doi.org/10.1002/9781119792529.ch10

Author Index

V. Sakalauskas et al. (Eds.): IBICA 2023, LNNS 1231, pp. 533–535, 2025.
https://doi.org/10.1007/978-3-031-78946-5

The manufacturer's authorised representative in the EU is Springer Nature Customer Service Centre GmbH, Europaplatz 3, 69115 Heidelberg, Germany. If you have any concerns regarding our products, please contact ProductSafety@springernature.com

Printed and bound by CPI Group (UK) Ltd, Croydon, CR0 4YY
30/09/2026
02225561-0017